Finite Element and Boundary Methods in Structural Acoustics and Vibration

Finite Element and Boundary Methods in Structural Acoustics and Vibration

Noureddine Atalla
Franck Sgard

CRC Press
Taylor & Francis Group
Boca Raton London New York

CRC Press is an imprint of the
Taylor & Francis Group, an **informa** business
A SPON PRESS BOOK

CRC Press
Taylor & Francis Group
6000 Broken Sound Parkway NW, Suite 300
Boca Raton, FL 33487-2742

First issued in paperback 2017

ISBN-13: 978-1-4665-9287-2 (hbk)
ISBN-13: 978-1-138-74917-7 (pbk)

Visit the Taylor & Francis Web site at
http://www.taylorandfrancis.com

and the CRC Press Web site at
http://www.crcpress.com

Thanks to all our loved ones for their support
and patience while writing this book.

Noureddine Atalla, Franck Sgard

My deepest thanks are reserved to Sonia, who provided me with her support and moral booster during the years of book preparation and who gave me the energy to complete this book.

Franck Sgard

Contents

Acknowledgments

This book is the fruit of a graduate-level class taught at the Department of Mechanical Engineering of l'Université de Sherbrooke. Thanks are due to the graduate students who through their questions and inputs enriched and influenced the content of this book. Several of the topics and examples discussed in the book were also prepared thanks to collaboration with past and current research associates and graduate students. Thanks in particular go to Mr Celse Kafui Amedin for his invaluable contribution to the development and validation of the FEM/BEM code (NOVAFEM) used in Chapters 6 through 8. Finally, the authors would like to acknowledge the use of compute Canada facilities.

MATLAB® is a registered trademark of The MathWorks, Inc. For product information, please contact:

The MathWorks, Inc.
3 Apple Hill Drive
Natick, MA 01760-2098 USA
Tel: 508 647 7000
Fax: 508-647-7001
E-mail: info@mathworks.com
Web: www.mathworks.com

Authors

Noureddine Atalla is a professor in the Department of Mechanical Engineering (Université de Sherbrooke). He is also a member and past director of GAUS (Group d'Acoustique et de vibration de l'Universite de Sherbrooke); one of the largest acoustic and vibration research group in North America (www.gaus.gme.usherbrooke.ca).

Professor Atalla received an MSc in 1988 from the Université de Technologie de Compiègne (France) and a PhD in 1991 in Ocean Engineering from Florida Atlantic University (USA). His core expertise is in computational vibroacoustics and modeling and characterization of acoustic materials. He has published more than 100 papers in acoustics and vibration, spanning different domains, including modeling poroelastic and viscoelastic materials, coupled fluid–structure problems, the acoustic and dynamic response of sandwich and composite structures, and modeling methods for industrial structures. He is also the co-author of a book on the modeling of sound porous materials.

Franck Sgard graduated from Ecole Nationale des Travaux Publics de l'Etat (ENTPE) in Vaulx en Velin (France) as a civil engineer in 1990. He then obtained his master's degree in mechanical engineering from the University of Washington (Seattle) in 1991. In 1992 and 1993, he worked as a research assistant in the acoustic group of the University of Sherbrooke (GAUS) on noise reduction inside aircraft cabins. He then started a joint PhD (University of Sherbrooke/Institut National des Sciences Appliquées in Lyon, France) in mechanical engineering (acoustics), which he completed in 1995. From 1995 till 2006, he worked as a professor at ENTPE, teaching acoustics and developing research activities in acoustics applied to transportation and building industries on topics such as the design of high efficiency acoustical materials for sound absorption and sound insulation, the numerical modeling of homogeneous and heterogeneous porous materials, the numerical modeling of complex multilayered structures at low frequencies, the characterization of mechanical parameters of porous materials. In 2006, he joined the Noise and Vibration group at the Institut Robert Sauvé en Santé et Sécurité du Travail (IRSST) in Montreal, Canada, first as an invited researcher and then as the Noise and Vibration Field leader. Since 2012, he has been the Mechanical and Physical Risk Prevention Research Field leader in the same institute. His current research activities deal with various problems related to noise and vibration in the field of occupational health and safety. He has published more than 100 papers and conference proceedings in acoustics and vibration.

Chapter 1

Introduction

1.1 COMPUTATIONAL VIBROACOUSTICS

Vibroacoustics deals with the interaction of sound waves and vibrating structures. Figure 1.1 provides a schematic of the problem and Figure 1.2 provides an example from the automotive industry. The generic structure is a homogeneous or a multilayered structure made up of various materials. It encompasses a cavity and is immersed in a fluid.

The structure is excited mechanically (point load), acoustically (plane waves, diffuse acoustic field, and point source in the interior or exterior fluids), or aerodynamically (e.g., turbulent boundary layer).

The vibroacoustic problem involves interaction between a vibrating structure and pressure fluctuations in the surrounding fluids. Figure 1.3 presents a diagram of the involved coupling mechanisms. The response of the structure is governed by the excitation, boundary conditions and coupling with the surrounding medium. This coupling depends on both the geometry of the problem and the properties of the structure and fluids.

The fluid–structure coupled problems can be classified into three main categories (see Figure 1.4).

Interior problems: In this case, the fluid domain is bounded. This is the case of problems involving a structure coupled with an internal cavity. Usually, the strength of the coupling depends on both the geometry of the cavity and the properties of the fluid (density, speed of sound). Still, for this class of problems, the fluid–structure coupling is always considered. The numerical formulations used to solve the coupled interior problem are generally based on the finite element method (FEM) (Zienkiewicz and Taylor 2000a, b) for both the fluid and the structure.

Exterior problems: In this case, the fluid domain is unbounded. The coupling is usually governed by the properties of the fluid and the mass of the structure. This is the case of free radiation and scattering problems. When the coupling is negligible, the problem can be solved

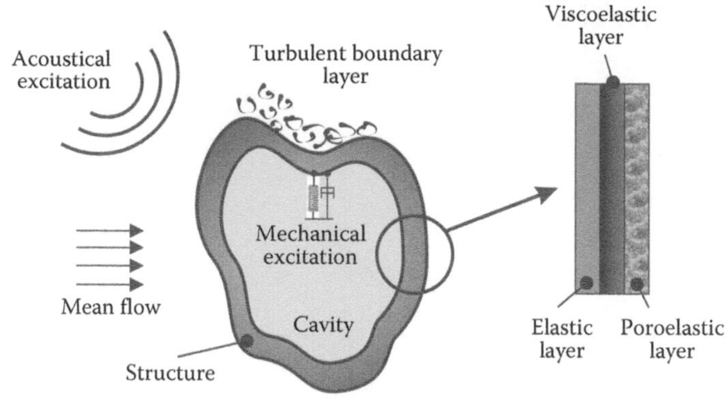

Figure 1.1 Schematic of the studied vibroacoustic problem.

in two steps. First, the response of the structure is calculated assuming it in vacuo. Then, the vibration field of the structure is used as a source for the acoustic problem. The formulations used to solve the coupled problem are generally based on a mixed FEM for the structure and a boundary element method (BEM) for the fluid (Ciskowski and Brebbia 1991). Modern methods use FEM-based methods for both the structure and unbounded domain. In particular, for the latter, perfectly matched layer (PML) absorbing boundaries are being widely used (e.g., Berenger 1994; Bermúdez et al. 2007; Zampolli et al. 2007).

Interior/exterior problems: Here, the structure is in contact with several fluid domains and at least one of them is unbounded. This is the configuration of Figure 1.1. The interior domain is bounded and

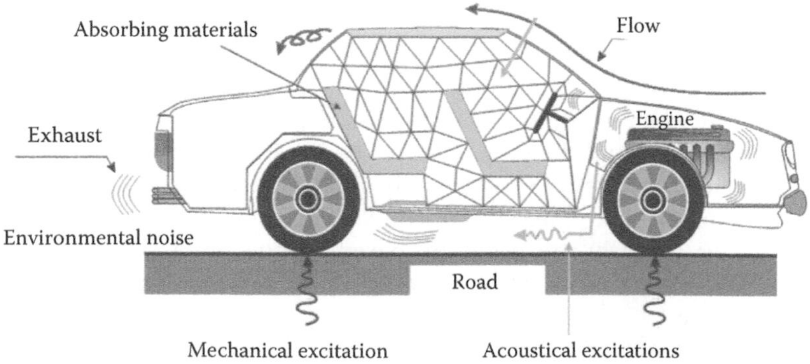

Figure 1.2 Example of a vibroacoustic problem. (Reproduced with permission from ESI/Group.)

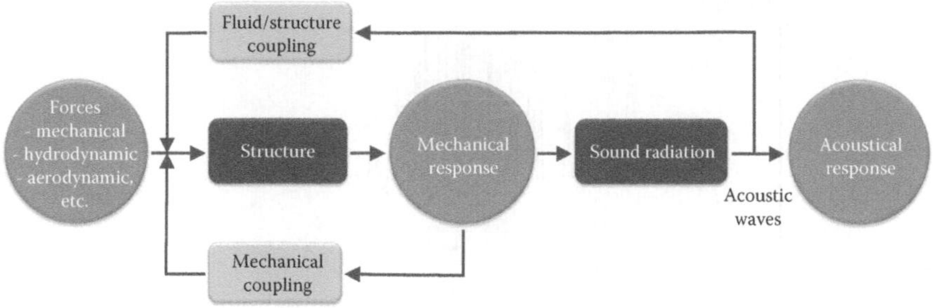

Figure 1.3 Typical coupling mechanisms in a vibroacoustic problem.

the exterior domain is unbounded. Here, depending on the nature of the exterior domain, fully coupled equations should be solved. Again, if the coupling with the exterior domain is negligible, the interior-coupled problem is first solved, followed by the radiation or scattering problem. It is common to use a combination of FEM for the structure and the interior fluid domain and BEM for the exterior fluid.

However, the selection of the solution method is not strictly defined by the class of the coupled vibroacoustics problem. It depends on the frequency domain of interest. Figure 1.5 presents the typical frequency response function (FRF) of a system. Three domains are clearly visible.

Low frequencies (LFs). This domain is characterized by clearly isolated and visible resonances hinting to a modal-controlled behavior. In this

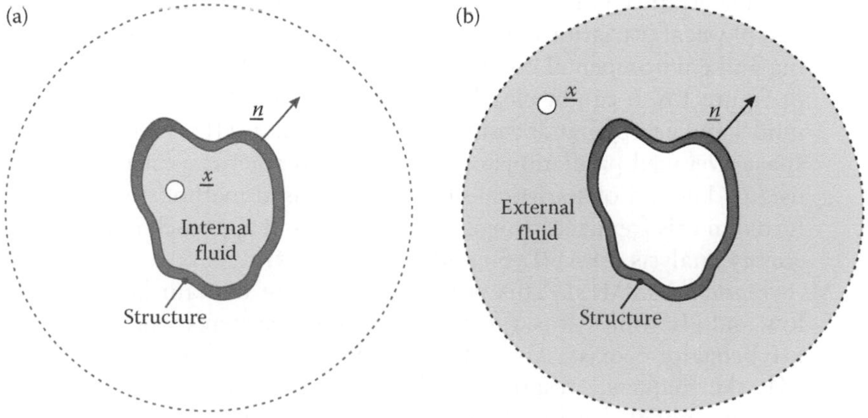

Figure 1.4 Interior (a) and exterior (b) coupled vibroacoustics problems.

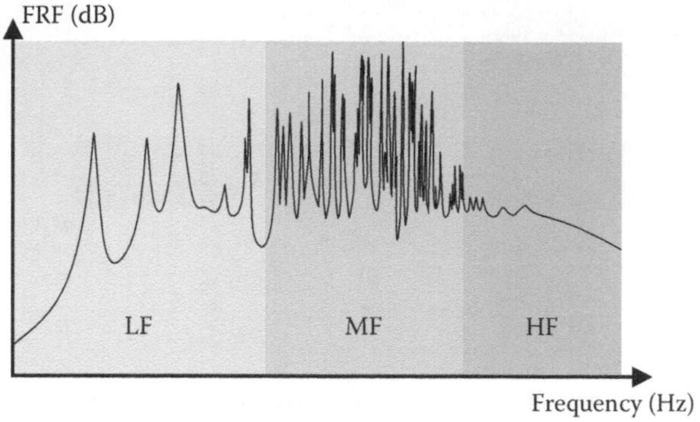

Figure 1.5 Typical FRF of a vibroacoustic system.

domain, the mode count (number of modes) is low and the wavelengths are long (global behavior). The system is deterministic and FEM/BEM is used. An accurate description of the geometry, properties of the fluid, and the structures, that is, the constitutive laws of the involved materials together with the boundary conditions, are needed for precision of the vibroacoustic response.

High frequencies (HFs). The frequency response of the system has no resonance or visible local strong variations. The behavior is very smooth indicating that the modal density of the system is uniform and the mode count is high. Wavelengths are short compared to the characteristic dimensions of the system (local behavior). In consequence, the size and thus the cost of the finite-element (FE)-based models become excessively high. More importantly, in this frequency range, the vibroacoustic response is extremely sensitive to variability and uncertainty in geometrical (fabrication and assembly tolerances) and physical (variations in material properties, joint behavior, operating and environmental conditions) parameters. An example is shown in Figure 1.6. It can be clearly seen that the use of a deterministic FE model and narrowbands results are of no value at HFs. Frequency and space-averaged predictions of the mean (average) response are more useful. The use of statistical- and energy-based methods is thus preferred in this frequency range. The widely used approach is statistical energy analysis (SEA) (Lyon and Dejong 1994).

Mid-frequencies (MFs). This is the intermediate domain between the low and HF limits. It is a feature of complex systems where low and HF behavior coexist. A simple example is provided in Figure 1.7. At MF, the frequency response depicts strong irregularities, indicating that the modal density of the associated conservative system is not uniform; it can be locally high or low. This is typically the case where

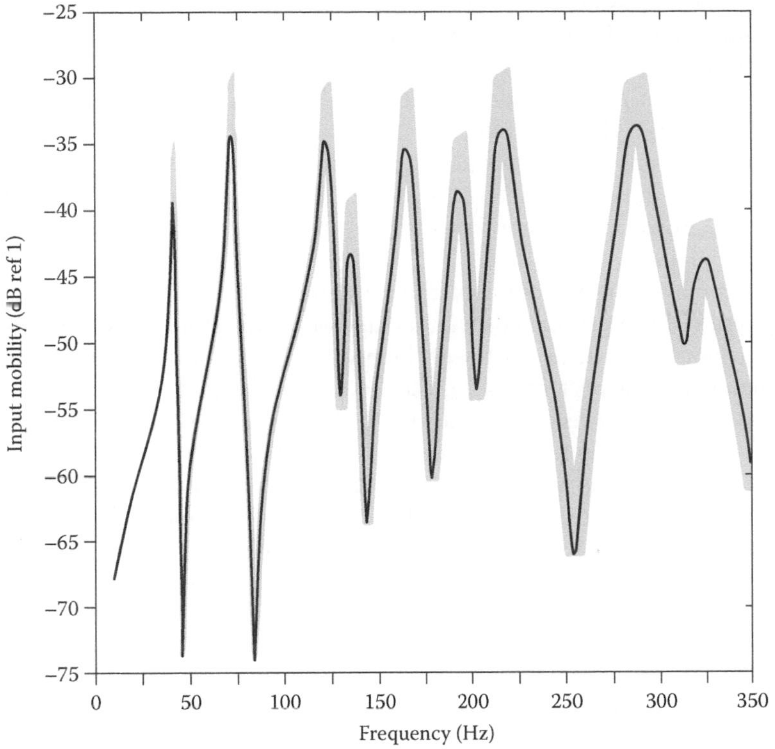

Figure 1.6 Example of the effect of structural uncertainty on the vibroacoustic response of a simply supported flat panel. The FRFs (input mobility) have been obtained by randomly varying the geometrical and physical parameters of the system. Mean value (black line) and spread (gray).

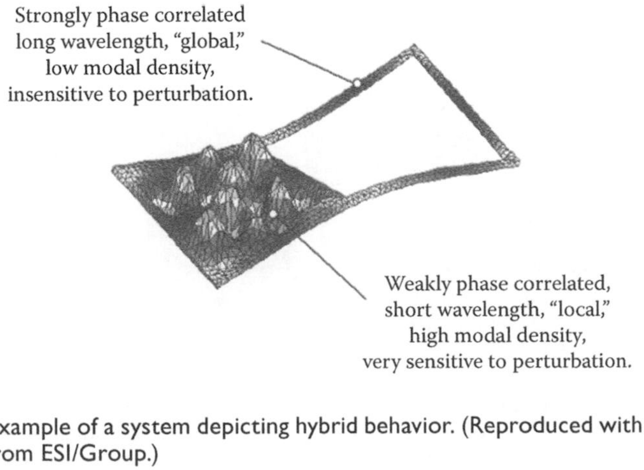

Figure 1.7 Example of a system depicting hybrid behavior. (Reproduced with permission from ESI/Group.)

part of the system (subsystem) depicts local behavior (HF) while other subsystems depict global behavior (LF). In this domain, the details of the boundary conditions, geometry, and constitutive laws of the materials are *a priori* needed for the low mode count subsystems (global behavior), while the effect of uncertainties is important for the high mode count subsystems (local behavior). This observation leads engineers to use hybrid deterministic–statistical methods for such systems (e.g., Soize 1993; Langley 2008).

In summary, the selection of the modeling methodologies depends on the physics of the problem (linear versus nonlinear; stationary, nonstationary, harmonic...), the geometry of the problem (shape, symmetry, periodicity, geometrical aspect ratios...), the frequency domain (low versus high), and the degree of precision or details needed in the solution (local versus global behavior; deterministic versus statistical...). Figure 1.8 provides a brief overview of the main methods used in vibroacoustics. This book will mainly concentrate on FEM and BEM. These two numerical methods are grouped together with analytical methods as shown in Figure 1.8, since both are based on the same fundamental principles and mathematical formulations. For FEM and BEM, additional approximations (e.g., weak formulations, discretization...) are usually added to solve a variety of problems with complex geometry, boundary conditions, load cases, and so on. Analytical methods, on the contrary, are only used when simple geometries (i.e., separable) and boundary conditions are considered. They are, however, more accurate and more insightful. Both methods should be used: analytical models for quick assessment of the vibroacoustic behavior together with quick parameters and "what if" studies and FE for more complex and real-life-like problems. The accuracy and application of FEM and BEM will also be discussed and illustrated through comparison with analytical methods.

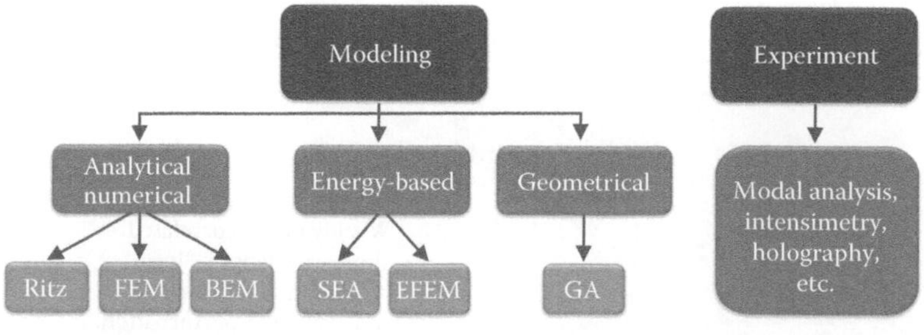

Figure 1.8 Typical methods used to solve vibroacoustic problems (FEM, finite-element method; BEM, boundary element method; SEA, statistical energy analysis; EFEM, energy finite element method; GA, geometrical acoustics.).

1.2 OVERVIEW OF THE BOOK

This book is written for engineering students as well as practicing engineers. Its goal is to present in a simple and straightforward manner the basic theory and implementation of the FE and boundary element (BE) methods applied to vibroacoustics problems.

Several advanced and current issues are not discussed in the book. This is done on purpose. The primary objective of the authors is rather to clearly present the basic concepts, allowing the students to correctly use the two covered computational methods in vibroacoustics applications.

The book is divided into eight chapters. The first two chapters introduce the basic equations governing linear acoustics, electrodynamics, and poroelastic-coupled problems (Chapter 2) and the associated integral formulations (Chapter 3). The next three chapters are dedicated to the interior problem in vibroacoustics. The one-dimensional wave equation is first considered to illustrate the basic concepts of the FE (Chapter 4). Then the three-dimensional (3-D) problem is used to detail the various steps in any classical FE implementation (Chapter 5). Emphasis has been given to the practical side rather than on mathematical justifications. The various steps are illustrated through examples. Next, the classical formulations for the interior fluid–structure coupled problem are presented (Chapter 6). In particular, the standard pressure–displacement formulation together with its numerical implementation are detailed and illustrated through examples. Finally, the last two chapters are devoted to the exterior problem in vibroacoustics. First, the classical BE formulations of the uncoupled exterior problem (acoustic radiation and scattering problems) are presented (Chapter 7). Direct, indirect, and variational formulations are discussed followed by their numerical implementations. Again, several examples are provided to illustrate the implementation and the applications of these methods. Next, the exterior coupled problem is discussed (Chapter 8). Here, an emphasis is given on the general variational formulation and its application.

REFERENCES

Berenger, J.-P. 1994. A perfectly matched layer for the absorption of electromagnetic waves. *Journal of Computational Physics* 114 (2): 185–200.

Bermúdez, A., L. Hervella-Nieto, A. Prieto, and R. Rodriguez. 2007. An optimal perfectly matched layer with unbounded absorbing function for time-harmonic acoustic scattering problems. *Journal of Computational Physics* 223 (2): 469–88.

Ciskowski, R. D. and C. A. Brebbia. 1991. *Boundary Element Methods in Acoustics*. Computational Mechanics Publications, Southampton, UK.

Langley, R. 2008. Recent advances and remaining challenges in the statistical energy analysis of dynamic systems. In *Proceedings of the 7th European Conference on Structural Dynamics*, Southampton, UK.

Lyon, R. H. and R. G. Dejong. 1994. *Theory and Application of Statistical Energy Analysis, Second Edition*. 2nd ed. Boston: Butterworth-Heinemann.

Soize, C. 1993. A model and numerical method in the medium frequency range for vibroacoustic predictions using the theory of structural fuzzy. *The Journal of the Acoustical Society of America* 94 (2): 849–65.

Zampolli, M., T. Alessandra, F. B. Jensen, N. Malm, and J. B. Blottman III. 2007. A computationally efficient finite element model with perfectly matched layers applied to scattering from axially symmetric objects. *The Journal of the Acoustical Society of America* 122 (3): 1472–85.

Zienkiewicz, O. C. and R. L. Taylor. 2000a. *The Finite Element Method*. 5th ed. Vol. 1. Woburn, MA, USA: Butterworth-Heinemann.

Zienkiewicz, O. C. and R. L. Taylor. 2000b. *The Finite Element Method*. 5th ed. Vol. 2. Woburn, MA, USA: Butterworth-Heinemann.

Chapter 2

Basic equations of structural acoustics and vibration

2.1 INTRODUCTION

As mentioned in Chapter 1, this book addresses classical numerical techniques to solve various vibroacoustics problems. In practice, a typical problem involves a structural domain coupled to bounded or unbounded fluid domains and sound-absorbing materials. We then seek to predict the coupled vibratory/acoustic response of the whole system subjected to the given excitations and boundary conditions. In this chapter, we introduce the fundamental equations governing the linear sound wave propagation in fluid domains (acoustics), in solid domains (elastodynamics), and in porous sound-absorbing materials (poroelasticity). We also establish the coupling equations between these domains. All these partial differential equations will be then solved using specific numerical methods, which will be presented in Chapters 3 and 4.

2.2 LINEAR ACOUSTICS

The acoustic pressure disturbance $p(\underline{x},t)$ in a perfect fluid volume Ω_f at rest with sound speed c_0, and density ρ_0, due to an acoustic source distribution $Q(x,t)$, satisfies the inhomogeneous wave equation:

$$\nabla^2 p(\underline{x},t) - \frac{1}{c_0^2}\ddot{p}(\underline{x},t) = -Q(\underline{x},t) \tag{2.1}$$

In the case of a mass source, $Q(x,t)$ represents the rate of mass injection and is related to the volume velocity $Q_s(x,t)$ by

$$Q(\underline{x},t) = \rho_0 \frac{dQ_s}{dt} \tag{2.2}$$

Mass injection is the typical model of monopoles. In this case, the volume velocity, also called source strength, is the product of the surface area and the normal surface velocity of the monopole. For a proof of Equation 2.1, the reader is invited to look at the excellent book of Pierce (1989). To the pressure disturbance, we can associate a particle velocity $\underline{V}(\underline{x},t)$ given by Euler's equation:

$$\frac{\partial \underline{V}(\underline{x},t)}{\partial t} = -\frac{\nabla p(\underline{x},t)}{\rho_0} \tag{2.3}$$

A particle displacement $\underline{U}(\underline{x},t)$ written as

$$\underline{V}(\underline{x},t) = \frac{\partial \underline{U}(\underline{x},t)}{\partial t} \tag{2.4}$$

A density fluctuation $\rho(\underline{x},t)$

$$\rho(\underline{x},t) = \frac{p(\underline{x},t)}{c_0^2} \tag{2.5}$$

Equation 2.1 must be associated to the boundary and initial conditions (Figure 2.1):

- *Neumann boundary condition*: Normal acoustic displacement is specified on the part $\partial\Omega_{f,N}$ of boundary $\partial\Omega_f$:

$$\underline{U} \cdot \underline{n} = \overline{\underline{U}} \cdot \underline{n} \tag{2.6}$$

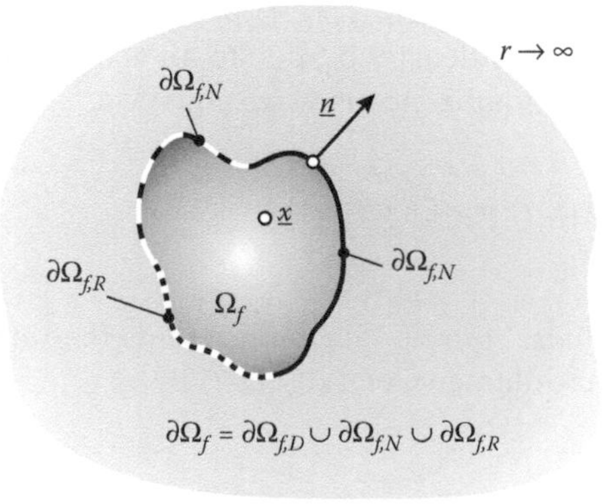

Figure 2.1 Fluid domain and boundary conditions.

- *Dirichlet boundary condition*: Acoustic pressure is specified on the part $\partial\Omega_{f,D}$ of boundary $\partial\Omega_f$:

$$p = \bar{p} \tag{2.7}$$

- *Mixed, Robin, or impedance boundary condition*:* Specific† normalized acoustic admittance is specified on the part $\partial\Omega_{f,R}$ of boundary $\partial\Omega_f$:

$$\beta = \frac{\rho_0 c_0}{Z_n} = \rho_0 c_0 \frac{\underline{V}.\underline{n}}{p} \tag{2.8}$$

where Z_n is the specific acoustic impedance applied on $\partial\Omega_{f,R}$.

In addition, for problems involving acoustic radiation in unbounded media (exterior problem, see Chapter 7), we must specify a condition to ensure that the wave amplitude vanishes at infinity.‡ This is the Sommerfeld radiation condition, which in the 3-D case, reads

$$\lim_{r\to\infty} r\left(\frac{\partial p}{\partial r} + \frac{1}{c_0}\frac{\partial p}{\partial t}\right) = 0 \tag{2.9}$$

Finally, the previous equations must be completed with initial conditions $p(\underline{x},t)\big|_{t=0}$ and $\left.\dfrac{\partial p(\underline{x},t)}{\partial t}\right|_{t=0}$ at all points of Ω_f.

In the following equation, we are interested in harmonic problems, that is, dynamic problems for which the temporal dependency of all variables is sinusoidal with circular frequency ω.§ A convenient mathematical description of the variables is to use the complex formalism

$$p(\underline{x},t) = \Re[\hat{p}(\underline{x})\exp(i\omega t)] = \Re[|\hat{p}(\underline{x})|\exp(i\varphi(\underline{x}))\exp(i\omega t)] \tag{2.10}$$

Solving for $p(\underline{x},t)$ is equivalent to solving for the complex-valued function $\hat{p}(\underline{x})$. There is an amplitude $|\hat{p}(\underline{x})|$ and a phase $\varphi(\underline{x})$ associated to $\hat{p}(\underline{x})$.

In the following equation, we systematically omit the factor $\exp(i\omega t)$ in all the equations. However, we have to keep in mind that the physical value of the acoustic pressure at point $\underline{x}$ and time t is recovered by multiplying $\hat{p}(\underline{x})$ by $\exp(i\omega t)$ and taking the real part.

---

* Also called radiation or absorption condition.

† The term specific will be omitted in the rest of the book.

‡ This condition assumes the absence of acoustic sources at infinity.

§ Note that if the disturbance spectrum is broadband, the temporal signal can be reconstructed using Fourier transform: $p(\underline{x},t) = \frac{1}{2\pi}\int_{-\infty}^{+\infty}\hat{p}(\underline{x},\omega)\exp(i\omega t)d\omega$, ω being the conjugate variable of time t.

For harmonic temporal dependence, Equation 2.1 can be rewritten as

$$\nabla^2 \hat{p}(\underline{x}) + k^2 \hat{p}(\underline{x}) = -\hat{Q}(\underline{x}) \tag{2.11}$$

where $k = \omega/c_0$ is the wavenumber Equation 2.11 is known as *Helmholtz equation*. Dirichlet boundary condition rewrites

$$\hat{p} = \bar{p} \tag{2.12}$$

Neumann boundary condition can be rewritten using Euler's equation (2.3):

$$\frac{\partial \hat{p}}{\partial n} = \rho_0 \omega^2 \underline{\bar{U}} \cdot \underline{n} \tag{2.13}$$

The impedance boundary condition rewrites

$$\frac{\partial \hat{p}}{\partial n} + ik\hat{\beta}\,\hat{p} = 0 \tag{2.14}$$

Finally, for exterior problems, Sommerfeld radiation condition becomes

$$\lim_{r\to\infty} r\left(\frac{\partial \hat{p}}{\partial r} + ik\hat{p}\right) = 0 \tag{2.15}$$

2.3 LINEAR ELASTODYNAMICS

Let us consider a linear elastic solid occupying volume Ω_s (see Figure 2.2). The fundamental equations governing its dynamic behavior are given by the conservation of mass equation, the conservation of momentum equation, and the behavioral law of the solid (the model defining how the solid deforms in response to an applied stress). In the framework of small disturbances around the solid equilibrium position, we can linearize these equations to come up with the following linear elastodynamics equations, relating the linearized displacement field $\underline{u}$, strain tensor $\underline{\underline{\varepsilon}}$, and stress tensor $\underline{\underline{\sigma}}$ at all points of Ω_s

$$\rho_s \frac{\partial^2 \underline{u}}{\partial t^2} = \underline{\nabla} \cdot \underline{\underline{\sigma}} + \rho_s \underline{F_b} \tag{2.16}$$

$$\underline{\underline{\sigma}} = \underline{\underline{\underline{\underline{C}}}}\,\underline{\underline{\varepsilon}} \tag{2.17}$$

$$\underline{\underline{\varepsilon}} = \frac{1}{2}\left(\underline{\nabla}\underline{u} + (\underline{\nabla}\underline{u})^T\right) \tag{2.18}$$

where ρ_s is the density of the solid, $\underline{\nabla}$ is the nabla operator (see Appendix 3B), $\underline{F_b}$ is the body force vector per unit volume and $\underline{\underline{\underline{\underline{C}}}}$ is the fourth-order stiffness (elasticity) tensor Equation 2.17 represents Hooke's law.

In Equation 2.16, the body force distribution per unit mass F_b can include various effects: gravity, thermal effects, initial deformation or pre-stress, etc. It is common to rewrite this equation according to d'Alembert as $\underline{\nabla} \cdot \underline{\underline{\sigma}} + \rho_s \underline{F_b} = 0$ where the body force $\rho_s \underline{F_b}$ accounts for the inertia pseudo-force $\underline{I} = -\rho_s(\partial^2 \underline{u}/\partial t^2)$.

The boundary $\partial\Omega_s$ of the solid is subjected to two types of boundary conditions (Figure 2.2):

- Specified contact forces $\underline{F}$ per unit area applied on $\partial\Omega_{s,N}$:

$$\underline{\underline{\sigma}} \cdot \underline{n} = \underline{F} \tag{2.19}$$

- Specified displacement over $\partial\Omega_{s,D} = \partial\Omega_s / \partial\Omega_{s,N}$:

$$\underline{u} = \underline{\bar{u}} \tag{2.20}$$

Finally, the previous equations are supplemented by two initial conditions providing the values of $\underline{u}(\underline{x},t)\big|_{t=0}$ and $(\partial \underline{u}(\underline{x},t)/\partial t)\big|_{t=0}$ at all points of Ω_s.

For harmonic problems, the linear elastodynamics equation writes

$$\underline{\nabla} \cdot \underline{\underline{\hat{\sigma}}} + \rho_s \underline{\hat{F}_b} + \rho_s \omega^2 \underline{\hat{u}} = 0 \tag{2.21}$$

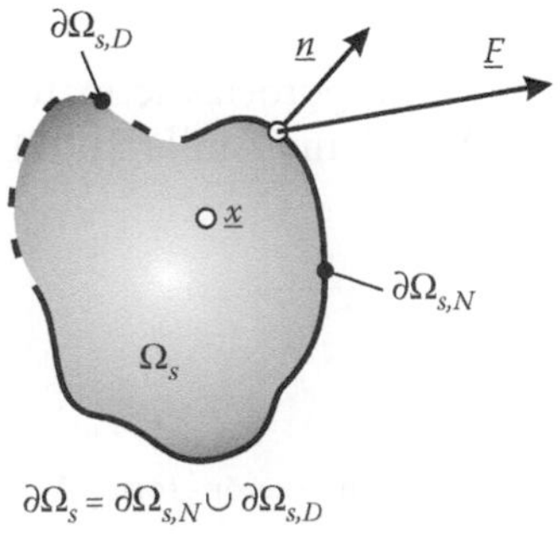

Figure 2.2 Solid domain and boundary conditions.

The boundary conditions of Equations 2.19 and 2.20 remain unchanged. For more details, the reader is invited to refer, for example, to Reddy's book (Reddy 2010).

2.4 LINEAR POROELASTICITY

From a qualitative point of view, a porous material is made up of a solid phase (the matrix or skeleton) and a fluid phase (pores) saturating its network of pores. The matrix can be continuous (e.g., plastic foams, porous ceramics) or not (fibrous or granular materials). The complexity of the microscopic geometry of such a medium makes it difficult to model it at this scale. The modeling is rather done at a macroscopic scale (defined by the wavelength in the medium), wherein this heterogeneous medium is seen as the superposition in time and space of two continuous coupled media, a solid and a fluid. This is the basis of the Biot theory. An extension of this theory, the Biot–Allard theory, is dedicated to the acoustics of porous media. It establishes partial differential equations involving macroscopic solid and fluid displacements $(\underline{u}^s, \underline{U}^f)$ averaged over a representative elementary volume. Alternatively, these equations can be rewritten in terms of solid-phase displacement and interstitial pressure $(\underline{u}^s, p^f)$.

Thus, the wave propagation in poroelastic materials is commonly described using either the classic displacement form $(\underline{u}^s, \underline{U}^f)$ or a mixed-displacement pressure $(\underline{u}^s, p^f)$ form. These are the most popular forms that have been implemented in the context of the FEM over the years. It is not the purpose of this book to cover all the various forms of poroelasticity equations. The reader can refer to Allard and Atalla's book (Allard and Atalla 2009) for details about the modeling of poroelastic materials. Instead, it has been chosen to focus on the mixed $(\underline{u}^s, p^f)$ displacement pressure form, which proves to be pretty efficient from the numerical point of view. Other forms can be implemented in a similar way (Allard and Atalla 2009). In addition, these equations will only be written in the frequency domain. Equations in the time domain can be found in Gorog et al. (1997) and Fellah et al. (2013).

The governing equations* of a poroelastic material in the framework of the mixed $(\underline{u}^s, p^f)$ read as (Allard and Atalla 2009)

$$\underline{\nabla} \cdot \underline{\underline{\tilde{\sigma}}}^s + \omega^2 \tilde{\rho} \underline{\hat{u}}^s + \tilde{\gamma} \underline{\nabla} \hat{p}^f = 0 \tag{2.22}$$

* These equations can be written in vector form $\underline{\nabla} \cdot \underline{\underline{\tilde{\sigma}}}^s + \omega^2 \tilde{\rho} \underline{u}^s + \tilde{\gamma} \underline{\nabla} \hat{p}^f = 0$ and $\nabla^2 \hat{p}^f + \frac{\tilde{\rho}_{22}}{\tilde{R}} \omega^2 \hat{p}^f - \frac{\tilde{\rho}_{22}}{\phi_p^2} \tilde{\gamma} \underline{\nabla} \cdot \underline{\hat{u}}^s = 0$.

$$\nabla^2\hat{p}^f + \frac{\tilde{\rho}_{22}}{\tilde{R}}\omega^2\hat{p}^f - \frac{\tilde{\rho}_{22}}{\phi_p^2}\tilde{\gamma}\underline{\nabla}\cdot\underline{\hat{u}}^s = 0 \tag{2.23}$$

where $\underline{\underline{\tilde{\sigma}}}^s$ is the in vacuo solid-phase stress tensor

$$\underline{\underline{\tilde{\sigma}}}^s = 2N(1+i\eta_p)\frac{\nu_p}{1-2\nu_p}\underline{\nabla}\cdot\underline{\hat{u}}^s\underline{\underline{I}} + 2N(1+i\eta_p)\underline{\underline{\varepsilon}}^s \tag{2.24}$$

where N, η_p, and ν_p are the solid-phase shear modulus, loss factor, and Poisson's ratio, respectively. $\underline{\underline{\varepsilon}}^s$ is the solid-phase strain tensor.

$\tilde{\gamma}$ is a coupling coefficient given by

$$\tilde{\gamma} = \phi_p\left(\frac{\tilde{\rho}_{12}}{\tilde{\rho}_{22}} - \frac{\tilde{Q}}{\tilde{R}}\right) \tag{2.25}$$

where ϕ_p denotes the porosity, $\tilde{Q}$ can be interpreted as a coupling coefficient between the deformation of the solid phase and the fluid phase, and $\tilde{R}$ is the dynamic bulk modulus of the fluid phase occupying a fraction ϕ_p of a unit volume of the porous material. $\tilde{Q}$ and $\tilde{R}$ are related to the dynamic bulk modulus of the air in the pores $\tilde{K}_e$. $\tilde{K}_e$ accounts for the dissipation due to the thermal exchanges between the two phases.

$\tilde{\rho}$ is a dynamic density given by

$$\tilde{\rho} = \tilde{\rho}_{11} - \frac{\tilde{\rho}_{12}^2}{\tilde{\rho}_{22}} \tag{2.26}$$

where $\tilde{\rho}_{11}$, $\tilde{\rho}_{12}$, and $\tilde{\rho}_{22}$ are Biot's complex-valued dynamic densities given by

$$\begin{aligned}\tilde{\rho}_{11} &= (1-\phi_p)\rho_{sk} + \phi_p\rho_0(\tilde{\alpha}(\omega)-1)\\ \tilde{\rho}_{12} &= -\phi_p\rho_0(\tilde{\alpha}(\omega)-1)\\ \tilde{\rho}_{22} &= \phi_p\rho_0\tilde{\alpha}(\omega)\end{aligned} \tag{2.27}$$

where ρ_0 is the density of the fluid in the pores, ρ_{sk} is the density of the material of the skeleton, and $\tilde{\alpha}(\omega)$ is the dynamic tortuosity. This coefficient accounts for the dissipation due to the viscous effects in the fluid phase. Note that $(1-\phi_p)\rho_{sk}$ represents the apparent mass of the porous material.

Generally, the Johnson–Champoux–Allard (JCA) model (Johnson et al. 1987; Champoux and Allard 1991) is chosen to describe the fluid phase of the porous material. In this case, five input parameters are required, namely the porosity ϕ_p, the flow resistivity σ_p, the tortuosity α_∞, the viscous characteristic length Λ, and the thermal characteristic length Λ'. In addition, if the vibration of the skeleton is accounted for and if the solid phase is assumed isotropic, Young's modulus E_p, loss factor η_p, and Poisson's ratio ν_p are required.

Using the JCA model, the dynamic tortuosity can be written as (Allard and Atalla 2009)

$$\tilde{\alpha}(\omega) = \alpha_\infty - i\frac{\phi_p \sigma_p}{\omega \rho_0}\tilde{G}(\omega) \tag{2.28}$$

where $\tilde{G}(\omega)$ is a viscous correction factor given by

$$\tilde{G}(\omega) = \sqrt{1 + i\frac{4\alpha_\infty^2 \eta \rho_0 \omega}{\sigma_p^2 \Lambda^2 \phi_p^2}} \tag{2.29}$$

and the dynamic bulk modulus $\tilde{K}_e(\omega)$ reads

$$\tilde{K}_e(\omega) = \frac{\gamma P_0}{\gamma - (\gamma - 1)\left[1 + \dfrac{8\eta}{i\Lambda'^2 B^2 \omega \rho_0}\tilde{G}'\right]^{-1}} \tag{2.30}$$

where $\tilde{G}'(\omega)$ is a thermal correction factor

$$\tilde{G}'(\omega) = \sqrt{1 + i\frac{\Lambda'^2 B^2 \rho_0 \omega}{16\eta}} \tag{2.31}$$

where $B^2 = \eta C_p / k_{tc}$ is Prandtl's number, η is the fluid dynamic viscosity, C_p is the heat capacity at constant pressure, k_{tc} is the thermal conductivity, and $\gamma = C_p/C_v$ is the heat capacity ratio.

Other expressions of dynamic tortuosity and dynamic bulk modulus can be used if additional porous material parameters are available (Pride et al. 1993; Wilson 1993; Lafarge and Lemarinier 1997).

Note that the fluid displacement is obtained from the interstitial pressure and the solid-phase displacement by

$$\underline{\hat{U}}^f = \frac{\phi_p}{\tilde{\rho}_{22}\omega^2}\underline{\nabla \hat{p}^f} - \frac{\tilde{\rho}_{12}}{\tilde{\rho}_{22}}\underline{\hat{u}}^s \tag{2.32}$$

The normal flux is given by

$$\phi_p\left(\underline{\hat{U}}^f - \underline{\hat{u}}^s\right) \cdot \underline{n} \tag{2.33}$$

and the total displacement of the porous material is written as

$$(1-\phi_p)\underline{\hat{u}}^s + \phi_p \underline{\hat{U}}^f \tag{2.34}$$

Also note that the total stress tensor of the porous material is

$$\underline{\underline{\hat{\sigma}}}^t = \underline{\underline{\tilde{\sigma}}}^s - \phi\left(1+\frac{\tilde{Q}}{\tilde{R}}\right)\hat{p}^f \underline{\underline{I}} \tag{2.35}$$

where $\phi_p(1 + (\tilde{Q}/\tilde{R}))$ is referred to as Biot–Willis coefficient. An important case of interest is when the porous material skeleton can be considered rigid and motionless or limp. In the case of a rigid motionless frame, $\underline{\hat{u}}^s = 0$ and the porous material is completely described by its interstitial pressure $\hat{p}^f$. Then, Equation 2.23 reduces to

$$\nabla^2 \hat{p}^f + \frac{\tilde{\rho}_{22}}{\tilde{R}}\omega^2 \hat{p}^f = 0 \tag{2.36}$$

which resembles Helmholtz equation. Using the analogy with a fluid, Equation2.36 can be rewritten as

$$\nabla^2 \hat{p}^f + \omega^2 \frac{\tilde{\rho}_e}{\tilde{K}_e}\hat{p}^f = 0 \tag{2.37}$$

where $\tilde{\rho}_e = \tilde{\rho}_{22}/\phi_p$ and $\tilde{K}_e = \tilde{R}/\phi_p$ correspond, respectively, to an effective complex dynamic density and an effective dynamic complex bulk modulus of an equivalent fluid occupying the totality of a unit volume of porous material.

Another important case is the one where the elasticity modulus of the matrix is weak. Then, the elastic force in the solid phase is negligible compared to the inertial and the pressure forces. The equation of motion of the solid phase reduces to

$$\omega^2 \tilde{\rho}\underline{\hat{u}}^s + \tilde{\gamma}\underline{\nabla}\hat{p}^f = 0 \tag{2.38}$$

Taking the divergence of Equation 2.38, the solid-phase dilatation can be substituted in Equation 2.23 to give

$$\nabla^2 \hat{p}^f + \omega^2 \frac{\tilde{\rho}'_e}{\tilde{K}_e}\hat{p}^f = 0 \tag{2.39}$$

where

$$\tilde{\rho}_e' = \left(\frac{1}{\tilde{\rho}_e} + \frac{\tilde{\gamma}^2}{\phi_p \tilde{\rho}} \right)^{-1} \tag{2.40}$$

where $\tilde{\rho}_e'$ is an apparent dynamic complex density of the fluid phase of the soft material. This equation is similar to the one of a rigid frame motionless porous material but accounts for the mass and the damping added by the solid phase.

2.5 ELASTO-ACOUSTIC COUPLING

When an elastic solid vibrates in the presence of a fluid, there is an interaction between the elastic and the acoustic waves. In this case, we must simultaneously solve the structural and the fluid equations subjected to the coupling conditions at the interface between the two domains.

For harmonic problems, the condition of stresses-continuity at the interface is written as

$$\underline{\underline{\hat{\sigma}}}\underline{n} + \hat{p}\underline{n} = 0 \tag{2.41}$$

In addition, the continuity of normal displacements at the interface gives

$$\frac{1}{\rho_0 \omega^2} \frac{\partial \hat{p}}{\partial n} = \underline{\hat{u}} \cdot \underline{n} \tag{2.42}$$

Finally, if the solid domain is coupled to an unbounded fluid domain (exterior problem, see Chapter 7), Sommerfeld condition Equation 2.15 must also be fulfilled.

2.6 PORO-ELASTO-ACOUSTIC COUPLING

Let us again consider the harmonic problems. At the interface between an elastic domain and a poroelastic domain, there is continuity of both the displacement vector and the total stress vector. Moreover, there is no relative displacement between the two phases (the flux is null). Thus

$$\begin{cases} \underline{\hat{u}}^s \cdot \underline{n} = \underline{\hat{u}} \cdot \underline{n} \\ \underline{\underline{\hat{\sigma}}}^t \cdot \underline{n} = \underline{\underline{\hat{\sigma}}} \cdot \underline{n} \\ \phi_p(\underline{\hat{U}}^f - \underline{\hat{u}}^s) \cdot \underline{n} = 0 \end{cases} \tag{2.43}$$

At the interface between a fluid domain and a poroelastic domain, there is continuity of the total normal displacement, the total stress vector, and the pressure. Thus

$$\begin{cases} (1-\phi_p)\underline{\hat{u}}^s \cdot \underline{n} + \phi \underline{\hat{U}}^f \cdot \underline{n} = \dfrac{1}{\rho_0 \omega^2} \dfrac{\partial \hat{p}}{\partial n} \\ \underline{\underline{\hat{\sigma}}}^t \cdot \underline{n} = -\hat{p}\underline{n} \\ \hat{p}^f = \hat{p} \end{cases} \tag{2.44}$$

Finally, at the interface between two poroelastic domains (described by superscripts (1) and (2)), there is continuity of the total stress vector, the interstitial pressure, the solid-phase displacement, and the fluxes. This can be written as

$$\begin{cases} \hat{p}^{f,(1)} = \hat{p}^{f,(2)} \\ \underline{\underline{\hat{\sigma}}}^{t,(1)} \cdot \underline{n} = \underline{\underline{\hat{\sigma}}}^{t,(2)} \cdot \underline{n} \\ \underline{\hat{u}}^{s,(1)} \cdot \underline{n} = \underline{\hat{u}}^{s,(2)} \cdot \underline{n} \\ \phi_p^{(1)} \left(\underline{\hat{U}}^{f,(1)} - \underline{\hat{u}}^{s,(1)} \right) \cdot \underline{n} = \phi_p^{(2)} \left(\underline{\hat{U}}^{f,(2)} - \underline{\hat{u}}^{s,(2)} \right) \cdot \underline{n} \end{cases} \tag{2.45}$$

2.7 CONCLUSION

This chapter described the fundamental governing equations for three classic problems of mechanics: linear acoustics, linear elastodynamics, and linear poroelasticity. The next chapter will introduce the associated integral forms necessary for their resolution using the finite and BE methods.

REFERENCES

Allard, J. F. and N. Atalla. 2009. *Propagation of Sound in Porous Media, Modelling Sound Absorbing Materials*. 2nd ed. Chichester, UK: Wiley-Blackwell.

Champoux, Y. and J.-F. Allard. 1991. Dynamic tortuosity and bulk modulus in air-saturated porous media. *Journal of Applied Physics* 70 (4): 1975–79.

Fellah, Z. E. A., M. Fellah, and C. Depollier. 2013. Transient acoustic wave propagation in porous media:Chapter 6. In *Modeling and Measurement Methods for Acoustic Waves and for Acoustic Microdevices*. Vol. 621. InTech. http://hal.archives-ouvertes.fr/docs/00/86/82/18/PDF/InTech-Transient_acoustic_wave_propagation_in_porous_media.pdf.

Gorog, S., R. Panneton, and N. Atalla. 1997. Mixed displacement–pressure formulation for acoustic anisotropic open porous media. *Journal of Applied Physics* 82 (9): 4192.

Johnson, D. L., J. Koplik, and R. Dashen. 1987. Theory of dynamic permeability and tortuosity in fluid-saturated porous media. *Journal of Fluid Mechanics* 176: 379–402.

Lafarge, D. and P. Lemarinier. 1997. Dynamic compressibility of air in porous structures at audible frequencies. *Journal of the Acoustical Society of America* 102: 1995–2006.

Pierce, A. D. 1989. *Acoustics, an Introduction to Its Physical Principles and Applications*. New York, USA: McGraw-Hill.

Pride, S. R., F. D. Morgan, and A. F. Gangi. 1993. Drag forces of porous-medium acoustics. *Physical Review B, Condensed Matter* 47 (9): 4964–78.

Reddy, J. N. 2010. *Principles of Continuum Mechanics: A study of Conservation Principles with Applications*. Cambridge, UK: Cambridge University Press.

Wilson, D. K. 1993. Relaxation-matched modeling of propagation through porous media, including fractal pore structure. *The Journal of the Acoustical Society of America* 94 (2): 1136–45.

Chapter 3

Integral formulations of the problem of structural acoustics and vibrations

3.1 INTRODUCTION

In the finite element method, we usually seek an approximate solution of a system of differential equations over a domain Ω with boundary $\partial\Omega$:

$$\begin{cases} \mathcal{L}(u) = f_v & \text{in}\,\Omega \\ \mathcal{C}(u) = f_s & \text{over}\,\partial\Omega \end{cases} \tag{3.1}$$

Here, $\mathfrak{L}$ and $\mathcal{C}$ denote differential operators characterizing the system; f_v and f_s are the associated loading terms. The unknown variable u is usually approximated in the form of a weighted sum:

$$u \approx \sum_{i=1}^{N} u_i \ \phi_{t,i} \tag{3.2}$$

The set of functions $\{\phi_{t,i}\}$ are called trial functions. They must satisfy all or part of the boundary conditions. Vector $\{u_i\}$ contains N unknowns to be determined. The direct substitution of Equation 3.2 into Equation 3.1 leads to an error or residual, given by

$$\mathcal{R}(u) = \mathcal{L}(u) - f_v \tag{3.3}$$

This residual is only equal to zero for the exact solution of Equation 3.1.

A classic way to solve for the unknown vector $\{u_i\}$ is to set the integral over domain Ω of the residual, weighted by N arbitrary but integrable functions $\psi_{w,i}$, to zero (Weighted residuals method):

$$\int_{\Omega} \psi_{w,i}\mathcal{R}(\phi_{t,i}, u_i)\,dV = 0 \tag{3.4}$$

This system of N equations can be solved directly for the unknown vector $\{u_i\}$. In this case, the trial functions need to satisfy all boundary conditions. It can also be transformed using an integral formulation to eliminate part of these boundary conditions. The accuracy of the approximation depends on the selection of the trial functions, the weight functions, and the number of terms in the approximation. This chapter summarizes, briefly, the main integral formulations used to solve the classical equations of structural acoustics and vibration. A thorough discussion can be found in several textbooks (e.g., Finlayson 1972; Lanczos 1986; Reddy 1993; Géradin and Rixen 1997). The different methods differ in the choice of test functions and the used integral formulation (=Rayleigh–Ritz, Petrov–Galerkin, Galerkin, least squares, collocations,...). These methods are illustrated through an example in Appendix 3A. In this chapter, we will however focus on describing a general method of construction of the integral formulation, called weak formulation. Then, we will discuss the link between this approach and the variational formulation of the problem. Finally, we will recall the fundamental energy-based principles commonly used in engineering for the direct construction of the weak integral formulation.

3.2 BASIC CONCEPTS

We start by recalling some basic concepts for the integral formulation of the classical problems of continuum mechanics. In an integral formulation, we usually encounter expressions of the form

$$\mathcal{W}(u) = \int_{\Omega} \mathcal{J}(u, \underline{\nabla} u, \ldots) dV = 0 \tag{3.5}$$

The integrand $\mathcal{J}$ is a function of u and its derivatives. $\mathcal{W}$ is in consequence a function whose arguments are themselves functions. It is called a functional.

For a given fixed value of the unknown variable u, $\mathcal{W}$ depends on the function and its derivatives. Suppose that the function u takes, at the same location $\underline{x}$, an arbitrary value $\hat{u}$. This value is different from the real value but is assumed consistent with the constraints of the problem; it must be compatible with the boundary conditions satisfied by u. By definition, the variation $\delta u = \hat{u} - u$ is called the virtual variation of function u. A virtual variation is an arbitrary (and imaginary) variation. However, since it must comply with the constraints of the problem, it must be admissible. Thus, an admissible virtual variation respects the homogeneous form of the boundary conditions imposed on the function. For example, $\delta u = 0$ at all points where the function u is given (Figure 3.1).

The virtual variation operator δ satisfies the classical properties of the derivation operator used in multivariable calculus:

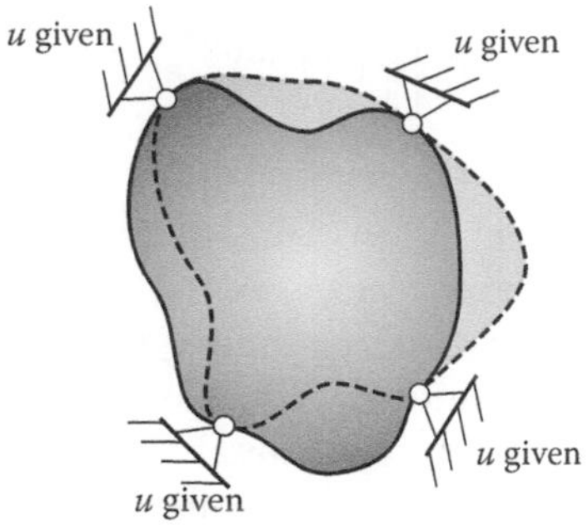

Figure 3.1 Notion of an admissible virtual variation.

$$\begin{cases} \delta(u+v) = \delta u + \delta v \\ \delta(uv) = v\delta u + u\delta v \\ \delta\left(\dfrac{\partial u}{\partial x}\right) = \dfrac{\partial \delta u}{\partial x} \\ \delta\left(\displaystyle\int_\Omega u d\Omega\right) = \displaystyle\int_\Omega \delta u d\Omega \end{cases} \tag{3.6}$$

In particular, the first virtual variation of a functional $\mathcal{J}$ is given by:

$$\delta\mathcal{J} = \frac{\partial \mathcal{J}}{\partial u}\delta u + \frac{\partial \mathcal{J}}{\partial(\partial u/\partial x_i)}\delta\left(\frac{\partial u}{\partial x_i}\right) + \cdots \tag{3.7}$$

where δu and $\delta(\partial u/\partial x_i)$ are virtual variations of functions u and $(\partial u/\partial x_i)$, respectively.

3.2.1 Variational statement: Stationarity of a functional

Consider the simple example of a function of three variables x_1, x_2, and x_3. A variational statement is a mathematical statement of the form: $\delta\mathcal{J} = 0$. This means that there is a point $\underline{x}_s$ in the vicinity of which the functional $\mathcal{J}$ does not change (=has a plateau). We say that $\mathcal{J}$ is stationary at point $\underline{x}_s$. Let $\underline{x}$ be an arbitrary point in the immediate vicinity of $\underline{x}_s$. One can write

$$\underline{x} - \underline{x}_s = \varepsilon \underline{d} = \delta \underline{x} \tag{3.8}$$

with ε an infinitesimal number and $\underline{d}$ an arbitrary orientation vector. The expansion of functional $\mathcal{J}$ in powers of ε in the vicinity of $\underline{x}_s$ leads to

$$\mathcal{J}(\underline{x}) = \mathcal{J}(\underline{x}_s) + \left[\left(\frac{\partial \mathcal{J}}{\partial x_i}\right)_s d_i\right]\varepsilon + O(\varepsilon^2) \tag{3.9}$$

The variational statement $\delta\mathcal{J} = 0$ means that for $\underline{x}$ given by Equation 3.8, the error is of the order of ε^2: $(\partial\mathcal{J}/\partial x_i)_s d_i = 0$. And, since the direction vector $\underline{d}$ is arbitrary, this also means that $(\partial\mathcal{J}/\partial x_i)_s = 0$.

We must distinguish variational principles where the functional is stationary from the ones where the functional is extremal (maximum or minimum). The two principles are not equivalent. For example, the function $\mathcal{J}(x,y,z) = x^2 - y^2 + z^2$ is stationary at point (0,0,0) but is not maximal at this point. Extremal principles contain more information (e.g., principle of minimum potential energy). However, in the variational formulations of structural acoustics and vibrations problems, stationarity principles are sufficient.

3.3 STRONG INTEGRAL FORMULATION

Let $(\mathcal{C})$ denote the space of integrable functions in Ω. Let u and ψ_w be two functions defined in $(\mathcal{C})$. The strong integral formulation, also known as the method of weighted residuals, consists in solving for functions u that cancel the weighted residual integral given in Equation 3.4. Note that we can always build the strong integral formulation of a given differential equation. The weighting functions should only be integrable. At this stage, their selection is arbitrary. In the general case, the weighting functions ψ_w satisfy weaker continuity and differentiability conditions compared to function u or trial functions ϕ_t for which an approximation of the form of Equation 3.2 is sought. A strong integral formulation is equivalent to the corresponding differential equation. It contains no information about the boundary conditions of the system.

EXAMPLE 3.1

Following Euler–Bernoulli's assumptions, and assuming a harmonic time dependence exp $(i\omega t)$ (see Chapter 2, Equation 2.10), the equations governing the flexural vibration of a thin simply supported beam are given by

$$\begin{cases} EI\dfrac{d^4\hat{u}}{dx^4} - m_s\omega^2\,\hat{u} = 0 \\ \hat{u}(0) = \dfrac{d^2\hat{u}(0)}{dx^2} = 0 \\ \hat{u}(L) = \dfrac{d^2\hat{u}(L)}{dx^2} = 0 \end{cases} \tag{3.10}$$

with EI: bending stiffness
m_s: mass per unit length
L: length of the beam
u: flexural displacement

Here, operator $\mathcal{L}$ of Equation 3.1, is given by $\mathcal{L} = EI\,(d^4/dx^4) - m_s\omega^2$. Its order is $m = 4$. The exact solution of Equation 3.10 must be differentiable up to the fourth order and satisfy all four boundary conditions. Let ψ_w be an arbitrary integrable function, the strong integral formulation associated to Equation 3.10 is given by

$$W(\hat{u}, \hat{\psi}_w) = \int_0^L \hat{\psi}_w(x)\left[EI\frac{d^4\hat{u}}{dx^4} - m_s\omega^2\hat{u}(x)\right]dx = 0 \tag{3.11}$$

3.4 WEAK INTEGRAL FORMULATION

In practice, it is desirable to lower the maximum order of differentiation appearing in the strong integral formulation and automatically satisfy certain boundary conditions called natural boundary conditions. The integral formulation achieving these two conditions is called the weak integral formulation. It is an integral in which the order of differentiability is distributed between the unknown variables and the weighting function and includes the natural boundary conditions of the system.

In the weak integral formulation, the order of the derivatives of the test functions is lower than in the strong formulation (which is equal to that in the initial differential equations). In particular, if the highest-order derivative is $m = 2n$ (a sufficient condition for a self-adjoint operator and therefore a symmetric system), the weak integral representation is obtained after n integration by parts. The maximum-order derivative in the weak formulation is then $m - n = n$. For this family of variational formulations, the boundary conditions can be divided into two classes:

- Essential boundary conditions (geometric or kinematic). They involve derivatives up to a maximum order of $n - 1$. These conditions are called essential because the weighting function must satisfy them to be admissible. In structural dynamics, these conditions involve displacements and rotations. In acoustics, they involve pressure, velocity potentials, and displacement potentials.
- Natural boundary conditions. The maximum order of the derivatives in these conditions is between n and m. They are called natural since they will be explicitly taken into account during the construction of the variational formulation. In structural dynamics, these involve

conditions on forces and moments. In acoustics, they involve the particle velocity or a relationship between the pressure and the velocity (impedance conditions).

3.5 CONSTRUCTION OF THE WEAK INTEGRAL FORMULATION

To build the weak integral formulation of a differential equation, we proceed in three steps:

Step 1: Write the strong integral formulation of the differential equation, Equation 3.4.

Step 2: Use the rules of integration by parts to distribute evenly, when possible, the order of differentiation between the unknown variable and the weighting function. Symbolically, we write

$$\int_{\Omega} \psi_w \, \mathcal{L}(u) dV = \int_{\Omega} \widehat{\mathcal{L}}^T(\psi_w) \widehat{\mathcal{L}}(u) dV + \int_{\partial\Omega} \widehat{\mathcal{C}}^T(\psi_w) \widehat{\mathcal{C}}(u) dS \tag{3.12}$$

The order of the differential operators $\widehat{\mathcal{L}}$ and $\widehat{\mathcal{L}}^T$ is less than the order m of the initial operator $\mathcal{L}$. Differential operators $\widehat{\mathcal{C}}$ and $\widehat{\mathcal{C}}^T$ are associated with boundary conditions. In the case of a self-adjoint operator, we can stop the process at an intermediate stage where the differential operator $\mathcal{L}$ is evenly distributed between ψ_w and u: $\widehat{\mathcal{L}}^T = \widehat{\mathcal{L}}$. In this case, Equation 3.12 reads as follows:

$$\int_{\Omega} \psi_w \, \mathcal{L}(u) dV = \int_{\Omega} \widehat{\mathcal{L}}(\psi_w) \widehat{\mathcal{L}}(u) dV + \int_{\partial\Omega} \tilde{\mathcal{C}}^T(\psi_w) \tilde{\mathcal{C}}(u) dS \tag{3.13}$$

Step 3: Evaluate the boundary integral terms by requiring the weighting functions to satisfy the homogeneous form of the essential boundary conditions and the unknown variables to satisfy the natural boundary conditions.

In summary, the solution of the system Equation 3.1 is approximated by the solution of the weak integral formulation, Equation 3.12 or 3.13. This solution has the advantage of requiring the trial functions to satisfy only part of the differentiability and boundary conditions of Equation 3.1. It represents, therefore, only an approximate (weak) solution. However, it is shown that this solution can be made as close as desired to the exact solution.

Note: The weak integral formulation obtained by choosing $\psi_w = \delta u$ is the so-called Galerkin weak formulation.

3.6 FUNCTIONAL ASSOCIATED WITH AN INTEGRAL FORMULATION: STATIONARITY PRINCIPLE

For certain class of problems defined by Equation 3.1, it is possible to construct a functional $\Pi(u, \underline{\nabla} u, \ldots)$ satisfying

$$\delta\Pi = \mathcal{W}(u, \delta u) = 0 \tag{3.14}$$

where $\mathcal{W}$ is a Galerkin weak formulation.

Statement (3.14) is called a variational statement and the corresponding weak formulation is called a variational formulation. In particular, this construction is possible if (i) operators $\mathcal{L}$ and $\mathcal{C}$ are linear and all derivatives in these operators have even orders and (ii) the forcing terms f_v and f_s are independent of u. These two conditions are sufficient but not necessary.

To clarify the variational statement (3.14), let us rewrite the system by separating the boundary conditions in two parts: (i) natural boundary conditions (automatically accounted for in the weak formulation):

$$\mathcal{C}_n(u) = f_n \quad \text{on } \partial\Omega_n \tag{3.15}$$

and essential boundary conditions (to be satisfied by the unknown variable u):

$$\mathcal{C}_e(u) = f_e \quad \text{on } \partial\Omega_e \tag{3.16}$$

In some cases (for conservative systems), we can construct from the Galerkin weak formulation, a functional Π satisfying:

$$\begin{cases} \delta\Pi = \mathcal{W}(u, \delta u) = 0 \\ \mathcal{C}_e(u) = f_e \quad \text{on } \partial\Omega_e \end{cases} \tag{3.17}$$

This variational equation, also known as the stationarity principle, states that among all admissible functions, the solution of the system makes the functional Π stationary.

Note: In continuum mechanics, the weak integral formulation or the associated variation statement can be constructed directly from the principles of analytical and Lagrangian mechanics (principle of virtual work, principle of minimum potential energy, Hamilton's principle…).

EXAMPLE 3.2

Consider Example 3.1. The strong integral formulation associated to the problem is given by Equation 3.11. Imposing on ψ_w to be twice differentiable, the weak integral formulation is obtained following two integration by parts:

$$\begin{aligned} W(\hat{u},\hat{\psi}_w) = &\int_0^L \left[EI \frac{d^2\hat{\psi}_w}{dx^2}\frac{d^2\hat{u}}{dx^2} - m_s\omega^2\hat{\psi}_w(x)\hat{u}(x) \right] dx \\ &+ \left[EI\hat{\psi}_w(x)\frac{d^3\hat{u}}{dx^3} \right]_0^L - \left[EI\frac{d\hat{\psi}_w}{dx}\frac{d^2\hat{u}}{dx^2} \right]_0^L \end{aligned} \tag{3.18}$$

Taking into account the boundary conditions simplifies the weak formulation into:

$$W(\hat{u},\hat{\psi}_w) = \int_0^L \left[EI \frac{d^2\hat{\psi}_w}{dx^2}\frac{d^2\hat{u}}{dx^2} - m_s\omega^2\hat{\psi}_w(x)\hat{u}(x) \right] dx + \left[EI\,\hat{\psi}_w(x)\frac{d^3\hat{u}}{dx^3} \right]_0^L \tag{3.19}$$

Thus, we see that the boundary condition $d^2\hat{u}/dx^2 = 0$ is automatically taken into account in the weak formulation. Such a condition is called a natural condition.

To obtain the Galerkin's weak formulation, we select $\hat{\psi}_w = \delta\hat{u}$ an arbitrary but admissible variation of $\hat{u}$. The variation is said to be admissible in the sense that it must satisfy the homogeneous form of the essential conditions of the problem. Here, it must satisfy $\delta\hat{u}(0) = \delta\hat{u}(L) = 0$. In consequence, the Galerkin's weak formulation is given by

$$\begin{cases} \displaystyle\int_0^L \left[EI \frac{d^2\delta\hat{u}}{dx^2}\frac{d^2\hat{u}}{dx^2} - m_s\omega^2\delta\hat{u}\,\hat{u} \right] dx = 0 \\ \forall\ (\hat{u},\delta\hat{u}) \Big/ \begin{array}{l} \hat{u}(0) = \hat{u}(L) = 0 \\ \delta\hat{u}(0) = \delta\hat{u}(L) = 0 \end{array} \end{cases} \tag{3.20}$$

The associated variational statement reads

$$\begin{cases} \delta\Pi = 0 / \hat{u}(0) = \hat{u}(L) = 0 \ \text{ with} \\ \Pi(\hat{u}) = \dfrac{1}{2}\displaystyle\int_0^L \left[EI\left(\frac{d^2\hat{u}}{dx^2}\right)^2 - m_s\omega^2\hat{u}^2 \right] dx \end{cases} \tag{3.21}$$

EXAMPLE 3.3

Variational formulation of the equation of linear acoustics

Recall the general Helmholtz equation (Chapter 2):

$$\begin{cases} \nabla^2\hat{p} + k^2\hat{p} = 0 & \text{in}\,\Omega_f \\ \hat{p} = \bar{p} & \text{over}\,\partial\Omega_{f,D} \\ \dfrac{\partial\hat{p}}{\partial n} = \rho_0\omega^2\underline{\bar{U}}\cdot\underline{n} & \text{over}\,\partial\Omega_{f,N} \\ \dfrac{\partial\hat{p}}{\partial n} + ik\hat{\beta}\hat{p} = 0 & \text{over}\,\partial\Omega_{f,R} \end{cases} \tag{3.22}$$

with $\partial\Omega_f = \partial\Omega_{f,D} \cup \partial\Omega_{f,N} \cup \partial\Omega_{f,R}$ (Figure 3.2).

The associated strong integral formulation reads

$$\int_{\Omega_f} \hat{q}_t(\nabla^2\hat{p} + k^2\hat{p})dV = 0 \tag{3.23}$$

with $\hat{q}_t$ an arbitrary but integrable function in Ω_f

Using the first Green's formula (see Appendix 3B), we obtain:

$$\int_{\Omega_f} \hat{q}_t\nabla^2\hat{p}\,dV = -\int_{\Omega_f} \underline{\nabla\hat{p}}\cdot\underline{\nabla\hat{q}_t}\,dV + \int_{\partial\Omega_f} \hat{q}_t\frac{\partial\hat{p}}{\partial n}dS \tag{3.24}$$

In consequence Equation 3.23 transforms into

$$\int_{\Omega_f} \left[\underline{\nabla}\hat{p}\cdot\underline{\nabla}\hat{q}_t - k^2\hat{p}\hat{q}_t\right]dV - \int_{\partial\Omega_f} \hat{q}_t\frac{\partial\hat{p}}{\partial n}dS = 0 \tag{3.25}$$

This time functions $\hat{q}_t$ and $\hat{p}$ must be at least differentiable.

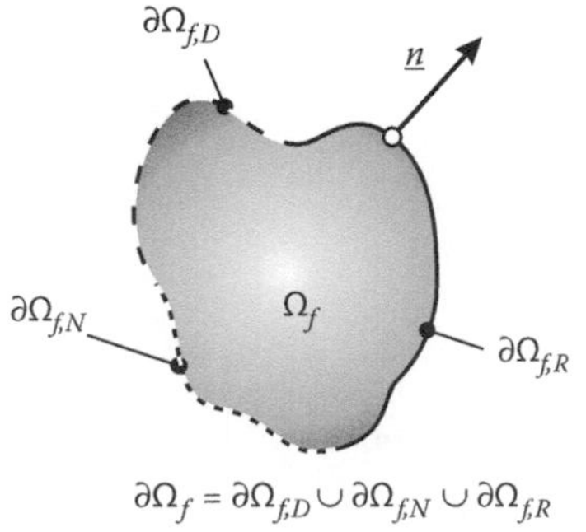

Figure 3.2 A schematic of the acoustic problem.

Using the boundary conditions, the integral over the boundary can be written as

$$\int_{\partial\Omega_f} \hat{q}_t \frac{\partial \hat{p}}{\partial n} dS = \int_{\partial\Omega_{f,D}} \hat{q}_t \frac{\partial \hat{p}}{\partial n} dS + \int_{\partial\Omega_{f,N}} \hat{q}_t (\rho_0 \omega^2 \underline{\bar{U}} \cdot \underline{n}) dS + \int_{\partial\Omega_{f,R}} \hat{q}_t(-ik\beta\hat{p})\, dS \tag{3.26}$$

Choosing, $\hat{q}_t = \delta\hat{p}$, an admissible variation of $\hat{p}$, we must have $\hat{q} = \delta\hat{p} = 0$ on $\partial\Omega_{f,D}$ since the pressure is known on $\partial\Omega_{f,D}$. In consequence, the Galerkin's weak formulation of the acoustic problem, Equation 3.22, reads:

$$\begin{cases} \int_{\Omega_f} [(\underline{\nabla}\delta\hat{p} \cdot \underline{\nabla}\hat{p}) - k^2 \delta\hat{p}\, \hat{p}]\, dV - \int_{\partial\Omega_{f,N}} \rho_0 \omega^2 \delta\hat{p} \underline{\bar{U}} \cdot \underline{n}\, dS + \int_{\partial\Omega_{f,R}} ik\beta\delta\hat{p}\, \hat{p}\, dS = 0 \\ \text{with } \hat{p} = \bar{p} \quad \text{over } \partial\Omega_{f,D} \end{cases} \tag{3.27}$$

The associated variational statement reads:

$$\begin{cases} \delta\Pi = 0 / \hat{p} = \bar{p} \text{ over } \partial\Omega_{f,D} \text{ with} \\ \Pi(\hat{p},\hat{p}) = \frac{1}{2}\left[\int_{\Omega_f} [\underline{\nabla}\hat{p} \,.\, \underline{\nabla}\hat{p} - k^2 \hat{p}^2]\, dV + \int_{\partial\Omega_{f,R}} ik\beta\, \hat{p}^2\, dS\right] - \int_{\partial\Omega_{f,N}} \rho_0 \omega^2 \underline{\bar{U}} \cdot \underline{n}\, \hat{p}\, dS \end{cases} \tag{3.28}$$

Note 1. Dividing the various terms of $\Pi(\hat{p},\hat{p})$ by $\rho_0\omega^2$, an energy interpretation of the functional associated with the acoustic problem can be given. The time-averaged energy associated to Equation 3.28 is obtained by taking $1/2\,\Re(\Pi(\hat{p},\hat{p}^*))$ namely

$$\frac{1}{2}\Re(\Pi(\hat{p},\hat{p}^*)) = T_{ac}(\hat{p},\hat{p}^*) - V_{ac}(\hat{p},\hat{p}^*) + D_{ac}(\hat{p},\hat{p}^*) + W_{ac}(\hat{p},\hat{p}^*) \tag{3.29}$$

with

$$\begin{cases} T_{ac}(\hat{p},\hat{p}^*) = \dfrac{1}{4\rho_o\omega^2} \displaystyle\int_{\Omega_f} \left|\underline{\nabla}\hat{p}\right|^2 dV \\ V_{ac}(\hat{p},\hat{p}^*) = \dfrac{1}{4\rho_o c^2} \displaystyle\int_{\Omega_f} \left|\hat{p}\right|^2 dV \\ D_{ac}(\hat{p},\hat{p}^*) = \Re\left(\dfrac{i\hat{\beta}}{4\rho_o c\omega} \displaystyle\int_{\partial\Omega_{f,R}} \left|\hat{p}\right|^2 dS\right) \\ W_{ac}(\hat{p}^*) = -\dfrac{1}{2}\Re\left(\displaystyle\int_{\partial\Omega_{f,N}} \hat{p}^* \underline{\bar{U}} \cdot \underline{n}\, dS\right) \end{cases} \tag{3.30}$$

T_{ac}, V_{ac}, D_{ac}, and W_{ac} represent the time averaged acoustic kinetic energy, time averaged potential energy, time averaged dissipated energy, and time averaged work done by the pressure in the prescribed normal displacement field, respectively.

Note 2. Consider the case where the displacement is prescribed over the total boundary $\partial\Omega_f$ of Ω_f. For $\omega = 0$, the pressure field reduces to the static pressure p^s in the cavity. This pressure is related to the particle displacement by

$$p^s = -\rho_0 c_0^2 \nabla \cdot \underline{U} \quad \text{in}\,\Omega_f \tag{3.31}$$

Moreover, the particle displacement is related to the prescribed displacement by

$$\underline{U} \cdot \underline{n} = \underline{\bar{U}} \cdot \underline{n} \quad \text{on}\,\partial\Omega_f \tag{3.32}$$

In consequence, using the divergence theorem (see Appendix 3B), the static pressure is related to the prescribed normal displacement by

$$p^s = -\frac{\rho_0 c_0^2}{\Omega_f} \int_{\partial\Omega_f} \underline{\bar{U}} \cdot \underline{n}\, dS \tag{3.33}$$

This relation represents a constraint that must be added to the weak formulation of the acoustic problem to be valid for $\omega = 0$.

EXAMPLE 3.4

Weak integral formulation of the poroelasticity equation

Recall Equations 2.21 and 2.22 governing the motion of a poroelastic material in the framework of the mixed $(\underline{u}^s, p^f)$ formulation

$$\begin{cases} \underline{\nabla} \cdot \underline{\underline{\tilde{\sigma}}}^s(\underline{\hat{u}}^s) + \omega^2 \tilde{\rho} \underline{\hat{u}}^s + \tilde{\gamma} \underline{\nabla} \hat{p}^f = 0 \\ \nabla^2 \hat{p}^f + \omega^2 \dfrac{\tilde{\rho}_{22}}{\tilde{R}} \hat{p}^f - \omega^2 \dfrac{\tilde{\rho}_{22}}{\phi_p^2} \tilde{\gamma} \underline{\nabla} \cdot \underline{\hat{u}}^s = 0 \end{cases} \tag{3.34}$$

Let $\delta\underline{\hat{u}}^s$ and $\delta\hat{p}^f$ be admissible variations of the solid phase displacement vector and the interstitial fluid pressure of the poroelastic medium, respectively. Using the methodology exposed in the previous examples, the weak integral formulation associated with these equations is given by

$$\int_{\Omega_p} \tilde{\sigma}^s_{ij}(\underline{\hat{u}}^s)\hat{\varepsilon}^s_{ij}(\delta\underline{\hat{u}}^s)\,dV - \omega^2\int_{\Omega_p} \tilde{\rho}\hat{u}^s_i\delta\hat{u}^s_i\,dV - \int_{\Omega_p} \tilde{\gamma}\frac{\partial \hat{p}^f}{\partial x_i}\delta\hat{u}^s_i\,dV$$
$$- \int_{\partial\Omega_p} \tilde{\sigma}^s_{ij}(\underline{\hat{u}}^s)n_j\delta\hat{u}^s_i\,dS = 0$$
$$\int_{\Omega_p}\left[\frac{\phi_p^2}{\omega^2\tilde{\rho}_{22}}\frac{\partial \hat{p}^f}{\partial x_i}\frac{\partial(\delta\hat{p}^f)}{\partial x_i} - \frac{\phi_p^2}{\tilde{R}}\hat{p}^f\delta\hat{p}^f\right]dV - \int_{\Omega_p}\tilde{\gamma}\frac{\partial(\delta\hat{p}^f)}{\partial x_i}\hat{u}^s_i\,dV$$
$$+ \int_{\partial\Omega_p}\left[\tilde{\gamma}\hat{u}_n - \frac{\phi_p^2}{\tilde{\rho}_{22}\omega^2}\frac{\partial \hat{p}^f}{\partial n}\right]\delta\hat{p}^f\,dS = 0 \quad \forall(\delta\underline{\hat{u}}^s,\delta\hat{p}^f) \tag{3.35}$$

where we have used Einstein summation notation.

Here, Ω_p and $\partial\Omega_p$ refer to the poroelastic domain and its bounding surface (Figure 3.3). $\underline{n}$ is the unit outward normal vector to the boundary surface $\partial\Omega_p$, and subscript n denotes the normal component of a vector. $\hat{u}^s_i$, $\tilde{\sigma}^s_{ij}$, ε^s_{ij} and n_i represent the components of solid phase displacement vector, in vacuo solid-phase stress tensor, solid-phase strain tensor and normal vector in a cartesian coordinate system (x_1, x_2, x_3).

To simplify the coupling conditions, with elastic and other poroelastic media, a total stress tensor with components $\hat{\sigma}^t_{ij} = \tilde{\sigma}^s_{ij} - \phi_p(1 + (\tilde{Q}/\tilde{R}))\hat{p}^f\delta_{ij}$ is introduced and the weak integral formulation transforms into:

$$\int_{\Omega_p}\left[\tilde{\sigma}^s_{ij}\delta\hat{\varepsilon}^s_{ij} - \omega^2\tilde{\rho}\hat{u}^s_i\delta\hat{u}^s_i\right]dV + \int_{\Omega_p}\left[\frac{\phi_p^2}{\tilde{\alpha}\rho_0\omega^2}\frac{\partial \hat{p}^f}{\partial x_i}\frac{\partial(\delta\hat{p}^f)}{\partial x_i} - \frac{\phi_p^2}{\tilde{R}}\hat{p}^f\,\delta\hat{p}^f\right]dV$$
$$- \int_{\Omega_p}\frac{\phi_p}{\tilde{\alpha}}\delta\left(\frac{\partial \hat{p}^f}{\partial x_i}\hat{u}^s_i\right)dV - \int_{\Omega_p}\phi_p\left(1 + \frac{\tilde{Q}}{\tilde{R}}\right)\delta(\hat{p}^f\hat{u}^s_{i,i})\,dV$$
$$- \int_{\partial\Omega_p}\hat{\sigma}^t_{ij}n_j\delta\hat{u}^s_i\,dS - \int_{\partial\Omega_p}\phi_p(\hat{u}^f_n - \hat{u}^s_n)\delta\hat{p}^f\,dS = 0 \quad \forall(\delta\hat{u}^s_i,\delta\hat{p}^f) \tag{3.36}$$

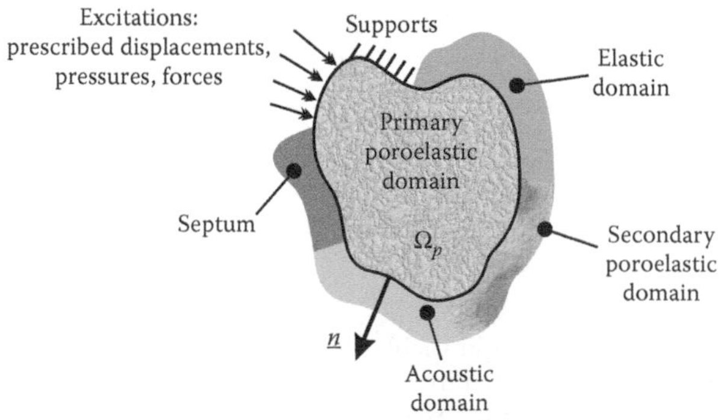

Figure 3.3 A schematic of the linear poroelastic problem with various coupling conditions.

In the above equation, we have used the identities $\tilde{\gamma} + \phi_p(1 + \tilde{Q}/\tilde{R}) = \phi_p/\tilde{\alpha}$ and $\tilde{\rho}_{22} = \tilde{\alpha}\rho_0$ where $\tilde{\alpha}$ is the dynamic tortuosity. Also, $(.)_{,i}$ denotes the partial derivative of the quantity in parenthesis with respect to x_i.

To illustrate the imposition of the boundary conditions, consider the case where the poroelastic domain is submitted to a prescribed displacement field $\underline{\bar{u}}$ on $\partial\Omega_{p,u}$ and a prescribed pressure field $\bar{p}$ on $\partial\Omega_{p,p} = \partial\Omega_p/\partial\Omega_{p,u}$ (this corresponds to a special case of the practical problem of a poroelastic treatment added to a structure-cavity problem, Figure 3.4).

The boundary conditions associated with the imposed displacement fields are

$$\begin{cases} \hat{u}_n^f - \hat{u}_n^s = 0 \\ \hat{u}_i^s = \bar{u}_i \end{cases} \quad \text{on } \partial\Omega_{p,u} \tag{3.37}$$

The first condition expresses the continuity of the normal displacements between the solid phase and the fluid phase. The second equation expresses the continuity between the imposed displacement vector and the solid phase displacement vector. Since the displacement is imposed, the admissible variation $\delta\underline{\hat{u}}^s = \underline{0}$ on $\partial\Omega_{p,u}$. Consequently, the surface integrals of Equation 3.36 simplify to

$$I^p = -\int_{\partial\Omega_{p,p}} \hat{\sigma}_{ij}^t n_j \delta\hat{u}_i^s \, dS - \int_{\partial\Omega_{p,p}} \phi(\hat{u}_n^f - \hat{u}_n^s)\delta\hat{p}^f \, dS \tag{3.38}$$

However, the boundary conditions associated with the imposed pressure field $\bar{p}$ on $\partial\Omega_{p,p}$ are

$$\begin{cases} \hat{\sigma}_{ij}^t n_j = -\bar{p} n_i \\ \hat{p}^f = \bar{p} \end{cases} \tag{3.39}$$

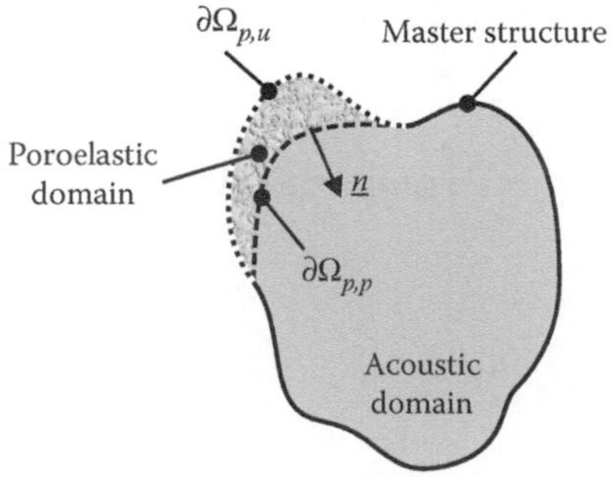

Figure 3.4 Example of a poroelastic domain coupled to an acoustic cavity and an elastic structure.

which express the continuity of the total normal stress and the continuity of the pressure through the interface $\partial\Omega_{p,p}$. Since the pressure is imposed, the admissible variation $\delta\hat{p}^f$ will fall to zero. As a consequence, Equation 3.36 reduces to

$$I^p = \int_{\partial\Omega_{p,p}} \bar{p}\delta\hat{u}_n^s \, dS \tag{3.40}$$

and the weak integral formulation of the problem reads

$$\begin{aligned}
&\int_{\Omega_p} [\tilde{\sigma}_{ij}^s \delta\hat{\varepsilon}_{ij}^s - \omega^2 \tilde{\rho}\hat{u}_i^s \delta\hat{u}_i^s] \, dV + \int_{\Omega_p} \left[\frac{\phi_p^2}{\tilde{\alpha}\rho_0\omega^2} \frac{\partial\hat{p}^f}{\partial x_i} \frac{\partial(\delta\hat{p}^f)}{\partial x_i} - \frac{\phi_p^2}{\tilde{R}} \hat{p}^f \, \delta\hat{p}^f \right] dV \\
&- \int_{\Omega_p} \frac{\phi_p}{\tilde{\alpha}} \delta\left(\frac{\partial\hat{p}^f}{\partial x_i} \hat{u}_i^s \right) dV - \int_{\Omega_p} \phi_p \left(1 + \frac{\tilde{Q}}{\tilde{R}} \right) \delta\left(\hat{p}^f \hat{u}_{i,i}^s \right) dV + \int_{\partial\Omega_{p,p}} \bar{p}\delta\hat{u}_n^s dS = 0 \\
&\forall \, (\delta\hat{u}_i^s, \delta\hat{p}^f) \text{ with } \underline{\hat{u}}^s = \underline{\bar{u}} \text{ on } \partial\Omega_{p,u}, \hat{p}^f = \bar{p} \text{ on } \partial\Omega_{p,p}
\end{aligned} \tag{3.41}$$

3.7 PRINCIPLE OF VIRTUAL WORK

The principle of virtual work is a variational model commonly used in solid mechanics. It states that for any virtual displacement $\underline{\breve{u}}$, the sum $W(\underline{u}, \underline{\breve{u}})$ of the works done by external forces, $W_{ext}(\underline{\breve{u}})$, internal forces, $W_{int}(\underline{u}, \underline{\breve{u}})$, and inertia pseudo-forces, $W_i(\underline{u}, \underline{\breve{u}})$ is null:

$$W(\underline{u}, \underline{\breve{u}}) = W_{ext}(\underline{\breve{u}}) + W_{int}(\underline{u}, \underline{\breve{u}}) + W_i(\underline{u}, \underline{\breve{u}}) = 0 \tag{3.42}$$

With

$$W_{ext}(\underline{\breve{u}}) = \int_{\Omega} \rho_s \underline{F_b} \cdot \underline{\breve{u}} \, dV + \int_{\partial\Omega} \underline{t} \cdot \underline{\breve{u}} \, dS \tag{3.43}$$

$$W_{int}(\underline{u}, \underline{\breve{u}}) = -V(\underline{u}, \underline{\breve{u}}) = -\int_{\Omega} \underline{\underline{\sigma}}(\underline{u}) : \underline{\underline{\varepsilon}}(\underline{\breve{u}}) \, dV \tag{3.44}$$

$$W_i(\underline{u}, \underline{\breve{u}}) = -\int_{\Omega} \rho_s \frac{d^2 u}{dt^2} \cdot \underline{\breve{u}} \, dV \tag{3.45}$$

In the above equation, V represents the strain energy associated with the virtual displacement $\underline{\breve{u}}$.

In the case where on a portion $\partial\Omega_D$ of the boundary $\partial\Omega$, the displacement field is prescribed, $\underline{u} = \underline{\bar{u}}$, the traction vector (stress vector) $\underline{t} = \underline{\underline{\sigma}} \cdot \underline{n}$ represents the reaction force needed to impose this displacement. In this case, one generally considers a kinematically admissible virtual displacement field, that is to say

$$\underline{\breve{u}} = 0 \quad \text{on } \partial\Omega_D \tag{3.46}$$

Equation 3.42 then reduces to

$$W(\underline{u}, \underline{\breve{u}}) = \int_{\Omega} \rho_s \underline{F_b} \cdot \underline{\breve{u}}\, dV - \int_{\Omega} \left(\rho_s \frac{d^2 \underline{u}}{dt^2} \right) \underline{\breve{u}}\, dV - \int_{\Omega} \underline{\underline{\sigma}} : \underline{\underline{\varepsilon}}(\underline{\breve{u}})\, dV + \int_{\partial\Omega/\partial\Omega_D} \underline{t} \cdot \underline{\breve{u}}\, dS = 0$$
$$\forall (\underline{u}, \underline{\breve{u}}) / \underline{\breve{u}} = 0, \ \underline{u} = \underline{\bar{u}} \text{ on } \partial\Omega_D \tag{3.47}$$

Similarly, when a force field is prescribed on $\partial\Omega/\partial\Omega_D$, $\underline{t} = \underline{F}$, Equation 3.47 reads

$$W(\underline{u}, \underline{\breve{u}}) = \int_{\Omega} \underline{\underline{\sigma}} : \underline{\underline{\varepsilon}}(\underline{\breve{u}})\, dV - \int_{\Omega} \rho_s \left[\underline{F_b} - \frac{d^2 \underline{u}}{dt^2} \right] \cdot \underline{\breve{u}}\, dV - \int_{\partial\Omega/\partial\Omega_D} \underline{F} \cdot \underline{\breve{u}}\, dS = 0$$
$$\forall (\underline{u}, \underline{\breve{u}}) / \underline{\breve{u}} = 0, \ \underline{u} = \underline{\bar{u}} \text{ on } \partial\Omega_D \tag{3.48}$$

Notes:

1. If $\underline{\breve{u}} = \delta \underline{u}$ is an admissible variation of $\underline{u}$, the principle of virtual work reduces to Galerkin's weak formulation of the linear elastodynamics problem (see Chapter 2):

$$\begin{cases} \underline{\nabla} \cdot \underline{\underline{\sigma}} + \rho_s \underline{F_b} - \rho_s \dfrac{d^2 \underline{u}}{dt^2} = 0 & \text{in } \Omega \\ \underline{u} = \underline{\bar{u}} & \text{on } \partial\Omega_D \\ \underline{\underline{\sigma}} \cdot \underline{n} = \underline{F} & \text{on } \partial\Omega/\partial\Omega_D \end{cases} \tag{3.49}$$

2. To simplify the writing of the principle of virtual work, one can follow D'Alembert and include the inertia pseudo-force per unit volume in $\underline{F_b}$:

$$\underline{F_b} \Rightarrow \underline{F_b} + \underline{I} = \underline{F_b} - \frac{d^2 \underline{u}}{dt^2} \tag{3.50}$$

3. For point loads $\underline{P}_i$ concentrated at locations $\underline{x} = \underline{x}_i$, the traction vector is given by $\underline{t} = \sum_i \underline{P}_i \delta(\underline{x} - \underline{x}_i)$ and the associated virtual work simplifies to

$$W_{ext} = \sum_i \underline{\breve{u}}(\underline{x}_i) \cdot \underline{P}_i \tag{3.51}$$

4. The principle of virtual work is applicable to arbitrarily moving reference systems as long as the inertia pseudo-forces are considered in relation to an absolute reference system (a Galilean system). In a moving system one must add, in the definition of the inertia pseudo-forces, correction terms (=pseudo-forces) taking into account the relativity of motion.

3.8 PRINCIPLE OF MINIMUM POTENTIAL ENERGY

The Galerkin's integral formulation of the linear elastodynamic problem is obtained from the principle of virtual work (Equation 3.48) by selecting $\underline{\breve{u}} = \delta\underline{u}$ as an admissible variation of the displacement field $\underline{u}$:

$$\begin{aligned} W(\underline{u},\delta\underline{u}) &= \int_\Omega \underline{\underline{\sigma}} : \underline{\underline{\varepsilon}}(\delta\underline{u})\,dV - \int_\Omega \rho_s \left[\underline{F}_b - \frac{d^2 u}{dt^2}\right] \cdot \delta\underline{u}\,dV \\ &- \int_{\partial\Omega/\partial\Omega_D} \underline{F} \cdot \delta\underline{u}\,dS = 0 \quad \forall(\underline{u},\delta\underline{u})/\delta\underline{u} = 0,\ \underline{u} = \underline{\bar{u}} \text{ on} \partial\Omega_D \end{aligned} \tag{3.52}$$

Using the constitutive material's law,

$$\underline{\underline{\sigma}} = \underline{\underline{C}} : \underline{\underline{\varepsilon}} \tag{3.53}$$

One obtains,

$$\begin{aligned} W(\underline{u},\delta\underline{u}) &= \int_\Omega \underline{\underline{\varepsilon}}(\underline{u}) : \underline{\underline{C}} : \underline{\underline{\varepsilon}}(\delta\underline{u})\,dV - \int_\Omega \rho_s \left[\underline{F}_b - \frac{d^2 u}{dt^2}\right] \cdot \delta\underline{u}\,dV \\ &- \int_{\partial\Omega/\partial\Omega_D} \underline{F} \cdot \delta\underline{u}\,dS = 0 \quad \forall(\underline{u},\delta\underline{u})/\delta\underline{u} = 0,\ \underline{u} = \underline{\bar{u}} \text{ on} \partial\Omega_D \end{aligned} \tag{3.54}$$

In statics, in the particular case where the stresses do not depend on the state of strain, we can define a functional Π called total potential energy such that

$$W(\underline{u},\delta\underline{u}) = \delta\Pi(\underline{u}) = 0 \tag{3.55}$$

with

$$\Pi(\underline{u}) = V_D - W_{ext} \tag{3.56}$$

V_D is the strain energy; a quadratic function of $\underline{\underline{\varepsilon}}$:

$$V = -W_{int} = \frac{1}{2}\int_{\Omega} \underline{\underline{\varepsilon}} : \underline{\underline{\underline{C}}} : \underline{\underline{\varepsilon}}\, dV \tag{3.57}$$

and $(-W_{ext})$ is the potential functional associated with the external forces, assumed all conservatives:

$$W_{ext}(\underline{u}) = \int_{\Omega} \rho_s F_b \cdot \underline{u}\, dV + \int_{\partial\Omega/\partial\Omega_D} \underline{F} \cdot \underline{u}\, dS \tag{3.58}$$

Equation 3.55 states that in static equilibrium, the principle of virtual work corresponds to the stationary condition of the total potential energy functional Π. In addition, the second-order variation of Π leads to

$$\delta^2\Pi(\underline{u}) = \int_{\Omega} \delta\underline{\underline{\varepsilon}} : \underline{\underline{\underline{C}}} : \delta\underline{\underline{\varepsilon}}\, dV \geq 0 \quad \forall\, \delta\underline{u} \neq 0 \tag{3.59}$$

since $\delta \underline{F_b} = \delta \underline{F} = \delta^2 \underline{u} = 0$.

Thus, we see that the solution of the static elasticity problem corresponds to a minimum of the total potential energy. This is the principle of minimum total potential energy which states that at static equilibrium, among all the kinematically admissible displacement fields, the actual displacement field makes the total potential energy minimum.

3.9 HAMILTON'S PRINCIPLE

Hamilton's principle is a time-integrated form of the principle of virtual work. It is obtained by an integral transformation of the expression (Appendix 3C):

$$\int_{t_1}^{t_2} W(\underline{u}, \delta\underline{u})\, dt = 0, \quad \delta\underline{u}(t_1) = \delta\underline{u}(t_2) = 0 \tag{3.60}$$

For conservative systems, it states that

> Among the possible trajectories between two fixed end points, subject to restrictive conditions $\delta \underline{u}(t_1) = \delta \underline{u}(t_2) = 0$ at the ends of the considered time interval $[t_1, t_2]$, the actual trajectory of the system is that which makes stationary the integral of the difference between the kinetic and total potential energy:

$$\delta \int_{t_1}^{t_2} (T - \Pi)\, dt = 0 \tag{3.61}$$

Hamilton's principle is often mathematically expressed as

$$\delta \int_{t_1}^{t_2} \mathfrak{L}(\underline{u})\, dt = 0 \tag{3.62}$$

where $\mathfrak{L}(\underline{u}) = T - \Pi$. The quantity $\int_{t_1}^{t_2} \mathfrak{L}(\underline{u})\, dt$ is called the action integral and the operator $\mathfrak{L}$ is called the Lagrangian of the system. The kinetic energy is given by

$$T(\underline{u}) = \frac{1}{2} \int_{\Omega} \rho_s \dot{\underline{u}} \cdot \dot{\underline{u}}\, dV \tag{3.63}$$

and the total potential energy Π is given by Equation 3.56.

In the case of nonconservative systems, it is easy to extend Hamilton's principle by adding a term for the nonconservative forces:

$$\delta \int_{t_1}^{t_2} (T - V)\, dt + \int_{t_1}^{t_2} \delta W_{ext}\, dt = 0 \tag{3.64}$$

with

$$\delta W_{ext} = \int_{\Omega} \rho_s \underline{F_b} \cdot \delta \underline{u}\, dV + \int_{\partial\Omega/\partial\Omega_D} \underline{F} \cdot \delta \underline{u}\, dS \tag{3.65}$$

Note 1. Hamilton's principle gives a direct variational statement for mechanical systems. However, its primary interest lies in the derivation of the equations of motion of these systems, and in particular through the use of the Lagrange equations. The latter are obtained directly from Hamilton's principle. However, in this book, for the construction of the variational form of a system, we prefer to proceed through the principle of virtual work or simply by direct construction of the weak formulation.

Note 2. To illustrate the use of Hamilton's principle, consider the one-dimensional case:

$$\mathfrak{L} = \int_0^L \mathcal{L}dx; \quad \mathcal{L} = \mathcal{L}(u, u', u'', \dot{u}, \dot{u}') \tag{3.66}$$

where in the following notations are used:

$$u' = (\partial u/\partial x); \quad u'' = (\partial^2 u/\partial x^2); \quad \dot{u} = (\partial u/\partial t); \quad \dot{u}' = (\partial^2 u/\partial x \partial t).$$

It is also assumed that the virtual work of the external forces $\mathcal{Q}$ and moments $\mathfrak{M}$ can be written as $\delta W_{ext} = [\mathcal{Q}\delta u + \mathfrak{M}\delta u']_0^L$, while the volumic term is assumed conservative and is thus included in the expression of the total potential energy. Algebraic manipulations of these equations (Appendix 3D) lead to the following equations, known as the Lagrange equations:

$$\frac{\partial \mathcal{L}}{\partial u} - \frac{\partial}{\partial x}\left(\frac{\partial \mathcal{L}}{\partial u'}\right) + \frac{\partial^2}{\partial x^2}\left(\frac{\partial \mathcal{L}}{\partial u''}\right) - \frac{\partial}{\partial t}\left[\frac{\partial \mathcal{L}}{\partial \dot{u}}\right] + \frac{\partial}{\partial t}\left(\frac{\partial}{\partial x}\frac{\partial \mathcal{L}}{\partial \dot{u}'}\right) = 0 \tag{3.67}$$

with the boundary (constraints) equations:

$$\left[\mathcal{Q} + \frac{\partial \mathcal{L}}{\partial u'} - \frac{\partial}{\partial x}\left(\frac{\partial \mathcal{L}}{\partial u''}\right) - \frac{\partial}{\partial t}\left(\frac{\partial \mathcal{L}}{\partial \dot{u}}\right)\right]\delta u = 0 \quad \text{at } x = 0 \text{ and } x = L \tag{3.68}$$

and

$$\left[\mathfrak{M} + \frac{\partial \mathcal{L}}{\partial u''}\right]\delta u' = 0 \quad \text{at } x = 0 \quad \text{and } x = L \tag{3.69}$$

Note 3. In some cases (e.g., in the Rayleigh-Ritz method), the displacement field can be expressed as a function of N generalized variables or degrees of freedom:

$$\underline{u} = \underline{u}(q_1, \ldots, q_N) \tag{3.70}$$

In this case, the Lagrange equations are reduced to a system of N coupled second-order differential equations:

$$\frac{d}{dt}\left(\frac{\partial \mathfrak{L}}{\partial \dot{q}_k}\right) - \frac{\partial \mathfrak{L}}{\partial q_k} - \mathcal{Q}_k = 0 \quad k = 1\ldots N \tag{3.71}$$

The Lagrangian can be written in the generic form:

$$\mathfrak{L}(q_1,\ldots,q_N,\dot{q}_1,\ldots,\dot{q}_N) = T - V = T - V_D - V_{ext} \tag{3.72}$$

V_D is the strain energy of the system while V_{ext} represents the potential energy associated with the conservative generalized forces:

$$\mathcal{Q}_{k,cons} = -\frac{\partial V_{ext}}{\partial q_k} \tag{3.73}$$

and $\mathcal{Q}_k$ represent the nonconservative generalized external forces, obtained from the expression of virtual work:

$$\delta W_{ext} = \sum_{k=1}^{N} \mathcal{Q}_k \delta q_k \tag{3.74}$$

In the case of a system with a viscous-type damping mechanism that can be represented by a quadratic dissipation functional involving generalized velocities

$$\mathfrak{D} = \frac{1}{2}\dot{q}_i c_{ij} \dot{q}_j \tag{3.75}$$

with constant coefficients c_{ij}. The virtual work of the generalized dissipation forces can then be written as

$$\delta W_{diss} = \sum_{k=1}^{N} \mathcal{Q}_{k,diss} \delta q_k \quad \text{with } \mathcal{Q}_{k,diss} = -\frac{\partial \mathfrak{D}}{\partial \dot{q}_k} = -c_{kj}\dot{q}_j \tag{3.76}$$

And consequently, Lagrange's equations take the form

$$\frac{d}{dt}\left(\frac{\partial T}{\partial \dot{q}_k}\right) - \frac{\partial T}{\partial q_k} + \frac{\partial V_d}{\partial q_k} + \frac{\partial \mathfrak{D}}{\partial \dot{q}_k} + \frac{\partial V_{ext}}{\partial q_k} = \mathcal{Q}_k \tag{3.77}$$

3.10 CONCLUSION

We have introduced in this chapter classical methods for obtaining the variational integral formulations associated with some common problems in continuum mechanics. The main objective of these methods is the construction of an approximate solution using a series of admissible test functions. In Chapters 4, 5, and 6 we will introduce the finite element method that operates through an approximation of the unknown variables on a spatial discretization of these integral formulations. This leads to a system of algebraic equations that provides an approximate solution to the problem. In Chapter 7, we will introduce other integral formulations, well adapted to unbounded domains.

APPENDIX 3A: METHODS OF INTEGRAL APPROXIMATIONS—EXAMPLE

3A.1 The problem

Let us consider the following problem:

$$\begin{cases} \dfrac{\partial^2 u}{\partial x^2} + u = -x & 0 < x < 1 \\ u(0) = 0 \quad \text{and} \quad u(1) = 0 \end{cases} \tag{3.78}$$

The exact solution is given by

$$u(x) = \frac{\sin(x)}{\sin(1)} - x \tag{3.79}$$

In this appendix, we seek an approximate solution to Equation 3.78 using different methods of weighted residuals.

3A.2 Method of weighted residuals

The first step in the method of weighted residuals consists in postulating an approximation using an expansion in terms of trial functions with unknown coefficients. These coefficients are determined later. It is preferable, when possible, to choose a trial function that satisfies the boundary conditions of the problem.

For our example, $u(0) = 0$ and $u(1) = 0$, we can then choose

$$\tilde{u}(x) = \sum_{i=1}^{N} a_i x^i (1 - x) \tag{3.80}$$

Table 3.1 Summary of the various approximation methods

Method	*Description*
Collocation	$\psi_{w,i} = \delta(x - x_i), i = 1,2,\ldots,n$ where x_i is a point of the domain
Least squares	$\psi_{w,i} = \frac{\partial \mathcal{R}}{\partial a_i}, i = 1,2,\ldots,n$ with $\mathcal{R}$ the residual and a_i is the ith coefficient of the trial function
Galerkin	$\psi_{w,i} = \frac{\partial \tilde{u}}{\partial a_i}, i = 1,2,\ldots,n$ with $\tilde{u}$ the trial function
Petrov–Galerkin	$\psi_{w,i} = x^i, i = 1,2,\ldots,n$

The a_i are unknown coefficients to be determined and N represents the order of the approximation (the number of trial functions). The symbol ~ indicates that the solution is approximate. The accuracy of the approximation depends on the choice and the number of trial functions.

One way to solve this problem is to set the integral of the residual,* weighted by arbitrary weighting functions, to zero.

$$\int_0^1 \psi_w(x)\mathcal{R}(\tilde{u}(x))dx = 0 \tag{3.81}$$

By choosing as many independent weighting functions as unknown coefficients in the approximation, we obtain the algebraic equations necessary to solve the system. Table 3.1 summarizes the choice of weighting functions for four common techniques.

The most popular methods are the Galerkin's method and the least-squares method. The Collocation and Petrov–Galerkin's approaches (also known as method of moments) are not used when high degrees of accuracy are required. The least-squares method always leads to a symmetric system. Galerkin's method does not usually lead to a symmetric system, except for self-adjoint operators. This method is particularly interesting in the case of the weak (or variational) form of the problem.

3A.2.1 Approximation by one term

The approximation is of the form

$$\tilde{u}(x) = ax(1 - x) \tag{3.82}$$

* The residual is defined for our problem as $\mathcal{R}(\tilde{u}(x)) = \frac{\partial^2 \tilde{u}}{\partial x^2} + \tilde{u} - (-x)$

The residual is given by the substitution of the approximation in the equation governing the problem to be solved minus the right-hand side, that is,

$$\mathcal{R} = \frac{\partial^2 \tilde{u}}{\partial x^2} + \tilde{u} + x = -2a + ax(1-x) + x \tag{3.83}$$

Since $\tilde{u}$ is different from u, the residual $\mathcal{R}$ does not vanish at all points of the domain.

The coefficient a is obtained from the integral form of the weighted residual:

$$I = \int_0^1 \psi_w \mathcal{R} dx = \int_0^1 \psi_w \left(-2a + ax(1-x) + x\right) dx = 0 \tag{3.84}$$

The corresponding approximate solution depends on the weighting function.

3A.2.1.1 Collocation method

The delta function $\delta(x - x_i)$ is used as a weighting function $\psi_w = \delta(x - x_i)$. The collocation point x_i belongs to the domain $]0,1[$. Obviously, the result depends on the choice of x_i. For $x_i = 0.5$, we obtain $a = 2/7$. The approximate solution becomes

$$\tilde{u}(x) = 0.2857x(1-x) \tag{3.85}$$

It is plotted in Figure 3.5 together with the exact solution.

3A.2.1.2 Least-squares method

The weighting function is obtained from the residual:

$$\psi_w = \frac{d\mathcal{R}}{da} \tag{3.86}$$

For our example, we have $\psi_w = x(1-x) - 2$, which brings us to the following result:

$$I = \int_0^1 \left(x(1-x) - 2\right)\left(-2a + ax(x-1) + x\right) dx = 0 \tag{3.87}$$

$$\frac{101}{30}a - \frac{11}{12} = 0 \Rightarrow a = \frac{55}{202} \tag{3.88}$$

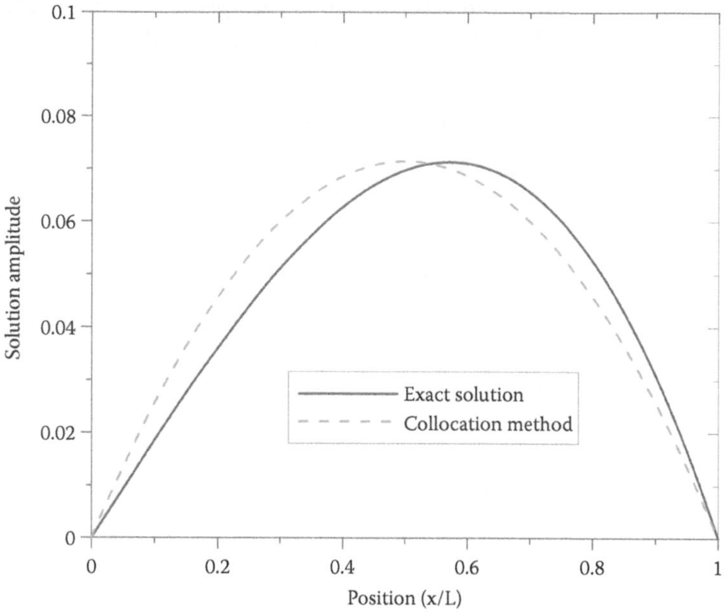

Figure 3.5 Comparison between the exact solution (plain) and that obtained with the collocation method (dashed).

$$\tilde{u}(x) = 0.2723x(1 - x) \tag{3.89}$$

The result is plotted in Figure 3.6 together with the exact solution.

3A.2.1.3 Galerkin's method

The weighting function is obtained by taking the derivative of the trial function with respect to the unknown amplitude:

$$\psi_w = \frac{d\tilde{u}}{da} = x(1 - x) \tag{3.90}$$

Therefore,

$$I = \int_0^1 x(1 - x)(-2a + ax(x - 1) + x)\,dx = 0 \tag{3.91}$$

$$\frac{3}{10}a - \frac{1}{12} = 0 \Rightarrow a = \frac{5}{18} \tag{3.92}$$

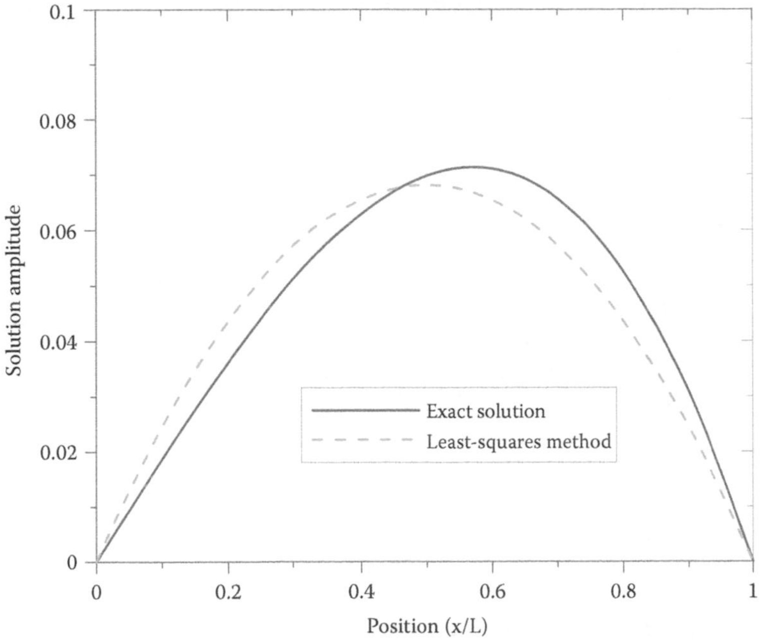

Figure 3.6 Comparison between the exact solution (plain) and that obtained with the least-squares method (dashed).

$$\tilde{u}(x) = 0.2778x(1 - x) \tag{3.93}$$

The result is plotted in Figure 3.7 together with the exact solution.

3A.2.1.4 Petrov–Galerkin's method

The weighting function is given by

$$\psi_w = x \tag{3.94}$$

$$\frac{11}{12}a - \frac{1}{3} \Rightarrow a = \frac{4}{11} \tag{3.95}$$

$$\tilde{u}(x) = 0.3636x(1 - x) \tag{3.96}$$

The result is plotted in Figure 3.8 together with the exact solution.

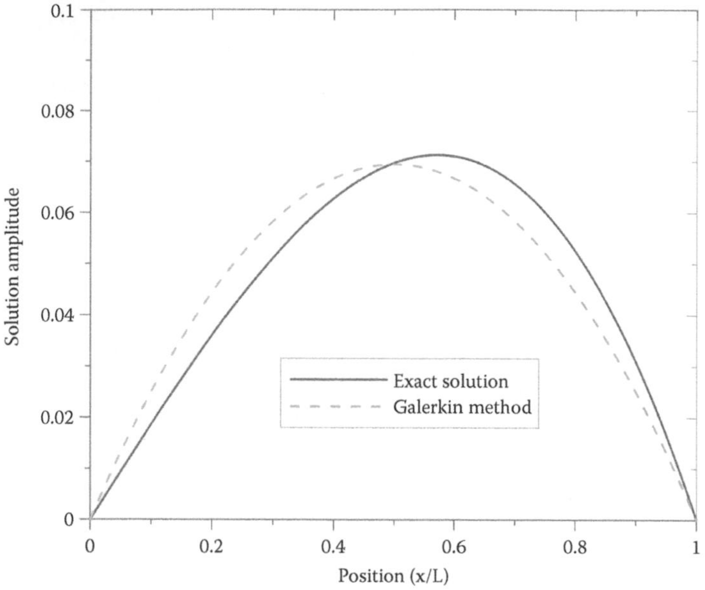

Figure 3.7 Comparison between the exact solution (plain) and that obtained with Galerkin method (dashed).

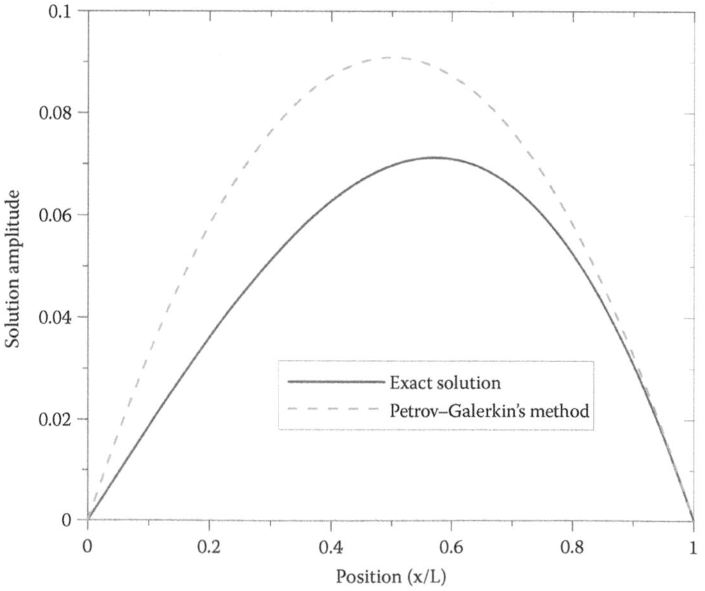

Figure 3.8 Comparison between the exact solution (plain) and that obtained with Petrov–Galerkin's method (dashed).

Table 3.2 Summary of the numerical solutions using two trial functions

Method	*Solution:* $\tilde{u}(x) = a_1x(l - x) + a_2x^2(l - x)$
Collocation	$x_1 = \frac{1}{3};\ x_2 = \frac{2}{3} \Rightarrow a_1 = \frac{81}{416};\ a_2 = \frac{9}{52}$
Least squares	$a_1 = \frac{46161}{246137};\ a_2 = \frac{413}{2437}$
Galerkin	$a_1 = \frac{71}{369};\ a_2 = \frac{7}{41}$
Petrov–Galerkin	$a_1 = \frac{35}{177};\ a_2 = \frac{85}{531}$

3A.2.2 Approximation using two terms

To improve the approximation, we can add more terms to the first trial function. For example, we can choose the following approximation using two trial functions:

$$\tilde{u}(x) = a_1x(1 - x) + a_2x^2(1 - x) \tag{3.97}$$

The residual becomes

$$\mathcal{R}(x, a_1, a_2) = x + a_1(-2 + x - x^2) + a_2(2 - 6x + x^2 - x^3) \tag{3.98}$$

For our example, we obtain the following results:

Collocation method: $\psi_{w,1} = \delta(x - x_1)$, $\psi_{w,2} = \delta(x - x_2)$
Least-squares method: $\psi_{w,1} = -2 + x - x^2$, $\psi_{w,2} = 2 - 6x + x^2 - x^3$
Galerkin's method: $\psi_{w,1} = x(1 - x)$, $\psi_{w,2} = x^2(1 - x)$
Petrov–Galerkin's method: $\psi_{w,1} = x$, $\psi_{w,2} = x^2$

The corresponding solutions are summarized in Table 3.2.

Figures 3.9 through 3.12 display the comparisons of the approximate solution with the exact solution for each method.

3A.3 Variational method

The previous formulation uses the strong integral form. The weak form is obtained from integration by parts. Its main advantage is that it reduces the conditions of differentiability of the trial function:

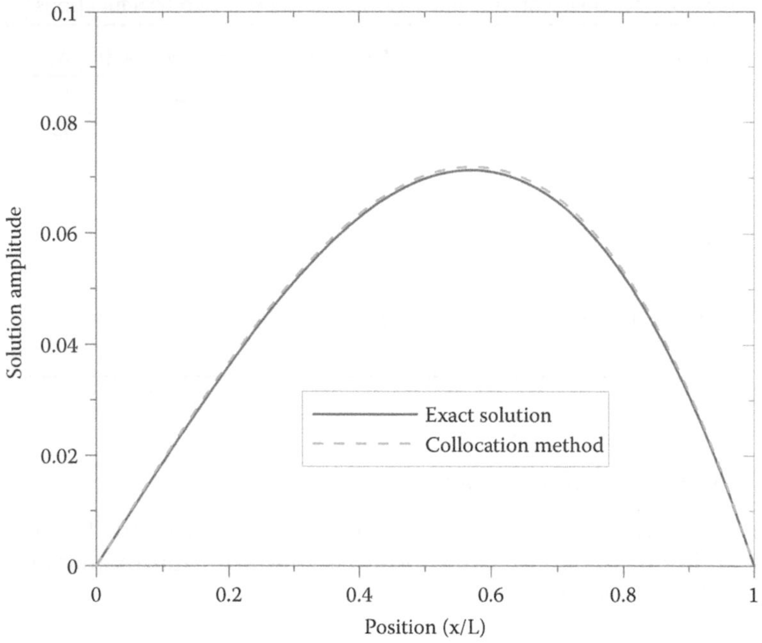

Figure 3.9 Comparison between the exact solution (plain) and that obtained with the collocation method with two terms (dashed).

$$\mathcal{J} = \int_0^1 \psi_w \left(\frac{\partial^2 \tilde{u}}{\partial x^2} + \tilde{u} + x \right) dx$$
$$= \int_0^1 \left(\frac{-\partial \psi_w}{\partial x} \frac{\partial \tilde{u}}{\partial x} + \psi_w \tilde{u} + \psi_w x \right) dx + \left[\psi_w \frac{\partial \tilde{u}}{\partial x} \right]_0^1 = 0 \tag{3.99}$$

The boundary term represents the Neumann's boundary condition. In this case, the trial function must be at least once differentiable.

In the special case of a self-adjoint operator, we can express the weak form in terms of the variation of a functional. This can be done with a specific choice of weighting function:

$$\psi_w = \delta \tilde{u} \tag{3.100}$$

where $\delta \tilde{u}$ is an admissible variation of $\tilde{u}$. Thus

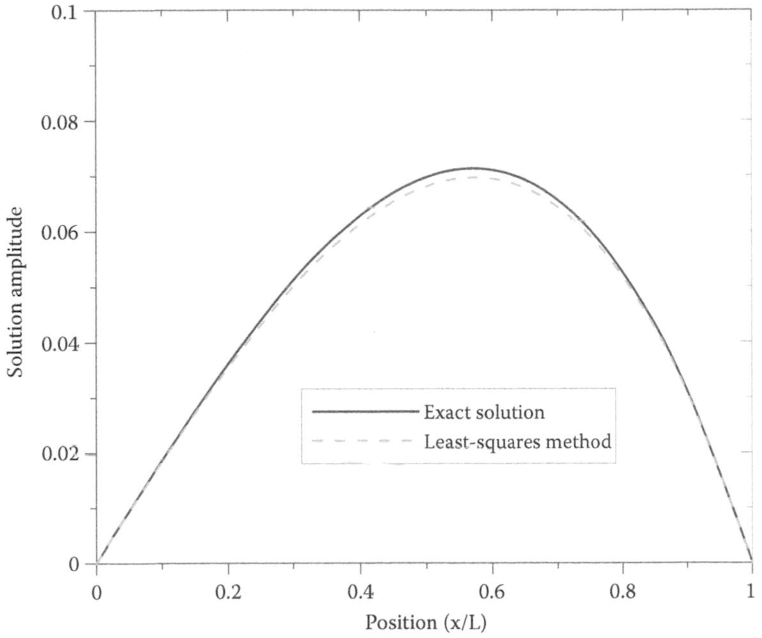

Figure 3.10 Comparison between the exact solution (plain) and that obtained with the least-squares method with two terms (dashed).

$$\delta\mathcal{J} = 0 \tag{3.101}$$

With

$$\delta\mathcal{J} = \int_0^1 \left(-\frac{\partial\delta\tilde{u}}{\partial x}\frac{\partial\tilde{u}}{\partial x} + \tilde{u}\delta\tilde{u} + x\delta\tilde{u} \right) dx + \left[\delta\tilde{u}\frac{\partial\tilde{u}}{\partial x} \right]_0^1 \tag{3.102}$$

Since $\tilde{u}$ is fixed on the boundary, we have

$$\delta\tilde{u} = 0 \quad \text{for } x = 0{,}1 \tag{3.103}$$

The weak form reduces to

$$\delta\mathcal{J} = \int_0^1 \left(\frac{\partial\delta\tilde{u}}{\partial x}\frac{\partial\tilde{u}}{\partial x} - \tilde{u}\delta\tilde{u} - x\delta\tilde{u} \right) dx = 0 \tag{3.104}$$

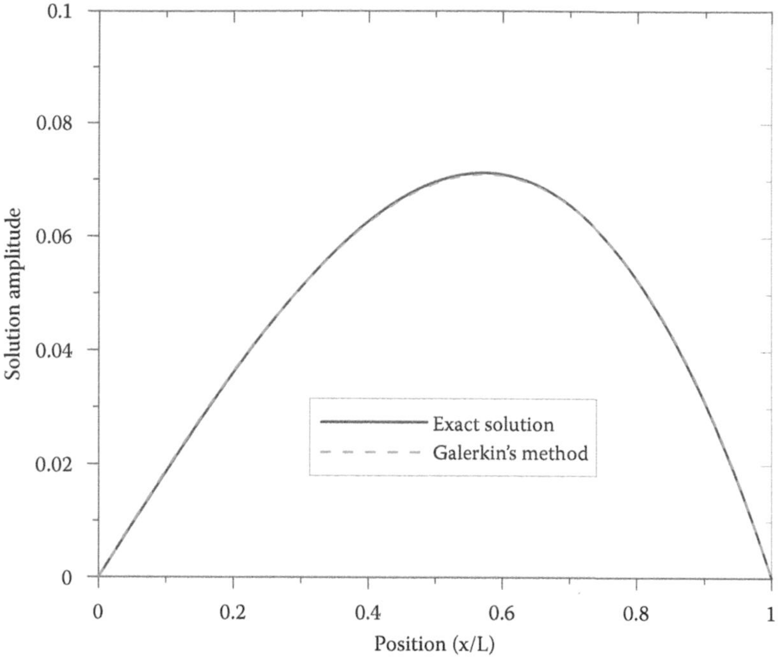

Figure 3.11 Comparison between the exact solution (plain) and that obtained with Galerkin's method with two terms (dashed).

Or, using the properties of the variation operator δ

$$\delta\mathcal{J} = \delta\int_0^1 \left(\frac{1}{2}\left(\frac{\partial\tilde{u}}{\partial x}\right)^2 - \frac{1}{2}\tilde{u}^2 - x\tilde{u}\right)dx = 0 \tag{3.105}$$

It can be deduced that the problem's functional is

$$\mathcal{J} = \int_0^1 \left(\frac{1}{2}\left(\frac{\partial\tilde{u}}{\partial x}\right)^2 - \frac{1}{2}\tilde{u}^2 - x\tilde{u}\right)dx \tag{3.106}$$

The variational statement of the problem is

$$\delta\mathcal{J} = 0, \quad \tilde{u}(0) = 0 \quad \text{and} \quad \tilde{u}(1) = 0 \tag{3.107}$$

which $\mathcal{J}$ is given by Equation 3.106.

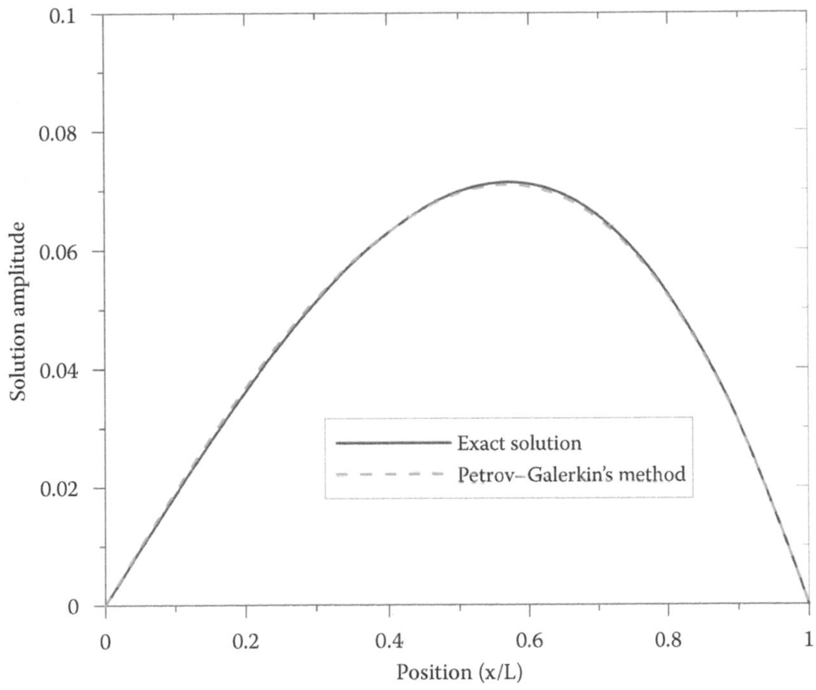

Figure 3.12 Comparison between the exact solution (plain) and that obtained with Petrov–Galerkin's method with two terms (dashed).

3A.4 Rayleigh–Ritz's method

This method leads to an approximate solution of a differential equation by operating directly on the functional when it exists. The method consists of two steps:

- Postulate an admissible solution as an expansion in terms of trial functions satisfying Dirichlet's boundary conditions of the problem (essential or geometrical conditions).
- Substitute the postulated solution in the functional and compute the unknown coefficients of the expansion from the variational statement.

In the case of our example with a single trial function:

$$\tilde{u}(x) = ax(1-x) \quad \tilde{u}(0) = 0 \quad \text{and} \quad \tilde{u}(1) = 0 \tag{3.108}$$

The functional can be written as

$$\mathcal{J} = \frac{1}{2}a^2\int_0^1\left[(1-2x)^2 - x^2(1-x)^2\right]dx - a\int_0^1 x^2(1-x)dx$$
$$= \frac{3}{20}a^2 - a\frac{1}{12} \tag{3.109}$$

$$\delta\mathcal{J} = 0 \Rightarrow \frac{d\mathcal{J}}{da} = 0 \Rightarrow a = \frac{5}{18} \cong 0.2778 \tag{3.110}$$

and

$$\tilde{u}(x) = 0.2278x(1-x) \tag{3.111}$$

We obtain the same solution as with Galerkin's approach (Figure 3.7). To improve the approximation, we can add another term in the expansion:

$$\tilde{u}(x) = a_1x(1-x) + a_2x^2(1-x) \tag{3.112}$$

The unknown coefficients are again obtained by substituting Equation 3.112 in the expression of the functional and by using

$$\delta\mathcal{J} = 0 \Rightarrow \frac{\partial\mathcal{J}}{\partial a_1} = 0 \quad \text{and} \quad \frac{\partial\mathcal{J}}{\partial a_2} = 0 \tag{3.113}$$

The calculation gives

$$a_1 = \frac{71}{369}; \quad a_2 = \frac{7}{41} \tag{3.114}$$

Again, this yields to the same solution as Galerkin's method, which is in good agreement with the exact solution (Figure 3.11).

APPENDIX 3B: VARIOUS INTEGRAL THEOREMS AND VECTOR IDENTITIES

3B.1 Nabla operator

Del, or Nabla represented by the symbol $\underline{\nabla}$, is an operator used in mathematics, in particular, in vector calculus, as a vector differential operator. Strictly speaking, $\underline{\nabla}$ is not a specific operator, but rather a convenient mathematical notation, that makes many equations easier to write and remember. It can be interpreted as a vector of partial derivative operators.

$$\underline{\nabla} = \left\langle \frac{\partial}{\partial x_1}, \frac{\partial}{\partial x_2}, \frac{\partial}{\partial x_3} \right\rangle^T \tag{3.115}$$

With this notation, we have

$$\begin{aligned} \underline{\text{grad}}\, g &= \underline{\nabla} g \\ \text{div}\underline{u} &= \underline{\nabla} \cdot \underline{u} \\ \underline{\text{rot}}\, \underline{u} &= \underline{\nabla} \times \underline{u} \\ \Delta g &= \nabla^2 g \\ \underline{\Delta}\, \underline{u} &= \underline{\nabla}^2 \underline{u} \end{aligned} \tag{3.116}$$

where the $\cdot$ denotes the dot or inner product and $\times$ denotes the cross product.

3B.2 Useful vector identities

$$\begin{aligned} \nabla^2 g &= \underline{\nabla} \cdot \underline{\nabla} g \\ \underline{\nabla} \times (\underline{\nabla} \times \underline{u}) &= \underline{\nabla}(\underline{\nabla} \cdot \underline{u}) - \nabla^2 \underline{u} \\ \underline{\nabla} \times (\underline{\nabla} g) &= 0 \\ \underline{\nabla} \times (g\underline{u}) &= g\underline{\nabla} \times \underline{u} + \underline{\nabla} g \times \underline{u} \\ \underline{\nabla} \cdot (\underline{\nabla} \times \underline{u}) &= 0 \\ \underline{\nabla} \cdot (g\underline{u}) &= g\underline{\nabla} \cdot \underline{u} + \underline{u} \cdot \underline{\nabla} g \end{aligned} \tag{3.117}$$

3B.3 Divergence theorem

The integral of the divergence of a vector field over some compact solid equals the integral of the flux through the closed surface bounding the solid of outward normal $\underline{n}$.

$$\int_{\Omega} \underline{\nabla} \cdot \underline{u}\, dV = \int_{\partial\Omega} \underline{u} \cdot \underline{n}\, dS \tag{3.118}$$

Applying the divergence theorem to the product of a scalar function g and a vector field $\underline{u}$, the result is

$$\int_{\Omega} (\underline{u} \cdot \underline{\nabla} g + g\underline{\nabla} \cdot \underline{u})\, dV = \int_{\partial\Omega} g\underline{u} \cdot \underline{n}\, dS \tag{3.119}$$

3B.4 Stoke's theorem

The integral of the curl of a vector field over a surface $\partial\Omega$ in $\mathbb{R}^3$ of outward normal $\underline{n}$ equals the line integral of the vector field over the closed curve $\partial^2\Omega$ bounding the surface

$$\int_{\partial\Omega} (\underline{\nabla} \times \underline{u}) \cdot \underline{n}\, dS = \int_{\partial^2\Omega} \underline{u} \cdot d\underline{s} \tag{3.120}$$

where $d\underline{s}$ is a line element.

3B.5 Green's first identity

If $\underline{u} = \underline{\nabla}\phi$, Equation 3.119 leads to Green's first identity

$$\int_{\Omega} (\underline{\nabla}\phi \cdot \underline{\nabla} g + g\nabla^2\phi)\, dV = \int_{\partial\Omega} g\underline{\nabla}\phi \cdot \underline{n}\, dS \tag{3.121}$$

3B.6 Green's second identity

If $\underline{u} = g\underline{\nabla}\phi - \phi\underline{\nabla}g$, Equation 3.119 leads to Green's second identity

$$\int_{\Omega} (g\nabla^2\phi - \phi\nabla^2 g)\, dV = \int_{\partial\Omega} (g\underline{\nabla}\phi \cdot \underline{n} - \phi\underline{\nabla}g \cdot \underline{n})\, dS \tag{3.122}$$

APPENDIX 3C: DERIVATION OF HAMILTON'S PRINCIPLE FROM THE PRINCIPLE OF VIRTUAL WORK

The principle of virtual work is

$$\delta W_{ext} + \delta W_i + \delta W_{int} = 0 \tag{3.123}$$

which is equivalent to

$$\delta W_{ext} + \delta W_i - \delta V = 0 \tag{3.124}$$

This can be written as

$$\delta W_{ext} = \int_{\Omega} \sigma_{ij}(\underline{u})\delta\varepsilon_{ij}\, dV + \int_{\Omega} \rho_s \frac{d^2 u_i}{dt^2} \delta u_i\, dV \tag{3.125}$$

where we have used Einstein summation notation.

Let us integrate Equation 3.125 between time t_1 and t_2 where the admissible variations δu_i satisfy the kinematic boundary conditions namely,

$$\delta u_i(t_1) = \delta u_i(t_2) = 0 \quad \text{on } \partial\Omega \tag{3.126}$$

We start from

$$\delta \int_{t_1}^{t_2} W_{ext}\, dt = \delta \int_{t_1}^{t_2} V\, dt + \int_{t_1}^{t_2} \int_{\Omega} \rho_s \frac{d^2 u_i}{dt^2} \delta u_i\, dV\, dt \tag{3.127}$$

Integrating by parts the last term of Equation 3.127, we get

$$\begin{aligned}\int_{t_1}^{t_2} \int_{\Omega} \rho_s \frac{\partial^2 u_i}{\partial t^2} \delta u_i\, dVdt &= -\int_{t_1}^{t_2} \int_{\Omega} \rho_s \frac{\partial u_i}{\partial t} \frac{\partial \delta u_i}{\partial t} dV\, dt + \underbrace{\left[\rho_s \frac{\partial u_i}{\partial t} \delta u_i \right]_{t_1}^{t_2}}_{=0} \\ &= -\delta \int_{t_1}^{t_2} T dt\end{aligned} \tag{3.128}$$

where T is the kinetic energy of the system:

$$T = \frac{1}{2} \int_{\Omega} \rho_s \frac{\partial u_i}{\partial t} \frac{\partial u_i}{\partial t} dV \tag{3.129}$$

Finally, using Equation 3.128, Equation 3.127 can be written as

$$\delta \int_{t_1}^{t_2} (W_{ext} + T - V)\, dt = 0 \tag{3.130}$$

Introducing the Lagrangian $\mathfrak{L} = T + W_{ext} - V = T - \Pi$, we end up with Hamilton's principle:

$$\delta \int_{t_1}^{t_2} \mathfrak{L}\, dt = 0 \tag{3.131}$$

This expression constitutes the variational formulation of the motion of a deformable body.

APPENDIX 3D: LAGRANGE'S EQUATIONS (1D)

Hamilton's principle writes:

$$\delta\int_{t_1}^{t_2} \mathfrak{L}\, dt = 0 \tag{3.132}$$

with $\mathfrak{L} = T + W_{ext} - V = T - \Pi$ is the Lagrangian of the system. T, W_{ext}, V are the kinetic energy, the work of external forces and the potential energy respectively.

For a 1D system,

$$\mathfrak{L} = \int_0^L \mathcal{L}(x,t)\, dx \tag{3.133}$$

$$T = \int_0^L \mathrm{T}(x,t)\, dx \tag{3.134}$$

$$W_{ext} = \int_0^L \mathrm{W}_{\mathrm{ext}}(x,t)\, dx \tag{3.135}$$

$$V = \int_0^L \mathrm{V}(x,t)\, dx \tag{3.136}$$

Let us assume $\mathrm{T} = \mathrm{T}(q,\dot{q},q',\dot{q}')$, $\mathrm{W}_{\mathrm{ext}} = \mathrm{W}_{\mathrm{ext}}(q,q')$, $V = V\,(q,q',q'')$ where q is the displacement field, $\dot{q} = (\partial q/\partial t)$, $q' = (\partial q/\partial x)$. Let us assume that the virtual work of external forces also comprises a boundary term $\delta W_{ext} = \left[\mathcal{Q}\delta q + \mathfrak{M}\delta q'\right]_0^L$. We can write $\mathcal{L} = \mathcal{L}(q,\dot{q},q',\dot{q}',q'')$. In consequence

$$\int_{t_1}^{t_2}\int_0^L \left[\frac{\partial \mathcal{L}}{\partial q}\delta q + \underbrace{\frac{\partial \mathcal{L}}{\partial \dot{q}}\delta \dot{q}}_{❶} + \underbrace{\frac{\partial \mathcal{L}}{\partial q'}\delta q'}_{❷} + \underbrace{\frac{\partial \mathcal{L}}{\partial \dot{q}'}\delta \dot{q}'}_{❸} + \underbrace{\frac{\partial \mathcal{L}}{\partial q''}\delta q''}_{❹} \right] dx\,dt$$
$$+ \int_{t_1}^{t_2} \left[\mathcal{Q}\delta q + \mathfrak{M}\delta q'\right]_0^L dt = 0 \tag{3.137}$$

$$❶ \int_{t_1}^{t_2}\int_{0}^{L} \frac{\partial \mathcal{L}}{\partial \dot{q}} \delta\dot{q}\, dx dt = -\int_{0}^{L}\left[\int_{t1}^{t_2} \frac{\partial}{\partial t}\left(\frac{\partial \mathcal{L}}{\partial \dot{q}}\right)\delta q dt + \underbrace{\left[\frac{\partial \mathcal{L}}{\partial \dot{q}}\delta q\right]_{t_1}^{t_2}}_{=0 \text{ because } \delta q(t_1)=\delta q(t_2)=0}\right] dx \tag{3.138}$$

$$❷ \int_{t_1}^{t_2}\int_{0}^{L} \frac{\partial \mathcal{L}}{\partial q'} \delta q'\, dx dt = -\int_{t_1}^{t_2}\left[\int_{0}^{L} \frac{\partial}{\partial x}\left(\frac{\partial \mathcal{L}}{\partial q'}\right)\delta q\, dx + \left[\frac{\partial \mathcal{L}}{\partial q'}\delta q\right]_{0}^{L}\right] dt \tag{3.139}$$

$$❸ \int_{t_1}^{t_2}\int_{0}^{L} \frac{\partial \mathcal{L}}{\partial \dot{q}'} \delta \dot{q}'\, dx dt = -\int_{t_1}^{t_2}\int_{0}^{L} \frac{\partial}{\partial t}\left(\frac{\partial \mathcal{L}}{\partial \dot{q}'}\right)\delta q'\, dx\, dt + \left[\int_{0}^{L} \frac{\partial \mathcal{L}}{\partial \dot{q}'}\delta q'\, dx\right]_{t_1}^{t_2}$$

$$= \int_{t_1}^{t_2}\int_{0}^{L} \frac{\partial}{\partial t}\frac{\partial}{\partial x}\left(\frac{\partial \mathcal{L}}{\partial \dot{q}'}\right)\delta q\, dx\, dt - \left[\int_{t_1}^{t_2} \frac{\partial}{\partial t}\left(\frac{\partial \mathcal{L}}{\partial \dot{q}'}\right)\delta q\, dt\right]_{0}^{L}$$

$$\underbrace{-\left[\int_{0}^{L} \frac{\partial}{\partial x}\left(\frac{\partial \mathcal{L}}{\partial \dot{q}'}\right)\delta q\, dx\right]_{t_1}^{t_2}}_{=0} + \left[\underbrace{\left[\frac{\partial \mathcal{L}}{\partial \dot{q}'}\delta q\right]_{t_1}^{t_2}}_{=0}\right]_{0}^{L} \tag{3.140}$$

$$❹ \int_{t_1}^{t_2}\int_{0}^{L} \frac{\partial \mathcal{L}}{\partial q''} \delta q''\, dx dt = -\int_{t_1}^{t_2}\left[\int_{0}^{L} \frac{\partial}{\partial x}\left(\frac{\partial \mathcal{L}}{\partial q''}\right)\delta q'\, dx + \left[\frac{\partial \mathcal{L}}{\partial q''}\delta q'\right]_{0}^{L}\right] dt$$

$$= \int_{t_1}^{t_2}\left[\int_{0}^{L} \frac{\partial^2}{\partial x^2}\left(\frac{\partial \mathcal{L}}{\partial q''}\right)\delta q\, dx - \left[\frac{\partial}{\partial x}\left(\frac{\partial \mathcal{L}}{\partial q''}\right)\delta q\right]_{0}^{L} + \left[\frac{\partial \mathcal{L}}{\partial q''}\delta q'\right]_{0}^{L}\right] dt \tag{3.141}$$

Finally for δq arbitrary, we get:

$$\frac{\partial \mathcal{L}}{\partial q} - \frac{\partial}{\partial x}\left(\frac{\partial \mathcal{L}}{\partial q'}\right) + \frac{\partial^2}{\partial x^2}\left(\frac{\partial \mathcal{L}}{\partial q''}\right) - \frac{\partial}{\partial t}\left(\frac{\partial \mathcal{L}}{\partial \dot{q}}\right) + \frac{\partial}{\partial t}\frac{\partial}{\partial x}\left(\frac{\partial \mathcal{L}}{\partial \dot{q}'}\right) = 0 \tag{3.142}$$

with the following boundary conditions:

$$\left[\mathcal{Q} + \frac{\partial \mathcal{L}}{\partial q'} - \frac{\partial}{\partial x}\left(\frac{\partial \mathcal{L}}{\partial q''}\right) - \frac{\partial}{\partial t}\left(\frac{\partial \mathcal{L}}{\partial \dot{q}'}\right)\right]\delta q = 0 \quad \text{in}\begin{cases} x = 0 \\ x = L \end{cases} \tag{3.143}$$

And

$$\left[\mathfrak{M} + \frac{\partial \mathcal{L}}{\partial q''}\right]\delta q' = 0 \quad \text{in}\begin{cases} x = 0 \\ x = L \end{cases} \tag{3.144}$$

Note: In several applications, the displacement field can be expressed in terms of N independent variables $(q_1, q_2, \ldots, q_N)$ called generalized coordinates. N is called the number of degrees of freedom. In this case, $\mathfrak{L} = \mathfrak{L}(q_1, q_2, \ldots, q_N, \dot{q}_1, \dot{q}_2, \ldots, \dot{q}_N)$ and Lagrange's equations reduce to a system of N equations:

$$\frac{\partial \mathfrak{L}}{\partial q_i} - \frac{\partial}{\partial t}\left(\frac{\partial \mathfrak{L}}{\partial \dot{q}_i}\right) = 0 \tag{3.145}$$

with $\mathfrak{L} = \mathrm{T} + \mathrm{W}_{ext} - \mathrm{V}$.

In the case where $\mathrm{T} = \mathrm{T}(\dot{q}_1, \dot{q}_2, \ldots, \dot{q}_N)$, $\mathrm{V} = \mathrm{V}(q_1, q_2, \ldots, q_N)$, and $\delta W_{ext} = \sum_{i=1}^{N} \mathcal{Q}_i \delta q_i$ with $\mathcal{Q}_i = \partial \mathrm{W}_{ext}/\partial q_i$. Lagrange's equations rewrite:

$$\frac{\partial}{\partial t}\left(\frac{\partial \mathrm{T}}{\partial \dot{q}_i}\right) + \frac{\partial \mathrm{V}}{\partial q_i} = \underbrace{\mathcal{Q}_i}_{\text{generalized force}} \tag{3.146}$$

EXAMPLE 3D.1: MASSES–SPRINGS SYSTEM (see Figure 3.13)

This is a two-degree-of-freedom system ($q_1 = x_1$, $q_2 = x_2$). We have $\mathfrak{L} = \mathfrak{L}(q_1, q_2, \dot{q}_1, \dot{q}_2)$ and Lagrange's equations reduce to

$$\frac{\partial}{\partial t}\left(\frac{\partial T}{\partial \dot{q}_i}\right) + \frac{\partial V}{\partial q_i} = \mathcal{Q}_i \quad i = 1,2 \tag{3.147}$$

We have

$$T = \frac{1}{2} m_1 \dot{x}_1^2 + \frac{1}{2} m_2 \dot{x}_2^2 \tag{3.148}$$

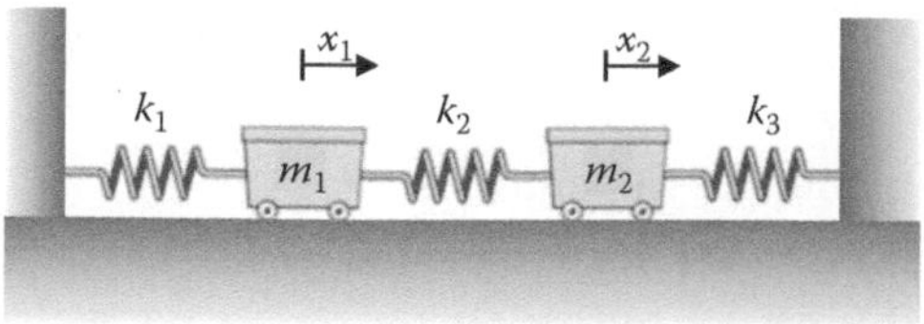

Figure 3.13 Masses–springs system.

$$V = \frac{1}{2}k_1x_1^2 + \frac{1}{2}k_2(x_1 - x_2)^2 + \frac{1}{2}k_3x_2^2 \qquad (3.149)$$

$$\delta W_{ext} = \underbrace{F_1}_{Q_1}\delta x_1 + \underbrace{F_2}_{Q_2}\delta x_2 \qquad (3.150)$$

This leads to

$$\begin{cases} m_1\ddot{x}_1 + (k_1 + k_2)x_1 - k_2x_2 = F_1 \\ m_2\ddot{x}_2 + (k_1 + k_2)x_2 - k_2x_2 = F_2 \end{cases} \qquad (3.151)$$

Or

$$\begin{pmatrix} m_1 & 0 \\ 0 & m_2 \end{pmatrix}\begin{Bmatrix} \ddot{x}_1 \\ \ddot{x}_2 \end{Bmatrix} + \begin{pmatrix} k_1 + k_2 & -k_2 \\ -k_2 & k_1 + k_2 \end{pmatrix}\begin{Bmatrix} x_1 \\ x_2 \end{Bmatrix} = \begin{Bmatrix} F_1 \\ F_2 \end{Bmatrix} \qquad (3.152)$$

EXAMPLE 3D.2: A TWO-DEGREE-OF-FREEDOM PENDULUM (see Figure 3.14)

This is a two-degree-of-freedom system ($q_1 = x$, $q_2 = \theta$). We have

$$\begin{aligned} T &= \frac{1}{2}m\left[\frac{d}{dt}(x + a\sin\theta)\right]^2 + \frac{1}{2}m\left[\frac{d}{dt}(a\cos\theta)\right]^2 + \frac{1}{2}I\dot{\theta}^2 \\ &= \frac{1}{2}m\dot{x}^2 + \frac{1}{2}ma^2\dot{\theta}^2\cos^2\theta + ma\dot{x}\dot{\theta}\cos\theta + \frac{1}{2}ma^2\dot{\theta}^2\sin^2\theta + \frac{1}{2}I\dot{\theta}^2 \end{aligned} \qquad (3.153)$$

where I is the moment of inertia of the rod around the pivot point. For small deformations, Equation 3.153 reduces to

$$T = \frac{1}{2}m\dot{x}^2 + \frac{1}{2}ma^2\dot{\theta}^2 + ma\dot{x}\dot{\theta} + \frac{1}{2}I\dot{\theta}^2 \qquad (3.154)$$

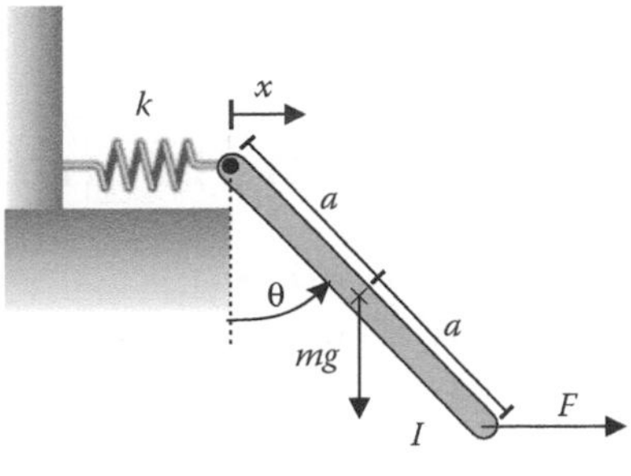

Figure 3.14 Two degrees-of-freedom pendulum.

We have

$$V = \frac{1}{2}kx^2 \tag{3.155}$$

and

$$\begin{aligned} \delta W_{ext} &= mga\delta(\cos\theta - 1) + F\delta(x + 2a\sin\theta) \\ &= -mga\sin\theta\delta\theta + F\delta x + 2aF\cos\theta\delta\theta \\ &= \underbrace{F}_{Q_1}\delta x + \underbrace{\left(2aF\cos\theta - mga\sin\theta\right)}_{Q_2}\delta\theta \end{aligned} \tag{3.156}$$

Lagrange's equations are

$$\begin{cases} \dfrac{\partial}{\partial t}\left(\dfrac{\partial T}{\partial \dot{x}}\right) + \dfrac{\partial V}{\partial x} = Q_1 \\ \dfrac{\partial}{\partial t}\left(\dfrac{\partial T}{\partial \dot{\theta}}\right) + \dfrac{\partial V}{\partial \theta} = Q_2 \end{cases} \tag{3.157}$$

Or

$$\begin{cases} ma\ddot{\theta} + m\ddot{x} + kx = F \\ ma^2\ddot{\theta} + ma\ddot{x} + I\ddot{\theta} = 2aF\cos\theta - mga\sin\theta \end{cases} \tag{3.158}$$

Assuming small deformations, we get

$$\begin{cases} ma\ddot{\theta} + m\ddot{x} + kx = F \\ (ma^2 + I)\ddot{\theta} + ma\ddot{x} + mga\theta = 2aF \end{cases} \tag{3.159}$$

If $x = 0$, we find the equation of a pendulum

$$(ma^2 + I)\ddot{\theta} + mga\theta = 2aF \tag{3.160}$$

EXAMPLE 3D.3: MASS–SPRING SYSTEM IN ROTATION (see **Figure 3.15**)

This is a one–degree-of-freedom system ($q_1 = x$). The reference system being non-Galilean we must add the inertia pseudo-force $m(a + x)\Omega_r^2$. We have

$$T = \frac{1}{2}m\dot{x}^2 \tag{3.161}$$

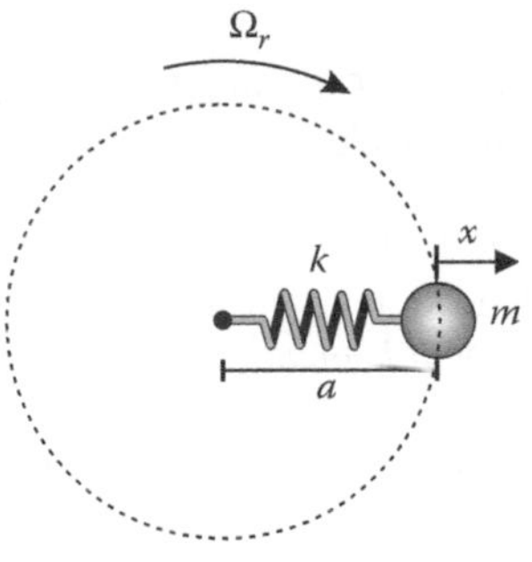

Figure 3.15 Mass–spring system in rotation.

$$V = \frac{1}{2}kx^2 \tag{3.162}$$

$$\delta W_{ext} = \underbrace{mg(a + x)\Omega_r^2}_{Q_1}\delta x \tag{3.163}$$

Lagrange's equations can be written as

$$\frac{\partial}{\partial t}\left(\frac{\partial T}{\partial \dot{x}}\right) + \frac{\partial V}{\partial x} = Q_1 \tag{3.164}$$

that is

$$m\ddot{x} + kx = ma\Omega_r^2 + mx\Omega_r^2 \tag{3.165}$$

Or equivalently

$$m\ddot{x} + \left(k - m\Omega_r^2\right)x = ma\Omega_r^2 \tag{3.166}$$

Let $X = x + x_{eq}$ with $kx_{eq} = m(a + x_{eq})\Omega_r^2$. We then get

$$m\ddot{X} + \left(k - m\Omega_r^2\right)X = \underbrace{ma\Omega_r^2 - \left(k - m\Omega_r^2\right)x_{eq}}_{=0} \tag{3.167}$$

The natural frequency of the system is

$$\omega_0 = \sqrt{\frac{k - m\Omega_r^2}{m}} \tag{3.168}$$

If $\Omega_r^2 \geq (k/m)$ the system becomes unstable.

REFERENCES

Finlayson, B. A. 1972. *The Method of Weighted Residuals and Variational Principles, with Application in Fluid Mechanics, Heat and Mass Transfer*, Vol. 87. New York: Academic Press.

Géradin, M. and D. Rixen. 1997. *Mechanical Vibrations: Theory and Application to Structural Dynamics*. Chichester, UK: John Wiley.

Lanczos, C. 1986. *The Variational Principles of Mechanics*. 4th ed. New York, USA: Dover Publications.

Reddy, J.N. 1993. *An Introduction to the Finite Element Method*. New York, USA: McGraw-Hill.

Chapter 4

The finite element method

An introduction

4.1 INTRODUCTION

This chapter introduces the fundamental concepts of the FEM through the study of a one-dimensional acoustic problem. The finite element discretization is based on a representation of the geometry and the unknown variables in terms of shape functions and nodal variables. This choice is mainly motivated by the requirements of generality and flexibility: to analyze, using a unified and systematic approach, boundary value problems related to domains of various shapes subjected to miscellaneous boundary conditions.

The analysis of a problem by FEM consists of seven basic steps. They are (see Figure 4.1)

Step 1: Writing the weak (or variational) integral form of the equations governing the problem.

Step 2: Meshing or geometry discretization. It consists in breaking up the geometry of the domain Ω into a set of subdomains Ω^e called "the elements." The subdomains are usually simple geometrical primitives (e.g., lines, triangles, quadrangles, tetrahedra ...).

Step 3: Approximation of variables and calculation of elementary matrices. It consists in defining on each subdomain Ω^e an approximate function p^e of the exact function p. This so-called nodal approximation function must have the following features:

- It must only involve the nodal variables attached to nodes located on Ω^e and its boundary;
- It must be continuous on Ω^e. In addition, all nodal approximation functions must satisfy the continuity conditions between subdomains.

The calculation of the elementary matrices consist then in evaluating, on each subdomain Ω^e, the nodal approximation of the integral form, determined in step 1.

Step 4: Assembling. In this step, the elementary matrices are combined to form global matrices. In other words, the integral form is evaluated over the entire domain Ω. By doing so, continuity conditions between elements are enforced.
Step 5: Imposition of boundary conditions or constraints.
Step 6: Invocation of stationarity condition. This step will lead to a system of algebraic equations. This system is solved using conventional numerical algorithms.
Step 7: Study of the convergence of the solution; calculation of physical indicators, and interpretation of results.
In the jargon of finite elements, steps (1) and (2) are part of the preprocessing phase, while step (7) is part of the postprocessing phase.

In this chapter, we will illustrate steps 1 through 7 with a simple one-dimensional example taken from acoustics. In Chapter 5, we will revisit, in the most general case, the essential ideas of these steps using this time the three-dimensional acoustic problem.

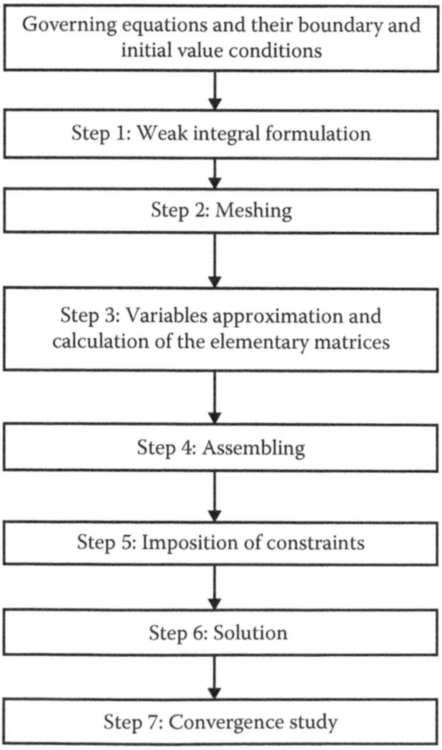

Figure 4.1 The seven steps of the FEM.

4.2 FINITE ELEMENT SOLUTION OF THE ONE-DIMENSIONAL ACOUSTIC WAVE PROPAGATION PROBLEM

4.2.1 Problem statement

Consider the following one-dimensional harmonic acoustic wave propagation problem:

$$\begin{cases} c_0^2 \dfrac{d^2\hat{p}}{dx^2} + \omega^2\hat{p} = 0 \quad x \in \Omega = \left]0, \mathrm{L}\right[\\ \hat{p}(0) = \hat{p}(L) = 0 \end{cases} \tag{4.1}$$

This is an eigenvalue problem. The associated natural frequencies and modal amplitudes are given by, respectively:

$$\omega_m = \frac{m\pi c_0}{L}, \quad p_m(x) = \sqrt{\frac{2}{L}}\sin\left(\frac{m\pi x}{L}\right), \quad m = 1,\cdots\infty \tag{4.2}$$

Note that the modal amplitudes have been normalized arbitrarily so that

$$\int_0^L p_m^2 dx = 1 \tag{4.3}$$

In the following, this eigenvalue problem is solved using the FEM*.

4.2.2 Step 1: Weak integral form

Let $\delta\hat{p}$ be an admissible variation of pressure $\hat{p}$, multiplying Equation 4.1 by $\delta\hat{p}$ and integrating over Ω gives

$$W(\hat{p},\delta\hat{p}) = \int_0^L \delta\hat{p}\left(c_0^2\frac{d^2\hat{p}}{dx^2} + \omega^2\hat{p}\right) dx = 0 \tag{4.4}$$

* This problem is a special case of the general form $\dfrac{d}{dx}\left(a\dfrac{du}{dx}\right) + b\,u = Q \quad 0 < x < L$ with essential boundary conditions at $x = 0$ and $x = L$ representing various problems in continuum mechanics (see Appendix IV.A).

An integration by parts leads to

$$W(\hat{p},\delta\hat{p}) = \int_0^L \left(-c_0^2 \frac{d\delta\hat{p}}{dx}\frac{d\hat{p}}{dx} + \omega^2 \hat{p}\delta\hat{p} \right) dx + \left[c_0^2 \delta\hat{p} \frac{d\hat{p}}{dx} \right]_0^L = 0 \tag{4.5}$$

$\delta\hat{p}$ being admissible, it must satisfy the homogenous form of the boundary conditions; in our example ($\delta\hat{p}(0) = \delta\hat{p}(L) = 0$). In consequence, the weak integral form associated with Equation 4.1 reads

$$W(\hat{p},\delta\hat{p}) = \int_0^L \left(c_0^2 \frac{d\delta\hat{p}}{dx}\frac{d\hat{p}}{dx} - \omega^2 \hat{p}\delta\hat{p} \right) dx = 0$$
$$\forall \delta\hat{p}/\delta\hat{p}(0) = \delta\hat{p}(L) = 0 \quad \text{and} \quad \hat{p}(0) = \hat{p}(L) = 0 \tag{4.6}$$

Defining

$$K(\hat{p},\delta\hat{p}) = c_0^2 \int_0^L \frac{d\delta\hat{p}}{dx}\frac{d\hat{p}}{dx} dx \tag{4.7}$$

and

$$M(\hat{p},\delta\hat{p}) = \int_0^L \delta\hat{p} \cdot \hat{p} dx \tag{4.8}$$

Equation 4.1 reads

$$\begin{cases} W(\hat{p},\delta\hat{p}) = K(\hat{p},\delta\hat{p}) - \omega^2 M(\hat{p},\delta\hat{p}) = 0 \\ \forall\ \delta\hat{p}/\delta\hat{p}(0) = \delta\hat{p}(L) = 0 \text{ and } \hat{p}(0) = \hat{p}(L) = 0 \end{cases} \tag{4.9}$$

Equivalently, the associated variational statement reads

$$\begin{cases} \delta\Pi(\hat{p}) = 0/\hat{p}(0) = \hat{p}(L) = 0 \text{ with,} \\ \Pi(\hat{p}) = \frac{1}{2}\left[K(\hat{p},\hat{p}) - \omega^2 M(\hat{p},\hat{p})\right] \end{cases} \tag{4.10}$$

Note that $K(\hat{p},\hat{p})$ is a positive semidefinite bilinear form while $M(\hat{p},\hat{p})$ is a positive definite bilinear form. The practical consequence of this will be made clear later in this chapter.

4.2.3 Step 2: Meshing

This is straightforward in this one-dimensional example. It consists in subdividing the domain Ω into a set of disjoint elements Ω^e. Denoting by n_e the number of elements, we write

$$\Omega = \bigcup_{e=1}^{n_e} \Omega^e \qquad e = 1,\ldots,n_e \tag{4.11}$$

Each element Ω^e is defined by a set of nn_e points (vertices) called nodes. Classically, two configurations are encountered (Figure 4.2):

1. Two-noded linear element: $nn_e = 2$. We will denote by x_1^e and x_2^e the two coordinates of the two nodes of the element in a cartesian coordinate system. Subscripts 1 and 2 refer here to a local numbering of the nodes of the element.
2. Three-noded quadratic element: $nn_e = 3$. We will denote by x_1^e, x_2^e, and x_3^e the coordinates of the three nodes of the element. Again subscripts 1, 2, and 3 refer to a local numbering of the nodes of the element.

4.2.4 Step 3: Approximation of the independent variable and calculation of the elementary matrices

4.2.4.1 *Nodal approximation of the variable*

Since the variable $\hat{p}$ (pressure) must be differentiable, we need to select at least a linear approximation to evaluate the form $K(\hat{p},\hat{p})$ and ensure inter-elements continuity. We will present two such approximations: the first is linear (order 1) and the second is quadratic (order 2).

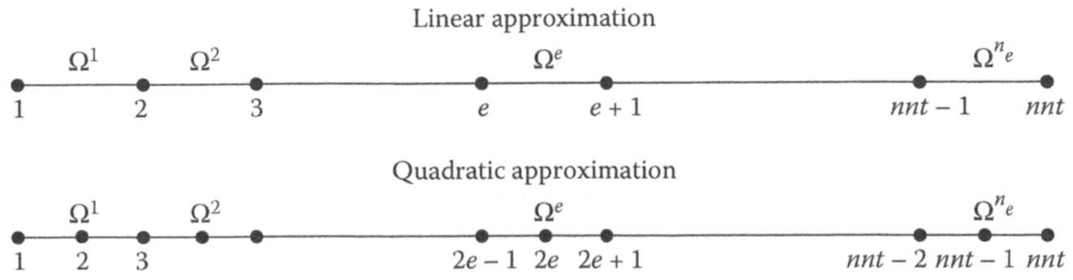

Figure 4.2 Two types of geometrical discretization of the one-dimensional problem.

4.2.4.1.1 Linear approximation

Let Ω^e represent a two-noded linear element. Knowing the values $\hat{p}_1^e$ and $\hat{p}_2^e$ of the pressure at the two nodes of the element, we seek to construct a linear approximation of the pressure within the element:

$$\hat{p}(x) = a\,x + b, \quad x \in \Omega^e \tag{4.12}$$

In consequence,

$$\hat{p}\left(x_1^e\right) = \hat{p}_1^e = a\,x_1^e + b, \quad \hat{p}\left(x_2^e\right) = \hat{p}_2^e = a\,x_2^e + b \tag{4.13}$$

This leads to

$$\hat{p}(x) = \frac{x_2^e - x}{h^e}\hat{p}_1^e + \frac{x - x_1^e}{h^e}\hat{p}_2^e \tag{4.14}$$

with $h^e = x_2^e - x_1^e$ the length of element Ω^e. Setting

$$N_1^e(x) = \frac{x_2^e - x}{h^e}, \quad N_2^e(x) = \frac{x - x_1^e}{h^e} \tag{4.15}$$

the nodal approximation of $\hat{p}$ can be written as

$$\hat{p}(x) = N_1^e(x)\,\hat{p}_1^e + N_2^e(x)\hat{p}_2^e = \sum_{i=1}^{2} N_i^e(x)\,\hat{p}_i^e \tag{4.16}$$

Or in matrix form*:

$$\hat{p}(x) = \left\langle N_1^e(x)\; N_2^e(x)\right\rangle \begin{Bmatrix} \hat{p}_1^e \\ \hat{p}_2^e \end{Bmatrix} = \left\langle N^e(x)\right\rangle\left\{\hat{p}^e\right\} \tag{4.17}$$

Remarks

1. Functions N_1^e and N_2^e are called nodal interpolation functions.

* In this book, we denote a line vector by symbol ⟨ ⟩ and a column vector by symbol { }.

2. The nodal interpolation functions must satisfy the following two conditions:
   - They should only involve the nodal variables attached to the nodes of the element.
   - They must be continuous across the element and differentiable as required by the weak integral form.
3. In general, polynomial functions are chosen to construct nodal interpolation functions since the process is simple and the numerical evaluation of the integrals involved in the weak integral form is easy.
4. When the interpolation functions are polynomials, these polynomials must be complete in order to represent all possible variations of the independent variable.
5. A nodal approximation requires nn_e interpolation functions. In symbolic form, we write

$$\hat{p}(x) = \sum_{i=1}^{nn_e} N_i^e(x)\, \hat{p}_i^e, \qquad x \in \Omega^e \tag{4.18}$$

   with $\hat{p}_i^e$, the value of the independent variable $\hat{p}$ at node i of the element and N_i^e the associated interpolation function. When N_i^e is polynomial, its order is equal to $(nn_e - 1)$.
6. The interpolation functions built solely from the nodal values of the independent variable (= functions ensuring C^0 continuity between elements) are called interpolation functions of Lagrange family.
7. Interpolation functions constructed from the nodal values of the independent variable and its derivatives are called interpolation functions of Hermite family.
8. The interpolation functions of the Lagrange family satisfy the following relationships:

$$\begin{cases} N_i^e(x_j^e) = \delta_{ij} \\ \\ \displaystyle\sum_{i=1}^{nn_e} N_i^e = 1 \Rightarrow \sum_{i=1}^{nn_e} \frac{dN_i^e}{dx} = 0 \end{cases} \tag{4.19}$$

   where δ_{ij} denotes the Kronecker symbol.

These two relations can be obtained easily from the nodal approximation $\hat{p}(x) = \sum_{i=1}^{nn_e} N_i^e(x)\, \hat{p}_i^e$. The first one ensures that $\hat{p}(x_i^e) = \hat{p}_i^e$ and the second one guarantees that the nodal approximation allows for representing a rigid state without deformation over the element ($\hat{p}$ constant over the element). The two conditions can be used to build the interpolation functions in a systematic way.

4.2.4.1.2 Quadratic approximation

Consider Ω^e a three-node line element. Let x_1^e, x_3^e denote the coordinates of its end-nodes, x_2^e the coordinate of its midnode, and h^e its length. We seek a quadratic approximation of the form:

$$\hat{p}(x) = a\,x^2 + bx + c, \quad x \in \Omega^e \tag{4.20}$$

knowing the nodal values of the independent variables:

$$\begin{cases} \hat{p}(x_1^e) = \hat{p}_1^e = a\,x_1^e + b(x_1^e)^2 + c \\ \hat{p}(x_2^e) = \hat{p}_2^e = a\,x_2^e + b(x_2^e)^2 + c \\ \hat{p}(x_3^e) = \hat{p}_3^e = a\,x_3^e + b(x_3^e)^2 + c \end{cases} \tag{4.21}$$

Making the change of variable $x = \xi + x_1^e$, the coordinates of the three nodes transforms into: $\xi_1^e = 0$, $\xi_2^e = h^e/2$ and $\xi_3^e = h^e$

The associated interpolation functions can be constructed directly from the properties given in Equation 4.19:

$$\begin{cases} N_1^e(\xi) = \left(1 - \dfrac{\xi}{h^e}\right)\left(1 - \dfrac{2\xi}{h^e}\right) \\ N_2^e(\xi) = \dfrac{4\xi}{h^e}\left(1 - \dfrac{\xi}{h^e}\right) \\ N_3(\xi) = \dfrac{-\xi}{h^e}\left(1 - \dfrac{2\xi}{h^e}\right) \end{cases} \tag{4.22}$$

4.2.4.2 Nodal approximation of the weak integral form

We start from Equation 4.9:

$$W(\hat{p},\delta\hat{p}) = K(\hat{p},\delta\hat{p}) - \omega^2 M(\hat{p},\delta\hat{p}) = 0 \tag{4.23}$$

with

$$K(\hat{p},\delta\hat{p}) = c_0^2 \int_0^L \frac{d\delta\hat{p}}{dx}\frac{d\hat{p}}{dx}\,dx \tag{4.24}$$

and

$$M(\hat{p},\delta\hat{p}) = \int_0^L \delta\hat{p}\hat{p}\;dx \tag{4.25}$$

Since $\Omega = \bigcup_{e=1}^{ne} \Omega^e$, the above integrals are written in terms of element integrals:

$$K(\hat{p},\delta\hat{p}) = \sum_{e=1}^{n_e} K^e(\hat{p},\delta\hat{p}) \tag{4.26}$$

with

$$K^e(\hat{p},\delta\hat{p}) = c_0^2 \int_{\Omega^e} \frac{d\delta\hat{p}}{dx}\frac{d\hat{p}}{dx}dx \tag{4.27}$$

and

$$M(\hat{p},\delta\hat{p}) = \sum_{e=1}^{n_e} M^e(\hat{p},\delta\hat{p}) \tag{4.28}$$

with

$$M^e(\hat{p},\delta\hat{p}) = \int_{\Omega^e} \delta\hat{p}\,\hat{p}\,dx \tag{4.29}$$

Using the nodal approximations of $\hat{p}$ and $\delta\hat{p}$

$$\hat{p} = \langle N^e\rangle\{\hat{p}^e\} \quad \text{and} \quad \delta\hat{p} = \langle N^e\rangle\{\delta\hat{p}^e\} \tag{4.30}$$

we write

$$K^e(\hat{p},\delta\hat{p}) = \langle\delta\,\hat{p}^e\rangle[K^e]\{\hat{p}^e\} \tag{4.31}$$

and

$$M^e(\hat{p},\delta\hat{p}) = \langle\delta\,\hat{p}^e\rangle[M^e]\{\hat{p}^e\} \tag{4.32}$$

with

$$[K^e] = c_0^2 \int_{\Omega^e} \left\{\frac{dN^e}{dx}\right\}\left\langle\frac{dN^e}{dx}\right\rangle dx \tag{4.33}$$

and

$$[M^e] = \int_{\Omega^e} \{N^e\}\langle N^e\rangle dx \tag{4.34}$$

The nodal approximation of the weak integral form reads finally

$$W(\hat{p},\delta\hat{p}) = \sum_{e=1}^{n_e} W^e(\hat{p},\delta\hat{p}) = 0 \tag{4.35}$$

with

$$W^e(\hat{p},\delta\hat{p}) = \langle\delta\hat{p}^e\rangle\big([K^e] - \omega^2[M^e]\big)\{\hat{p}^e\} \tag{4.36}$$

Matrices $[K^e]$ and $[M^e]$ are called **elementary matrices** of the system.

4.2.4.3 Evaluations of the elementary matrices

This corresponds to evaluating integrals Equations 4.33 and 4.34. In general, the evaluation is done numerically (see Chapter 5). In the simple case of our one-dimensional problem, an exact integration is possible. We present calculations for both types of elements.

For the linear element, using Equation 4.15 in Equations 4.33 and 4.34, we obtain

$$[K^e] = c_0^2 \int_{x_1^e}^{x_2^e} \begin{Bmatrix} -\dfrac{1}{h^e} \\ 1/h^e \end{Bmatrix} \left\langle -\frac{1}{h^e} \;\; \frac{1}{h^e} \right\rangle dx \tag{4.37}$$

$$[M^e] = \int_{x_1^e}^{x_2^e} \begin{bmatrix} \left(\dfrac{x_2^e - x}{h^e}\right)^2 & \left(\dfrac{x_2^e - x}{h^e}\right)\left(\dfrac{x - x_1^e}{h^e}\right) \\ \left(\dfrac{x_2^e - x}{h^e}\right)\left(\dfrac{x - x_1^e}{h^e}\right) & \left(\dfrac{x - x_1^e}{h^e}\right)^2 \end{bmatrix} dx \tag{4.38}$$

which leads to

$$[K^e] = \frac{c_0^2}{h^e} \begin{bmatrix} 1 & -1 \\ -1 & 1 \end{bmatrix} \tag{4.39}$$

and

$$[M^e] = \frac{h^e}{6}\begin{bmatrix} 2 & +1 \\ +1 & 2 \end{bmatrix} \tag{4.40}$$

For the quadratic element, using Equation 4.22 in Equations 4.33 and 4.34, we obtain:

$$[K^e] = \frac{c_0^2}{3h^e}\begin{bmatrix} 7 & -8 & 1 \\ -8 & 16 & -8 \\ 1 & -8 & 7 \end{bmatrix} \tag{4.41}$$

and

$$[M^e] = \frac{h^e}{30}\begin{bmatrix} 4 & 2 & -1 \\ 2 & 16 & 2 \\ -1 & 2 & 4 \end{bmatrix} \tag{4.42}$$

Remarks

1. Matrix $[K^e]$ is square and symmetric. Its diagonal elements are positive. In addition, it is singular since without constraints, the element can exhibit a rigid behavior (constant pressure here). In particular, note that the sum of its elements is zero since $K^e(\hat{p},\delta\hat{p}) = 0$ for $\hat{p} = \delta\hat{p} = 1$; see Equation 4.27.
2. Matrix $[M^e]$ is square and symmetric. The sum of its elements is equal to the length of the element since $M^e(\hat{p},\delta\hat{p}) = h^e$ for $\hat{p} = \delta\hat{p} = 1$; see Equation 4.29.
3. In the case of an acoustic source with volume density Q in domain Ω, the one-dimensional acoustic problem (4.1), becomes

$$\begin{cases} c_0^2 \dfrac{d^2\hat{p}}{dx^2} + \omega^2\hat{p} = -\hat{Q} \qquad x \in \Omega = \left]\, 0, L \,\right[\\ \hat{p}(0) = \hat{p}(L) = 0 \end{cases} \tag{4.43}$$

The associated weak integral form is given by

$$\begin{cases} W(\hat{p},\delta\hat{p}) = K(\hat{p},\delta\hat{p}) - \omega^2 M(\hat{p},\delta\hat{p}) - R(\delta\hat{p}) = 0 \\ \forall\ \delta\hat{p} / \delta\hat{p}(0) = \delta\hat{p}(L) = 0 \quad \text{and} \quad \hat{p}(0) = \hat{p}(L) = 0 \end{cases} \tag{4.44}$$

with

$$R(\delta\hat{p}) = \int_{\Omega} \delta\hat{p}\hat{Q}\,d\Omega \tag{4.45}$$

In discretized form, we obtain

$$R(\delta\hat{p}) = \sum_{e=1}^{n_e} R^e(\delta\hat{p}) \tag{4.46}$$

with

$$R^e(\delta\hat{p}) = \int_{\Omega^e} \delta\hat{p}\hat{Q}\,dx = \langle\delta\hat{p}^e\rangle\{\hat{f}^e\} \tag{4.47}$$

and

$$\{\hat{f}^e\} = \int_{\Omega^e} \{N^e\}\hat{Q}\,dx \tag{4.48}$$

Vector $\{\hat{f}^e\}$ is called source or load vector.
For a constant source density ($\hat{Q}(x) = \hat{Q}^e, x \in \Omega^e$), $\{\hat{f}^e\}$ is given by

$$\{\hat{f}^e\} = \frac{\hat{Q}^e\ h^e}{2}\begin{Bmatrix}1\\1\end{Bmatrix} \tag{4.49}$$

for a linear element and

$$\{\hat{f}^e\} = \frac{\hat{Q}^e h^e}{6}\begin{Bmatrix}1\\4\\1\end{Bmatrix} \tag{4.50}$$

for a quadratic element.

4. In the case of a point source located at $x = x_0$, the element load vector $\{\hat{f}^e\}$ becomes

$$\{\hat{f}^e\} = \int_{\Omega^e} \{N^e\}\hat{Q}\delta(x - x_0)dx = \begin{cases}\{0\} & x_0 \notin \Omega^e\\ \hat{Q}\{N^e(x_0)\} & x_0 \in \Omega^e\end{cases} \tag{4.51}$$

5. Reconsider problem (4.1), with a natural Neumann* type boundary condition:

$$\begin{cases} c_0^2 \dfrac{d^2\hat{p}}{dx^2} + \omega^2\hat{p} = 0 & x \in \Omega = \left]\, 0, L\, \right[\\ \hat{p}(0) = 0 \\ \dfrac{d\hat{p}}{dx}(L) = \rho_0\omega^2\overline{U}_n \end{cases} \tag{4.52}$$

The associated integral form is given by

$$\begin{cases} W(\hat{p},\delta\hat{p}) = K(\hat{p},\delta\hat{p}) - \omega^2 M(\hat{p},\delta\hat{p}) - R(\delta\hat{p}) = 0 \\ \forall \delta\hat{p} / \delta\hat{p}(0) = 0 \quad \text{and} \quad \hat{p}(0) = 0 \end{cases} \tag{4.53}$$

with

$$R(\delta\hat{p}) = \delta\hat{p}(L)\ \rho_0 c_0^2 \omega^2 \overline{U}_n \tag{4.54}$$

In discretized form, we obtain

$$R^e(\delta p) = \begin{cases} 0 & \text{if } e \neq n_e \\ \delta\hat{p}_{nnt}\ \rho_0 c_0^2 \omega^2 \overline{U}_n & \text{if } e = n_e \end{cases} \tag{4.55}$$

The only nonzero elemental load vector $\{\hat{f}^e\}$ is associated with element Ω^{n_e}. It is given by

$$\{\hat{f}^e\} = \rho_0\omega^2 c_0^2 \overline{U}_n \begin{Bmatrix} 0 \\ 1 \end{Bmatrix} \tag{4.56}$$

for the linear element and

$$\{\hat{f}^e\} = \rho_0\omega^2 c_0^2 \overline{U}_n \begin{Bmatrix} 0 \\ 0 \\ 1 \end{Bmatrix} \tag{4.57}$$

for the quadratic element.

* For example, this is the case of an acoustic medium driven by a piston, with displacement $\overline{U}_n \exp(i\omega t)$, located at $x = L$.

4.2.5 Step 4: Assembling

In discretized form, the weak integral form (4.35) reads

$$W(\hat{p},\delta\hat{p}) = \sum_{e=1}^{n_e} \langle\delta\hat{p}^e\rangle\left([K^e] - \omega^2[M^e]\right)\{\hat{p}^e\} \tag{4.58}$$

The assembly procedure consists in writing this sum in terms of the vectors of nodal unknowns. Doing so requires the continuity of the variable between elements. To introduce the basic idea, consider the case of approximation by linear elements. For n_e elements, let us sequentially number the geometric nodes from 1 to nnt. The associated global vector of nodal unknowns is given by

$$\langle\hat{p}\rangle = \langle\hat{p}_1, \hat{p}_2, \ldots, \hat{p}_{nnt}\rangle \tag{4.59}$$

In this sequential numbering, the nodes ID of element Ω^e are given by e and $e+1$. In the particular case of our problem, the assembly procedure is reduced to the imposition of the following continuity conditions:

$$\hat{p}_2^e = \hat{p}_1^{e+1} = \hat{p}_{e+1} \tag{4.60}$$

Using this global numbering, an element localization matrix, $[L^e]$ of dimension $(nn_e \times nnt)$, relating local and global numbering is introduced:

$$\{\hat{p}^e\} = [L^e]\{\hat{p}\} \tag{4.61}$$

For the two-noded linear element, $[L^e]$ is given by

$$\begin{matrix} & \cdots & e & e+1 & & \\ [L^e] = \Big[& \begin{matrix} 0 & \cdots & 1 & 0 & \cdots & 0 \\ 0 & \cdots & 0 & 1 & \cdots & 0 \end{matrix} & \Big] \end{matrix} \tag{4.62}$$

Using Equation 4.61, the discretized weak integral form, Equation 4.58, becomes

$$W(\hat{p},\delta\hat{p}) = \sum_{e=1}^{n_e} \langle\delta\hat{p}\rangle[L^e]^T([K^e] - \omega^2[M^e])[L^e]\{\hat{p}\} \tag{4.63}$$

In compact form

$$W(\hat{p},\delta\hat{p}) = \langle\delta\hat{p}\rangle([K] - \omega^2[M])\{\hat{p}\} \tag{4.64}$$

with

$$[K] = \sum_{e=1}^{n_e}[L^e]^T[K^e][L^e] \tag{4.65}$$

and

$$[M] = \sum_{e=1}^{n_e}[L^e]^T[M^e][L^e] \tag{4.66}$$

Matrices $[K]$ and $[M]$ resulting from the assembling of elementary matrices $[K^e]$ and $[M^e]$ are called global stiffness and mass matrix of the system, respectively. In the case of the two-nodes linear element, it is easy to verify that $[K]$ and $[M]$ are given by*

$$[K] = \frac{c_0^2}{h}\begin{bmatrix} 1 & -1 & & & \\ & 2 & -1 & & \\ & & \ddots & \ddots & \\ & sym. & & 2 & -1 \\ & & & & 1 \end{bmatrix} \tag{4.67}$$

$$[M] = \frac{h}{6}\begin{bmatrix} 2 & 1 & & & \\ & 4 & \ddots & & \\ & & \ddots & \ddots & \\ & sym. & & 4 & 1 \\ & & & & 2 \end{bmatrix} \tag{4.68}$$

In summary, the assembly process is used to write the weak integral form in a discrete form involving the nodal unknowns of the problem. Knowing that $\hat{p}(0) = \hat{p}_1$ and $\hat{p}(L) = \hat{p}_{nnt}$, the weak integral form (4.9) is written in discrete form:

$$\begin{cases} W(\hat{p},\delta\hat{p}) = \langle\delta\hat{p}\rangle([K] - \omega^2[M])\{\hat{p}\} = 0 \\ \forall\langle\delta\hat{p}\rangle/\delta\hat{p}_1 = \delta\hat{p}_{nnt} = 0 \quad \text{and} \quad \hat{p}_1 = \hat{p}_{nnt} = 0 \end{cases} \tag{4.69}$$

* For ease of presentation, same length elements are assumed: $h^e = h$.

Remarks

1. Matrices $[K]$ and $[M]$ are symmetric* and band-limited matrices. This is a general property of matrices derived from the FEM.
2. For the problem with an acoustic source defined in Equation 4.43, the global load vector is of the form:

$$\{\hat{f}\} = \sum_{e=1}^{n_e} [L^e]^T \{\hat{f}^e\} \tag{4.70}$$

For the two-noded element it is given by†

$$\{\hat{f}\} = \frac{\hat{Q}h}{2} \begin{Bmatrix} 1 \\ 2 \\ \vdots \\ 2 \\ 1 \end{Bmatrix} \tag{4.71}$$

3. For the Neumann problem defined in Equation 4.52, the global load vector is of the form

$$\{\hat{f}\} = \rho_0 c_0^2 \omega^2 \overline{U}_n \begin{Bmatrix} 0 \\ 0 \\ \vdots \\ 0 \\ 1 \end{Bmatrix} \tag{4.72}$$

4.2.6 Step 5: Imposition of constraints

Before invoking the stationarity in Equation 4.69 to solve the system, we must impose the essential boundary conditions of the problem. That is, $\delta\hat{p}_1 = \delta\hat{p}_{nnt} = 0$ and $\hat{p}_1 = \hat{p}_{nnt} = 0$. For the approximation by linear elements Equation (4.69) becomes

* This is a consequence of the symmetry of the weak integral form.

† We assume elements of the same length and a constant source density: $\hat{Q}^e = \hat{Q}$.

$$W(\hat{p},\delta\hat{p}) = \langle 0,\ \delta\hat{p}_2,\dots,\ \delta\hat{p}_{n-1},\ 0\rangle\ ([K] - \omega^2[M]) \begin{Bmatrix} 0 \\ \hat{p}_2 \\ \vdots \\ \hat{p}_{n-1} \\ 0 \end{Bmatrix} \tag{4.73}$$

or explicitly using Equations 4.67 and 4.68

$$\left\langle \delta\hat{p}_2,\dots,\ \delta\hat{p}_{nnt-1}\right\rangle \left(\frac{c_0^2}{h} \begin{bmatrix} 2 & -1 & & \\ & 2 & \ddots & \\ & & \ddots & -1 \\ & sym. & & 2 \end{bmatrix} - \omega^2 \frac{h}{6} \begin{bmatrix} 4 & 1 & & \\ & 4 & \ddots & \\ & & \ddots & 1 \\ & sym. & & 4 \end{bmatrix} \right) \times \begin{Bmatrix} \hat{p}_2 \\ \vdots \\ \vdots \\ \hat{p}_{nnt-1} \end{Bmatrix} = 0 \tag{4.74}$$

for any arbitrary vector $\left\langle \delta\hat{p}_2,\dots,\ \delta\hat{p}_{nnt-1}\right\rangle$. In consequence, the following eigenvalue problem is obtained:

$$\frac{c_0^2}{h} \begin{bmatrix} 2 & -1 & & \\ & 2 & \ddots & \\ & & \ddots & -1 \\ & sym. & & 2 \end{bmatrix} \begin{Bmatrix} \hat{p}_2 \\ \hat{p}_3 \\ \vdots \\ \hat{p}_{nnt-1} \end{Bmatrix} = \omega^2 \frac{h}{6} \begin{bmatrix} 4 & 1 & & \\ & 4 & \ddots & \\ & & \ddots & 1 \\ & sym. & & 4 \end{bmatrix} \begin{Bmatrix} \hat{p}_2 \\ \hat{p}_3 \\ \vdots \\ \hat{p}_{nnt-1} \end{Bmatrix} \tag{4.75}$$

For the problem with an acoustic source with constant density $\hat{Q}$, defined in Equation 4.43, the obtained global system is given by

$$\left(\frac{c_0^2}{h} \begin{bmatrix} 2 & -1 & & \\ & 2 & \ddots & \\ & & \ddots & -1 \\ & sym. & & 2 \end{bmatrix} - \omega^2 \frac{h}{6} \begin{bmatrix} 4 & 1 & & \\ & 4 & \ddots & \\ & & \ddots & 1 \\ & sym. & & 4 \end{bmatrix} \right) \begin{Bmatrix} \hat{p}_2 \\ \hat{p}_3 \\ \vdots \\ \hat{p}_{nnt-2} \\ \hat{p}_{nnt-1} \end{Bmatrix} = \frac{\hat{Q}h}{2} \begin{Bmatrix} 1 \\ 2 \\ \vdots \\ 2 \\ 1 \end{Bmatrix} \tag{4.76}$$

For the Neumann problem defined in Equation 4.52, the global system is given by

$$\left(\frac{c_0^2}{h}\begin{bmatrix} 2 & -1 & & \\ & 2 & \ddots & \\ & & \ddots & -1 \\ sym. & & & 2 \end{bmatrix} - \omega^2\frac{h}{6}\begin{bmatrix} 4 & 1 & & \\ & 4 & \ddots & \\ & & \ddots & 1 \\ sym. & & & 4 \end{bmatrix}\right)\begin{Bmatrix} \hat{p}_2 \\ \hat{p}_3 \\ \vdots \\ \hat{p}_{nnt-2} \\ \hat{p}_{nnt-1} \end{Bmatrix} = \rho_0 c_0^2\omega^2\overline{U}_n\begin{Bmatrix} 0 \\ 0 \\ \vdots \\ 0 \\ 1 \end{Bmatrix} \tag{4.77}$$

4.2.7 Steps 6 and 7: Solution and convergence study

Let us consider first the solution resulting from the use of two linear elements: $n_e = 2$; $nnt = 3$; $h = L/2$ and $\langle \hat{p} \rangle = \langle \hat{p}_1, \hat{p}_2, \hat{p}_3 \rangle$ with $\hat{p}_1 = \hat{p}_3 = 0$. The eigenvalue problem (4.75), reduces to

$$\left(-\omega^2\frac{h}{6}4 + \frac{c_0^2}{h}2\right)\hat{p}_2 = 0 \tag{4.78}$$

Its solution is given by

$$\omega_1 = \sqrt{3}\,\frac{c_0}{h} = 2\sqrt{3}\,\frac{c_0}{L} \tag{4.79}$$

To compare the associated mode shape with the analytical solution Equation 4.2, we should normalize the amplitude following Equation 4.3:

$$\int_0^L (\hat{p}(x))^2\,dx = \langle \hat{p} \rangle [M] \{\hat{p}\} = 1 \tag{4.80}$$

leading to $\hat{p}_2 = \sqrt{3/L}$. The exact solution for the first mode ($m = 1$) is obtained from $\omega_m = m\pi c_0/L$ and $\hat{p}_m(x) = \sqrt{2/L}\sin(m\pi x/L)$. The comparison is illustrated in Figure 4.3. The relative error on the first circular frequency is given by

$$\frac{\omega_1 - \omega_{ex}}{\omega_{ex}} = \frac{2\sqrt{3} - \pi}{\pi} \simeq 10\% \tag{4.81}$$

And the error on the amplitude of $\hat{p}_2$ is

$$\frac{\sqrt{3/L} - \sqrt{2/L}\sin\pi/2}{\sqrt{2/L}\sin\pi/2} = \frac{\sqrt{3} - \sqrt{2}}{\sqrt{2}} \simeq 22\% \tag{4.82}$$

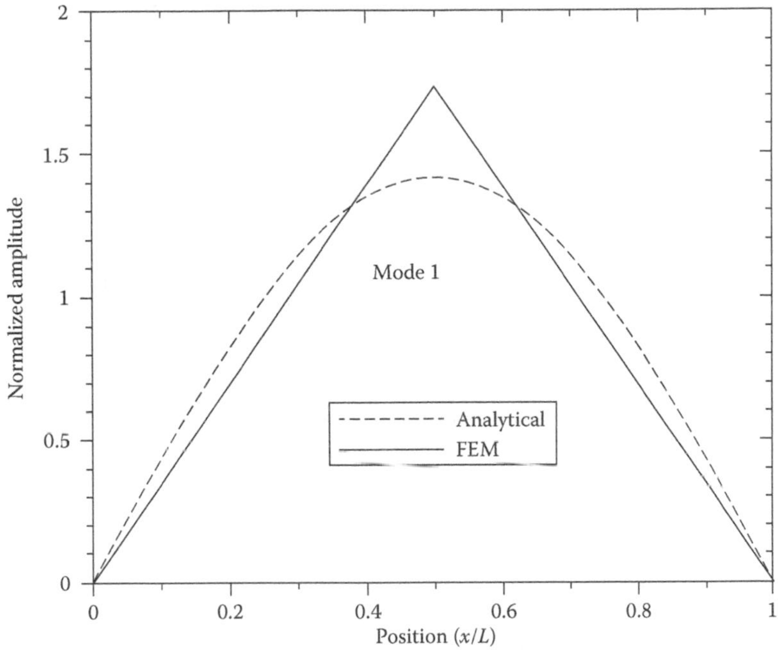

Figure 4.3 Convergence of the first mode. Exact solution versus FEM solution using two linear elements.

Note that the calculated frequency is greater than the exact frequency (why is this?). To improve the convergence more elements should be used. For instance, using three linear elements we have: $n_e = 3$, $nnt = 4$, $h = L/3$, $\langle\hat{p}\rangle = \langle\hat{p}_1, \hat{p}_2, \hat{p}_3, \hat{p}_4\rangle$ with $\hat{p}_1 = \hat{p}_4 = 0$. The associated eigenvalue problem is given by

$$\left(-\omega^2 \frac{h}{6}\begin{bmatrix} 4 & 1 \\ 1 & 4 \end{bmatrix} + \frac{c_0^2}{h}\begin{bmatrix} 2 & +1 \\ -1 & 2 \end{bmatrix}\right)\begin{Bmatrix} \hat{p}_2 \\ \hat{p}_3 \end{Bmatrix} = \begin{Bmatrix} 0 \\ 0 \end{Bmatrix} \tag{4.83}$$

Its solution is given by

$$\begin{cases} \omega_1 = \sqrt{\dfrac{6}{5}}\,\dfrac{c_0}{h} = 3\sqrt{\dfrac{6}{5}}\,\dfrac{c_0}{L} \\ \omega_2 = \sqrt{6}\,\dfrac{c_0}{h} = 3\sqrt{6}\,\dfrac{c_0}{L} \end{cases} \tag{4.84}$$

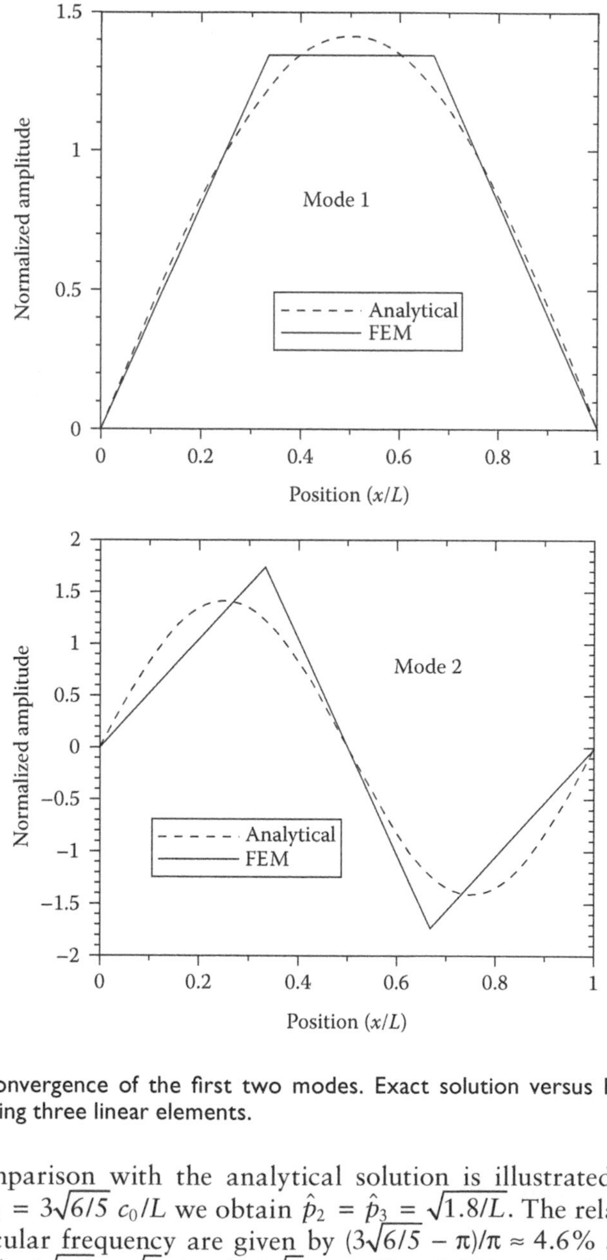

Figure 4.4 Convergence of the first two modes. Exact solution versus FEM solution using three linear elements.

The comparison with the analytical solution is illustrated in Figure 4.4. For $\omega_1 = 3\sqrt{6/5}\, c_0/L$ we obtain $\hat{p}_2 = \hat{p}_3 = \sqrt{1.8/L}$. The relative errors on the circular frequency are given by $(3\sqrt{6/5} - \pi)/\pi \approx 4.6\%$ and on the amplitude by $(\sqrt{1.8} - \sqrt{2}\,\sin\pi/3)/(\sqrt{2}\,\sin\pi/3) \approx 9.5\%$ at $x = L/3$.

For $\omega_2 = 3\sqrt{6}\, c_0/L$, we obtain $\hat{p}_2 = -\hat{p}_3 = \sqrt{3/L}$. The relative errors on the circular frequency is $(3\sqrt{6} - 2\pi)/2\pi \approx 17\%$ and on the amplitude $(\sqrt{3} - \sqrt{2}\,\sin 2\pi/3)/(\sqrt{2}\,\sin 2\pi/3) \approx 41\%$ at $x = 2L/3$.

Thus, as the mesh is refined, the FEM solution approaches the analytical solution (see Appendix 4C for MATLAB scripts to study the convergence). The convergence is reached from higher values. The FEM overestimates the natural frequencies. A qualitative explanation lies in the fact that for a given mesh, we constrain the variable to vary along a fixed set of shape functions. Thus, an additional numerical constraint is imposed, which increases the rigidity compared to its natural value. This observation can be exploited to study the convergence of the solution. This propriety is, however, lost when consistent implementations are not used (e.g., use of diagonal lumped mass matrices).

Several examples illustrating the FEM solution of the one-dimensional acoustic problem for various types of boundary conditions and excitations, together with the associated MATLAB scripts, are given in Appendix 4C.

4.3 CONCLUSION

We have introduced in this chapter and its appendixes the fundamental concepts of FEM through the study of a one-dimensional acoustic problem. The seven-step-based methodology was introduced and illustrated via several examples. In Chapter 5, the same steps are revisited and detailed in the context of the uncoupled three-dimensional acoustic and structural problems. The study of the coupled problem is deferred to Chapter 6 for the interior problem and to Chapter 8 for the exterior problem.

APPENDIX 4A: DIRECT STIFFNESS APPROACH: SPRING ELEMENTS

4A.1 Linear springs

EXAMPLE 4A.1

The equilibrium equation of a spring element (see Figure 4.5).

$$\begin{bmatrix} k & -k \\ -k & k \end{bmatrix} \begin{Bmatrix} u_1 \\ u_2 \end{Bmatrix} = \begin{Bmatrix} F_1 \\ F_2 \end{Bmatrix} \tag{4.85}$$

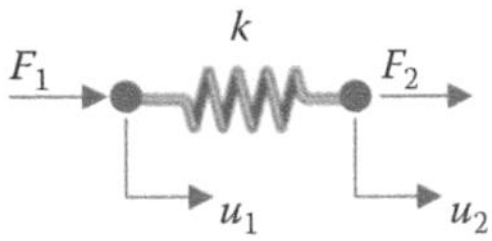

Figure 4.5 Linear spring.

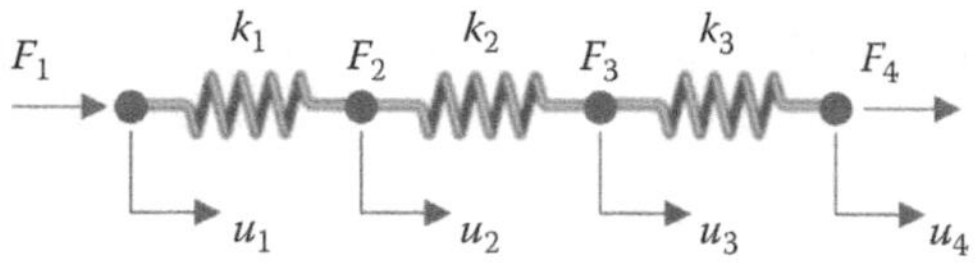

Figure 4.6 Assembling and continuity.

EXAMPLE 4A.2

Illustration of the assembly of spring elements (see Figure 4.6).

$$\begin{bmatrix} k_1 & -k_1 & 0 & 0 \\ -k_1 & k_1+k_2 & -k_2 & 0 \\ 0 & -k_2 & k_2+k_3 & -k_3 \\ 0 & 0 & -k_3 & k_3 \end{bmatrix} \begin{Bmatrix} u_1 \\ u_2 \\ u_3 \\ u_4 \end{Bmatrix} = \begin{Bmatrix} F_1 \\ F_2 \\ F_3 \\ F_4 \end{Bmatrix} \tag{4.86}$$

The FEM works as well for isostatic problems (Ex: $u_1 = 0$) as for hyperstatic problems (Ex: $u_1 = u_4 = 0$).

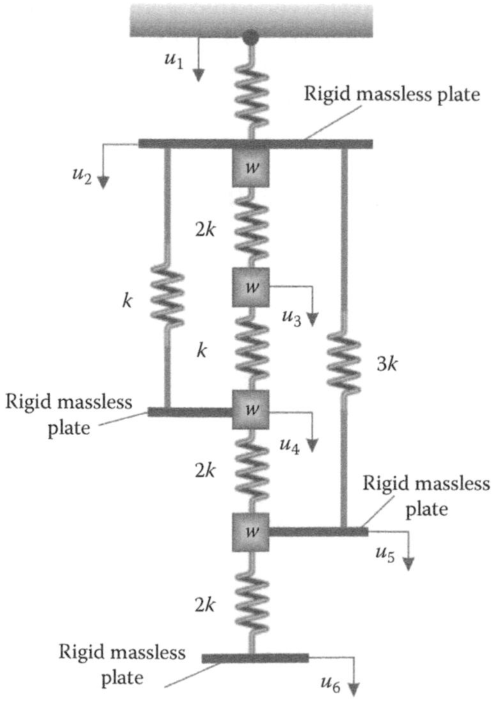

Figure 4.7 Assembling of spring elements.

EXAMPLE 4A.3

Illustration of the assembly of spring elements (see Figure 4.7).

$$\begin{bmatrix} k & -k & 0 & 0 & 0 & 0 \\ -k & 7k & -2k & -3k & -3k & 0 \\ 0 & -2k & 3k & -k & 0 & 0 \\ 0 & -k & -k & 4k & -2k & 0 \\ 0 & -3k & 0 & -2k & 7k & -2k \\ 0 & 0 & 0 & 0 & 2k & 2k \end{bmatrix} \begin{Bmatrix} u_1 \\ u_2 \\ u_3 \\ u_4 \\ u_5 \\ u_6 \end{Bmatrix} = \begin{Bmatrix} 0 \\ w \\ w \\ w \\ w \\ 0 \end{Bmatrix} \tag{4.87}$$

with the kinematic conditions $u_1 = 0$.

4A.2 Uniaxial traction–compression

EXAMPLE 4A.4

Illustration of the modeling of uniaxial bars using springs (see Figure 4.8).

$$K_i = \frac{A_i E_i}{L_i}$$

A is the cross-section area
E is Young's modulus
L is the length

$$\begin{bmatrix} k_1 & -k_1 & 0 & 0 \\ -k_1 & k_1 + k_2 & -k_2 & 0 \\ 0 & -k_2 & k_2 + k_3 & -k_3 \\ 0 & 0 & -k_3 & k_3 \end{bmatrix} \begin{Bmatrix} u_1 \\ u_2 \\ u_3 \\ u_4 \end{Bmatrix} = \begin{Bmatrix} F_1 \\ 0 \\ 0 \\ P \end{Bmatrix} \tag{4.88}$$

F_1 is the reaction force to be determined via condition $u_1 = 0$.

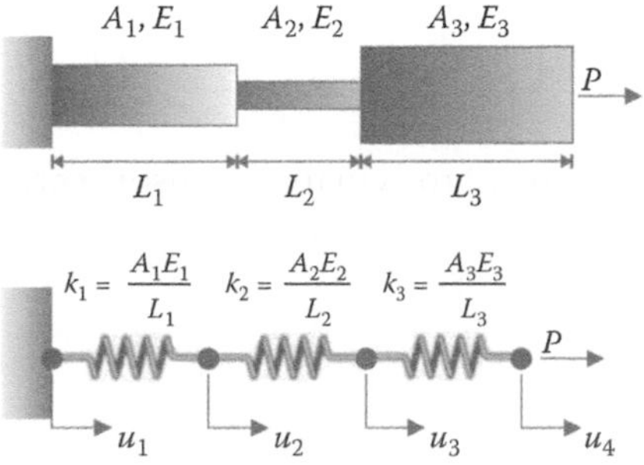

Figure 4.8 Uniaxial traction–compression.

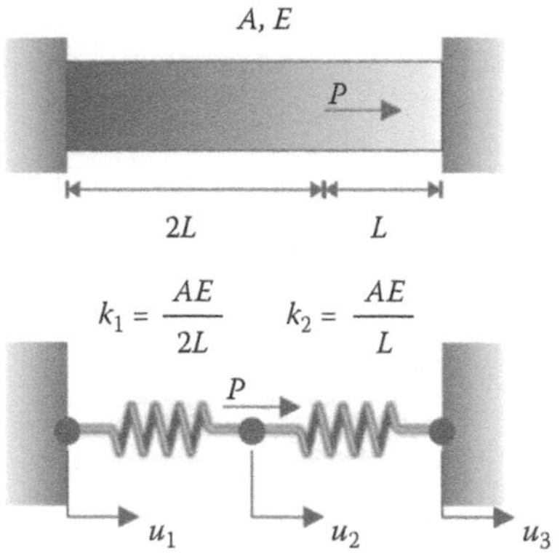

Figure 4.9 Example 1 of hyperstatic problem.

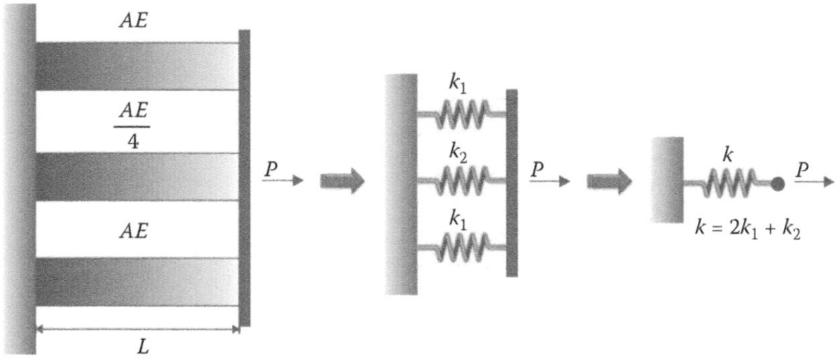

Figure 4.10 Example 2 of hyperstatic problem.

Figures 4.9 and 4.10 illustrate the modeling of two hyperstatic problems involving bar elements using springs.

4A.3 Torsion

Illustration of the modeling of rods in torsion using springs (see Figure 4.11).

$$K_{eq} = \frac{GJ}{L} \tag{4.89}$$

G is the shear modulus
J is the polar moment of inertia of the cross section (torsion constant)
L is the element length.

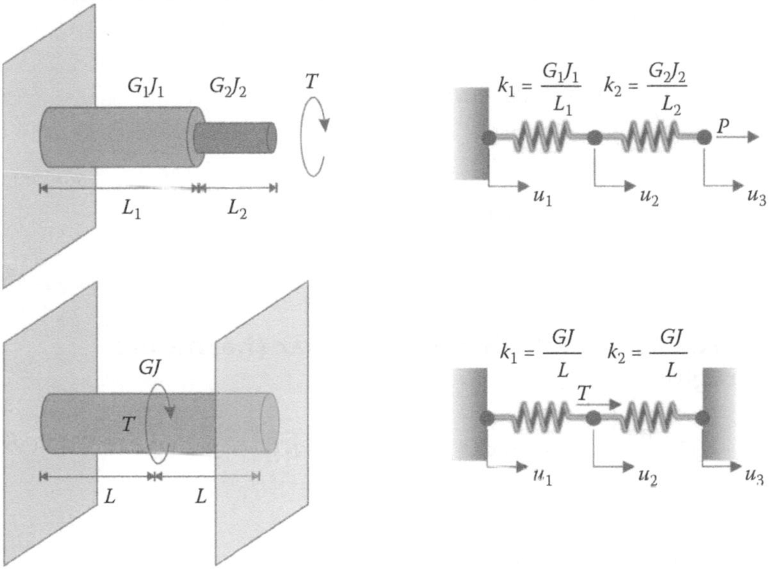

Figure 4.11 Torsion problems.

APPENDIX 4B: EXAMPLES OF TYPICAL 1D PROBLEMS

This appendix gives examples of typical 1D problems that can be solved using the approach (matrices) developed for the 1D acoustic wave propagation problem.

4B.1 Forced response of a bar in traction (harmonic regime)

$$\frac{\partial}{\partial x}\left(ES\frac{\partial \hat{u}}{\partial x}\right) + \rho_s\omega^2\hat{u} = \hat{F}_s \qquad + B.C. \tag{4.90}$$

with

ES: product of the material Young's modulus and the bar cross-section area (traction stiffness)
$\hat{u}$: axial displacement
ρ_s: mass per unit length
$\hat{F}_s$: force per unit length.
$B.C.$: Boundary conditions

4B.2 Free transverse vibration of a string (harmonic regime)

$$\frac{\partial}{\partial x}\left(T_s(x)\frac{\partial \hat{u}}{\partial x}\right) + m_s(x)\omega^2 \hat{u}(x) = 0 \qquad + B.C. \tag{4.91}$$

with

$\hat{u}$: transverse displacement of the string
T_s: tension in the string
m_s: mass per unit length.

4B.3 Free torsional vibration of a bar (harmonic regime)

$$\frac{\partial}{\partial x}\left[GJ(x)\frac{\partial \hat{\theta}(x)}{\partial x}\right] + \omega^2 I(x)\hat{\theta}(x) = 0 \qquad + B.C. \tag{4.92}$$

with

$\hat{\theta}$: section rotation angle
GJ: product of material's shear modulus and section torsional constant (torsional stiffness)
I: second polar moment of the section.

4B.4 Heat conduction (stationary regime)

$$\frac{\partial}{\partial x}\left(\kappa\frac{\partial \hat{T}}{\partial x}\right) + \hat{\gamma} = 0 \qquad + B.C. \tag{4.93}$$

with

κ: isotropic thermal diffusivity
$\hat{T}$: temperature
$\hat{\gamma}$: internal distributed heat source.

4B.5 Incompressible laminar flow under the action of a constant pressure gradient (stationary regime)

$$\eta\frac{\partial^2 \hat{u}}{\partial y^2} + \frac{\partial \hat{p}}{\partial x} = 0 \qquad + B.C. \tag{4.94}$$

with

η: dynamic viscosity of the fluid
$\hat{u}$: axial component of the velocity
$\hat{p}$: pressure.

4B.6 Hagen-Poiseuille laminar pipe flow equation

$$\frac{\partial}{\partial x}\left(k_p \frac{d\hat{p}}{dx}\right) + \hat{\upsilon} = 0 \qquad + B.C. \tag{4.95}$$

with

$k_p = \pi D^4/128\eta$ with η dynamic viscosity of the fluid and D the pipe diameter

$\hat{\upsilon}$: volume flow rate

$\hat{p}$: pressure.

APPENDIX 4C: APPLICATION TO ONE-DIMENSIONAL ACOUSTIC WAVE PROPAGATION PROBLEM

The following examples illustrate the steps described in this chapter through various problems solved using MATLAB.

EXAMPLE 4C.1

This example describes the solution of Equation 4.1 using linear elements. It can be used to study the convergence of these elements. The comparison with the analytical solution, Equation 4.2, for the first four modes is given in Figure 4.12 for a mesh of 10 elements.

```
%=================================================================
% Example 4C.1
%
% Solve a 1D Acoustic Eigenvalue Problem using Linear elements
% Boundary conditions: Fixed pressure at both ends%
%
%=================================================================
clear all; close all; clc
%
% Problem data
%
Lx=1;             % Length of the domain; fixed arbitrary to 1
c0=342;           % Speed of sound (m/s)
%
% Step 1: Mesh
%
ne= input(' Number of linear elements: '); % number of elements
nnt = ne+1;                       % Total number of nodes
h=Lx/ne;                          % Length of the elements
x=[0:h:Lx];                       % Coordinates table
%
```

```
% Step 2: Compute Elementary matrices
%
Ke=c0^2* [1,-1;-1,1]*1/h;
Me=[2,1;1,2]*h/6;
%
% Step 3: Assembling
%
I=eye(2,2);
K=zeros(nnt,nnt); M = zeros(nnt,nnt);
for ie=1: ne
  L=zeros(2,nnt); L(:,ie:ie+1)=I; % Location matrix for element ie
  K = K + L'*Ke*L;
  M = M + L'*Me*L;
end
%
% Step 4: Boundary conditions: p(1) = p(nnt) = 0
%
K = K(2:nnt-1,2:nnt-1);
M = M(2:nnt-1,2:nnt-1);
ndof = nnt-2;           % Final number of unknown (equations)
%
% Step 5: Compute the eigenvalues and eigenvectors
%
[V,D]=eig(K,M); D= sqrt(D);
Nor=V'*M*V; V=V*sqrt(inv(Nor));   % Normalization of the eigenvectors
for idof = 1: ndof
  w(idof) = D(idof,idof);
end
[w,ind]=sort(w);            % Sort the eigenvalues in ascending order
%
% Step 6: Comparison with the exact solution for a selected number
   of modes
%
nm = min(4, ndof);          %  Here we select the first modes up to
   the fourth
theo = c0*[1:ndof]*pi / Lx;
Err = abs(theo-w)./theo;

fprintf('\n ************Results for %d lin elements ************', ne);
fprintf('\n ================================================');
fprintf('\n Mode  FEM (Hz) Exact (Hz) Error(%%)');
fprintf('\n ---------------------------------');
for m=1:nm
  fprintf('\n %d %7.2f %7.2f  %3.1f', m, w(m)/2/pi, theo(m)/2/pi,
   Err(m)*100);
end
fprintf('\n ================================================\n');
fprintf('\n');
%
% comparison of the mode shapes
%
xt=[0:0.025:1] * Lx;        % Mesh to visualize the mode shapes
for m=1:nm
  figure;
  yt= sqrt(2)*sin(m*pi*xt);
```

```
  plot(xt,yt,'LineWidth',2,'Color',[0 0 0])
  hold on, % Theoretical solution for
  y=[0,real(V(:,ind(m))'),0]; if (yt(2) >0 && y(2) < 0), y = -y; end;
   % Recall : p(0)=p(L)=0
  plot(x,y,'-*','LineWidth',2,'Color',[0.5 0.5 0.5])    % FE solution
   title(['Mode' num2str(m)]);
   xlabel('Position x/L');
   ylabel(' Normalized Amplitude ');
   legend('Analytical', 'FEM Linear');
end
```

```
************ Results for 10 lin elements ************
====================================================
Mode  FEM (Hz) Exact (Hz) Error(%)
----------------------------------------------------
  1    171.70    171.00      0.4
  2    347.65    342.00      1.7
  3    532.13    513.00      3.7
  4    729.36    684.00      6.6
====================================================
```

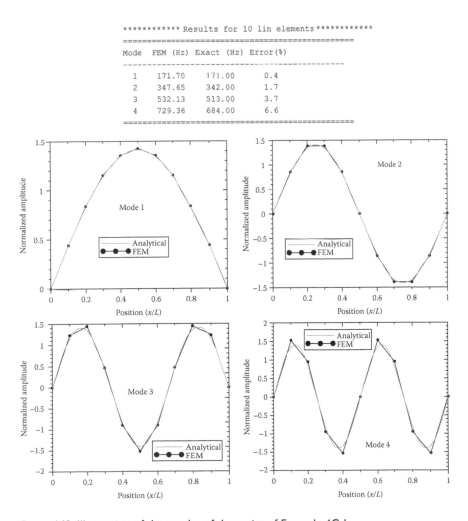

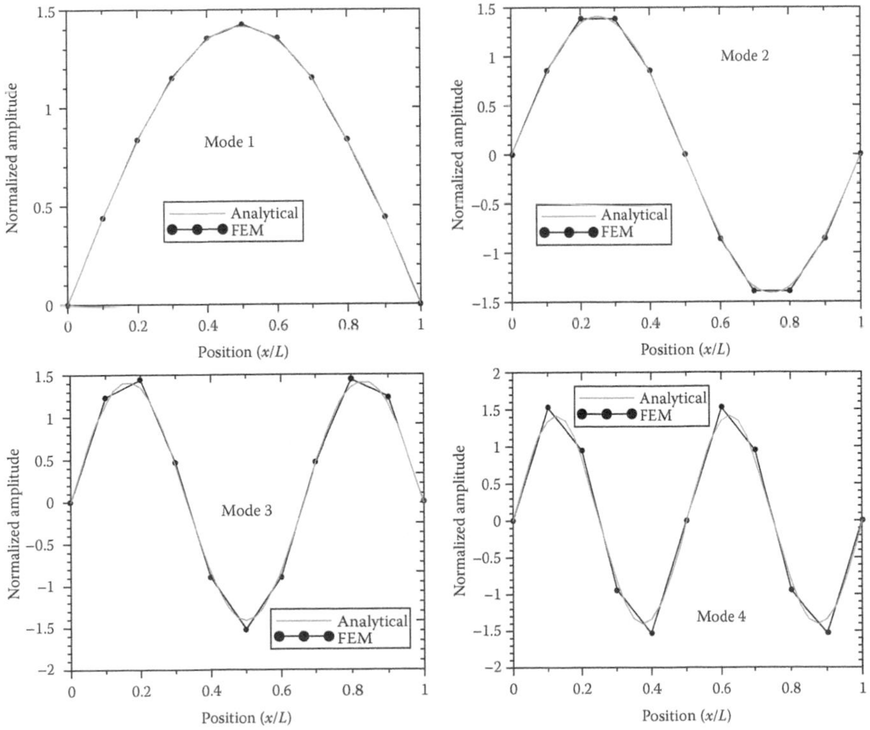

Figure 4.12 Illustration of the results of the script of Example 4C.1.

```
************Results for 5 quadratic elements ************
=================================================
Mode  FEM (Hz) Exact (Hz) Error(%)
-----------------------------------------------
  1    171.02    171.00      0.0
  2    342.55    342.00      0.2
  3    516.81    513.00      0.7
  4    698.03    684.00      2.1
=================================================
```

Figure 4.13 Illustration of the results of the script of Example 4C.2.

EXAMPLE 4C.2

Repeat Example 4C.1 using quadratic elements. It can be used to study the convergence of these elements. The comparison for the first four natural frequencies is shown in Figure 4.13 using a mesh of 5 elements (11 nodes). How does the convergence compare for the mode shapes?

```
%===============================================================
% Example 4C.2
%
% Solve a 1D Acoustic Eigenvalue Problem using quadratic elements
% Boundary conditions: Fixed pressure at both ends%
%
%===============================================================
clear all; close all; clc
%
% Problem data
%
Lx=1;                     % Length of the domain; fixed arbitrary to 1
c0=342;                   % Speed of sound (m/s)
%
% Step 1: Mesh
%
ne= input(' Number of quadratic element: ');  % number of elements
nnt = 2*ne+1;                          % Total number of nodes
h=Lx/ne;                               % Length of the elements
x=[0:h/2:Lx];                          % Coordinates table
%
% Step 2: Compute Elementary matrices
%
Ke=c0^2*[7,-8,1;-8,16,-8;1,-8,7]/(3*h);
Me=[4,2,-1;2,16,2;-1,2,4]*h/30;
%
% Step 3: Assembling
%
I=eye(3,3);
K=zeros(nnt,nnt); M = zeros(nnt,nnt);
for ie=1: ne
  L=zeros(3,nnt); L(:,2*ie-1:2*ie+1)=I;     % Location matrix for
  element ie
```

```
  K = K + L'*Ke*L;
  M = M + L'*Me*L;
end
%
% Step 4: Boundary conditions: p(1) = p(nnt) = 0
%
K = K(2:nnt-1,2:nnt-1);
M = M(2:nnt-1,2:nnt-1);
ndof = nnt-2;          % Final number of unknown (equations)
%
% Step 5: Compute the eigenvalues and eigenvectors
%
[V,D]=eig(K,M); D= sqrt(D);
%
% Normalisation et ordre des vecteurs propres
%
Nor=V'*M*V; V=V*sqrt(inv(Nor)); % Normalization of the
   eigenvectors
for idof = 1: ndof
  w(idof) = D(idof,idof);
end
[w,ind]=sort(w);          % Sort the eigenvalues in ascending order
%
% Step 6: Comparison with the exact solution for a selected number
   of modes
%
%
nm = min(4, ndof);  %  Here we select the first modes up to the
   fourth
theo = c0*[1:ndof]*pi/Lx;
Err = abs(theo-w)./theo;


fprintf('\n ************Results for %d quadratic elements
   ************', ne);
fprintf('\n =================================================');
fprintf('\n Mode  FEM (Hz) Exact (Hz) Error(%%)');
fprintf('\n -----------------------------------------------');
for m=1:nm
  fprintf('\n   %d   %7.2f   %7.2f       %3.1f ', m, w(m)/2/pi,
   theo(m)/2/pi, Err(m)*100);
end
fprintf('\n =================================================\n');
fprintf('\n');
%
% comparison of the mode shapes
%
xt=[0:0.025:1] * Lx;          % Mesh to visualize the mode shapes
for m=1:nm
  figure;
  yt= sqrt(2)*sin(m*pi*xt); plot(xt,yt), hold on, % Theoretical
   solution for
  y=[0,real(V(:,ind(m))'),0]; if (yt(2) >0 && y(2) < 0), y = -y; end;
   % Recall : p(0)=p(L)=0
  plot(x,y,'r-*')          % solution par element finis
```

```
    title(['Mode ' num2str(m)]);
    xlabel('Position x/L');
    ylabel(' Normalized Amplitude ');
    legend('Analytical', 'FEM Quadratic ');
end
```

EXAMPLE 4C.3

This example describes the solution of Equation 4.52 using quadratic elements. The analytical solution is given by

$$\hat{p}(x,\omega) = \rho_0 c_0^2 (1 + i\eta_a) \frac{\omega \sin(k_0 x)}{\cos(k_0 L)}, \quad k_0 = \frac{\omega}{c_0 \sqrt{1 + i\eta_a}} \tag{4.96}$$

with η_a the fluid structural damping.

Derive this equation and use this script to study the convergence in terms of number of elements per wavelength; study also the effect of damping in the fluid. An example of the comparison is given in Figure 4.14 for the quadratic pressure in the cavity.

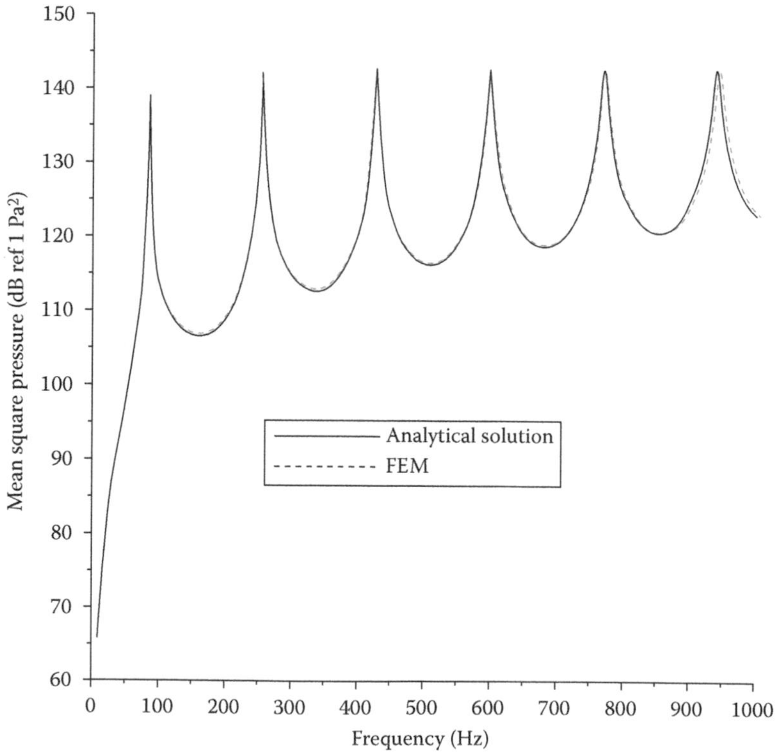

Figure 4.14 Comparison between analytical solution and FEM for the script of Example 4C.3—10 quadratic elements, loss factor = 1%.

```
%=====================================================================
% Example 4C.3
%
% Solve a 1D Acoustic cavity excited by a piston at one and closed at
% the other end.
%
% Use of three-noded 1D finite element
%
%=====================================================================
clear all; close all; clc
%
% Problem data
%
Lx=1;                    % Length of the domain; fixed arbitrary to 1
c0=342;                  % Speed of sound (m/s)
rho0=1.2;                % Fluid density (kg/m^3))
neta_a= input(' loss factor (damping) in the fluid: '); % Play with
   this parameter

freq=[10:2:1000];         % Frequency domain
omega=2*pi*freq;
%
% Step 1: Mesh
%
ne= input(' Number of quadratic element: ');   % number of elements
nnt = 2*ne+1;                       % Total number of nodes
h=Lx/ne;                            % Length of the elements
x=[0:h/2:Lx];                       % Coordinates table
%
% Step 2: Compute Elementary matrices
%
Ke=c0^2*[7,-8,1;-8,16,-8;1,-8,7]/(3*h);
Me=[4,2,-1;2,16,2;-1,2,4]*h/30;
%
% Step 3: Assembling
%
I=eye(3,3);
K=zeros(nnt,nnt); M = zeros(nnt,nnt);
for ie=1: ne
  L=zeros(3,nnt); L(:,2*ie-1:2*ie+1)=I;    % Location matrix for
   element ie
   K = K + L'*Ke*L;
   M = M + L'*Me*L;
end
%
% Step 4: Boundary conditions: p(1) = 0
%
K = K(2:nnt,2:nnt);
M = M(2:nnt,2:nnt);
ndof = nnt-1;    % Final number of unknown (equations)
%
% Step 5: Solving the system with the Force vector : Piston at a x=L
%
R=zeros(1,ndof); % Initialize the force vector
Pquad=zeros(1,ndof);
```

```
Pquad_exact=zeros(1,ndof);
for n=1:length(omega)
  w=omega(n);
  R(ndof) = w^2*rho0*c0^2*(1+1i*neta_a);  % Piston (fixed
   displacement) at x=L
  MK=K*(1+1i*neta_a)-w^2*M;
  P=inv(MK)*R';                            % Solve the linear system
  Pquad(n) = real((P'*M*P)/2);             % Compute the space
   avergaed quadratic pressure
  % Analytical solution (to be derived as an exercise)
  k0=w/c0/sqrt(1+1i*neta_a);
  amp=sin(k0*x);
  Pquad_exact(n) = real(rho0*c0*sqrt(1+1i*neta_a)*w/cos(k0*Lx))^2
   .* norm(amp)^2/length(x)/2;
end
%
% Step 6: Comparison with the exact solution
%

plot(freq,10*log10(Pquad_exact),freq,10*log10(Pquad),'r-');
xlabel('Frequency (Hz)');
ylabel(' Quadratic pressure (dB)');
legend('Analytical', 'FEM ');
sprintf('Analytical vs. FEM: %d quadratic elements; /eta=%f', ne,
neta_a);
text=sprintf('Analytical vs. FEM using %d quadratic elements;
damping =%4.2f %%', ne, neta_a);
Title(text)
```

EXAMPLE 4C.4

This example discusses the solution of the following system using quadratic elements:

$$\begin{cases} c_0^2 \dfrac{d^2\hat{p}}{dx^2} + \omega^2\hat{p} = 0 \qquad x \in \Omega = \left]0, L\right[\\ \dfrac{d\hat{p}}{dx}(0) = -i\rho_0\omega v_0 \\ \dfrac{d\hat{p}}{dx}(L) + ik_0\hat{\beta}\hat{p}(L) = 0 \end{cases} \tag{4.97}$$

This problem describes the sound propagation in a waveguide excited by a piston at $x = 0$ with an acoustic impedance condition at $x = L$.

The analytical solution is given by

$$\hat{p}(x,\omega) = \hat{A}\left(\exp(-ik_0x) + \hat{\Re}\exp(ik_0x)\right), \quad k_0 = \frac{\omega}{c_0\sqrt{1+i\eta_a}}$$

$$\hat{A} = \frac{\rho_0 c_0 v_0}{\exp(ik_0L) - \hat{\Re}\exp(-ik_0L)}, \quad \hat{\Re} = \frac{(\hat{Z}-1)}{(\hat{Z}+1)} \tag{4.98}$$

with η_a the fluid structural damping.

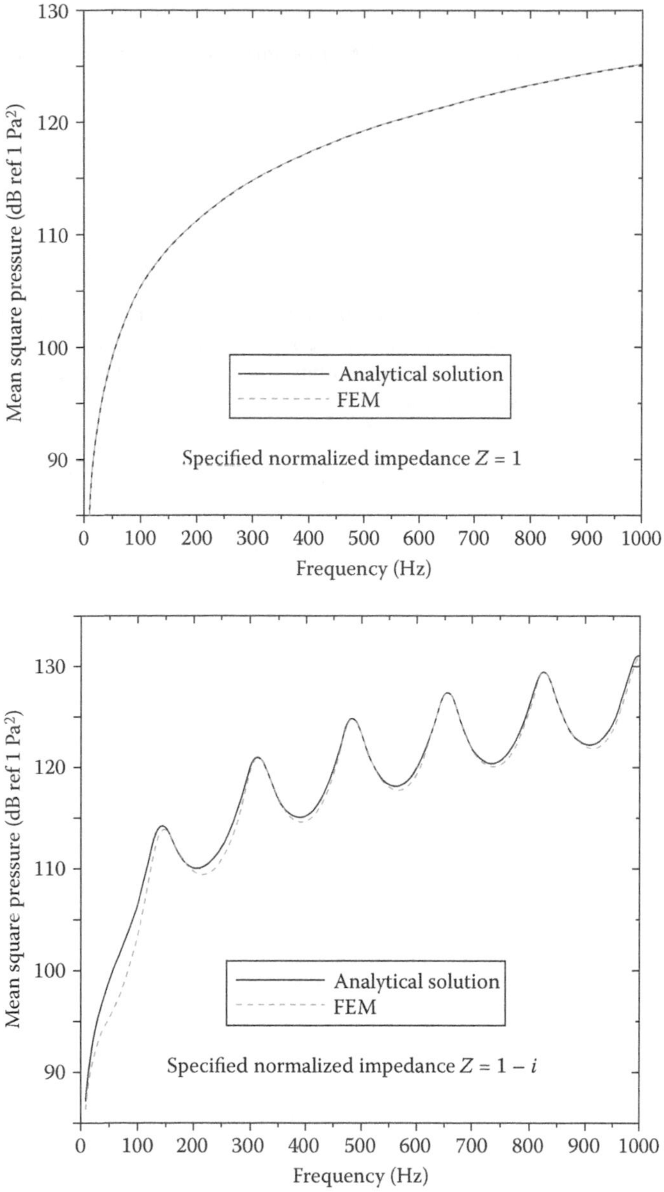

Figure 4.15 Comparison between analytical solution and FEM for the script of Example 4C.4 using 12 quadratic elements (4 elements per wavelength)—mean square pressure in the cavity.

The weak form associated with Equation 4.97 is given by

$$W(\hat{p},\delta\hat{p}) \;=\; K(\hat{p},\delta\hat{p}) - \omega^2 M(\hat{p},\delta\hat{p}) + i\omega A(\hat{p},\delta\hat{p}) - R(\delta\hat{p}) = 0\ ,\ \forall\delta\hat{p} \tag{4.99}$$

with

$$A(\hat{p},\delta\hat{p}) = \hat{\beta}\, c_0 \delta\hat{p}(L)\hat{p}(L) \quad , \; R(\delta\hat{p}) = \delta\hat{p}(0)(-i\rho_0\omega c_0^2 v_0) \tag{4.100}$$

Use the following script to study the convergence in terms of (i) number of elements per wavelength and (ii) complex value of the normalized impedance. An example of the comparison is given in Figure 4.15 for the quadratic pressure in the cavity.

```
%======================================================================
% Example 4C.4
%
% Piston excited 1D cavity with an impedance condition at the other
  end.
%
% Mesh: using of quadratic elements
%
% Illustration of the convergence with absorption in the cavity
%
% Note the following limiting cases :
% Z=infinity (a large number) ==> rigid termination
% Z=0 ==> pressure release)
% Z=1 ==>  Impedance matching (semi-infinite tube)
%======================================================================
clc; clear all; close all;
%% Input data
rho0=1.2;         % Fluid density
c0=342.2;         % Speed of sound
neta_a=0.00;      % Damping in the cavity
U0=1;             % Piston displacement
L=1;              % Length of the cavity

%% Impedance properties
Z0=rho0 * c0;
Z= input('\n Input the normalized complex impedance at one end : ');
if (Z~=0), beta= 1/Z;else beta=1e6; end    % convert to admittance

%% Frequency domain
freq=[10:2:1000];           % Chosen arbitrarily
omega=2*pi*freq;
lamda_min= c0/max(freq);    % Smallest wavelength in the fluid

%% Step 1: Mesh
nel=input ('\n Number of quadratic element / wavelength:');
ne= ceil(nel* L/lamda_min);
```

```
fprintf('\n Number of quadratic elements : %d \n', ne')

nnt = 2*ne+1;                         % Total number of nodes
h=L/ne;                               % Length of the elements
x=[0:h/2:L];                          % Coordinates table

%% Step 2: Compute Elementary matrices
He=[7,-8,1;-8,16,-8;1,-8,7]/(3*h)/rho0;
Qe=[4,2,-1;2,16,2;-1,2,4]*h/30/(rho0*c0^2);

%% Step 3: Assembling
I=eye(3,3);
H=zeros(nnt,nnt); Q = zeros(nnt,nnt);
for ie=1: ne
  LM=zeros(3,nnt); LM(:,2*ie-1:2*ie+1)=I;
  H = H + LM'*He*LM;
  Q = Q + LM'*Qe*LM;
end

%% Step 4: Impedance condition at x=0 : dPdn(0)+j*k*beta*P(0) = 0
A = zeros(nnt,nnt);
A(1,1) = beta/(rho0*c0);

%% Step 5: Solving the system with the Force vector : Piston at a x=L
ndof = nnt;
Pquad_direct=zeros(1,ndof);
Pquad_modal=zeros(1,ndof);
Pquad_exact=zeros(1,ndof);
for n=1:length(omega)
  w=omega(n);
  % FE solution
  R=zeros(ndof,1);      % Initialize force vector
  R(ndof) = w^2*U0;     % displacement of the piston fixed to 1
  P = (H - w^2*Q/(1+1i*neta_a)+1i*w*A)\R;        % Solve the system
  Pquad_direct(n) = (rho0*c0^2)* real(P'*Q*P)/(2*L);    % Compute the
   space avergaed quadratic pressure

   % Analytical solution
   k0=w/c0/sqrt(1+1i*neta_a);
   r=(Z-1)/(Z+1);
   a=(1j*w*U0)*Z0/(exp(1j*k0*L)-r*exp(-1j*k0*L));
   x=[0:L/100:L]; p=a*(exp(-1i*k0*x)+r*exp(1i*k0*x));
   Pquad_exact(n) = real(norm(p)^2/length(x)/2);
end
% Step 6: Comparison with the exact solution

plot(freq,10*log10(Pquad_direct),'k','LineWidth',2)
hold on
plot(freq,10*log10(Pquad_exact),':','LineWidth',2,'Color',[0.5 0.5
   0.5])
xlabel('Frequency (Hz)');
ylabel(' Quadratic pressure (dB ref.1)');
legend('Analytical', 'FEM ');
```

```
text(100 , 95,['Analytical vs. FEM using ', num2str(ne), 'quadratic
elements'],'Color','k','FontSize',12)
text(100 , 90,['Specified normalized Impedance Z = ',
num2str(Z)],'Color','k','FontSize',12)
```

EXAMPLE 4C.5

This example discusses the modeling using one-dimensional-FEM, the procedure used in an impedance tube to measure the surface impedance and absorption coefficient of an acoustic material (Figure 4.16). The tube is of length L. A loudspeaker is installed at one end ($x = 0$) and an acoustic material of thickness h and surface impedance $\hat{Z}$ is bonded onto the other rigid and impervious end ($x = L$). Below the cut-off frequency of the tube, only plane waves propagate and the problem is amenable to a one-dimensional analysis.

In this case, the surface impedance of the material can be obtained from the measurement of the transfer function $\hat{H}_{12} = \hat{P}_2/\hat{P}_1$ between two microphones adequately placed in the tube (e.g., standard ASTM E-1050):

$$\frac{\hat{Z}}{\rho_0 c_0} = \frac{1+\hat{\Re}}{1-\hat{\Re}}, \quad \hat{\Re} = \frac{\hat{H}_{12} - \exp(-ik_0 s)}{\exp(ik_0 s) - \hat{H}_{12}} \exp\left(2ik_0(d+s)\right) \tag{4.101}$$

where d is the distance from microphone 2 to the sample and s is the spacing between the two microphones. The normal incidence absorption coefficient α is directly obtained from the reflection coefficient $\hat{\Re}$: $\alpha = 1 - |\hat{\Re}|^2$.

This example compares the estimation of Equation 4.101 using FEM to its exact value (known input here).

In the FE model, the loudspeaker can be simply replaced by an oscillating rigid piston and the governing equations are given by Equation 4.97.

The following script demonstrates the accuracy of FE solution for the case of a cylindrical tube with diameter $D = 100$ mm (and thus cut-off frequency

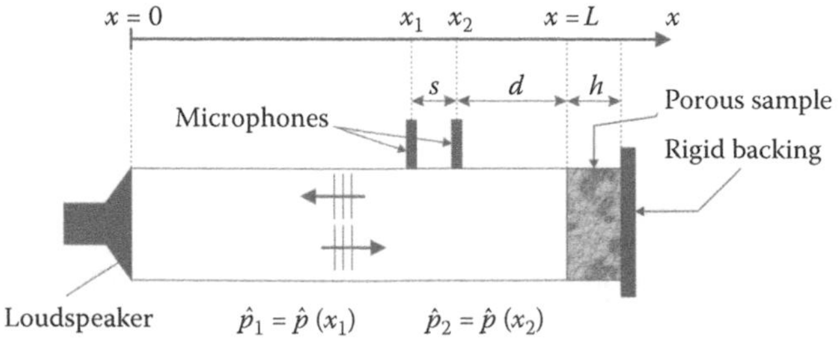

Figure 4.16 A schematic of a two-microphones impedance tube.

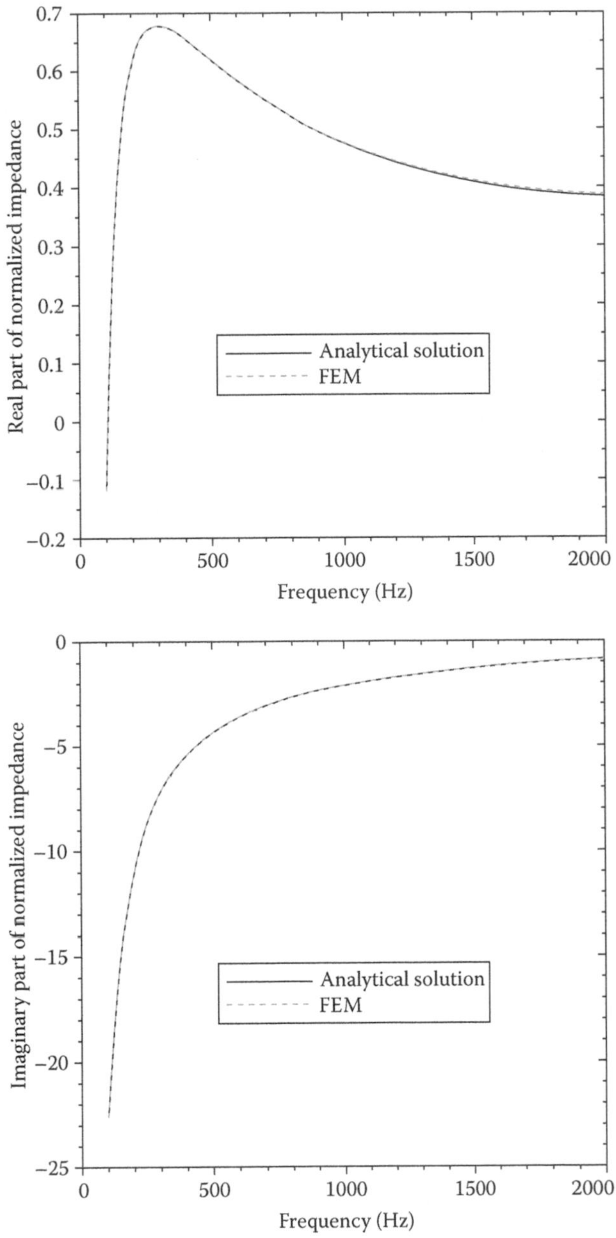

Figure 4.17 Comparison between analytical solution and FEM for the script of Example 4C.5—normal incidence surface impedance.

$0.59c_0/D = 2006$ Hz), length $L = 40D$, microphone spacing $s = 0.5D$ and distance between microphone 2 and the surface of the material $d = D/2$. The material of thickness: $h = 2$ cm is assumed fibrous and its characteristic impedance $\hat{Z}_c$ and wavenumber $\hat{k}_c$ are given by the Delany-Bazley model (Delany and Bazley 1970):

$$\begin{aligned}\hat{Z}_c &= \rho_0 c_0\left[1 + 0.057X^{-0.754} - i0.087X^{-0.732}\right]\\ \hat{k}_c &= \frac{\omega}{c_0}\left[1 + 0.0978X^{-0.700} - i0.189X^{-0.595}\right]\end{aligned} \tag{4.102}$$

with $X = \rho_0 f/\sigma_p$, f the frequency, $\rho_0 = 1.2$ kg/m^3, $c_0 = 340$ m/s, and $\sigma_p = 10{,}000$ Rayls/m the flow resistivity of the material.

Note that the impedance of a material of thickness h bonded onto a rigid wall and excited by a plane wave with incidence angle θ is given by (Allard and Atalla, 2009): $\hat{Z} = -\hat{Z}_c\hat{k}_c/\hat{k}_3\, i\cot(\hat{k}_3 h)$; $\hat{k}_3 = \hat{k}_c\cos\theta$. In our example, $\theta = 0$ and thus $\hat{Z} = -i\hat{Z}_c\cot(\hat{k}_c h)$.

An example of the comparison, using 30 quadratic elements (approximately five elements per wavelength) is given in Figure 4.17.

Note that the real part of the surface impedance is negative at low frequencies, which is not physical. This defect in the Delany and Bazley model has been noticed by Miki (1990) who proposed modifications to Equations 4.102.

```
%==================================================================
% Example: Impedance tube.
%
% Mesh: using of quadratic elements
%
% Illustration of the procedure for the measurement of an impedance
  of a
% material using the two-microphone method
%
%==================================================================
clc; clear all; close all;


%% Input data
%
rho0=1.2;           % Fluid density
c0=342.2;           % Speed of sound
neta_a=0.000;       % Damping in the cavity
U0=1;               % Piston displacement
D=0.1;              %100 mm diameter
L=10 *D;            % Length of the cavity
s=0.5*D;            % microphone spacing
d=D/2;              % distance between mic 2 and sample
%% Frequency domain
fc=floor(1.84*c0/D/pi); % Cut off frequency
```

```
freq=[100:2:fc];    % Choose correctly the lowest frequency ( a
   funtion of mics spacing)
omega=2*pi*freq;
k0=omega/c0;

%% Impedance properties
Z0=rho0 * c0;
h=0.02;          % thickness of the material
sigma=10000;     % flow resitivity
X=rho0*freq/sigma;
Zc=Z0*(1+0.057*X.^(-0.754)-1i*0.087.*X.^(-0.732));
k=k0 .*(1+0.0978*X.^(-0.700)-1i*0.189.*X.^(-0.595));
Z=-1i.*Zc.*cot(k*h) /Z0;
beta= 1./Z;  % convert to admittance

%% Step 1: Mesh
ne= 30;     % Number of quadratic elements (selected to make sure
   that there are nodes at the two mics locations)
nnt = 2*ne+1;                       % Total number of nodes
h=L/ne;                             % Length of the elements
x=[0:h/2:L];                        % Coordinates table

in2=find(abs(x-d)<1e-6);            % Location of mic 2
in1=find(abs(x-(d+s))<1e-6);        % Location of mic 1
s=abs(x(in1)-x(in2));               % recalculate microphones separation
d=x(in2);                       % and distanmce between microphone 2
   and the sample
%% Step 2: Compute Elementary matrices
He=[7,-8,1;-8,16,-8;1,-8,7]/(3*h)/rho0;
Qe=[4,2,-1;2,16,2;-1,2,4]*h/30/(rho0*c0^2);

%% Step 3: Assembling
I=eye(3,3);
H=zeros(nnt,nnt); Q = zeros(nnt,nnt);
for ie=1: ne
  LM=zeros(3,nnt); LM(:,2*ie-1:2*ie+1)=I;
  H = H + LM'*He*LM;
  Q = Q + LM'*Qe*LM;
end

%% Step 4 & 5: specify frquency dependent impedance and Solve the
system with the Force vector : Piston at a x=L
ndof = nnt;
Pquad_direct=zeros(1,ndof);
Pquad_modal=zeros(1,ndof);
Pquad_exact=zeros(1,ndof);
P_mic1=zeros(1,ndof);
P_mic2=zeros(1,ndof);
A = zeros(nnt,nnt);
R=zeros(ndof,1);               % Initialize force vector
for n=1:length(omega)
  w=omega(n);
  R(ndof) = w^2*U0;              % displacement of the piston fixed to 1
  A(1,1) = beta(n)/(rho0*c0);
```

```
  P = (H - w^2*Q/(1+1i*neta_a)+1i*w*A)\R;        % Solve the system
  P_mic1(n)=P(in1);
  P_mic2(n)=P(in2);
end

% calculate the normalized impedance
clear R
k=omega/c0;
H12=P_mic2./P_mic1;
R=(H12-exp(-1i*k*s))./(exp(1i*k*s)-H12) .*exp(1i*2*k*(d+s));
Z_num=(1+R)./(1-R);

%% Step 6: Comparison with the exact solution
figure(1)
subplot(1,2,1)
plot(freq,real(Z),'k','LineWidth',2)
hold on
plot(freq,real(Z_num),':','LineWidth',2,'Color',[0.5 0.5 0.5])
xlim([100 2000])
grid on
xlabel('Frequency (Hz)');
ylabel(' Normalized Impedance - Real part');
legend('Analytical', 'FEM ');
subplot(1,2,2)
plot(freq,imag(Z),'k','LineWidth',2)
hold on
plot(freq,imag(Z_num),':','LineWidth',2,'Color',[0.5 0.5 0.5]);
xlim([100 2000])
grid on
xlabel('Frequency (Hz)');
ylabel(' Normalized Impedance - Imaginary part');

figure (2)
R_theo=(Z-1)./(Z+1); alpha_theo=1-abs(R_theo).^2;
R_num=(Z_num-1)./(Z+1); alpha_num=1-abs(R_num).^2;
plot(freq,alpha_theo,'k','LineWidth',2)
hold on
plot(freq,alpha_num,':','LineWidth',2,'Color',[0.5 0.5 0.5])
xlim([100 2000])
grid on
xlabel('Frequency (Hz)');
ylabel(' Absorption coefficient');
legend('Analytical', 'FEM ');
```

Chapter 5

Solving uncoupled structural acoustics and vibration problems using the finite-element method

5.1 INTRODUCTION

In this chapter, we present the main ideas behind the FEM based on a general three-dimensional acoustic problem. Section 5.2 describes the various steps of the FEM. Section 5.3 discusses convergence considerations. Section 5.4 presents several examples to illustrate the concepts discussed in Section 5.2. Finally, Section 5.5 formulates the acoustic and vibratory indicators commonly used in the field of vibroacoustics. The reader may refer to several references (Batoz and Dhatt 1990; Petyt 1990; Imbert 1991; Bathe 1996; Craveur 1997; Géradin and Rixen 1997; Hughes 2000; Zienkiewicz and Taylor 2005) for more details about the FEM in general.

5.2 THREE-DIMENSIONAL WAVE EQUATION: GENERAL CONSIDERATIONS

Let us consider the resolution of the three-dimensional Helmholtz equation in a given closed volume Ω using the FEM (see Figure 5.1). The boundary $\partial\Omega$ comprises three parts over which the acoustic pressure $\hat{p}$ $(\partial\Omega_{f,D})$, its normal gradient $(\partial\hat{p}/\partial n)(\partial\Omega_{f,N})$, or an admittance boundary condition $(\partial\Omega_{f,R})$ is prescribed:

$$\begin{cases} \nabla^2\hat{p} + k^2\hat{p} = 0 & \text{in } \Omega_f \\ \hat{p} = \overline{p} & \text{on } \partial\Omega_{f,D} \\ \dfrac{\partial\hat{p}}{\partial n} = \rho_0\,\omega^2\overline{U}_n & \text{on } \partial\Omega_{f,N} \\ \dfrac{\partial\hat{p}}{\partial n} + ik\hat{\beta}\hat{p} = 0 & \text{on } \partial\Omega_{f,R} \end{cases} \tag{5.1}$$

where $\overline{p}$ and $\overline{U}_n = \underline{\overline{U}} \cdot \underline{n}$ are a prescribed pressure and normal particle displacement, respectively, and $\hat{\beta}$ is the normalized acoustic admittance.

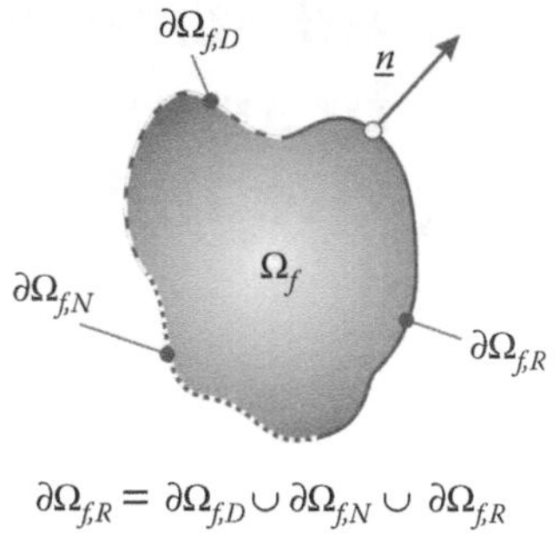

Figure 5.1 Acoustics problem.

5.2.1 Step 1: Integral formulation of the problem

The weak integral Galerkin's formulation associated with Equation 5.1 is written as (see Chapter 3)

$$\begin{cases} W\left(\hat{p},\delta\hat{p}\right) = \int\limits_{\Omega_f}\left[\underline{\nabla}\hat{p}\cdot\underline{\nabla}\delta\hat{p} - k^2\hat{p}\delta\hat{p}\right]dV - \int\limits_{\partial\Omega_{f,N}}\rho_0\omega^2\bar{U}_n\delta\hat{p}\,dS + \int\limits_{\partial\Omega_{f,R}} ik\hat{\beta}\hat{p}\delta\hat{p}\,dS \\ \text{with } \delta\hat{p} = 0 \quad \text{and} \quad \hat{p} = \bar{p} \quad \text{on } \partial\Omega_{f,D} \end{cases} \tag{5.2}$$

Let

$$H(\hat{p},\delta\hat{p}) = \int\limits_{\Omega_f}\frac{1}{\rho_0}\left(\underline{\nabla}\hat{p}\cdot\underline{\nabla}\delta\hat{p}\right)dV \tag{5.3}$$

$$Q(\hat{p},\delta\hat{p}) = \int\limits_{\Omega_f}\frac{1}{\rho_0 c_0^2}\,\hat{p}\delta\hat{p}\,dV \tag{5.4}$$

$$A(\hat{p},\delta\hat{p}) = \int\limits_{\partial\Omega_{f,R}}\frac{\hat{\beta}}{\rho_0 c_0}\,\hat{p}\delta\hat{p}\,dS \tag{5.5}$$

$$R(\delta\hat{p}) = \omega^2\int\limits_{\partial\Omega_{f,N}}\delta\hat{p}\bar{U}_n\,dS \tag{5.6}$$

The weak form Equation 5.2 can be rewritten as

$$\begin{cases} W\left(\hat{p},\delta\hat{p}\right) = H(\hat{p},\delta\hat{p}) - \omega^2 Q(\hat{p},\delta\hat{p}) + i\omega A(\hat{p},\delta\hat{p}) - R(\delta\hat{p}) = 0 \\ \text{with } \delta\hat{p} = 0 \quad \text{and} \quad \hat{p} = \bar{p} \quad \text{on } \partial\Omega_{f,D} \end{cases} \tag{5.7}$$

In the following section, we present the general ideas behind the FEM approximation of this equation.

5.2.2 Step 2: Discretization and approximation of geometry

The domain Ω is approximated by a set of N_e disjoint elements $\Omega^1,\Omega^2,\ldots,\Omega^{n_e}$:

$$\Omega \cong \tilde{\Omega} = \bigcup_{e=1}^{n_e} \Omega^e \tag{5.8}$$

To make the analytical development of shape functions (also called interpolation functions) and the calculation of elementary integrals easier, we use a local dimensionless coordinate system associated with each element. Thus, to each FE Ω^e, there corresponds a reference element Ω^r defined by the transformation:

$$T:\underline{\xi} \in \Omega^r \longrightarrow \underline{x}(\underline{\xi}) \in \Omega^e \tag{5.9}$$

This transformation is bijective: to a point pertaining to the reference element Ω^r, corresponds a unique point of Ω^e. The function T transforms the local coordinates (ξ_1,ξ_2,ξ_3) into global coordinates (x_1,x_2,x_3) (Figure 5.2):

$$\begin{pmatrix} x_1 \\ x_2 \\ x_3 \end{pmatrix} = T \begin{pmatrix} \xi_1 \\ \xi_2 \\ \xi_3 \end{pmatrix} \tag{5.10}$$

The construction of this transformation relies on the representation by nodes and shape functions. Thus, each element Ω^e of the approximated domain $\tilde{\Omega}$ is described by a mapping such as:

$$\underline{x}(\underline{\xi}) = \sum_{i=1}^{nn_e} N_i^e(\underline{\xi})\, \underline{x}_i^e = \langle N^e(\underline{\xi}) \rangle \, \{\underline{x}^e\} \tag{5.11}$$

where $\{\underline{x}^e\}$ is a vector containing the (x_1,x_2,x_3) coordinates of each node of element Ω^e, and $\langle N^e \rangle = \langle N_1^e,\ldots,N_{nn_e}^e \rangle$ is the vector of local shape functions of the element. More explicitly

$$x_1 = \langle N^e(\xi_1,\xi_2,\xi_3) \rangle \{x_1^e\} \tag{5.12}$$

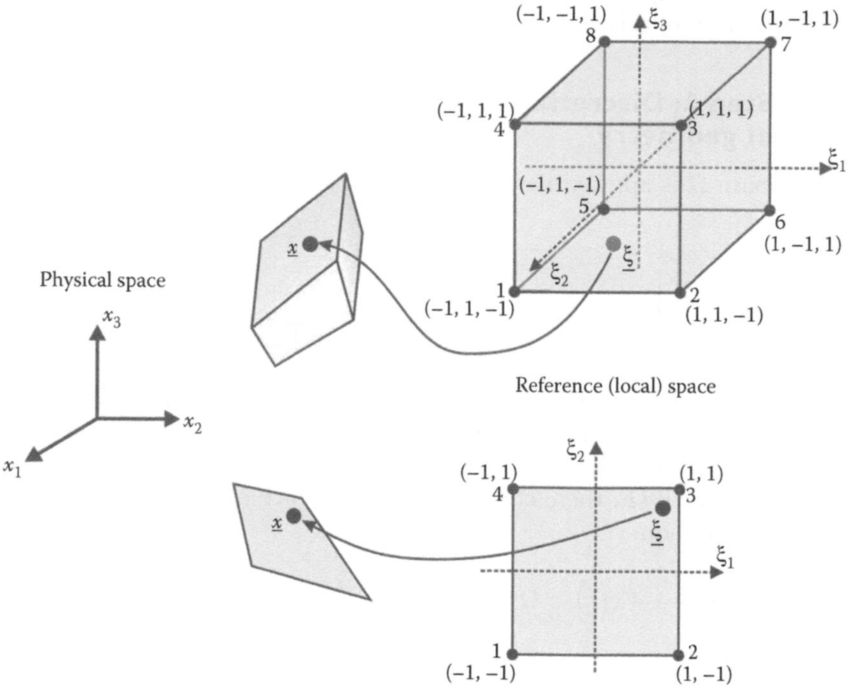

Figure 5.2 Transformation between the reference and the physical space. Example of a surface quadrilateral element and a volume hexahedron element.

$$x_2 = \langle N^e(\xi_1,\xi_2,\xi_3)\rangle\{x_2^e\} \tag{5.13}$$

$$x_3 = \langle N^e(\xi_1,\xi_2,\xi_3)\rangle\{x_3^e\} \tag{5.14}$$

with

$$\{x_1^e\} = \begin{Bmatrix} x_{1,1}^e \\ \vdots \\ x_{1,nn_e}^e \end{Bmatrix}; \quad \{x_2^e\} = \begin{Bmatrix} x_{2,1}^e \\ \vdots \\ x_{2,nn_e}^e \end{Bmatrix}; \quad \{x_3^e\} = \begin{Bmatrix} x_{3,1}^e \\ \vdots \\ x_{3,nn_e}^e \end{Bmatrix} \tag{5.15}$$

the vectors containing the nodal coordinates of element Ω^e along each direction i, $i = 1,2,3$ and

$$\langle N^e(\xi_1,\xi_2,\xi_3)\rangle = \langle N_1^e(\xi_1,\xi_2,\xi_3),\ N_2^e(\xi_1,\xi_2,\xi_3),\ldots,\ N_{nn_e}^e(\xi_1,\xi_2,\xi_3)\rangle \tag{5.16}$$

are shape functions (usually polynomials).

Every discretization $\tilde{\Omega} = \bigcup_{e=1}^{n_e} \Omega^e$ of Ω should be such that two neighboring elements interconnect exactly along their common boundary (no overlapping or hole). In other words, the nodal and shape function-based geometrical approximation of Ω must be such that

- The tangent plane and the normal exist and are continuous at all points interior to the element faces.
- The interface between two elements must be continuous and there must be no overlapping of the elements' boundaries.

Thus, the shape functions $N_i^e(\xi_1, \xi_2, \xi_3)$ must satisfy

$$\begin{cases} N_i^e\left(\underline{\xi}_j\right) = \delta_{ij} \\ \sum_{i=1}^{nn_e} N_i^e\left(\underline{\xi}\right) = 1 \end{cases} \tag{5.17}$$

The first condition expresses the correspondence between the nodes of the physical element and those of the reference element. The second condition expresses the validity of the nodal approximation when the element size tends to zero.

Remark: Similarly, to calculate the elementary matrices on the boundary $\partial\Omega$ of volume Ω, $\partial\Omega$ is approximated with a set of distinct surface elements $\partial\Omega^e$. In practice, these elements are built from the trace of the volume elements in contact with boundary $\partial\Omega$ (i.e., the skin of the volume mesh used to approximate the domain Ω).

5.2.2.1 Building a mesh in practice

We choose a partition of domain Ω constituted of n_e elements Ω^e. The nodes of the elements are numbered sequentially from 1 to nnt where nnt is the total number of nodes. To define the mesh completely, two tables are built: a table of coordinates denoted as XYZ (:,:) and a connectivity table referred to as IEN (:,:). Table XYZ (:,:) of dimension $n_d \times nnt$ contains the coordinates of the nodes (n_d is the geometrical dimension of the problem; $n_d = 3$ in the tridimensional case). Table IEN (:,:) of dimension $nn_e \times n_e$ establishes the correspondence between the local and the global numbering of the nodes. It contains the global indexes of the nn_e nodes of each element.

EXAMPLE 5.1

To illustrate the construction of these two tables, consider the following two-dimensional (2D) mesh (see Figure 5.3):

Here, we have: $n_d = 2$, $n_e = 6$, $nn_e = 4$, $nnt = 12$

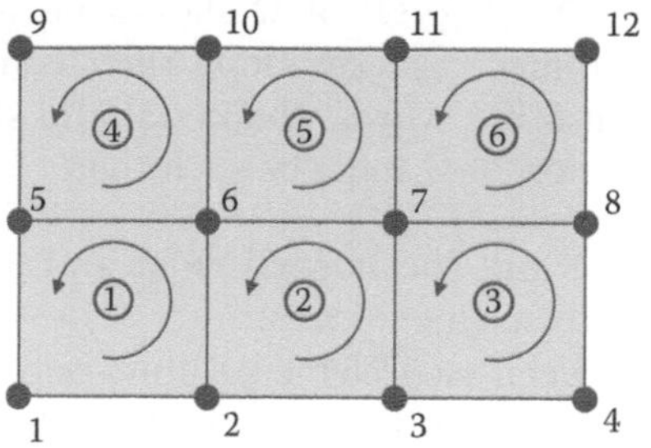

Figure 5.3 Bidimensional mesh.

- Table of coordinates: $XYZ(1,i) = x_{1,i}$; $XYZ(2,i) = x_{2,i}$ for $i = 1, \ldots, nnt$
- Connectivity table: *IEN* $(i,e) \Rightarrow i^e$ is the global index of the local node *i* of element *e*:

		Element e					
		1	2	3	4	5	6
nodes (local numbering)	1	1	2	3	5	6	7
	2	2	3	4	6	7	8
	3	6	7	8	10	11	12
	4	5	6	7	9	10	11

<= global numbering

5.2.3 Step 3: Approximation of variables and calculation of elementary matrices

5.2.3.1 *Nodal approximation of variables*

Two types of interpolation occur in the construction of a finite element: the interpolation of the geometry (= approximation of the geometry by shape functions N_i) and the interpolation of the variables (= approximation of the variable using shape functions $\bar{N}_i$):

$$\underline{x} = \sum_{i=1}^{nn_e} N_i^e(\underline{\xi})\, \underline{x}_i^e \tag{5.18}$$

$$\hat{p} = \sum_{i=1}^{nn_e} \bar{N}_i^e(\underline{\xi})\, \hat{p}_i^e \tag{5.19}$$

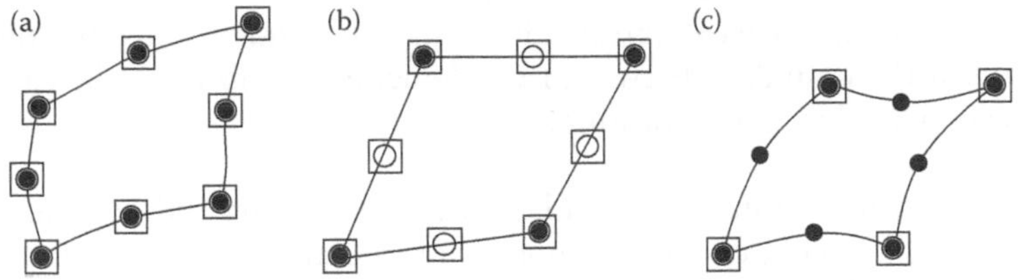

Figure 5.4 Illustration of the different types of FEs: (a) isoparametric, (b) subparametric, and (c) superparametric. Symbol □ is used to represent the solution nodes and • is associated with the geometrical nodes.

In practice, three cases are encountered (see Figure 5.4):

- *Isoparametric elements:* The same shape functions are used for the geometry and the solution ($N_i = \bar{N}_i$). These elements are the most commonly used.
- *Subparametric elements:* The degree of interpolation for the geometry is smaller than for the solution. This is for example the case of Hermite elements.
- *Superparametric elements:* The degree of interpolation for the geometry is larger than for the solution. These elements are rarely used in the scope of FE applications because of their poor convergence. However, they are commonly used in BE applications (see Chapter 7).

In the following section, we are going to limit the presentation to isoparametric elements. We will, therefore, use the same symbol N_i for the shape functions associated with the geometry and the solution.

5.2.3.1.1 Choice of nodal shape functions

To ensure the convergence by refining the mesh (elements become smaller), the nodal shape functions must satisfy the following conditions (Shames and Dym 1995):

1. In the case where the nodal displacement is compatible with a rigid body motion, the shape function must not allow for any deformation.
2. In the case where the nodal variables together with their derivatives up to the maximum order appearing in the weak integral formulation are compatible with uniform states of deformation, these states must be satisfied at all points inside the element when its dimensions tend to zero. In this case, the interpolation is said to be complete and we have $\sum_{i=1}^{nn_e} N_i = 1$.

3. The nodal variables and their derivatives up to the maximum order appearing in the weak integral formulation minus 1, must be continuous at the boundary between two elements. In this case, the elements are referred to as conforming.

Thus, for a weak integral formulation containing partial derivatives up to order m, the condition for a complete interpolation requires a C^m continuity, whereas the conformity requires a C^{m-1} continuity.

5.2.3.2 Transformation of gradients and integrals

To evaluate the elementary integral over the reference element, we must express the physical space gradients and integrals in terms of local coordinates. The classic differential geometry provides the tools to carry out this transformation.

5.2.3.2.1 Case of a volume element

Starting from

$$\begin{cases} dx_1 = \dfrac{\partial x_1}{\partial \xi_1} d\xi_1 + \dfrac{\partial x_1}{\partial \xi_2} d\xi_2 + \dfrac{\partial x_1}{\partial \xi_3} d\xi_3 \\ dx_2 = \dfrac{\partial x_2}{\partial \xi_1} d\xi_1 + \dfrac{\partial x_2}{\partial \xi_2} d\xi_2 + \dfrac{\partial x_2}{\partial \xi_3} d\xi_3 \\ dx_3 = \dfrac{\partial x_3}{\partial \xi_1} d\xi_1 + \dfrac{\partial x_3}{\partial \xi_2} d\xi_2 + \dfrac{\partial x_3}{\partial \xi_3} d\xi_3 \end{cases} \tag{5.20}$$

We can write

$$d\underline{x} = \underline{a}_1^e d\xi_1 + \underline{a}_2^e d\xi_2 + \underline{a}_3^e d\xi_3 \tag{5.21}$$

where

$$\begin{cases} \underline{a}_1^e = \dfrac{\partial \underline{x}}{\partial \xi_1} \\ \underline{a}_2^e = \dfrac{\partial \underline{x}}{\partial \xi_2} \\ \underline{a}_3^e = \dfrac{\partial \underline{x}}{\partial \xi_3} \end{cases} \tag{5.22}$$

Thus, $\left(\underline{a}_1^e, \underline{a}_2^e, \underline{a}_3^e\right)$ constitute the vectors of the natural basis of the reference element Ω^r associated with Ω^e.

Furthermore, we have

$$\begin{cases} \dfrac{\partial}{\partial \xi_1} = \dfrac{\partial}{\partial x_1}\dfrac{\partial x_1}{\partial \xi_1} + \dfrac{\partial}{\partial x_2}\dfrac{\partial x_2}{\partial \xi_1} + \dfrac{\partial}{\partial x_3}\dfrac{\partial x_3}{\partial \xi_1} = \underline{a}_1 \cdot \dfrac{\partial}{\partial \underline{x}} \\ \dfrac{\partial}{\partial \xi_2} = \dfrac{\partial}{\partial x_1}\dfrac{\partial x_1}{\partial \xi_2} + \dfrac{\partial}{\partial x_2}\dfrac{\partial x_2}{\partial \xi_2} + \dfrac{\partial}{\partial x_3}\dfrac{\partial x_3}{\partial \xi_2} = \underline{a}_2 \cdot \dfrac{\partial}{\partial \underline{x}} \\ \dfrac{\partial}{\partial \xi_3} = \dfrac{\partial}{\partial x_1}\dfrac{\partial x_1}{\partial \xi_3} + \dfrac{\partial}{\partial x_2}\dfrac{\partial x_2}{\partial \xi_3} + \dfrac{\partial}{\partial x_3}\dfrac{\partial x_3}{\partial \xi_3} = \underline{a}_3 \cdot \dfrac{\partial}{\partial \underline{x}} \end{cases} \tag{5.23}$$

Or under matrix form

$$\left\{\frac{\partial}{\partial \underline{\xi}}\right\} = \begin{bmatrix} \langle \underline{a}_1^e \rangle \\ \langle \underline{a}_2^e \rangle \\ \langle \underline{a}_3^e \rangle \end{bmatrix} \left\{\frac{\partial}{\partial \underline{x}}\right\} = [J^e] \left\{\frac{\partial}{\partial \underline{x}}\right\} \tag{5.24}$$

Matrix $[J^e]$ is the Jacobian matrix of the mapping:

$$[J^e] = \begin{bmatrix} \langle \underline{a}_1^e \rangle \\ \langle \underline{a}_2^e \rangle \\ \langle \underline{a}_3^e \rangle \end{bmatrix} = \begin{bmatrix} \dfrac{\partial x_1}{\partial \xi_1} & \dfrac{\partial x_2}{\partial \xi_1} & \dfrac{\partial x_3}{\partial \xi_1} \\ \dfrac{\partial x_1}{\partial \xi_2} & \dfrac{\partial x_2}{\partial \xi_2} & \dfrac{\partial x_3}{\partial \xi_2} \\ \dfrac{\partial x_1}{\partial \xi_3} & \dfrac{\partial x_2}{\partial \xi_3} & \dfrac{\partial x_3}{\partial \xi_3} \end{bmatrix}^e \tag{5.25}$$

More precisely, it reads

$$[J^e] = \begin{bmatrix} \left\langle \dfrac{\partial N^e}{\partial \xi_1} \right\rangle \{x_1^e\} & \left\langle \dfrac{\partial N^e}{\partial \xi_1} \right\rangle \{x_2^e\} & \left\langle \dfrac{\partial N^e}{\partial \xi_1} \right\rangle \{x_3^e\} \\ \left\langle \dfrac{\partial N^e}{\partial \xi_2} \right\rangle \{x_1^e\} & \left\langle \dfrac{\partial N^e}{\partial \xi_2} \right\rangle \{x_2^e\} & \left\langle \dfrac{\partial N^e}{\partial \xi_2} \right\rangle \{x_3^e\} \\ \left\langle \dfrac{\partial N^e}{\partial \xi_3} \right\rangle \{x_1^e\} & \left\langle \dfrac{\partial N^e}{\partial \xi_3} \right\rangle \{x_2^e\} & \left\langle \dfrac{\partial N^e}{\partial \xi_3} \right\rangle \{x_3^e\} \end{bmatrix} \tag{5.26}$$

The Jacobian of the mapping is given by

$$|J^e| = \det [J^e] = \underbrace{\left(\underline{a}_1^e, \underline{a}_2^e, \underline{a}_3^e\right)}_{\text{mixed product}} = \left(\underline{a}_1^e \times \underline{a}_2^e\right) \cdot \underline{a}_3^e \tag{5.27}$$

The mapping (5.9) being bijective, the Jacobian is not singular and we can express the physical coordinates gradient in terms of local coordinates gradient:

$$\underline{\nabla}_x(\cdots) = [J^e]^{-1}\underline{\nabla}_\xi(\cdots) \tag{5.28}$$

namely

$$\left\{\frac{\partial}{\partial \underline{x}}\right\} = [J^e]^{-1}\left\{\frac{\partial}{\partial \underline{\xi}}\right\} \tag{5.29}$$

with

$$[J^e]^{-1} = \frac{1}{\left|J^e\right|}\left[\underline{a}_2^e \times \underline{a}_3^e \quad \underline{a}_3^e \times \underline{a}_1^e \quad \underline{a}_1^e \times \underline{a}_2^e\right] \tag{5.30}$$

Finally, the elementary volume dV at point $\underline{x}$ of the physical element Ω^e is linked to the elementary volume $d\xi_1 d\xi_2 d\xi_3$ of the reference element Ω^r by

$$d\Omega = \left|J^e\right| d\xi_1 d\xi_2 d\xi_3 \tag{5.31}$$

Consequently,

$$\int_{\Omega^e}(\cdots)\,d\Omega = \int_{\Omega^r}(\cdots)\left|J^e\right| d\xi_1\, d\xi_2\, d\xi_3 \tag{5.32}$$

5.2.3.2.2 Case of a surface element

A point $\underline{x}(x_1, x_2, x_3)$* in the physical space is defined by its parametric coordinates (ξ_1, ξ_2) in the reference space. Classic formulas of the theory of surfaces allow for relating gradients and integrals expressed in either system of coordinates together with calculating the normal.

The vectors defining the tangent plane at point $\underline{x}(x_1, x_2, x_3)$ of the physical element are given by

$$\begin{cases} \underline{a}_1^e = \dfrac{\partial \underline{x}}{\partial \xi_1} \\[2ex] \underline{a}_2^e = \dfrac{\partial \underline{x}}{\partial \xi_2} \end{cases} \tag{5.33}$$

* Since the point belongs to a surface, there exists a relation between its physical coordinates of the form $f(x_1, x_2, x_3) = 0$.

The unit normal $\underline{n}^e$ at point $\underline{x}$ of element Ω^e is given by

$$\underline{n}^e = \frac{\underline{a}_1^e \times \underline{a}_2^e}{\left|\underline{a}_1^e \times \underline{a}_2^e\right|} = \frac{\underline{a}_1^e \times \underline{a}_2^e}{\left|j^e\right|} \tag{5.34}$$

with

$$\left|j^e\right| = \left|\underline{a}_1^e \times \underline{a}_2^e\right| = \left(\left|\underline{a}_1^e\right|^2 \left|\underline{a}_2^e\right|^2 - \left(\underline{a}_1^e \cdot \underline{a}_2^e\right)^2\right)^{\frac{1}{2}} \tag{5.35}$$

The area element dS reads

$$dS = \left|j^e\right| d\xi_1 d\xi_2 \tag{5.36}$$

Consequently,

$$\int_{S^e} (\cdots)\, dS = \int_{S^r} (\cdots)\left|j^e\right| d\xi_1\, d\xi_2 \tag{5.37}$$

Let

$$\underline{a}_3^e = \frac{\partial \underline{x}}{\partial \xi_3} = \underline{a}_1^e \times \underline{a}_2^e = \left|j^e\right| \underline{n}^e \tag{5.38}$$

Then, using Equations 5.27 and 5.35, $[J^e]^{-1}$ in Equation 5.28 becomes

$$[J^e]^{-1} = \frac{1}{\left|j^e\right|^2}\left[\underline{a}_2^e \times \underline{a}_3^e \quad \underline{a}_3^e \times \underline{a}_1^e \quad \underline{a}_3^e\right] \tag{5.39}$$

5.2.3.2.3 Case of a line element

A point $\underline{x}(x_1, x_2, x_3)$ in the physical space is defined by its parametric coordinate (ξ_1) in the reference space. Classic formulas of differential geometry provide the unit tangent together with the curvilinear abscissa element ds at point $\underline{x}(x_1, x_2, x_3)$ of the curvilinear element Ω^e.

The vector of the element natural basis is given by

$$\underline{a}_1^e = \frac{\partial \underline{x}}{\partial \xi_1} \tag{5.40}$$

The unit tangent $\underline{\tau}^e$ at point $\underline{x}$ of element Ω^e is written as

$$\underline{\tau}^e = \frac{\underline{a}_1^e}{\left|\underline{a}_1^e\right|} = \frac{1}{\left|J^e\right|}\,\underline{a}_1^e \tag{5.41}$$

with

$$\left|J^e\right| = \left|\underline{a}_1^e\right| \tag{5.42}$$

The curvilinear abscissa element ds is

$$ds = \left|J^e\right| d\xi_1 \tag{5.43}$$

Remark: The shape functions for usual three- and two-dimensional elements together with the properties of the corresponding mapping are given in Appendix 5A and 5B, respectively.

5.2.3.3 Calculation of elementary matrices

We must evaluate over each element the weak form of the problem in Equation 5.7.

5.2.3.3.1 Calculation of $Q(\hat{p}, \delta\hat{p})$ over element Ω^e

Applying Equation 5.4 over element Ω^e, we obtain

$$Q^e(\hat{p},\delta\hat{p}) = \int_{\Omega^e} \frac{1}{\rho_0 c_0^2}\,\hat{p}\delta\hat{p}\,dV \tag{5.44}$$

Using Equations 5.19 and 5.32, we get

$$Q^e(\hat{p},\delta\hat{p}) = \left\langle \delta\hat{p}^e \right\rangle [Q^e] \left\{\hat{p}^e\right\} \tag{5.45}$$

where

$$[Q^e] = \int_{\Omega^r} \frac{1}{\rho_0 c_0^2} \left\{N(\xi_1,\xi_2,\xi_3)\right\}\left\langle N(\xi_1,\xi_2,\xi_3)\right\rangle \left|J^e(\xi_1,\xi_2,\xi_3)\right| d\xi_1 d\xi_2 d\xi_3 \tag{5.46}$$

Note that $[Q^e]$ is a square, symmetric, and positive semidefinite matrix of size $(nn_e \times nn_e)$.

5.2.3.3.2 *Calculation of* $H(\hat{p},\delta\hat{p})$ *over element* $(nn_e \times nn_e)$

Applying Equation 5.3 over element Ω^e, we obtain

$$H^e(\hat{p},\delta\hat{p}) = \int_{\Omega^e} \frac{1}{\rho_0}\left(\underline{\nabla}\delta\hat{p} \cdot \underline{\nabla}\hat{p}\right) dV \tag{5.47}$$

Using Equations 5.19 and 5.28, we can write

$$\underline{\nabla}\hat{p} = \nabla_x \langle N(\underline{\xi})\rangle \{\hat{p}^e\} = [B^e]\{\hat{p}^e\} \tag{5.48}$$

with

$$[B^e] = [J^e]^{-1} \underline{\nabla}_\xi \langle N(\xi)\rangle \tag{5.49}$$

In other words, matrix $[B^e]$ of dimensions $(3 \times nn_e)$ is

$$[B^e] = \begin{bmatrix} \left\langle \dfrac{\partial \underline{x}}{\partial \xi_1} \right\rangle \\ \left\langle \dfrac{\partial \underline{x}}{\partial \xi_2} \right\rangle \\ \left\langle \dfrac{\partial \underline{x}}{\partial \xi_3} \right\rangle \end{bmatrix}^{-1} \begin{bmatrix} \left\langle \dfrac{\partial N}{\partial \xi_1} \right\rangle \\ \left\langle \dfrac{\partial N}{\partial \xi_2} \right\rangle \\ \left\langle \dfrac{\partial N}{\partial \xi_3} \right\rangle \end{bmatrix} \tag{5.50}$$

Substituting Equation 5.49 and using Equation 5.32 in Equation 5.47, we obtain

$$H^e(\hat{p},\delta\hat{p}) = \langle \delta\hat{p}^e \rangle [H^e]\{\hat{p}^e\} \tag{5.51}$$

with

$$[H^e] = \int_{\Omega^r} \frac{1}{\rho_0} [B^e(\xi_1,\xi_2,\xi_3)]^T [B^e(\xi_1,\xi_2,\xi_3)] \left| J^e(\xi_1,\xi_2,\xi_3) \right| d\xi_1\, d\xi_2\, d\xi_3 \tag{5.52}$$

Note that $[H^e]$ is a square, symmetric, and positive semidefinite matrix of size $(nn_e \times nn_e)$.

5.2.3.3.3 *Calculation of* $A(\hat{p},\delta\hat{p})$ *over element* $\partial\Omega^e_{f,R}$

Applying Equation 5.5 over element $\partial\Omega^e_{f,R}$, we obtain

$$A^e(\hat{p},\delta\hat{p}) = \int_{\partial\Omega^e_{f,R}} \frac{\hat{\beta}}{\rho_0 c_0}\, \delta\hat{p}\,\hat{p}\, dS \tag{5.53}$$

Using Equations 5.19 and 5.37, we get

$$A^e(\hat{p},\delta\hat{p}) = \langle \delta\hat{p}^e \rangle \left[\hat{A}^e\right]\left\{\hat{p}^e\right\} \tag{5.54}$$

with*

$$\left[\hat{A}^e\right] = \int_{\partial\Omega^r_{f,R}} \frac{\hat{\beta}}{\rho_0 c_o} \{N(\xi_1,\xi_2,\xi_3)\}\langle N(\xi_1,\xi_2,\xi_3)\rangle \left|j^e(\xi_1,\xi_2,\xi_3)\right| d\xi_1\, d\xi_2 \tag{5.55}$$

Note that $[\hat{A}^e]$ is a square, symmetric matrix of size $(nn_e \times nn_e)$. Note that here, nn_e denotes the number of nodes of the surface element $\partial\Omega^e_{f,N}$ that is different from the number of nodes of volumic elements involved in Equations 5.46 and 5.52.

5.2.3.3.4 *Calculation of* $R(\delta\hat{p})$ *over element* $\partial\Omega^e_{f,N}$

Applying Equation 5.6 over element $\partial\Omega^e_{f,N}$, we obtain

$$R^e(\delta\hat{p}) = \omega^2 \int_{\partial\Omega^e_{f,N}} \delta\hat{p}\,\bar{U}_n\, dS \tag{5.56}$$

Using Equations 5.19 and 5.37, we get

$$R^e(\hat{p},\delta\hat{p}) = \langle \delta\hat{p}^e \rangle \left\{\hat{f}^e\right\} \tag{5.57}$$

with

$$\left\{\hat{f}^e\right\} = \omega^2 \left[C^e_{up}\right]\left\{\bar{U}^e_n\right\} \tag{5.58}$$

* Note that the shape functions N_i used to approximate $A^e(\hat{p},\delta\hat{p})$ are associated with the nodal approximation over the element $\partial\Omega^e_{f,R}$ and the corresponding reference element $\partial\Omega^r_{f,R}$.

and

$$\left[C_{up}^e\right] = \int_{\partial\Omega_{f,N}^r} \{N(\xi_1,\xi_2,\xi_3)\}\langle N(\xi_1,\xi_2,\xi_3)\rangle \left|j^e(\xi_1,\xi_2,\xi_3)\right| d\xi_1\, d\xi_2 \quad (5.59)$$

$[C_{up}^e]$ is a square, symmetric matrix of size $(nn_e \times nn_e)$ where nn_e denotes the number of nodes of the surface element $\partial\Omega_{f,N}^e$.

Remark: In the case where the nodal displacement vector $\{\bar{U}^e\}$ is given,* we must first calculate the nodal approximation of the normal displacement $\bar{U}_n = \underline{\bar{U}} \cdot \underline{n}$ at all points of $\partial\Omega_{f,N}^e$. Let

$$\left\{\bar{U}^e\right\} = \begin{Bmatrix} \{\underline{\bar{U}}_1\} \\ \vdots \\ \{\underline{\bar{U}}_{nn_e}\} \end{Bmatrix} = \begin{Bmatrix} \bar{U}_{1,1} \\ \bar{U}_{2,1} \\ \bar{U}_{3,1} \\ \vdots \\ \bar{U}_{1,nn_e} \\ \bar{U}_{2,nn_e} \\ \bar{U}_{3,nn_e} \end{Bmatrix} \quad (5.60)$$

The nodal approximation of $\bar{U}_n = \underline{\bar{U}} \cdot \underline{n}$ is given by (see Equation 5.38):

$$\left\{\bar{U}_n^e\right\} = \frac{1}{\left|j^e\right|} \begin{bmatrix} \langle \underline{a}_3^e \rangle & 0 & 0 \\ 0 & \ddots & 0 \\ 0 & 0 & \langle \underline{a}_3^e \rangle \end{bmatrix} \begin{Bmatrix} \{\underline{\bar{U}}_1\} \\ \vdots \\ \{\underline{\bar{U}}_{nn_e}\} \end{Bmatrix} \quad (5.61)$$

Or more compactly,

$$\left\{\bar{U}_n^e\right\} = \frac{1}{\left|j^e\right|}\left[a_3^e\right]\left\{\bar{U}^e\right\} \quad (5.62)$$

where $[a_3^e]$ is a block diagonal matrix of dimension $(nn_e \times 3nn_e)$.

Consequently, the vector $\left\{\hat{f}^e\right\}$ in Equation 5.58 can be rewritten as

$$\left\{\hat{f}^e\right\} = \omega^2\left[C_{up}^e\right]\left\{\bar{U}^e\right\} \quad (5.63)$$

* This is the case for fluid-structure coupling (see Chapter 6).

with this time

$$\left[C_{up}^e\right] = \int_{\partial\Omega_{f,N}^r} \{N(\xi_1,\xi_2,\xi_3)\}\langle N(\xi_1,\xi_2,\xi_3)\rangle\left[a_3^e(\xi_1,\xi_2,\xi_3)\right] d\xi_1\, d\xi_2 \qquad (5.64)$$

which is a rectangular matrix of size $(nn_e \times 3nn_e)$ that can be rewritten explicitly as

$$\left[C_{up}^e\right] = \int_{\partial\Omega_{f,N}^r} \begin{Bmatrix} N_1 \\ \vdots \\ N_{nn_e} \end{Bmatrix} \left\langle N_1\left\langle \underline{a}_3^e\right\rangle \quad \ldots \quad N_{nn_e}\left\langle \underline{a}_3^e\right\rangle\right\rangle d\xi_1\, d\xi_2 \qquad (5.65)$$

5.2.3.4 *Discretized weak integral formulation*

By substituting Equations 5.46, 5.52, 5.55, and 5.63 in Equation 5.7, we obtain the discretized weak integral formulation of the problem

$$\begin{cases} W(\hat{p},\delta\hat{p}) = \sum_{e/\Omega^e\in\Omega} \langle\delta\hat{p}^e\rangle\left(\left[H^e\right] - \omega^2\left[Q^e\right]\right)\{\hat{p}^e\} \\ \qquad + i\omega \sum_{e/\partial\Omega^e\in\partial\Omega_{f,R}} \langle\delta\hat{p}^e\rangle\left[\hat{A}^e\right]\{\hat{p}^e\} - \sum_{e/\partial\Omega^e\in\partial\Omega_{f,N}} \langle\delta\hat{p}^e\rangle\left\{\hat{f}^e\right\} \\ \text{with } \delta\hat{p} = 0 \quad \text{and} \quad \hat{p} = \bar{p} \quad \text{on } \partial\Omega_{f,D} \end{cases} \qquad (5.66)$$

where the summation is taken in the assembling sense.

The discretized weak integral Galerkin's formulation associated with the linear elastodynamic equation can be found following the same procedure.

5.2.3.5 *Calculation of elementary integrals*

The calculation of elementary integrals appearing in the discretized weak integral formulation is carried out numerically in practice. Indeed, except for simple cases where the Jacobian is constant, the analytical integration is impossible. The numerical integration scheme usually relies on the use of Gauss quadrature rules (see Appendix 5C). Let us just recall the principle here.

5.2.3.5.1 Consider an interval [−1;1]

For a nonsingular function $f(x)$, the approximated value of the integral

$$I = \int_{-1}^{1} f(x)\,dx \qquad (5.67)$$

is given by

$$I \cong \sum_{i=1}^{npg} W_i^g f\left(x_i^g\right) \tag{5.68}$$

where x_i^g and W_i^g are, respectively, the abscissae and the weights of the *npg* Gauss points. x_i^g correspond to the zeroes of Legendre polynomial of order *npg* (see Appendix 5C). Gauss quadrature rule for a given order *npg* exactly integrates monomials of degree $\leq 2npg - 1$.

5.2.3.5.2 Consider a square [−1;1]² and a cube [−1;1]³

The corresponding formulae are obtained from the one-dimensional case, namely

$$I_2 = \int_{-1}^{1}\int_{-1}^{1} f(x_1, x_2)\, dx_1 dx_2 \cong \sum_{i=1}^{npg}\sum_{j=1}^{npg} f\left(x_{1,i}^g, x_{2,j}^g\right) W_i^g W_j^g \tag{5.69}$$

$$I_3 = \int_{-1}^{1}\int_{-1}^{1}\int_{-1}^{1} f(x_1, x_2, x_3)\, dx_1 dx_2 dx_3 \cong \sum_{i=1}^{npg}\sum_{j=1}^{npg}\sum_{k=1}^{npg} f\left(x_{1,i}^g, x_{2,j}^g, x_{3,k}^g\right) W_i^g W_j^g W_k^g \tag{5.70}$$

There also exist numerical quadrature rules for other elementary shapes such as triangles, tetrahedra, and wedges (see Appendix 5C). Appendix 5D provides examples of Fortran routines for the calculation of matrices $[Q^e]$ and $[H^e]$ using Gauss quadrature rules.

5.2.4 Step 4: Assembling

Consider the vector $\langle \hat{p} \rangle = \langle \hat{p}_1, \hat{p}_2, \ldots, \hat{p}_{nnt} \rangle$ of nodal unknowns of the problem and the elementary localization matrix $[L^e]$ of dimensions $(n_e \times nnt)$ allowing for the correspondence between the local numbering and the global numbering $\{\hat{p}^e\} = [L^e]\{\hat{p}\}$. The discretized formulation Equation 5.66 can be rewritten

$$\begin{cases} W\left(\hat{p}, \delta\hat{p}\right) = \langle \delta\hat{p} \rangle \left(\left([H] - \omega^2 [Q] + i\omega \left[\hat{A}\right] \right) \{\hat{p}\} - \left\{\hat{f}\right\} \right) = 0 \\ \text{with } \delta\hat{p} = 0 \quad \text{and} \quad \hat{p} = \bar{p} \quad \text{on } \partial\Omega_{f,D} \end{cases} \tag{5.71}$$

where

$$[H] = \sum_{e/\Omega^e \in \Omega} [L^e]^T [H^e][L^e] \tag{5.72}$$

$$[Q] = \sum_{e/\Omega^e \in \Omega} [L^e]^T [Q^e][L^e] \tag{5.73}$$

$$[\hat{A}] = \sum_{e/\partial\Omega^e \in \partial\Omega_{f,R}} [L^e]^T [\hat{A}^e][L^e] \tag{5.74}$$

$$\{\hat{f}\} = \sum_{e/\partial\Omega^e \in \partial\Omega_{f,N}} [L^e]^T \{\hat{f}^e\} \tag{5.75}$$

where matrices $[H]$,$[Q]$, and $[\hat{A}]$ are obtained by assembling of elementary matrices $[H^e]$, $[Q^e]$, and $[\hat{A}^e]$. These matrices are square, band, and symmetric of size $(nnt \times nnt)$. Similarly, vector $\{\hat{f}\}$ is the global acoustical load vector. It is obtained by assembling the elementary load vectors $\{\hat{f}^e\}$.

5.2.4.1 Computer implementation procedure

Practically, the assembling procedure is intimately linked to the storage technique of global matrices (band format, skyline format, sparse format, etc.) and to the technique used to account for boundary conditions. Appendix 5E provides an example of Fortran routines in the case where matrices are stored in full format.

5.2.5 Step 5: Constraints and boundary conditions

There are three basic techniques to impose constraints and boundary conditions (Cook et al. 2002):

- Lagrange multipliers
- Global matrices partitioning (= reduction of matrices)
- Penalty method

To introduce the different techniques, we consider the general canonical form of the functional governing the problem:

$$\begin{cases} \Pi(\hat{p}) = \dfrac{1}{2}\langle\hat{p}\rangle[\hat{Z}]\{\hat{p}\} - \langle\hat{p}\rangle\{\hat{f}\} \\ \text{with } \hat{p} = \bar{p} \quad \text{on } \partial\Omega_{f,D} \end{cases} \tag{5.76}$$

The dynamic matrix $[\hat{Z}]$ is assumed to be symmetric and of dimensions $(n_{eq} \times n_{eq})$, where n_{eq} represents the number of equations of the system (i.e., total number of degrees of freedom).

5.2.5.1 Lagrange's multiplier method

Let us assume, in the general case, a constraint of the form

$$\Gamma(\hat{p}) = \hat{h} \quad \text{on} \quad \partial\Omega_{f,D} \tag{5.77}$$

The problem consists in making the functional $\Pi(\hat{p})$ subjected to the constraint Equation 5.77 stationary. In Lagrange's multiplier method, a modified functional Π^* is introduced:

$$\Pi^* = \Pi + \int_{\partial\Omega_{f,D}} \hat{\lambda} C(\hat{p})\, dS \tag{5.78}$$

where $\hat{\lambda}$ is an arbitrary integrable function called Lagrange's multiplier. This function can be approximated on the elements of $\partial\Omega_{f,D}$ using the associated shape functions and nodal values

$$\hat{\lambda} = \langle N \rangle \{\hat{\lambda}\} \tag{5.79}$$

where $\{\hat{\lambda}\}$ represents the vector of nodal values of $\hat{\lambda}$ that is, the vector of nodal Lagrange's multipliers.

In the case of linear constraints*

$$\Gamma(\hat{p}) = C\hat{p} - \hat{h} = 0 \quad \text{on} \quad \partial\Omega_{f,D} \tag{5.80}$$

We can then write

$$\int_{\partial\Omega_{f,D}} \hat{\lambda}\Gamma(\hat{p})\, dS = \langle \hat{\lambda} \rangle \int_{\partial\Omega_{f,D}} \{N^{\lambda}\} C \langle N^{\lambda} \rangle dS \{\hat{p}\} - \langle \hat{\lambda} \rangle \int_{\partial\Omega_{f,D}} \{N^{\lambda}\} \hat{h}\, dS \tag{5.81}$$

Let

$$[C_{\lambda}] = \int_{\partial\Omega_{f,D}} \{N^{\lambda}\} C \langle N^{\lambda} \rangle dS \tag{5.82}$$

* The linear form is a case frequently encountered in practice. For the acoustical problem ($\hat{p} = \bar{p}$) we have $C = 1$; $\hat{h} = \bar{p}$.

$$\left\{\hat{h}_\lambda\right\} = \int_{\partial\Omega_{f,D}} \left\{N^\lambda\right\} \hat{h}\, dS \tag{5.83}$$

The discretized modified functional is

$$\Pi^* = \Pi + \left\langle\hat{\lambda}\right\rangle \left[C_\lambda\right] \left\{\hat{p}\right\} - \left\langle\hat{\lambda}\right\rangle \left\{\hat{h}_\lambda\right\} \tag{5.84}$$

Or

$$\Pi^* = \frac{1}{2}\langle\hat{p}\rangle\left[\hat{Z}\right]\{\hat{p}\} - \langle\hat{p}\rangle\left\{\hat{f}\right\} + \langle\hat{p}\rangle \left[C_\lambda\right]^T \left\{\hat{\lambda}\right\} - \left\langle\hat{\lambda}\right\rangle \left\{\hat{h}_\lambda\right\} \tag{5.85}$$

The invocation of the stationarity of Π^* leads to

$$\begin{aligned}\delta\Pi^* = {} & \langle\delta\hat{p}\rangle\left[\hat{Z}\right]\{\hat{p}\} - \langle\delta\hat{p}\rangle\left\{\hat{f}\right\} + \langle\delta\hat{p}\rangle \left[C_\lambda\right]^T \left\{\hat{\lambda}\right\} \\ & + \left\langle\delta\hat{\lambda}\right\rangle \left[C_\lambda\right] \{\hat{p}\} - \left\langle\delta\hat{\lambda}\right\rangle \left\{\hat{h}_\lambda\right\} = 0\end{aligned} \tag{5.86}$$

This expression being valid for all arbitrary variation $\left(\delta\hat{p}, \delta\hat{\lambda}\right)$, there follows that

$$\begin{cases}\left[\hat{Z}\right]\{\hat{p}\} + \left[C_\lambda\right]^T \left\{\hat{\lambda}\right\} = \left\{\hat{f}\right\} \\ \left[C_\lambda\right]\{\hat{p}\} = \left\{\hat{h}_\lambda\right\}\end{cases} \tag{5.87}$$

Or in matrix form

$$\underbrace{\begin{bmatrix}\left[\hat{Z}\right] & \left[C_\lambda\right]^T \\ \left[C_\lambda\right] & [0]\end{bmatrix}}_{\text{band symmetric matrix}} \begin{Bmatrix}\{\hat{p}\} \\ \left\{\hat{\lambda}\right\}\end{Bmatrix} = \begin{Bmatrix}\left\{\hat{f}\right\} \\ \left\{\hat{h}_\lambda\right\}\end{Bmatrix} \tag{5.88}$$

We thus end up with a symmetric system of larger size but with a band form.

Remarks

1. The numerical implementation of Lagrange's multipliers method is easy.

2. Note that the final matrix is nonpositive definite. This prevents the use of several efficient direct resolution algorithms.
3. Lagrange's multipliers method is not efficient for simple boundary conditions. However, being general, it is well adapted to take into account complex boundary conditions (e.g., links between two structures, transfer impedance conditions, etc.).
4. In the case of boundary conditions such as kinematic relationships between nodal values, Lagrange's multipliers technique can be introduced easily. Let us assume that we have m linear relations of the form

$$\underbrace{[C_\lambda]}_{m\times n}\underbrace{\{\hat{p}\}}_{n} = \underbrace{\{\hat{h}_\lambda\}}_{m} \tag{5.89}$$

The modified functional is given by Equation 5.84 where $\langle\hat{\lambda}\rangle = \langle\hat{\lambda}_1,\ldots,\hat{\lambda}_m\rangle$ are the m Lagrange's multipliers.

Invoking the stationarity of Π^* leads to matrix system Equation 5.88 of dimensions $(n_{eq}+m)\times(n_{eq}+m)$.
5. In the case where $[\hat{Z}] = [K]$ is the stiffness matrix of an elasticity problem, the first line of Equation 5.88 leads to

$$[K]\{\hat{p}\} = \{\hat{F}\} - [C_\lambda]^T\{\hat{\lambda}\} \tag{5.90}$$

where $\{\hat{F}\}$ is the force nodal vector acting on the structure. Thus, vector $\{\hat{r}_\lambda\} = -[C_\lambda]^T\{\hat{\lambda}\}$ represents the nodal reaction forces required to satisfy the constraints $[C_\lambda]\{\hat{p}\} = \{\hat{h}_\lambda\}$.

5.2.5.2 *Partitioning method*

Let us assume that we have to impose kinematic boundary conditions of the form of Equation 5.89. We can partition the vector of nodal unknowns $\{\hat{p}\}$ under the form $\left\{\begin{matrix}\{\hat{p}_k\}\\ \{\hat{p}_c\}\end{matrix}\right\}$ where $\{\hat{p}_k\}$ is the vector of degrees of freedom that are kept (i.e., free or master degrees of freedom) and $\{\hat{p}_c\}$ is the vector of the m condensed degrees of freedom (i.e., imposed or slave degrees of freedom).

Similarly, let us partition vector $\{\hat{h}_\lambda\}$ under the form $\left\{\begin{matrix}\{\hat{h}_k\}\\ \{\hat{h}_c\}\end{matrix}\right\}$.

Equation 5.89 becomes

$$\begin{bmatrix} [C_{kk}] & [C_{kc}] \\ [C_{ck}] & [C_{cc}] \end{bmatrix} \begin{Bmatrix} \{\hat{p}_k\} \\ \{\hat{p}_c\} \end{Bmatrix} = \begin{Bmatrix} \{\hat{h}_k\} \\ \{\hat{h}_c\} \end{Bmatrix} \tag{5.91}$$

The second line of this system leads to

$$\{\hat{p}_c\} = [C_{cc}]^{-1}\{\hat{h}_c\} - [C_{cc}]^{-1}[C_{ck}]\{\hat{p}_k\} \tag{5.92}$$

Using this relation, we can express the vector of nodal unknowns $\{\hat{p}\}$ in terms of the master degrees of freedom $\{\hat{p}_k\}$:

$$\{\hat{p}\} = [G]\{\hat{p}_k\} + \{\hat{h}_r\} \tag{5.93}$$

with

$$[G] = \begin{bmatrix} [I] \\ \hline [R] \end{bmatrix}; \quad [R] = -[C_{cc}]^{-1}[C_{ck}] \tag{5.94}$$

and

$$\{\hat{h}_r\} = \begin{Bmatrix} \{0\} \\ \hline [C_{cc}]^{-1}\{\hat{h}_c\} \end{Bmatrix} \tag{5.95}$$

Matrix $[G]$ is a reduction matrix. It is rectangular of dimension $(n_{eq} \times (n_{eq} - m))$.

The system functional is

$$\begin{cases} \Pi(\hat{p}) = \dfrac{1}{2}\langle\hat{p}\rangle\left[\hat{Z}\right]\{\hat{p}\} - \langle\hat{p}\rangle\{\hat{f}\} \\ [C_\lambda]\{\hat{p}\} = \{\hat{h}\} \end{cases} \tag{5.96}$$

and can be expressed in terms of $\{\hat{p}_k\}$

$$\Pi(\hat{p}) = \frac{1}{2}\langle \hat{p}_k \rangle [G]^T [\hat{Z}][G]\{\hat{p}_k\} + \frac{1}{2}\left(\langle \hat{p}_c \rangle [G]^T [\hat{Z}]\{\hat{h}_r\} + \langle \hat{h}_r \rangle [\hat{Z}][G]\{\hat{p}_k\}\right)$$
$$- \langle \hat{p}_k \rangle [G]^T \{\hat{f}\} + \frac{1}{2}\langle \hat{h}_r \rangle [\hat{Z}]\{\hat{h}_r\} - \langle \hat{h}_r \rangle \{\hat{f}\} \tag{5.97}$$

Let

$$[\hat{Z}_r] = [G]^T [\hat{Z}][G] \tag{5.98}$$

and

$$\{\hat{f}_r\} = [G]^T \{\hat{f}\} \tag{5.99}$$

$[\hat{Z}_r]$ is the reduced dynamic matrix of size $(n_{eq} - m) \times (n_{eq} - m)$ and $\{\hat{f}_r\}$ is the reduced loading vector of dimension $(n_{eq} - m)$. Assuming that matrix $[\hat{Z}]$ is symmetric, the algebraic system obtained by invoking the stationarity principle is

$$[\hat{Z}_r]\{\hat{p}_k\} = \{\hat{f}_r\} - \{\hat{r}\} \tag{5.100}$$

with

$$\{\hat{r}\} = [G]^T [\hat{Z}]\{\hat{h}_r\} \tag{5.101}$$

Remarks:

1. In the case of boundary conditions of type $\left(\hat{p}_i = \bar{p}_i \text{ for } i = 1,\ldots,m\right)$, we have

$$\{\hat{p}\} = \begin{Bmatrix} \{\hat{p}_k\} \\ \{\hat{p}_c\} \end{Bmatrix}; \quad \{\hat{h}\} = \begin{Bmatrix} \{0\} \\ \{\bar{p}\} \end{Bmatrix}; \quad [C_\lambda] = \begin{bmatrix} [0] & [0] \\ [0] & [I] \end{bmatrix} \tag{5.102}$$

since $[C_{ck}] = [0]$ and $[C_{cc}]^{-1} = [I]$. In addition,

$$[G] = \begin{bmatrix} [I] \\ [0] \end{bmatrix} \tag{5.103}$$

and

$$\left\{\hat{h}_r\right\} = \left\{\frac{\{0\}}{[C_{cc}]^{-1}\left\{\hat{h}_c\right\}}\right\} = \left\{\frac{\{0\}}{\{\bar{p}\}}\right\} \tag{5.104}$$

Denoting

$$\left[\hat{Z}\right] = \begin{bmatrix} \left[\hat{Z}_{kk}\right] & \left[\hat{Z}_{kc}\right] \\ \left[\hat{Z}_{ck}\right] & \left[\hat{Z}_{cc}\right] \end{bmatrix}; \quad \left\{\hat{f}\right\} = \left\{\begin{matrix} \hat{f}_k \\ \hat{f}_c \end{matrix}\right\} \tag{5.105}$$

we get

$$\left[\hat{Z}_r\right] = [G]^T\left[\hat{Z}\right][G] = \left[\hat{Z}_{kk}\right] \tag{5.106}$$

and

$$\{\hat{r}\} = [G]^T\left[\hat{Z}\right]\left\{\hat{h}_r\right\} = \left[\hat{Z}_{kc}\right]\{\bar{p}\} \tag{5.107}$$

Consequently, the reduced system is

$$\left[\hat{Z}_{kk}\right]\{\hat{p}_k\} = \left\{\hat{f}_k\right\} - \left[\hat{Z}_{kc}\right]\{\bar{p}\} \tag{5.108}$$

2. The partitioning method leads to a reduced system. Moreover, it preserves the symmetry and the properties of matrix $[\hat{Z}]$. However, it requires matrix operations that can be complex to implement depending on the type of constraints and used storage format.
3. In the case of boundary conditions of type ($\hat{p}_i = 0$ for $i = 1,\ldots,m$) are not accounted for at the assembling step, they can be taken into account at this stage of the FEM using the partitioning method. The analysis of system Equation 5.108 shows that it is equivalent to discarding the lines and columns corresponding to the m imposed degrees of freedom in the final system.
4. For small values of m, a simple method to account for conditions of the type ($\hat{p}_i = \bar{p}_i$) while avoiding matrix operations related to the partitioning method is to write

$$
\text{line i}\begin{bmatrix} \hat{Z}_{11} & & 0 & & \\ & \ddots & \vdots & & \\ 0 & \cdots & 1 & \cdots & 0 \\ & & \vdots & \ddots & \\ & & 0 & & \hat{Z}_{n_{eq}n_{eq}} \end{bmatrix} \begin{Bmatrix} \hat{p}_1 \\ \vdots \\ \hat{p}_i \\ \vdots \\ \hat{p}_{n_{eq}} \end{Bmatrix} = \begin{Bmatrix} \hat{f}_1 - \hat{Z}_{1i}\,\bar{p}_i \\ \vdots \\ \bar{p}_i \\ \vdots \\ \hat{f}_{n_{eq}} - \hat{Z}_{n_{eq}i}\,\bar{p}_i \end{Bmatrix} \tag{5.109}
$$

column i

5. The partitioning method uses a transformation matrix to impose the constraints. Transformation matrices are commonly used in FE; for example, to transform elementary matrices between multiple coordinate systems.

5.2.5.3 Penalty method

This simple technique is used when boundary conditions of the type ($\hat{p}_i = \bar{p}_i$ for $i = 1,\ldots,m$) have to be imposed while avoiding the matrix operations of the partitioning method.

For the sake of clarity, let us consider a single-node constraint ($\hat{p}_i = \bar{p}_i$). Let ϖ be a "large" number whose value is to be defined. Let us introduce the modified functional Π^* of the problem:

$$
\Pi^* = \frac{1}{2}\langle\hat{p}\rangle\left[\hat{Z}\right]\{\hat{p}\} - \langle\hat{p}\rangle\left\{\hat{f}\right\} + \frac{\varpi}{2}\left(\hat{p}_i - \bar{p}_i\right)^2 \tag{5.110}
$$

Writing the stationarity for Equation 5.110 leads to

$$
\delta\Pi^* = \langle\delta\hat{p}\rangle\left(\left[\hat{Z}\right]\{\hat{p}\} - \left\{\hat{f}\right\}\right) + \varpi\delta\hat{p}_i\left(\hat{p}_i - \bar{p}_i\right) = 0 \tag{5.111}
$$

Introducing the localization vector $\{e_i\}$ of dimension equal to the total number of degrees of freedom, defined by

$$
\hat{p}_i = \langle e_i\rangle\left\{\hat{p}\right\};\quad \langle e_i\rangle = \langle 0,\ldots,1,\ldots 0\rangle \tag{5.112}
$$

where the number "1" in vector $\{e_i\}$ is in the position of degree of freedom "i." We then have

$$
\delta\Pi^* = \langle\delta\hat{p}\rangle\left(\left[\hat{Z}\right]\{\hat{p}\} - \left\{\hat{f}\right\}\right) + \varpi\{e_i\}\langle e_i\rangle\left\{\hat{p}\right\} - \varpi\{e_i\}\,\bar{p}_i = 0 \tag{5.113}
$$

The vector $\{\delta\hat{p}\}$ being arbitrary, we get

$$\left(\left[\hat{Z}\right] + \varpi\{e_i\}\langle e_i\rangle\right)\{\hat{p}\} = \{\hat{f}\} + \varpi\{e_i\}\bar{p}_i \tag{5.114}$$

That is,

$$\begin{bmatrix} \hat{Z}_{11} & \hat{Z}_{12} & & \cdots & \hat{Z}_{1n_{eq}} \\ & \ddots & & & \\ & & \hat{Z}_{ii} + \varpi & & \vdots \\ & sym & & \ddots & \\ & & & & \hat{Z}_{n_{eq}n_{eq}} \end{bmatrix} \begin{Bmatrix} \hat{p}_1 \\ \vdots \\ \hat{p}_i \\ \vdots \\ \hat{p}_{n_{eq}} \end{Bmatrix} = \begin{Bmatrix} \hat{f}_1 \\ \vdots \\ \hat{f}_i + \varpi\bar{p}_i \\ \vdots \\ \hat{f}_{n_{eq}} \end{Bmatrix} \tag{5.115}$$

For the ith equation, we have

$$\sum_{\substack{j=1 \\ j\neq i}}^{n_{eq}} \hat{Z}_{ij}\hat{p}_j + \hat{Z}_{ii}\hat{p}_i + \varpi\hat{p}_i = \hat{f}_i + \varpi\bar{p}_i \tag{5.116}$$

If a very large number ϖ is chosen (= largest machine number), we obtain $\varpi\hat{p}_i \approx \varpi\bar{p}_i$. In other words, the condition $\hat{p}_i = \bar{p}_i$ is satisfied approximately.

Thus, in this technique, to impose $\hat{p}_i = \bar{p}_i$, it is sufficient to add the value ϖ to the diagonal element $\hat{Z}_{ii}$ and to add $\varpi\bar{p}_i$ to the ith element of the right-hand-side load vector, without modifying the assembled system.

Remarks

1. The main issue associated with the penalty method lies in the choice of value ϖ. A too large value of ϖ may lead to ill-conditioned systems. However, numerical experiments can be used to determine an optimal choice for ϖ depending on the problem and the computer.
2. The penalty method can be easily generalized for kinematic boundary conditions of the form $\left(\sum_{i=1}^{m} c_i\hat{p}_i = cst\right)$. For example, to impose the boundary condition ($\hat{p}_i = \hat{p}_j$), the assembled system $[\hat{Z}]\{\hat{p}\} = \{\hat{f}\}$ needs to be modified accordingly:

$$\begin{cases} \hat{Z}_{ii} \Rightarrow \hat{Z}_{ii} + \varpi \\ \hat{Z}_{jj} \Rightarrow \hat{Z}_{jj} + \varpi \\ \hat{Z}_{ij} \Rightarrow \hat{Z}_{ij} - \varpi \\ \hat{Z}_{ji} \Rightarrow \hat{Z}_{ji} - \varpi \end{cases} \tag{5.117}$$

The remaining terms are left unchanged.

5.2.6 Step 6: Stationarity

Once the boundary conditions have been taken into account, the invocation of the stationarity of the functional leads to a matrix system of the form

$$\left(-\omega^2[Q] + i\omega\left[\hat{A}\right] + [H]\right)\{\hat{p}\} = \left\{\hat{f}\right\} \tag{5.118}$$

where the expressions of the different matrices and vectors depend on the technique used to account for the kinematic boundary condition $\hat{p} = \bar{p}$ on $\partial\Omega_{f,D}$.

5.2.7 Step 7: Resolution of linear systems

In general, the time-domain general system of equations governing a dynamic system in linear acoustics and elastodynamics takes the form

$$[M]\{\ddot{u}\}+[C_d]\{\dot{u}\}+[K]\{u\} = \{F\} \tag{5.119}$$

where $[K]$ and $[M]$ denote the stiffness and mass matrix, respectively. Both are symmetric. Matrix $[M]$ is definite positive and matrix $[K]$ is semi-definite positive. $[C_d]$ is a matrix describing the damping in the system (here we select a viscous model), and $\{F\}$ is the nodal load vector. When the excitation changes rapidly with time and the time history of the response is sought, a transient solution is required. The solution depends on the initial conditions: u_0 and $\dot{u}_0$. The time response can be calculated with a step-by-step integration of the equations of motion in the time domain (direct response) or by reduction of the system using the modal basis followed by integration of the modal equations (modal superposition method). The latter is only possible when the frequency content of the excitation is limited to low frequencies (i.e., only the first modes of the system are excited). For excitation spectra with a high-frequency content, such as impulses and shocks, a direct integration is required. The main difficulties of the latter method lie in the control of the stability of the method and the selection of the time step for the time integration. An excellent discussion of these issues is given in Imbert (1991), Craveur (1997), and Géradin and Rixen (1997).

When the excitation is harmonic (or decomposable into harmonic components), a frequency response analysis is performed. This is the case discussed in this book. Let us rewrite the matrix system, Equation 5.118, under the classic form found for structural problems:

$$\left(-\omega^2[M] + i\omega\left[\hat{C}_d\right] + [K]\right)\{\hat{u}\} = \left\{\hat{F}\right\} \tag{5.120}$$

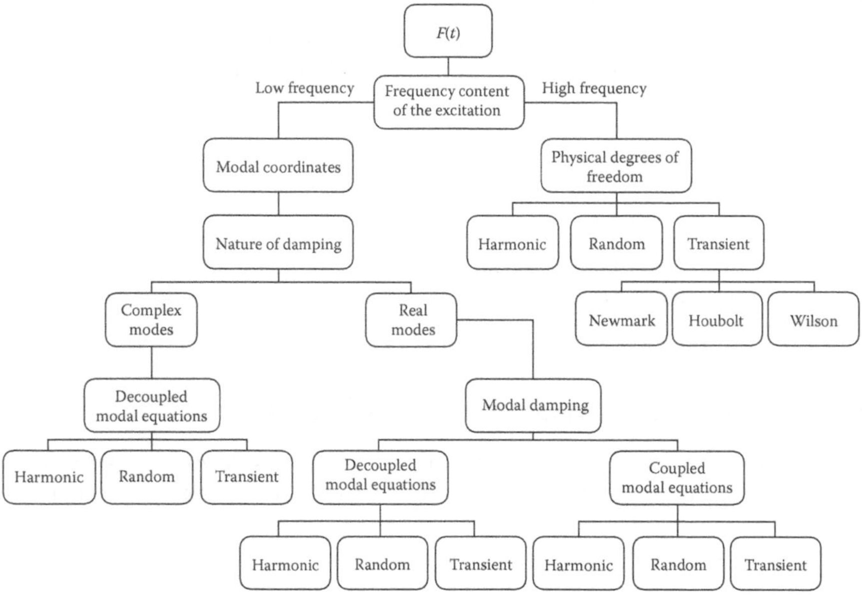

Figure 5.5 Summary of the strategies used to solve dynamic analysis problems. (Adapted from Imbert, J. F. 1991. *Analyse des structures par éléments finis*. Editions Cépaduès, reproduced with permission of Cépaduès.)

where $[K]$ and $[M]$ are real symmetric definite positive matrices, $[\hat{C}_d]$ is a complex-valued symmetric matrix, and $\{\hat{F}\}$ is the nodal load vector.

Again, the solution can be obtained from the resolution of Equation 5.120 in terms of nodal degrees of freedom or most commonly using a modal superposition method. The choice of the technique can be oriented based on Figure 5.5.

5.2.7.1 Resolution methods

The previous linear system Equation 5.120 looks like

$$\left[\hat{Z}(\omega)\right]\{\hat{u}\} = \left\{\hat{F}\right\} \tag{5.121}$$

This system is complex valued and symmetric. It can be solved using either direct or iterative algorithms. The direct algorithms are the most general. However, they are costly because the algorithm operates on the full system of degrees of freedom. Table 5.1 provides an overview of the different numerical algorithms used to solve the aforementioned system together with their advantages and drawbacks. For more details, the reader can refer to Imbert (1991).

Table 5.1 Comparison between the different methods to solve linear system of equations

Method = f(size, loading, etc.)	
Iterative methods	*Direct methods*
• Gauss–Seidel	• Gauss elimination and variants
• Relaxation	• Cholesky factorization
• Conjugate gradients	☺ Excellent reliability
☺ Fast and efficient for high n_{eq}	☺ Multiple loadings
☹ Problem of convergence and stability	☺ Efficient (provided that it is suitable to the problem topology: sparse, band, etc.)
☹ Choice of the preconditioner	☹ High numerical cost $\cong n_{eq}^3$
☹ For complex problems, the number of operations can be unpredictable	

5.2.7.2 *Modal methods*

In the field of linear dynamics, the resolution is generally carried out using the modal superposition. This method allows for replacing the initial system by a reduced size system, by truncating the modal basis. This assumes that only the first eigenmodes contribute to the dynamic behavior of the structure. In particular, the excitation frequency spectrum must be limited to the first eigenmodes of the structure which is not always the case when shocks occur.

In the case of an undamped structure, the eigenmodes are real. For a damped structure, eigenmodes are complex. In the case of weakly damped structures, the eigenvalues and eigenvectors associated with the nondissipative structure are sufficiently close to the complex eigenmodes and can therefore be used to represent the solution. In general, the results of a classic modal analysis can only be used to calculate the dynamic response when the structure is weakly damped and when the behavior of the system is linear. Several methods can be used to solve the eigenvalue problem. The selection of a method depends on the size of the system (i.e., number of degrees of freedom of the system), desired number of eigenvalues, and the capacity of the algorithm to differentiate between close-by eigenvalues and determination of degenerate eigenvalues. In practice, subspace iteration and Lanczos-based methods are the two widely used and in particular block Lanczos due to its performance, accuracy, and robustness. A thorough discussion can be found in several textbooks (e.g., Craveur 1997).

5.2.7.2.1 *Undamped system*

Let us write the undamped problem as

$$\left(-\omega^2[M]+[K]\right)\{u\}=\{F\} \tag{5.122}$$

This system is real valued and symmetric. Its size is $n_{eq} \times n_{eq}$ where n_{eq} is the number of equations or degrees of freedom of the problem. This system, therefore, has n_{eq} real eigenvalues* $\omega_1^2,...,\omega_{n_{eq}}^2$. However, only the first ones have a physical sense. Moreover, the corresponding eigenvectors, $\{\Phi_{s,1}\},\ldots,\{\Phi_{s,n_{eq}}\}$,satisfy the following orthogonal equations:

$$\langle\Phi_{s,i}\rangle[M]\{\Phi_{s,j}\} = M_{m,i}\delta_{ij}; \quad \langle\Phi_{s,i}\rangle[K]\{\Phi_{s,j}\} = K_{m,i}\delta_{ij} \tag{5.123}$$

$M_{m,i}$ and $K_{m,i}$ are the generalized (modal) mass and stiffness of mode i. They are related by the Rayleigh quotient

$$\omega_i^2 = \frac{K_{m,i}}{M_{m,i}} \tag{5.124}$$

In general, the eigenvectors are orthonormalized with respect to matrix $[M]$ and thus, $M_{m,i}$ =1 and $K_{m,i} = \omega_i^2$.

If an eigenvalue has multiplicity m, the associated m eigenvectors are linearly independent and can always be orthogonalized.† In consequence, the eigenvectors of the system form a complete basis of order n_{eq} and any vector can be written as a unique linear superposition of these vectors. This is the basis of the modal superposition method.

For a given mesh, eigenfrequencies ω_i and eigenvectors $\{\Phi_{s,i}\}$ are just approximations of the exact values $\omega_{i,ex}$ and $\{\Phi_{s,i,ex}\}$. In consequence, the difference between the two vectors (approximation error) can be written in terms of the "exact" eigenvectors of the system:

$$\{\Phi_{s,i}\} = \{\Phi_{s,i,ex}\} + \varepsilon\sum_{i\neq j} a_j\{\Phi_{s,j,ex}\} \tag{5.125}$$

with a_j constants (participation coefficients) and ε, a small number that can be made as small as possible with the refinement of the mesh. Using the orthogonality property $\langle\Phi_{s,i}\rangle[K]\{\Phi_{s,j}\} = \omega_i^2\delta_{ij}$, it can be shown that (exercise):

$$\omega_i^2 = \omega_{i,ex}^2 + \varepsilon^2\sum_{i\neq j} a_j^2\omega_{j,ex}^2 \tag{5.126}$$

* We solve the associated free vibration system $(-\omega^2[M] + [K])\{u\} = \{0\}$.

† This is a consequence of the following theorem: "To a repeated eigenvalue ω_i^2 of system $[K]\{u\} - \omega^2[M]\{u\} = 0$ there correspond as many linearly independent eigenvectors as the degree of multiplicity of the eigenvalue."

This equation shows that for a given mesh, the approximation of the eigenfrequencies is more accurate compared to the associated mode shapes. It also shows that, in general, the eigenfrequencies are larger or equal than the exact values.

When only the first eigenmodes are excited, the resolution based on the modal superposition becomes interesting. Instead of working with the n_{eq} modes, it is sufficient to keep only a reduced number of modes n_m small compared to n_{eq} ($n_m \ll n_{eq}$) that allow for obtaining a converged solution. For better accuracy, all modes up to one-half or two times the highest forcing frequency should be retained.

Using classic eigenvectors calculation algorithms, we calculate the in vacuo normalized modal basis $\left([\Omega_s^2],[\Phi_s]\right)$ that is truncated at order n_m ($n_m \ll n_{eq}$):

$$\{u\} = [\Phi_s]\{u_m\}; \quad [\Phi_s]^T[K][\Phi_s] = \left[\Omega_s^2\right]; \quad [\Phi_s]^T[M][\Phi_s] = \left[I_{n_m}\right] \quad (5.127)$$

where $\{u_m\}$ is the vector of modal coordinates and

$$[\Phi_s] = \begin{bmatrix} \Phi_{s,1} & \Phi_{s,2} & \dots & \Phi_{s,n_m} \end{bmatrix} \quad (5.128)$$

the matrix containing the n_m first eigenmodes of the system, normalized with respect to $[M]$. It is a rectangular matrix of dimensions $n_{eq} \times n_m$ and

$$\left[\Omega_s^2\right] = \begin{bmatrix} \omega_{s,1}^2 & & & \\ & \omega_{s,2}^2 & & \\ & & \ddots & \\ & & & \omega_{s,n_m}^2 \end{bmatrix} \quad (5.129)$$

is a diagonal matrix of dimensions $n_m \times n_m$ whose coefficients are the squared circular eigenfrequencies of the system. Finally, $[I_{n_m}]$ denotes the identity matrix of dimensions $n_m \times n_m$.

To solve for the forced problem, we use a projection of Equation 5.122 on the truncated modal basis $\left([\Omega_s^2],[\Phi_s]\right)$. Using the orthogonality relationships between the modes, we get

$$\left[\left[\Omega_s^2\right] - \omega^2\left[I_{n_m}\right]\right]\{u_m\} = \{F_m\} \quad (5.130)$$

with

$$\{F_m\} = [\Phi_s]^T\{F\} \quad (5.131)$$

the generalized load vector acting on the structure. The solution is given by

$$\{u_m\} = \left[\left[\Omega_s^2\right] - \omega^2\left[I_{n_m}\right]\right]^{-1}\{F_m\} \tag{5.132}$$

or

$$\{u_m\} = \begin{Bmatrix} \dfrac{F_1}{\omega_1^2 - \omega^2} \\ \dfrac{F_2}{\omega_2^2 - \omega^2} \\ \cdots \\ \dfrac{F_{nm}}{\omega_{n_m}^2 - \omega^2} \end{Bmatrix} \tag{5.133}$$

The nodal vector $\{u\}$ can be recovered using Equation 5.127.

Remark

When the system has n_r rigid body modes, the previous steps still hold. Indeed, it is always possible to construct n_r rigid body modes that are mutually orthogonal with respect to $[M]$. Let us denote

$$\left[\Phi_{s,r}\right] = \left[\Phi_{s,r,1} \quad \Phi_{s,r,2} \quad \ldots \quad \Phi_{s,r,n_r}\right] \tag{5.134}$$

the matrix containing the rigid body modes. We have

$$\left[\Phi_{s,r}\right]^T[K]\left[\Phi_{s,r}\right] = 0; \quad \left[\Phi_{s,r}\right]^T[M]\left[\Phi_{s,r}\right] = \left[I_{n_r}\right] \tag{5.135}$$

In addition, the rigid body modes $[\Phi_{s,r}]$ and the elastic modes $[\Phi_{s,e}]$ are associated with distinct eigenvalues and are orthogonal to each other:

$$\left[\Phi_{s,r}\right]^T[K]\left[\Phi_{s,e}\right] = 0; \quad \left[\Phi_{s,r}\right]^T[M]\left[\Phi_{s,e}\right] = 0 \tag{5.136}$$

Let us split the contribution, to the response, of the rigid body modes and elastic modes:

$$\{u\} = \left[\Phi_{s,r}\right]\{u_r\} + \left[\Phi_{s,e}\right]\{u_m\} \tag{5.137}$$

where $\{u_r\}$ is the vector containing the modal contributions of the rigid body modes.

Equation 5.122 becomes

$$\left(-\omega^2[M][\Phi_{s,r}]+[K][\Phi_{s,r}]\right)\{u_r\}+\left(-\omega^2\,[M][\Phi_{s,e}]+[K][\Phi_{s,e}]\right)\{u_m\}=\{F\} \tag{5.138}$$

Premultiplying successively by $[\Phi_{s,e}]^T$ and $[\Phi_{s,r}]^T$ and taking into account the orthogonality relationships between modes, we get

$$\begin{cases}\left([\Omega_s^2]-\omega^2[I_{n_m-n_r}]\right)\{u_m\}=\{F_m\}\\ -\omega^2\{u_r\}=\{F_r\}\end{cases} \tag{5.139}$$

with

$$\{F_r\}=[\Phi_{s,r}]^T\{F\} \tag{5.140}$$

We clearly see that the equations corresponding to the elastic modes remain unchanged, whereas the contribution of the rigid body modes amounts to an inertial effect.

5.2.7.2.2 Damped system: Structural modal damping

In the case of a structural modal damping model for the system* $\{\eta_1,\eta_2,\ldots,\eta_{n_m}\}$, the solution Equation 5.132 is

$$\{\hat{u}_m\}=\left[[\hat{\Omega}_s^2]-\omega^2[I_{n_m}]\right]^{-1}\{\hat{F}_m\} \tag{5.141}$$

with

$$[\hat{\Omega}_s^2]=\begin{pmatrix}\omega_1^2(1+i\eta_1) & & & \\ & \omega_2^2(1+i\eta_2) & & \\ & & \ddots & \\ & & & \omega_{n_m}^2(1+i\eta_{n_m})\end{pmatrix} \tag{5.142}$$

* This is a classic damping model for an elastic structure. It consists in including the dissipation in Young's modulus for a structural problem $\hat{E}=E(1+i\eta_s)$ where η_s is the loss factor of the structure. For an acoustic problem, the bulk modulus of the fluid $\rho_0c_0^2$ is modified similarly: $\widehat{\rho_0c_0^2}=\rho_0c_0^2(1+i\eta_a)$. ρ_0, c_0, and η_a denote the fluid density, sound speed, and loss factor, respectively. Note that the loss factor can be either affected to the squared sound speed $\hat{c}_0^2=c_0^2(1+i\eta_a)$ (includes also dissipation due to thermal effects) or to the density (to specifically account for dissipation due to viscous effects).

or

$$\{\hat{u}_m\} = \begin{Bmatrix} \dfrac{\hat{F}_1}{\omega_1^2(1+i\eta_1) - \omega^2} \\ \dfrac{\hat{F}_2}{\omega_2^2(1+i\eta_2) - \omega^2} \\ \cdots \\ \dfrac{\hat{F}_{nm}}{\omega_{nm}^2(1+i\eta_{nm}) - \omega^2} \end{Bmatrix} \tag{5.143}$$

5.2.7.2.3 Damped system: General damping model

Usually, when damping is high, a direct solution strategy is used. This is, for instance, the case when a sound package is added to a structure or a cavity. However, when damping is low to moderate, a modal reduction method is still efficient. Starting from the general system with a general frequency-dependent damping function (cast in the form of viscous damping):

$$\left(-\omega^2[M] + i\omega\left[\hat{C}_d\right] + [K]\right)\{\hat{u}\} = \left\{\hat{F}\right\} \tag{5.144}$$

Two reduction methods can be used to solve such a system:

- Projection on the damped modal basis
- Projection on the undamped modal basis

Regarding the first method, the complex modal basis must be determined. The reader may refer to Géradin and Rixen (1997), for example, to obtain details about this method. Here, we limit ourselves to the second method.

The second approach assumes that the damping is small. Generally, the damping matrix coefficients are at least of an order of magnitude smaller than the stiffness and mass matrices coefficients (viscous damping, structural damping, radiation damping, etc.). The eigenvalues of the damped structure are then hardly different from the undamped structure eigenvalues. It is, therefore, relevant to seek the solution as a linear combination of undamped modes. The procedure is as follows. We project the system on the undamped normalized modal basis truncated at order n_m. Using Equation 5.127, we end up with a compact system of the form:

$$\left[\left[\Omega_s^2\right] + i\omega\left[\hat{C}_{d,m}(\omega)\right] - \omega^2\left[I_{n_m}\right]\right]\{\hat{u}_m\} = \left\{\hat{F}_m\right\} \tag{5.145}$$

where

$$\left[\hat{C}_{d,m}(\omega)\right] = [\Phi_s]^T \left[\hat{C}_d(\omega)\right][\Phi_s] \tag{5.146}$$

is the projected damping matrix. Note that in the general case, matrix $[\hat{C}_{d,m}(\omega)]$ is nondiagonal.*

System Equation 5.145 is a symmetric system of small size $(n_m \times n_m)$. It can be solved using a classic Gauss elimination algorithm.

The use of a truncated modal basis may lead to two types of errors:

- An insufficient number of modes kept in the truncated modal basis leading to an incomplete frequency response.
- An incomplete decomposition of the applied loading.

To improve the convergence of the solution, the modal acceleration method is often used. It consists in adding to the "dynamic" solution, the static contribution of unkept modes (Petyt 1990; Craveur 1997). Indeed, the modes beyond the last one kept in the truncated modal basis (omitted modes) respond in a quasi-static manner to an excitation with frequency components below their eigenfrequency. To improve the convergence, the modal solution must be able to reproduce the pseudo-static response:

$$\{\hat{u}_s\} = [K]^{-1}\{\hat{F}\} \tag{5.147}$$

Algebraic manipulations of Equation 5.145 show that the pseudo-static contribution of the kept modes is given by $[\Phi_s][\Omega_s^2]^{-1}\{\hat{F}_m\}$ (recall that rigid body modes do not contribute to stiffness and thus, only elastic modes are included here). Consequently, the corrected solution is given by

$$\{\hat{u}\} = [\Phi_s]\{\hat{u}_m\} + [K]^{-1}\{\hat{F}\} - [\Phi_s]\left[\Omega_s^2\right]^{-1}\{\hat{F}_m\} \tag{5.148}$$

The added correction (second and third terms) represents the static contribution of the omitted modes; it is referred to as the residual flexibility. It can be shown that this correction improves the convergence of the modal basis but with increased cost. An example is given in Section 5.5.2.

5.3 CONVERGENCE CONSIDERATIONS

There are basically three types of errors that can affect the quality of the solution: modeling error, discretization error, and numerical error.

* It becomes diagonal for proportional damping models where $[\hat{C}_d]$ can be written as $[\hat{C}_d] = \alpha[M] + \beta[K]$.

The modeling error arises from the description of the problem (geometry, material behavioral law, load and boundary conditions, and type of analysis). The discretization error is induced by the creation of the mesh and the approximation used for the geometry and the solution (type of elements used, mesh density, and accuracy of the geometry). The numerical error includes integration, round-off, or truncation and matrix-conditioning errors. These different types of errors have been discussed in detail in various FE books (Reddy 1993; Bathe 1996; Zienkiewicz and Taylor 2000) and hence will not be the subject of this book. We rather focus here on the discretization error in the context of vibroacoustics applications and more specifically on the selection of the mesh.

A question that arises in the FEM is how do we mesh a structure, a fluid cavity, or a porous medium to ensure that the solution is converged? Unfortunately, there is no universal answer to this question. It depends on the physical problem of interest, the formulation of the FE, mesh quality, excitation, frequency, type of desired vibroacoustic indicators, and so on. However, practically, if the mesh is kept refined until the change in the result of interest (displacement, stresses, strain, pressure, etc.) is less than a specified tolerance value, the optimal mesh will be achieved. Convergence refers to this process. In vibroacoustics, we are mainly interested in calculating structural displacements, acoustic pressure, or quadratic indicators such as mean square structural velocity or pressure, energies, injected powers, radiated powers, and dissipated powers. This involves two types of convergence: nodal values convergence and energy convergence. In both cases, the mesh is refined until the percentage of variation in the variable or the associated energy is smaller than a given accuracy.

Actually, the mesh chosen to represent the system is a matter of engineering experience and compromise between the solution accuracy and computational resources. The elements size should be kept small enough to yield good results but not too small to avoid prohibitive computational time and exceed the available memory size. Smaller elements should be used where the variables change rapidly. Larger elements can be used where the solution is expected not to vary much.

In practice, the choice of the number of elements is dictated by the type of element (linear, quadratic, etc.) and the maximum frequency of interest. Typically, in vibroacoustics application, a commonly accepted convergence criterion is that for structural and fluid domains, at least six elements (respectively four elements) per wavelength should be used for linear elements (respectively quadratic) (see also Section 7.9). Everything being equal, quadratic elements should be preferred to linear elements. For a structural domain, the question of which wavelength should be selected for evaluating the convergence criterion arises, since a structure can support many kinds of waves (bending, compression, shear, etc.). This requires an examination of the physics of the problem prior to building the mesh. When poroelastic materials are involved, the physics is even more complicated because the material is biphasic and dissipative. There is no single criterion and trial-and-error meshing should be considered. A recommendation would be to start with 10–12 elements per wavelength where the wavelength is chosen as the smallest of the three Biot waves (P1, P2, or S) if nothing is known about the physical behavior of the system. The practice is to start with a mesh that has been chosen based on the aforementioned convergence criterion and to refine the mesh to check that it is converged in the sense defined by the user. Note that in vibroacoustics, for a given frequency of calculation, the mesh density of structures, fluid cavities, and poroelastic media can be very different since the wavelengths in each medium are different. Then, the question is how to connect meshes of various densities at interfaces? This is discussed in Section 8.7.

5.4 CALCULATION OF ACOUSTIC AND VIBRATORY INDICATORS

This section provides the reader with the expressions of the acoustic and vibratory indicators that are commonly used in the field of vibroacoustics. Both the definitions together with the expression of the indicator in terms of physical nodal values are presented.

5.4.1 Kinetic energy

5.4.1.1 Case of a structure

The time-averaged kinetic energy of a structure occupying a domain Ω is given by

$$\begin{aligned} E_c(\omega) &= \frac{1}{4}\int_{\Omega} \rho_s \underline{v} \cdot \underline{v}^* \, dV \\ &= \frac{\omega^2}{4}\langle \hat{u}^* \rangle [M] \{\hat{u}\} \end{aligned} \tag{5.149}$$

where $[M]$ is the structural mass matrix.

$$[M] = \int_{\Omega} \rho_s \left[N^s\right]^T \left[N^s\right] dV \tag{5.150}$$

$[N^s]$ is a matrix of nodal shape functions used to interpolate displacement $\underline{u}$. The dimension of $[N^s]$ is $3 \times nnt$.

For a thin structure, this is

$$E_c(\omega) = \frac{1}{4}\int_S m_s \underline{v} \cdot \underline{v}^* dS = \frac{\omega^2}{4}\langle \hat{u}^* \rangle [M]\{\hat{u}\} \tag{5.151}$$

where $m_s = \rho_s h_s$ is the mass per unit area of the structure of thickness h_s and density ρ_s, and $[M]$ is the structural mass matrix.

$$[M] = \int_S m_s \left[N^s\right]^T \left[N^s\right] dS \tag{5.152}$$

$[N^s]$ is a matrix of surface nodal shape functions used to interpolate displacement $\underline{u}$ on surface S. The dimension of $[N^s]$ is $3 \times nntS$ where $nntS$ is the number of nodes on surface S.

5.4.1.2 Case of a fluid

Assume that $\hat{p}$ is the acoustic pressure inside a fluid domain Ω. The time-averaged kinetic energy stored in the fluid reads as

$$\begin{aligned} E_c(\omega) &= \frac{1}{4\rho_0\omega^2}\int_\Omega \underline{\nabla}\hat{p} \cdot \underline{\nabla}\hat{p}^* dV \\ &= \frac{1}{4\omega^2}\langle \hat{p}^* \rangle [H]\{\hat{p}\} \end{aligned} \tag{5.153}$$

where $[H]$ is the kinetic energy matrix defined in Equation 5.72.

5.4.1.3 Case of a poroelastic material

Using the notations of Section 2.4, the time-averaged kinetic energy of the porous material associated with the solid phase is given by

$$\begin{aligned} E_c^s(\omega) &= \frac{1}{4}\Re\left(\int_\Omega \tilde{\rho}\hat{\underline{u}}^s \cdot \hat{\underline{u}}^{s*} dV\right) \\ &= \frac{1}{4}\Re\left(\langle \hat{u}^{s*} \rangle [\tilde{M}]\{\hat{u}^s\}\right) \end{aligned} \tag{5.154}$$

where $\hat{\underline{u}}^s$ is the solid phase displacement field and $[\tilde{M}]$ is the solid phase mass matrix:

$$[\tilde{M}] = \int_{\Omega} \tilde{\rho}\left[N^s\right]^T\left[N^s\right] dV \tag{5.155}$$

where $[N^s]$ is a matrix of nodal shape functions used to interpolate the solid-phase displacement $\underline{\hat{u}}^s$. Its size is $3 \times nnt$.

Let $\hat{p}^f$ be the fluid phase interstitial pressure field of the porous material; then, the time-averaged kinetic energy associated with the fluid phase is given by

$$\begin{aligned} E_c^f(\omega) &= \frac{1}{4\omega^2}\Re\left(\int_{\Omega} \frac{\phi^2}{\tilde{\rho}_{22}} \underline{\nabla}\hat{p}^{f*} \cdot \underline{\nabla}\hat{p}^f \, dV\right) \\ &= \frac{1}{4\omega^2}\left(\left\langle \hat{p}^{f*}\right\rangle [\tilde{H}]\left\{\hat{p}^f\right\}\right) \end{aligned} \tag{5.156}$$

where $[\tilde{H}]$ is the fluid phase kinetic energy matrix.

$$[\tilde{H}] = \int_{\Omega} \frac{\phi^2}{\tilde{\rho}_{22}}\left[\underline{\nabla}\left\langle N^f\right\rangle\right]^T\left[\underline{\nabla}\left\langle N^f\right\rangle\right] dV \tag{5.157}$$

$\{N^f\}$ is a vector of nodal shape functions used to interpolate the fluid phase interstitial pressure $\hat{p}^f$. The size of $\{N^f\}$ is nnt.

5.4.2 Strain energy

5.4.2.1 *Case of a structure*

The time-averaged strain energy of a structure occupying a domain Ω is given by

$$\begin{aligned} E_d(\omega) &= \frac{1}{4}\Re\left(\int_{\Omega} \underline{\underline{\sigma}}(\underline{\hat{u}}) : \underline{\underline{\varepsilon}}\left(\underline{\hat{u}}^*\right) dV\right) \\ &= \frac{1}{4}\Re\left(\left\langle \hat{u}^*\right\rangle [K]\{\hat{u}\}\right) \end{aligned} \tag{5.158}$$

where $[K]$ is the structural stiffness matrix.

$$[K] = \int_{\Omega} \left(\left[L^s\right]\left[N^s\right]\right)^T \left[D^s\right]\left[L^s\right]\left[N^s\right] dV \tag{5.159}$$

where $[L^s]$ is a differential operator allowing to obtain the strain from the displacement (dimension $6 \times nnt$) and $[D^s]$ is the constitutive matrix of the material (dimension 6×6). For an isotropic material, it is given by

$$[D^s] = \begin{bmatrix} 2G\frac{1-\nu}{1-2\nu} & 2G\frac{\nu}{1-2\nu} & 2G\frac{\nu}{1-2\nu} & 0 & 0 & 0 \\ 2G\frac{\nu}{1-2\nu} & 2G\frac{1-\nu}{1-2\nu} & 2G\frac{\nu}{1-2\nu} & 0 & 0 & 0 \\ 2G\frac{\nu}{1-2\nu} & 2G\frac{\nu}{1-2\nu} & 2G\frac{1-\nu}{1-2\nu} & 0 & 0 & 0 \\ 0 & 0 & 0 & G & 0 & 0 \\ 0 & 0 & 0 & 0 & G & 0 \\ 0 & 0 & 0 & 0 & 0 & G \end{bmatrix} \tag{5.160}$$

where G is the shear modulus and ν is Poisson's ratio.

5.4.2.2 *Case of a fluid*

The time-averaged compression energy of a fluid is given by

$$\begin{aligned} E_d(\omega) &= \frac{1}{4\rho_0 c_0^2} \int_\Omega |\hat{p}|^2 \, dV \\ &= \frac{1}{4} \langle \hat{p}^* \rangle [Q] \{\hat{p}\} \end{aligned} \tag{5.161}$$

where $[Q]$ is the compression energy matrix defined in Equation 5.73.

5.4.2.3 *Case of a poroelastic material*

Using the notations of Section 5.2.4, the time-averaged strain energy of the solid phase is

$$\begin{aligned} E_d^s(\omega) &= \frac{1}{4} \int_\Omega \underline{\underline{\tilde{\sigma}}}^s\left(\underline{\hat{u}}^s\right) : \underline{\underline{\varepsilon}}^s\left(\underline{\hat{u}}^{s*}\right) dV \\ &= \frac{1}{4}\left(\langle \hat{u}^{s*} \rangle [K^p] \{\hat{u}^s\}\right) \end{aligned} \tag{5.162}$$

$[K^p]$ is the stiffness matrix of the solid phase of the porous material. It is similar to Equation 5.159.

The time-averaged compression energy of the fluid phase is given by

$$E_d^f(\omega) = \frac{1}{4}\int_\Omega \frac{\phi_p^2}{\tilde{R}}\left|\hat{p}^f\right|^2 dV$$
$$= \frac{1}{4}\left(\langle \hat{p}^{f*}\rangle [\tilde{Q}]\{\hat{p}^f\}\right) \tag{5.163}$$

where $\hat{p}^f$ is the fluid phase pressure field and $[\tilde{Q}]$ is the compression energy matrix of the fluid phase:

$$[\tilde{Q}] = \int_\Omega \frac{\phi_p^2}{\tilde{R}}\{N^f\}\langle N^f\rangle dV \tag{5.164}$$

5.4.3 Dissipated power

5.4.3.1 Case of the structure

The time-averaged power dissipated in a structure occupying a domain Ω is given by

$$\Pi_d(\omega) = \frac{1}{2}\Im\left(\omega\int_\Omega \underline{\underline{\sigma}}(\hat{\underline{u}}) : \underline{\underline{\varepsilon}}(\hat{\underline{u}}^*)dV\right)$$
$$= \frac{1}{2}\Im\left(\omega\langle \hat{u}^*\rangle [K]\{\hat{u}\}\right) \tag{5.165}$$

The above expression assumes a structural damping model. In the case of viscous damping, the dissipated power is given by

$$\Pi_d(\omega) = \frac{1}{2}\Re\left(\omega^2\langle \hat{u}^*\rangle [\hat{C}_d]\{\hat{u}\}\right) \tag{5.166}$$

5.4.3.2 Case of a fluid

When the dissipation is included in the squared sound speed ($c_0^2 \to \hat{c}_0^2 = c_0^2(1 + i\eta_a)$) where η_a is the structural loss factor of the fluid, the time-averaged power dissipated in a fluid occupying a domain Ω is given by

$$\Pi_d(\omega) = \frac{1}{2}\Im\left(-\omega\int_\Omega \frac{\left|\hat{p}\right|^2}{\rho_0\hat{c}_0^2}dV\right)$$
$$= \frac{1}{2}\Im\left(-\omega\langle \hat{p}^*\rangle \left[\hat{Q}\right]\{\hat{p}\}\right) \tag{5.167}$$

If the dissipation is included in the fluid density while keeping the compressibility real valued ($\rho_0 \rightarrow \hat{\rho}_0 = \rho_0(1 - i\eta_a)$), the time-averaged power dissipated in the fluid is given by

$$\Pi_d(\omega) = -\frac{1}{2}\Im\left(\int_\Omega \frac{|\nabla \hat{p}|^2}{\hat{\rho}_0\omega} dV\right)$$
$$= -\frac{1}{2}\Im\left(\frac{1}{\omega}\langle \hat{p}^*\rangle[\hat{H}]\{\hat{p}\}\right) \tag{5.168}$$

5.4.3.3 Case of a poroelastic material

In a porous material, the power can be dissipated by structural damping in the skeleton, by thermal effects, and by viscous effects. The time-averaged power dissipated by structural damping in a porous medium (Dazel et al. 2008) occupying a domain Ω is given by

$$\Pi_d(\omega) = \frac{1}{2}\Im\left(\omega\int_\Omega \underline{\underline{\tilde{\sigma}}}^s\left(\underline{\hat{u}}^s\right) : \underline{\underline{\varepsilon}}^s\left(\underline{\hat{u}}^{s*}\right) dV\right)$$
$$= \frac{1}{2}\Im\left(\omega\langle \hat{u}^{s*}\rangle[K^p]\{\hat{u}^s\}\right) \tag{5.169}$$

The time-averaged power dissipated by thermal effects in a porous medium occupying a domain Ω is given by

$$\Pi_{d,t}(\omega) = \frac{1}{2}\Im\left(-\omega\int_\Omega \frac{\phi_p^2}{\tilde{R}}\left|\hat{p}^f\right|^2 dV\right)$$
$$= \frac{1}{2}\Im\left(\omega\langle \hat{p}^{f*}\rangle[\tilde{Q}]\{\hat{p}^f\}\right) \tag{5.170}$$

The time-averaged power dissipated by viscous effects in a porous medium occupying a domain Ω is given by

$$\Pi_{d,v}(\omega) = \frac{1}{2}\Im\left(\frac{1}{\omega}\int_\Omega \frac{\phi_p^2}{\tilde{\rho}_{22}}\underline{\nabla p}^{f*}\cdot\underline{\nabla p}^f dV - \omega^3\int_\Omega \tilde{\rho}\underline{u}^s\cdot\underline{u}^{s*} dV - \omega\int_\Omega \frac{2\phi_p^2}{\tilde{\alpha}}\Re\left(\underline{\nabla p}^f\cdot\underline{\hat{u}}^{s*}\right) dV\right)$$
$$= \frac{1}{2}\Im\left(\frac{1}{\omega}\langle \hat{p}^{f*}\rangle[\tilde{H}]\{\hat{p}^f\} - \omega^3\langle \hat{u}^{s*}\rangle[\tilde{M}]\{\hat{u}^s\} - 2\tilde{\gamma}\omega\langle \hat{u}^{s*}\rangle\left[C^{(1)}_{u^s p^f}\right]\{\hat{p}^f\}\right) \tag{5.171}$$

where $\left[C^{(1)}_{u^s p^f}\right]$ is a volume-coupling matrix between the solid phase and the fluid phase of the porous material given by

$$\left[C^{(1)}_{u^s p^f}\right] = \int_{\Omega} \left[N^s\right]^T \underline{\nabla}\left[N^f\right] dV \tag{5.172}$$

5.4.4 Mean square velocity

5.4.4.1 *Case of a structure*

Assume that the structure occupying volume Ω vibrates with a displacement field $\underline{\hat{u}}$ (velocity field $\underline{\hat{v}} = i\omega\underline{\hat{u}}$). The mean (i.e., time and space averaged) square velocity of the structure in direction i is given by

$$\begin{aligned}
\left\langle V_i^2 \right\rangle(\omega) &= \frac{1}{2\Omega}\int_{\Omega} |\hat{v}_i|^2 \, dV \\
&= \frac{1}{2\Omega}\left(\left\langle \hat{v}^* \right\rangle [L_i]^T \left[C^{(1)}_{uu}\right][L_i]\{\hat{v}\}\right) \\
&= \frac{\omega^2}{2\Omega}\left(\left\langle \hat{u}^* \right\rangle [L_i]^T \left[C^{(1)}_{uu}\right][L_i]\{\hat{u}\}\right)
\end{aligned} \tag{5.173}$$

where $\{\hat{u}\}$ is the nodal displacement vector, $[L_i]$ is a localization matrix for the degrees of freedom in direction i, and $\left[C^{(1)}_{uu}\right]$ is a volume-coupling matrix given by

$$\left[C^{(1)}_{uu}\right] = \int_{\Omega} \left[N^s\right]^T \left[N^s\right] dV \tag{5.174}$$

The mean square velocity in direction i can also be defined on a surface S as

$$\begin{aligned}
\left\langle V_i^2 \right\rangle(\omega) &= \frac{1}{2S}\int_{S} |\hat{v}_i|^2 \, dS \\
&= \frac{\omega^2}{2S}\left(\left\langle \hat{u}^* \right\rangle [L_i]^T \left[C^{(2)}_{uu}\right][L_i]\{\hat{u}\}\right)
\end{aligned} \tag{5.175}$$

$[C^{(2)}_{uu}]$ is a surface-coupling matrix given by

$$\left[C^{(2)}_{uu}\right] = \int_{S} \left[N^s\right]^T \left[N^s\right] dS \tag{5.176}$$

where $[N^s]$ is a matrix of surface nodal shape functions used to interpolate displacement $\underline{u}$ on surface S. Its size is $3 \times nnt_s$.

We can also define a mean square normal velocity of surface S as

$$\begin{aligned}\left\langle V_n^2 \right\rangle(\omega) &= \frac{1}{2S}\int_S \left|\underline{\hat{v}} \cdot \underline{n}\right|^2 dS \\ &= \frac{\omega^2}{2S}\left(\left\langle \hat{u}^* \right\rangle \left[C_{uu}^{(3)}\right]\{\hat{u}\}\right)\end{aligned} \tag{5.177}$$

where $\underline{n}$ is the normal to surface S and $\left[C_{uu}^{(3)}\right]$ is a surface-coupling matrix given by

$$\left[C_{uu}^{(3)}\right] = \int_S \left[N^s\right]^T \langle n \rangle \{n\} \left[N^s\right] dS \tag{5.178}$$

5.4.4.2 *Case of a poroelastic material*

For a poroelastic material, the mean square velocity of the solid phase in direction i is given by

$$\begin{aligned}\left\langle V_i^2 \right\rangle(\omega) &= \frac{1}{2\Omega}\int_\Omega \left|\hat{v}_i^s\right|^2 dV \\ &= \frac{\omega^2}{2\Omega}\left(\left\langle \hat{u}^{s*} \right\rangle [L_i]^T \left[C_{u^s u^s}\right][L_i]\{\hat{u}^s\}\right)\end{aligned} \tag{5.179}$$

with

$$\left[C_{u^s u^s}\right] = \int_\Omega \left[N^s\right]^T \left[N^s\right] dV \tag{5.180}$$

5.4.5 Mean square pressure

5.4.5.1 *Case of a fluid*

The mean square pressure inside Ω is given by

$$\begin{aligned}\left\langle p^2 \right\rangle(\omega) &= \frac{1}{2\Omega}\int_\Omega \left|\hat{p}\right|^2 dV \\ &= \frac{1}{2\Omega}\left(\left\langle \hat{p}^* \right\rangle \left[C_{pp}\right]\{\hat{p}\}\right)\end{aligned} \tag{5.181}$$

where $[C_{pp}]$ is a volume-coupling matrix given by

$$[C_{pp}] = \int_{\Omega} \{N^f\}\langle N^f\rangle dV \tag{5.182}$$

5.4.5.2 Case of a poroelastic material

Using the notations of Section 2.4, the mean square interstitial pressure inside a poroelastic domain Ω is given by

$$\begin{aligned}\langle p^2\rangle(\omega) &= \frac{1}{2\Omega}\int_{\Omega} \left|\hat{p}^f\right|^2 dV \\ &= \frac{1}{2\Omega}\left(\langle \hat{p}^{f*}\rangle\left[C_{p^f p^f}\right]\{\hat{p}^f\}\right)\end{aligned} \tag{5.183}$$

Note that for a homogeneous material $[C_{p^f p^f}]$ can be expressed in terms of the compression energy matrix of the fluid phase $[\tilde{Q}]$ (see Equation 5.164) as

$$\left[C_{p^f p^f}\right] = \frac{\tilde{R}}{\phi_p^2}\left[\tilde{Q}\right] \tag{5.184}$$

5.4.6 Power injected to a structure

5.4.6.1 Case of a mechanical excitation

The power injected by a force field per unit area applied on the structure is given by

$$\begin{aligned}\Pi_{in} &= \frac{1}{2}\Re\left(\int_S \underline{\hat{F}} \cdot \underline{\hat{v}}^* dS\right) \\ &= \frac{1}{2}\Re\left(-i\omega\langle \hat{F}\rangle\left[C_{uu}^{(1)}\right]\{u^*\}\right)\end{aligned} \tag{5.185}$$

5.4.6.2 Case of an acoustical excitation

The power injected by an acoustic pressure field $\bar{p}$ applied on the structure is given by

$$\Pi_{in} = \frac{1}{2}\Re\left(\int_S \bar{p}\underline{n}\cdot\hat{\underline{v}}^* dS\right)$$
$$= \frac{1}{2}\Re\left(-i\omega\langle\bar{p}\rangle[C_{up}]^T\{\hat{u}^*\}\right) \quad (5.186)$$

where $[C_{up}]$ is a surface fluid-structure-coupling matrix given by

$$[C_{up}] = \int_S [N^u]\{n\}\langle N^p\rangle dS \quad (5.187)$$

5.4.7 Radiated power

The sound power radiated by a vibrating structure that generates a radiated sound pressure $\hat{p}$ is given by (see Chapter 8):

$$\Pi_{rad} = \frac{1}{2}\Re\left(\int_S \hat{p}(\hat{\underline{v}}.\underline{n})^* dS\right)$$
$$= \frac{1}{2}\Re\left(-i\omega\langle\hat{p}\rangle[C_{up}]^T\{\hat{u}^*\}\right)$$
$$= \frac{1}{2}\Re\left(\omega^2\langle\hat{u}\rangle[\hat{Z}_{rad}]\{\hat{u}^*\}\right) \quad (5.188)$$

where $[\hat{Z}_{rad}]$ is the radiation impedance matrix of the structure (see Chapter 8).

5.4.8 Radiation efficiency

The radiation efficiency is a measure of the ability of the structure to radiate sound. It is defined as

$$\sigma_{rad} = \frac{\Pi_{rad}}{\rho_0 c_0 S\langle V_n^2\rangle} \quad (5.189)$$

$\rho_0 cS\langle V_n^2\rangle$ corresponds to the acoustic power radiated by a piston of normal mean square velocity $\langle V_n^2\rangle$. Note that σ_{rad} can be >1.

5.4.9 Sound transmission loss

The sound transmission loss of a structure denoted as STL is an indicator of its sound insulation capacity when it is subjected to an acoustic excitation. It is defined as

$$STL = 10\log\left(\frac{1}{\tau}\right) \tag{5.190}$$

where τ is the transmission factor given by

$$\tau = \frac{\Pi_t}{\Pi_{inc}} \tag{5.191}$$

with Π_t the transmitted sound power. This power can be calculated from Equation 5.188. Π_{inc}* is the incident sound power. The incident sound power is the intensity that flows through the surface of the structure as if the latter was absent. For an incident plane wave of complex amplitude $\hat{A}_i$ and incident angles (θ_i,ϕ_i) impinging on a plane surface S with an angle θ_i with respect to the normal to the surface

$$\Pi_{inc}(\theta_i,\phi_i) = \frac{|\hat{A}_i|^2 S\cos\theta_i}{2\rho_0 c_0} \tag{5.192}$$

For a diffuse field,

$$\Pi_{inc,d} = \int_0^{2\pi}\int_0^{\pi/2} \Pi_{inc}(\theta_i,\phi_i)\sin\theta_i \, d\theta_i \, d\phi_i \tag{5.193}$$

For a plane surface, we have

$$\begin{aligned}\Pi_{inc,d} &= \int_0^{2\pi}\int_0^{\pi/2} \frac{|A_i|^2 S\cos\theta_i}{2\rho_0 c_0}\sin\theta_i \, d\theta_i \, d\phi_i \\ &= \frac{|\hat{A}_i|^2 S\pi}{2\rho_0 c_0}\end{aligned} \tag{5.194}$$

5.5 EXAMPLES OF APPLICATIONS

5.5.1 Imposition of the boundary conditions for a cantilever bar

Let us consider the static equilibrium of a cantilever bar of cross section S subjected to axial load (see Figure 5.6). The bar is discretized with three two-noded bar elements. Let $K = ES/L$ be the axial rigidity of the bar.

* It should not be confused with the injected power that accounts for the presence of the structure.

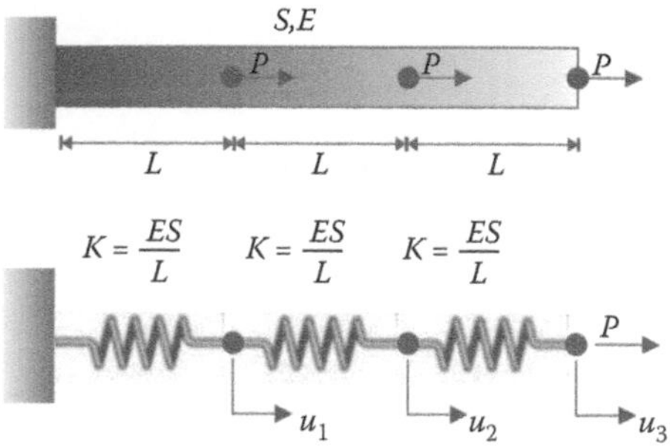

Figure 5.6 Example of a cantilever bar subjected to axial load.

The system governing the problem after imposition of the boundary condition $u(0) = 0$ is

$$\begin{bmatrix} 2K & -K & 0 \\ -K & 2K & -K \\ 0 & -K & K \end{bmatrix} \begin{Bmatrix} u_1 \\ u_2 \\ u_3 \end{Bmatrix} = \begin{Bmatrix} P \\ P \\ P \end{Bmatrix} \tag{5.195}$$

Let us assume that we want to impose $u_2 = u_3$ (the case where the element 2–3 is a rigid link) using the three methods discussed in Section 5.2.5.

In Lagrange's multiplier technique, the condition $u_2 = u_3$ is equivalent to

$$[C_\lambda]\{u\} = \langle 0, 1, -1 \rangle \begin{Bmatrix} u_1 \\ u_2 \\ u_3 \end{Bmatrix} = 0 \tag{5.196}$$

System Equation 5.88 becomes

$$\begin{bmatrix} 2K & -K & 0 & 0 \\ -K & 2K & -K & 1 \\ 0 & -K & K & -1 \\ 0 & 1 & -1 & 0 \end{bmatrix} \begin{Bmatrix} u_1 \\ u_2 \\ u_3 \\ \lambda \end{Bmatrix} = \begin{Bmatrix} P \\ P \\ P \\ 0 \end{Bmatrix} \tag{5.197}$$

The solution is given by

$$\begin{Bmatrix} u_1 \\ u_2 \\ u_3 \\ \lambda \end{Bmatrix} = \begin{Bmatrix} 3P/K \\ 5P/K \\ 5P/K \\ -P \end{Bmatrix} \tag{5.198}$$

We can verify that

$$-[C_\lambda]^T\{\lambda\} = -\begin{Bmatrix} 0 \\ 1 \\ -1 \end{Bmatrix}(-P) = \begin{Bmatrix} 0 \\ +P \\ -P \end{Bmatrix} \tag{5.199}$$

is the force vector that must be exerted at the nodes to have a rigid link between nodes 2 and 3.

Using the partitioning method, let

$$\{u\} = \begin{Bmatrix} u_1 \\ u_2 \\ u_3 \end{Bmatrix}; \quad \{u_k\} = \begin{Bmatrix} u_1 \\ u_2 \end{Bmatrix}; \quad \{u_c\} = u_3 \tag{5.200}$$

The reduction matrix is given by

$$\{u\} = \begin{Bmatrix} u_1 \\ u_2 \\ u_3 \end{Bmatrix} = \underbrace{\begin{bmatrix} 1 & 0 \\ 0 & 1 \\ 0 & 1 \end{bmatrix}}_{[G]} \begin{Bmatrix} u_1 \\ u_2 \end{Bmatrix} \tag{5.201}$$

Furthermore, vector $\{\hat{h}\}=\{0\}$ and consequently $\{\hat{h}_r\}=\{0\}$ and $\{\hat{r}\}=\{0\}$. Therefore,

$$\left[\hat{Z}_r\right] = \begin{bmatrix} 1 & 0 & 0 \\ 0 & 1 & 1 \end{bmatrix} \begin{bmatrix} 2K & -K & 0 \\ -K & 2K & -K \\ 0 & -K & K \end{bmatrix} \begin{bmatrix} 1 & 0 \\ 0 & 1 \\ 0 & 1 \end{bmatrix} = \begin{bmatrix} 2K & -K \\ -K & K \end{bmatrix} \tag{5.202}$$

and

$$\{\hat{f}_r\} = [G]^T\{\hat{f}\} = \begin{bmatrix} 1 & 0 & 0 \\ 0 & 1 & 1 \end{bmatrix} \begin{bmatrix} P \\ P \\ P \end{bmatrix} = \begin{Bmatrix} P \\ 2P \end{Bmatrix} \tag{5.203}$$

The reduced system then is

$$\begin{bmatrix} 2K & -K \\ -K & K \end{bmatrix} \begin{Bmatrix} u_1 \\ u_2 \end{Bmatrix} = \begin{Bmatrix} P \\ 2P \end{Bmatrix} \tag{5.204}$$

We verify that the solution is $\begin{Bmatrix} u_1 \\ u_2 \end{Bmatrix} = \begin{Bmatrix} 3P/K \\ 5P/K \end{Bmatrix}$ with $u_3 = u_2$.

Finally, in the penalty method, the imposition of the constraint $u_2 = u_3$ leads to the system:

$$\begin{bmatrix} 2K & -K & 0 \\ -K & 2K+\varpi & -K-\varpi \\ 0 & -K-\varpi & K+\varpi \end{bmatrix} \begin{Bmatrix} u_1 \\ u_2 \\ u_3 \end{Bmatrix} = \begin{Bmatrix} P \\ P \\ P \end{Bmatrix} \tag{5.205}$$

The resolution of this system yields

$$\begin{Bmatrix} u_1 \\ u_2 \\ u_3 \end{Bmatrix} = \begin{Bmatrix} 3P/K \\ 5P/K \\ 5P/K + P/(K+\varpi) \end{Bmatrix} \tag{5.206}$$

Thus, by choosing ϖ very large compared to the stiffness K, the exact solution of the problem is recovered.

5.5.2 Application to the flexural vibrations of a beam: Pseudo-static corrections

This example considers the flexural vibration of a clamped-free beam of length L, bending stiffness EI, density ρ_s, and height h_b excited at its center to illustrate the modal acceleration method. Figure 5.7 depicts the geometry of the problem.

An analytical calculation shows that the input mobility of the system is given by

$$\hat{H} = \frac{\hat{v}(x_1 = L)}{\hat{F}(x_1 = L)} = i\omega \frac{\sinh(kL)\cos(kL) - \cosh(kL)\sin(kL)}{\big(1 + \cosh(k\mathrm{L})\cos(k\mathrm{L})\big)k^3 EI} \tag{5.207}$$

where $k^4 = \rho_s h_b \omega^2 / EI$ is the wavenumber in the beam.

Figure 5.7 A clamped-free beam excited at its tip.

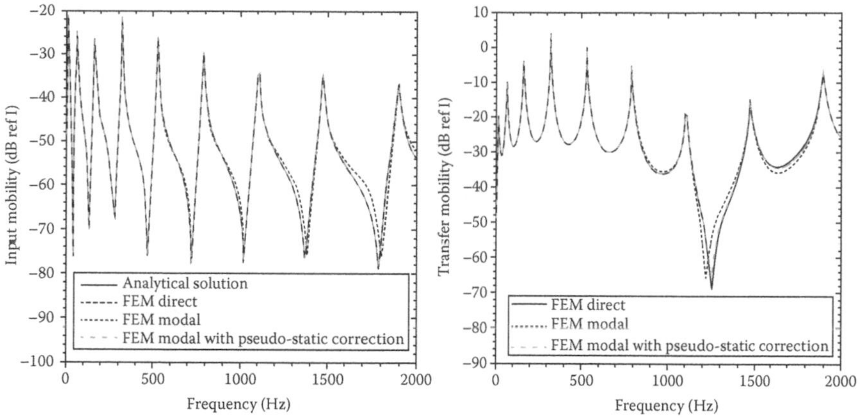

Figure 5.8 Comparisons between the analytical approach and the FE model: effect of the pseudo-static correction.

The solution is compared here to the FEM solution. Following Euler–Bernoulli's assumptions, the elementary mass and stiffness matrices $[M_e]$ and $[K_e]$ of a thin beam (mass per unit length m_s, length L, and flexural rigidity EI) are given by (Petyt 1990)

$$[M_e] = \frac{m_s L}{420}\begin{bmatrix} 156 & 22L & 54 & -13L \\ & 4L^2 & 13L & -3L^2 \\ & & 156 & -22L \\ & & & 4L^2 \end{bmatrix} \quad [K_e] = \frac{EI}{L^3}\begin{bmatrix} 12 & 6L & -12 & 6L \\ & 4L^2 & -6L & 2L^2 \\ & & 12 & -6L \\ & & & 4L^2 \end{bmatrix} \tag{5.208}$$

Figure 5.8 compares the obtained response for two locations: the excitation position (input mobility) and a location on the beam (transfer mobility). The studied beam is made from steel with length $L = 30$ cm and a height $h_b = 1$ mm. In the FE solution, six elements per flexural wavelength were used (38 elements) and all modes up to 1.2 times the maximum frequency of interest were kept (10 modes). It is observed that, in this example, the pseudo-static correction mainly improves the convergence of the response at the antiresonances of the system.

5.5.3 Application to the flexural vibrations of a beam: Lagrange's multipliers

This second example illustrates the use of Lagrange multipliers. It considers a beam with an imposed normal displacement at its center. Figure 5.9

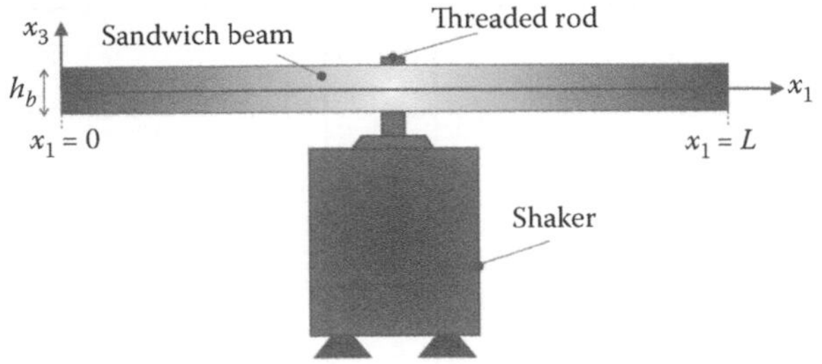

Figure 5.9 Sandwich beam excited at its center by a mechanical shaker.

depicts the geometry of a typical practical configuration. This setup is used to identify the mechanical properties of materials (a variant of Oberst's beam technique).

An analytical calculation shows that the ratio of the transverse displacement of a point x of the beam to the imposed displacement is given by

$$\hat{H}(x_1,\omega) = \frac{1}{2}\frac{\cosh(kL/2)+\cos(kL/2)}{1+\cosh(kL/2)\cos(kL/2)}\left[\cosh(kx_1)+\cos(kx_1)\right] + \frac{1}{2}\frac{\sinh(kL/2)-\sin(kL/2)}{1+\cosh(kL/2)\cos(kL/2)}\left[\sinh(kx_1)+\sin(kx_1)\right] \quad (5.209)$$

where $k^4 = \rho_s h_b \omega^2 / EI$ is the wavenumber in the beam.

The resonance frequencies of the system are given by the solutions of

$$1 + \cosh(kL/2)\cos(kL/2) = 0 \quad (5.210)$$

They are identical to those of a clamped-free beam of half-length $L/2$. The natural frequencies and modes are given by

$$\omega_n = \frac{\lambda_n^2}{L^2}\sqrt{\frac{Eh_b^3}{12\rho_s h_b}}$$
$$\phi_n(x) = \left[\cos(k_n x) + \cosh(k_n x)\right] + \gamma_n\left[\sin(k_n x) + \sinh(k_n x)\right] \quad (5.211)$$

where

$$k_n = \frac{\lambda_n}{L},\ \cos(\lambda_n)\cosh(\lambda_n) = 0 \quad \text{and} \quad \gamma_n = \frac{\sin(\lambda_n)+\sinh(\lambda_n)}{\cos(\lambda_n)-\cosh(\lambda_n)}$$

Table 5.2 Parameters of the first elastic modes of the beam

n	3	4	5	6	7	8	9
λ_n	4.730	7.853	10.996	14.137	17.279	20.420	23.562
γ_n	−0.9825	−1.008	−1	−1	−1	−1	−1

λ_n are coefficients that are given in Table 5.2 for the first nine elastic modes. For large values of n, $\lambda_n \approx (n-1.5)\pi$.

There are two rigid body modes. They correspond to the translation and rotation body modes. Their normalized values $\left(\int_0^L \phi_n^2 dx = 1\right)$ are given by

$$\begin{cases} \phi_1(x) = 1 \\ \phi_2(x) = 2\sqrt{3}\left(\dfrac{x}{L} - 1\right) \end{cases} \tag{5.212}$$

The rigid body displacement can be written as $w_0(x) = a_1\phi_1(x) + a_2\phi_2(x)$.

Using the modal superposition technique, the displacement field of the beam is

$$\hat{w}(x) = \sum_m \hat{a}_m \phi_m(x) \tag{5.213}$$

The imposition of the boundary condition $\hat{w}(0) = \hat{w}_0$ implies that

$$\sum_m \hat{a}_m \phi_m(0) = \hat{w}_0 \Leftrightarrow \langle \Phi(0) \rangle \{\hat{a}_m\} = \hat{w}_0 \tag{5.214}$$

Before imposing the boundary conditions, the equation of motion is given by

$$\left[\left[\hat{\Omega}_s^2\right] - \omega^2 \left[I_{nm}\right]\right]\{\hat{a}_m\} = \{0\} \text{ with}$$

$$\left[\hat{\Omega}_s^2\right] = \begin{pmatrix} \omega_1^2(1+i\eta_1) & & & \\ & \omega_2^2(1+i\eta_2) & & \\ & & \ddots & \\ & & & \omega_{nm}^2(1+i\eta_{nm}) \end{pmatrix} \tag{5.215}$$

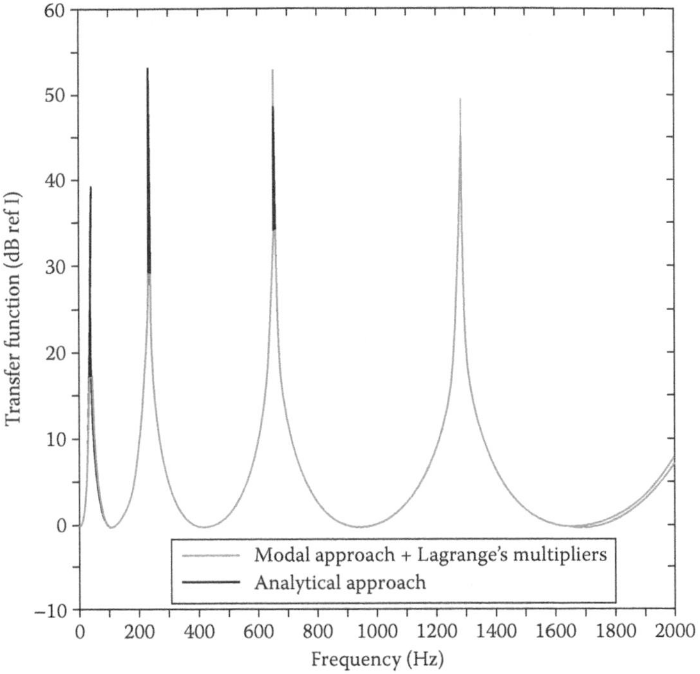

Figure 5.10 Comparisons between the analytical approach and the modal + Lagrange's multipliers technique.

where n_m is the number of kept modes and $[I_{n_m}]$ denotes the identity matrix of order n_m.

Using Lagrange's multiplier technique, the system after imposition of the boundary conditions is

$$\begin{bmatrix} [\hat{\Omega}_s^2] - \omega^2 [I_{n_m}] & \langle \Phi(0) \rangle^T \\ \langle \Phi(0) \rangle & 0 \end{bmatrix} \begin{Bmatrix} \{\hat{a}_m\} \\ \hat{\lambda} \end{Bmatrix} = \begin{Bmatrix} 0 \\ \hat{w}_0 \end{Bmatrix} \tag{5.216}$$

Figure 5.10 presents an example of comparisons between the analytic method and the modal calculation.

5.5.4 Acoustic response of a rectangular cavity with a porous absorber

This example considers a monopole in a rectangular cavity with one wall covered by a porous material. Figure 5.11 depicts the geometry of the problem. The dimensions of the air-filled cavity (speed of sound 342.20 m/s,

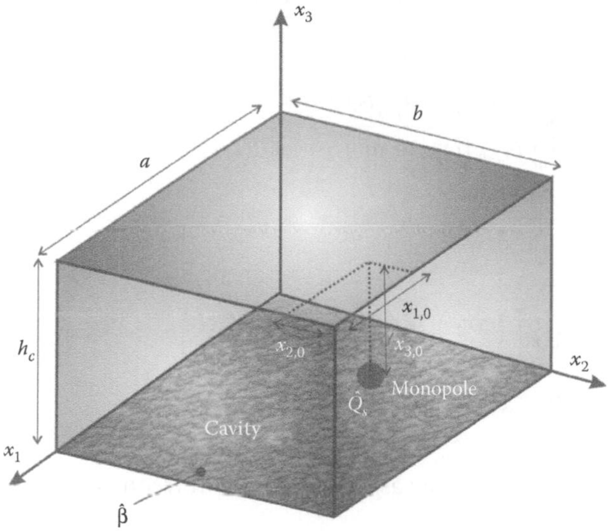

Figure 5.11 Rectangular cavity with a point source and one wall covered with a porous material.

density: 1.21 kg/m) are $a = 0.8$ m, $b = 1.7$ m, and $h_c = 1$ m. Structural damping in the cavity is taken equal to 1%. The foam (melamine) is 4 cm thick and covers the bottom wall of the cavity (0.8 × 1.7 m). It is modeled as a limp porous material (see Chapter 2). Its properties are given in Table 5.3. The monopole is located at position (0, 1.7, 0.96) and has a unit volume velocity. Two models are compared. The first is analytical and is based on Rayleigh-Ritz method. The second uses FE.

In the Rayleigh-Ritz approach, the foam is represented by a locally reacting impedance condition. The foam normal impedance $\hat{Z}_s$ can be calculated using an analytical model or a transfer matrix method (see Allard and Atalla 2009). The pressure in the cavity is calculated using (see Chapter 7)

$$\hat{p}(\underline{x}) = -ik \int_{\partial\Omega} \hat{\beta}(\underline{y}) \hat{G}(\underline{x}, \underline{y}) dS_y + i\omega\rho_0 \hat{Q}_s \hat{G}(\underline{x}, \underline{x}_0) \tag{5.217}$$

where $\hat{\beta}$ is the normalized acoustic admittance of the foam ($\hat{\beta} = \rho_0 c_0 / \hat{Z}_s$), $\hat{Q}_s$ is the volume source strength of the monopole in m³/s, and $\underline{x}_0$ is its

Table 5.3 Properties of the foam

ϕ_p *[1]*	σ_p *[Rayls]*	α_∞ *[1]*	Λ *[μm]*	Λ' *[μm]*	ρ_l *[kg/m]*	η_p *[1]*	ρ_0 *[kgm⁻³]*	c_0 *[ms⁻¹]*
0.99	10,900	1.02	100	130	8.8	0.17	1.213	342.2

location. $\hat{G}(\underline{x},\underline{y})$ denotes Green's function of the hard-walled rectangular cavity:

$$\hat{G}(\underline{x},\underline{y}) = \sum_{p,q,r} \frac{c^2 \Psi_{pqr}(\underline{x}) \Psi_{pqr}(\underline{y})}{N_{pqr}\left(\omega_{pqr}^2 - \omega^2\right)} \tag{5.218}$$

with Ψ_{pqr} the mode shapes:

$$\Psi_{pqr}(\underline{x}) = \cos\left(\frac{p\pi x_1}{a}\right)\cos\left(\frac{q\pi x_2}{b}\right)\cos\left(\frac{r\pi x_3}{h_c}\right)$$
$$\left(0 \le x_1 \le a; 0 \le x_2 \le b; 0 \le x_3 \le h_c\right) \tag{5.219}$$

ω_{pqr} and $k_{pqr} = \omega_{pqr}/c_0$ are the associated natural frequencies and wavenumbers, respectively

$$\omega_{pqr} = c_0\pi\sqrt{\frac{p^2}{a^2} + \frac{q^2}{b^2} + \frac{r^2}{h_c^2}} \tag{5.220}$$

Finally,

$$N_{pqr} = \int_{\Omega} \Psi_{pqr}^2 dV = \frac{abh_c}{8}\varepsilon_p\varepsilon_q\varepsilon_r; \quad \varepsilon_i = \begin{cases} 2 & i = 0 \\ 1 & i > 0 \end{cases} \tag{5.221}$$

Modes up to 1.5 times the maximum frequency of interest are kept in the Rayleigh-Ritz approach.

In the FE case, the cavity mesh consists of 24 × 50 × 30 Hexa8 elements (10 elements/wavelength), and the melamine mesh uses five Hexa8 elements through the thickness. A modal superposition method is used, and modes up to 1.5 times the maximum frequency of interest are kept (570 modes).

The results of the comparison are given in Figure 5.12. Good agreement is observed, keeping in mind the difference between the two approaches (analytical modes vs. FE-calculated modes and the locally reacting model for the foam vs. FE modeling of the foam). In particular, it is observed that for this configuration, the locally reacting impedance model is sufficient to capture the effect of the sound package. However, this is not true for more complicated sound packages or in the presence of vibrating structures. An example is given in Chapter 6.

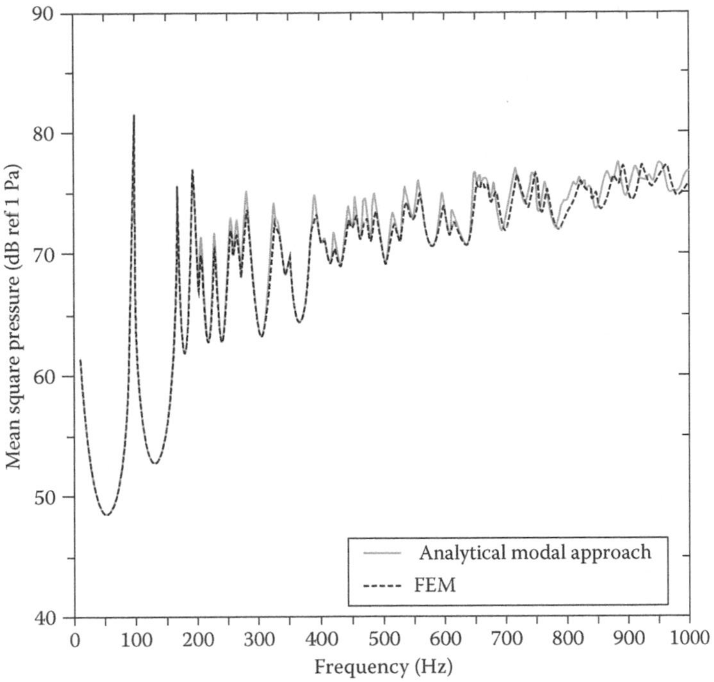

Figure 5.12 Comparisons between the semianalytical (Rayleigh-Ritz) approach and FE for a monopole in an acoustically treated rigid cavity.

5.5.5 Response of a structure to a random excitation

This section describes two methods to model the vibroacoustic response of a structure subjected to a random excitation (e.g., turbulent boundary layer [TBL]) using the FEM. The formulas are given without proof.

5.5.5.1 Method 1: Classic calculation

5.5.5.1.1 Response of the structure in terms of physical degrees of freedom (direct response)

The discretized response of a structure subjected to a given excitation is given by

$$i\omega\left[\hat{Z}_s\right]\{\hat{u}\} = \left\{\hat{F}\right\} \tag{5.222}$$

where $\{\hat{u}\}$ is the nodal displacement vector and

$$\left[\hat{Z}_s\right] = i\omega\left[M\right] + \frac{[K]}{i\omega} + [C_a] \tag{5.223}$$

is the mechanical impedance matrix of the structure (it can include the coupling with the surrounding fluid or the effect of an acoustic treatment) and $\{\hat{F}\}$ is the loading vector. In the case of a distributed parietal pressure, this vector is written

$$\left\{\hat{F}\right\} = \left[C_{up}\right]\left\{\hat{p}\right\} \tag{5.224}$$

where $[C_{up}]$ is a fluid-structure-coupling matrix coming out from the discretization of the term associated with the work of pressure forces in the weak formulation of the problem:

$$\int_S \hat{p} w_n dS = \langle\hat{u}\rangle\left[C_{up}\right]\left\{\hat{p}\right\} \tag{5.225}$$

Using Equation 5.222, the displacement nodal vector is given by

$$\{\hat{u}\} = \frac{1}{i\omega}\left[\hat{H}_s\right]\left[C_{up}\right]\left\{\hat{p}\right\} \tag{5.226}$$

with $[\hat{H}_s] = [\hat{Z}_s]^{-1}$ is a mobility matrix.

For a random excitation, only the power spectral density (PSD) of the excitation is known and is written as

$$S_{pp}(\underline{x}, \underline{y}, \omega) = S_p(\omega)\hat{\psi}_c(\underline{x}, \underline{y}, \omega) \tag{5.227}$$

where S_p is the power autospectrum and $\hat{\psi}_c$ is the spatial correlation function of the excitation. Assuming a spatially homogeneous and stationary field, the PSD only depends on the distance between two points $(\underline{x}, \underline{y})$ of the structure. For instance, for a diffuse acoustic field (DAF), it is given by (Pierce 1989)

$$\hat{\psi}_c\left(\underline{x}, \underline{y}, \omega\right) = \frac{\sin kr}{kr} \tag{5.228}$$

where $k = \omega/c_0$ is the acoustic wavenumber and $r = |\underline{x} - \underline{y}|$ is the distance between two points of the structure.

For classic turbulent boundary layer excitation models (Corcos, Efimtsov, and Cockburn), the spatial correlation function $\hat{\psi}_c(\underline{x},\underline{y},\omega)$ takes the form:

$$\hat{\psi}_c\left(\underline{x},\underline{y},\omega\right) = \exp\left(-|x_1 - y_1|/\delta_1\right)\exp\left(-|x_2 - y_2|/\delta_2\right)\exp\left(-ik_c\left(x_1 - y_1\right)\right) \tag{5.229}$$

where

$$k_c = \frac{\omega}{U_c}, \quad \delta_1 = \frac{1}{\alpha_1 k_c}, \quad \delta_2 = \frac{1}{\alpha_2 k_c} \tag{5.230}$$

U_c denotes the convective velocity and the spatial decay coefficients α_1 and α_2 depend on the model. The power autospectrum $S_p(\omega)$ is generally measured or given by empirical models (CockBurn and Jolly 1968; Efimtsov 1982).

Using Equation 5.226, the spectral density of the structural displacement field is given by

$$\{\hat{u}\}\left\langle\hat{u}^*\right\rangle = \omega^2\left[\hat{H}_s\right]\left[C_{up}\right]\{\hat{p}\}\left\langle\hat{p}^*\right\rangle\left[C_{up}\right]^T\left[\hat{H}_s^*\right]^T \tag{5.231}$$

Using Equation 5.227, this yields

$$\left[S_{uu}\right] = \omega^2\left[\hat{H}_s\right]\left[C_{up}\right]\left[S_{pp}\right]\left[C_{up}\right]^T\left[\hat{H}_s^*\right]^T \tag{5.232}$$

where $[S_{pp}] = \{\hat{p}\}\langle\hat{p}^*\rangle$ and $\left[S_{uu}\right] = \{\hat{u}\}\left\langle\hat{u}^*\right\rangle$ are the spectral density nodal matrices of the excitation (see Figure 5.13) and of the response, respectively.

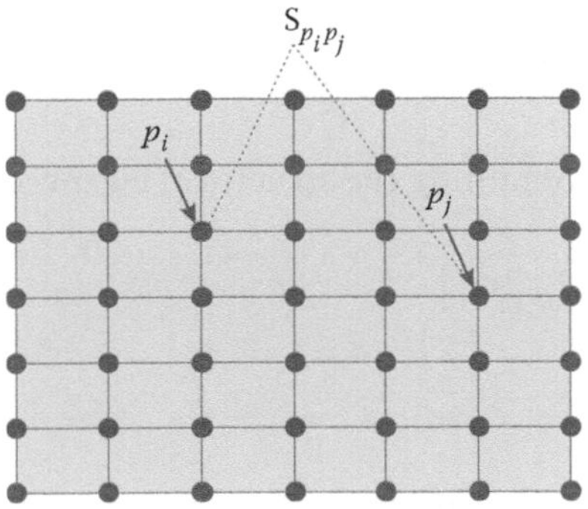

Figure 5.13 Illustration of the calculation of the excitation PSD on the mesh nodes.

Let

$$[S_{xx}] = [C_{up}][S_{pp}][C_{up}]^T \tag{5.233}$$

$[S_{xx}]$ is the matrix arising from the discretization of

$$\langle\hat{u}\rangle[S_{xx}]\{\hat{u}^*\} = \int_S\int_S \hat{w}_n(x_1,x_2)S_{pp}(x_1,x_2,x_1',x_2')\hat{w}_n^*(x_1',x_2')\,dS\,dS' \tag{5.234}$$

Equation 5.232 can be written as

$$[S_{uu}] = \omega^2[\hat{H}_s][S_{xx}][\hat{H}_s^*]^T \tag{5.235}$$

This direct approach is general. It allows, for example, to account for high dampings and frequency dependency of matrices. It is commonly recommended for structures with attached damping and absorbing treatment. However, it is costly from a numerical point of view. Another method that is more efficient from a numerical standpoint is described in the next section.

5.5.5.1.2 Response of the structure in terms of modal coordinates

In its modal form, Equation 5.235 can be written as

$$[S_{uu}] = [\Phi_s][S_{uu,m}][\Phi_s]^T \tag{5.236}$$

with

$$[S_{uu,m}] = \omega^2[\hat{H}_m][S_{xx,m}][\hat{H}_m^*]^T \tag{5.237}$$

$[\Phi_s]$ is the matrix containing the structural eigenvectors

$$[S_{xx,m}] = [\Phi_s]^T[S_{xx}][\Phi_s] \tag{5.238}$$

and

$$[\hat{H}_m] = \left([\Phi_s]^T[\hat{Z}_s][\Phi_s]\right)^{-1} \tag{5.239}$$

is the inverse of the modal impedance matrix.

Equation 5.237 can be rewritten as

$$[S_{uu,m}] = \omega^2[\hat{H}_m][C_{up,m}][S_{pp}][C_{up,m}]^T[\hat{H}_m^*]^T \quad (5.240)$$

with

$$[C_{up,m}] = [\Phi_s]^T[C_{up}] \quad (5.241)$$

This method is very efficient for slightly damped structures. It is, however, not suitable to account for high dampings and frequency-dependent matrices.

5.5.5.1.3 *Calculation of vibroacoustic indicators*

Kinetic energy spectral density

The kinetic energy spectral density is

$$S_{E_c}(\omega) = \frac{1}{2}tr([S_{vv}][M]) \quad (5.242)$$

where $tr([A]) = \sum_i A_{ii}$.*

In its modal form, it becomes

$$S_{E_c}(\omega) = \frac{\omega^2}{2}tr([S_{ww,m}]) \quad (5.243)$$

Mean square velocity spectral density

The mean square velocity spectral density is given by

$$S_{\langle v^2\rangle}(\omega) = \frac{1}{2S}\int_S \hat{v}_n\hat{v}_n^*\,dS = \frac{\omega^2}{2S}tr([S_{uu}][C_{up}]) \quad (5.244)$$

In the case of a planar structure (one degree of freedom per node), $S_{\langle v^2\rangle}(\omega)$ can also be rewritten approximately as

$$S_{\langle v^2\rangle}(\omega) = \frac{\omega^2 S}{2N_s}tr([S_{uu}]) \quad (5.245)$$

where N_s is the number of structural nodes.

* Indeed, we have $S_{E_c}(\omega) = \frac{1}{2}\langle\hat{v}^*\rangle[M]\{\hat{v}\} = \frac{1}{2}\hat{v}_i^*M_{ij}\hat{v}_j = \frac{1}{2}\hat{v}_i^*\hat{v}_jM_{ji} = \frac{1}{2}S_{vv,ij}M_{ji}$ that corresponds to Equation 5.242.

In modal coordinates, using $[C_{up,m}] = [\Phi_s]^T[C_{up}][\Phi_s]$, we have

$$S_{\langle v^2\rangle}(\omega) = \frac{\omega^2}{2S} tr([S_{uu,m}][C_{up,m}]) \tag{5.246}$$

Injected power spectral density
The power spectral density injected by the excitation into the structure is given by

$$S_{\Pi_{in}}(\omega) = \frac{1}{2}\Re\left(tr\left(\{\hat{v}\}\langle\hat{p}^*\rangle[C_{up}]^T\right)\right) \tag{5.247}$$

Using Equation 5.226, we have

$$S_{\Pi_{in}}(\omega) = \frac{1}{2}\Re\left(tr\left(\omega^2\left[\hat{H}_s\right][C_{up}][S_{pp}][C_{up}]^T\right)\right) \tag{5.248}$$

Namely,

$$S_{\Pi_{in}}(\omega) = \frac{1}{2}\Re\left(tr\left(\omega^2\left[\hat{H}_s\right][S_{xx}]\right)\right) \tag{5.249}$$

The equivalent modal equation is given by

$$S_{\Pi_{in}}(\omega) = \frac{1}{2}\Re\left(tr\left(\omega^2\left[\hat{H}_m\right][S_{xx,m}]\right)\right) \tag{5.250}$$

where we have used the fact that $tr([\Phi_s][\hat{H}_m][\Phi_s]^T[S_{xx}]) = tr([\hat{H}_m][\Phi_s]^T[S_{xx}][\Phi_s])$.

Radiated power spectral density (see Chapter 7 for details)
The power spectral density radiated by the structure is given by

$$S_{\Pi_{rad}}(\omega) = \frac{\omega^2}{2}\Re\left(tr\left(\left[\hat{\mathcal{Z}}_{rad}\right][S_{uu}]\right)\right) \tag{5.251}$$

where $[\hat{\mathcal{Z}}_{rad}]$ is the radiation impedance matrix of the structure.
In terms of modal coordinates, we have

$$S_{\Pi_{rad}}(\omega) = \frac{\omega^2}{2}\Re\left(tr\left([S_{uu,m}]\left[\hat{\mathcal{Z}}_{rad,m}\right]\right)\right) \tag{5.252}$$

where $[\hat{\mathcal{Z}}_{rad,m}] = [\Phi_s]^T[\hat{\mathcal{Z}}_{rad}][\Phi_s]$ is the modal impedance radiation matrix.

5.5.5.2 Method 2: Approximation of the solution by deterministic forces

The principle of this method is to represent the solution by an approximation under deterministic excitations (Witting and Sinha 1975; Coyette et al. 2008; Coyette and Meerbergen 2008). The procedure is the following:

1. For a given frequency ω, generate the nodal PSD matrix ($[S_{pp}]$)
2. Perform a Cholesky factorization of $[\hat{S}_{pp}]$*

$$\left[S_{pp}\right] = \left[\hat{L}\right]^{H}\left[\hat{L}\right] \tag{5.253}$$

where superscript H refers to the Hermitian transpose.

3. Generate a sample of N random phases $\langle \varphi_1,\ldots,\varphi_N \rangle$ with $0 \le \varphi_i \le 2\pi$

$$\varphi_i = 2\pi\,\text{rand}(0,1) \tag{5.254}$$

together with the phase vector

$$\left\langle e^{j\varphi_i}\right\rangle = \left\langle e^{j\varphi_1},\ldots,e^{j\varphi_N}\right\rangle \tag{5.255}$$

4. Calculate the nodal pressure-loading equivalent to this sample:

$$\left\{\hat{p}\right\} = \left[\hat{L}\right]^{H}\left\{e^{j\varphi_i}\right\} \tag{5.256}$$

and its discretized form (nodal force vector created from the parietal pressure field)

$$\left\{\hat{F}\right\} = \left[C_{up}\right]\left\{\hat{p}\right\} \tag{5.257}$$

5. Solve system Equation 5.222:

$$\left\{\hat{u}\right\} = \frac{1}{i\omega}\left[\hat{H}_s\right]\left[\hat{F}\right] \tag{5.258}$$

and calculate the associated vibroacoustic indicators.

* It can be shown that this factorization is always possible for a TBL or a DAF since $[S_{pp}]$ is complex, Hermitian for the TBL and positive definite, real, and symmetric for the DAF.

6. Repeat steps 3–5 M times (M outcomes) and average the indicators. For example, for the mean square velocity:

$$S_{\langle v^2 \rangle} = \frac{1}{M} \sum_{i=1}^{M} S_{\langle v^2 \rangle_i} \tag{5.259}$$

The convergence of the method depends on parameter M (= number of outcomes or trials). Practically, a value of $M = 5$ can be sufficient to reach the convergence. This method has the advantage of using the classic solution (excitation by a nodal force distribution equivalent to the parietal pressure field created by a DAF or a TBL) to deal with a problem with a random excitation. Furthermore, it is compatible with classic methods to account for damping and frequency-dependent material properties (either modal or direct).

5.5.5.3 *Application example*

In this section, we compare the FEM calculations based on method 2 with a Rayleigh-Ritz based analytical calculation in the case of a panel excited by a TBL.

The studied panel is 0.5 m long by 0.7 m wide and 2 mm thick. It is made up of aluminum ($E = 2 \times 10^{11}$ Pa; $\nu = 0.33$; $\rho_s = 7780$ kgm^{-3}; $\eta_s = 0.01$). It is simply supported and assumed baffled for acoustic radiation (see Chapter 7). Two configurations are considered: the bare panel and the panel with an added two-layer sound package made up of a foam and a heavy layer (HL).

The Corcos model (Corcos 1963) is used to model the TBL. Assuming the flow along the x_1-axis, the model of the correlation function, in space-frequency domain, is given by

$$\begin{aligned} C(x_1 - x_1', x_2 - x_2', \omega) &= \exp\left(-|x_1 - x_1'|/\delta_1\right) \exp\left(-|x_2 - x_2'|/\delta_2\right) \\ &\quad \times \exp\left(-ik_c (x_1 - x_1')\right) \end{aligned} \tag{5.260}$$

with $\delta_1 = 1/\alpha_1 k_c$, $\delta_2 = 1/\alpha_2 k_c$ and $k_c = \omega/U_c$. U_c represents the convective velocity, α_1 and α_2 the coherence decay coefficients and δ_1 and δ_2 the correlation lengths. In the presented example, the mean square velocity and the transmission loss of the panel are calculated using a convection velocity of 140 m/s and decay coefficients $\alpha_1 = 0.125$, $\alpha_2 = 0.83$, and the number of outcomes $M = 8$.

In the FEM simulation, the mesh must be selected to capture both the smallest structural wavelength and the correlation lengths of the TBL. Figures 5.14 and 5.15 show an example of the comparison using 50×35 quad 4 plate elements. The mesh along the x_1-axis corresponds to a criterion of 6 elements per correlation length δ_1. It is observed that FEM and the modal approach provide comparable results.

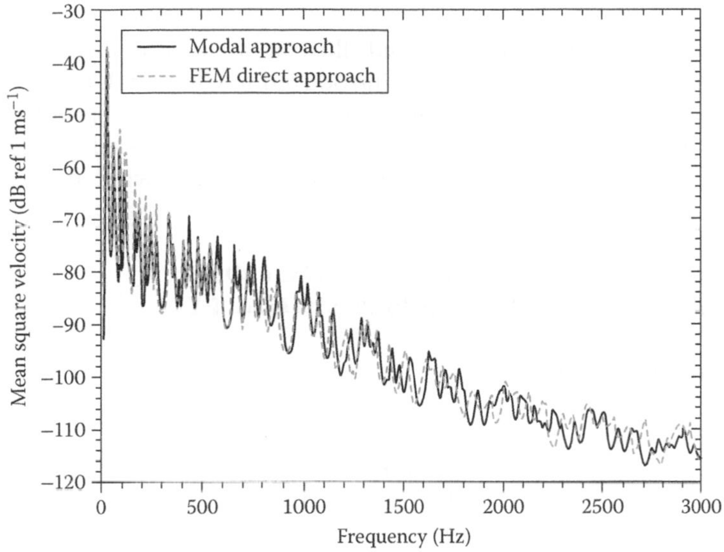

Figure 5.14 Mean square velocity spectral density: comparisons between FEM direct approach and modal approach – 50 × 35 mesh.

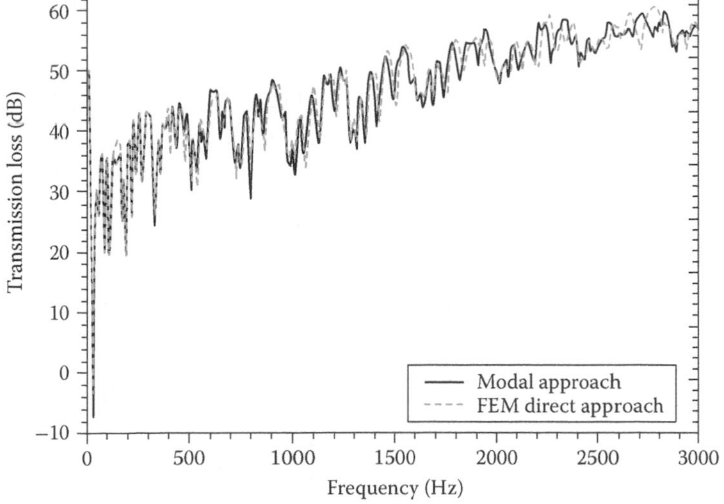

Figure 5.15 Sound transmission loss: comparisons between the direct approach and modal approach – 50 × 35 mesh.

5.6 CONCLUSION

This chapter revisited, detailed, and illustrated via several examples, the seven-step-based FE methodology, introduced in Chapter 4, in the context of the uncoupled three-dimensional acoustic and structural problems. The coupled interior (structure–cavity) problem will be discussed in Chapter 6. The study of the exterior problem (radiation or coupling with an unbounded fluid) is deferred to Chapters 7 and 8.

APPENDIX 5A: USUAL SHAPE FUNCTIONS FOR LAGRANGE THREE-DIMENSIONAL FEs

Tetrahedron element

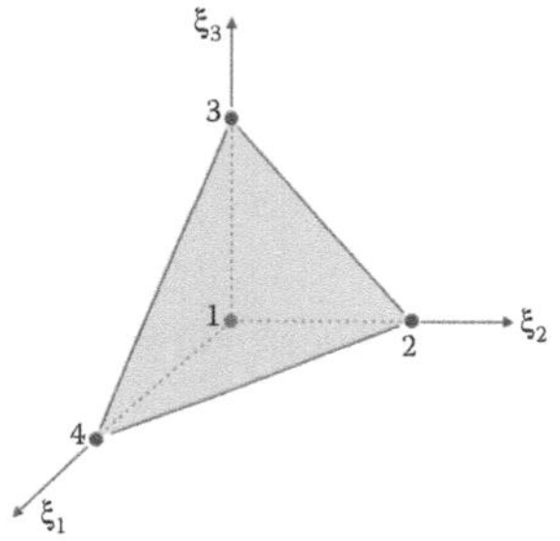

$n_e = 4$

Node id	Coordinates	Shape functions
1	(0,0,0)	$N_1 = 1 - \xi_1 - \xi_2 - \xi_3$
2	(0,1,0)	$N_2 = \xi_2$
3	(0,0,1)	$N_3 = \xi_3$
4	(1,0,0)	$N_4 = \xi_1$

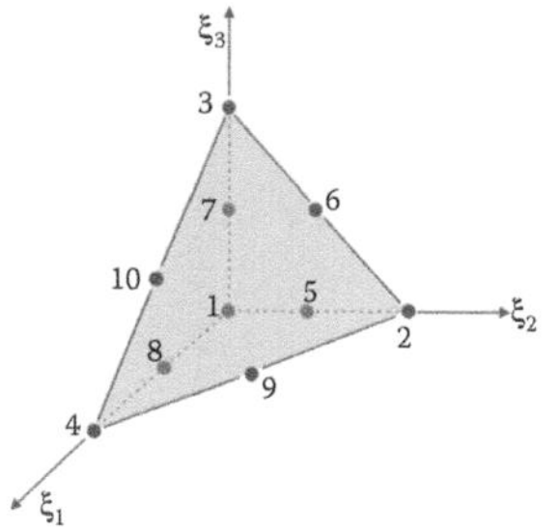

$n_e = 10$ $\quad \xi_4 = 1 - \xi_1 - \xi_2 - \xi_3$

Node id	Coordinates	Shape functions
1	(0,0,0)	$N_1 = (2\xi_4 - 1)\xi_4$
2	(0,1,0)	$N_2 = (2\xi_2 - 1)\xi_2$
3	(0,0,1)	$N_3 = (2\xi_3 - 1)\xi_3$
4	(1,0,0)	$N_4 = (2\xi_1 - 1)\xi_1$
5	$\left(0, \frac{1}{2}, 0\right)$	$N_5 = 4\xi_4\xi_2$
6	$\left(0, \frac{1}{2}, \frac{1}{2}\right)$	$N_6 = 4\xi_2\xi_3$
7	$\left(0, 0, \frac{1}{2}\right)$	$N_7 = 4\xi_3\xi_4$
8	$\left(\frac{1}{2}, 0, 0\right)$	$N_8 = 4\xi_1\xi_4$
9	$\left(\frac{1}{2}, \frac{1}{2}, 0\right)$	$N_9 = 4\xi_1\xi_2$
10	$\left(\frac{1}{2}, 0, \frac{1}{2}\right)$	$N_{10} = 4\xi_1\xi_3$

Hexahedron element

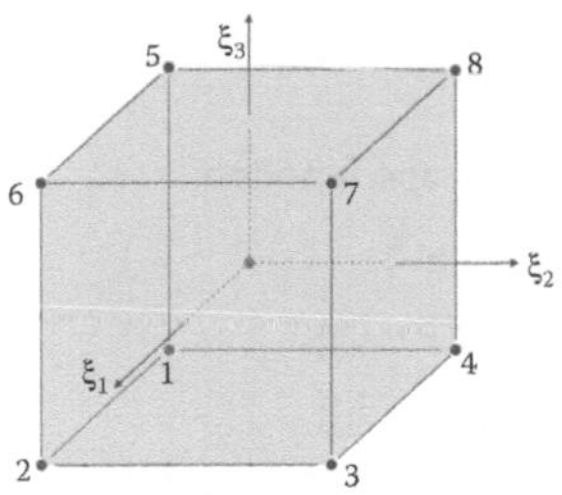

$n_e = 8$

Node id	Coordinates	Shape functions
1	(−1,−1,−1)	$N_1 = 0.125(1-\xi_1)(1-\xi_2)(1-\xi_3)$
2	(1,−1,−1)	$N_2 = 0.125(1+\xi_1)(1-\xi_2)(1-\xi_3)$
3	(1,1,−1)	$N_3 = 0.125(1+\xi_1)(1+\xi_2)(1-\xi_3)$
4	(−1,1,−1)	$N_4 = 0.125(1-\xi_1)(1+\xi_2)(1-\xi_3)$
5	(−1,−1,1)	$N_5 = 0.125(1-\xi_1)(1-\xi_2)(1+\xi_3)$
6	(1,−1,1)	$N_6 = 0.125(1+\xi_1)(1-\xi_2)(1+\xi_3)$
7	(1,1,1)	$N_7 = 0.125(1+\xi_1)(1+\xi_2)(1+\xi_3)$
8	(−1,1,1)	$N_8 = 0.125(1-\xi_1)(1+\xi_2)(1+\xi_3)$

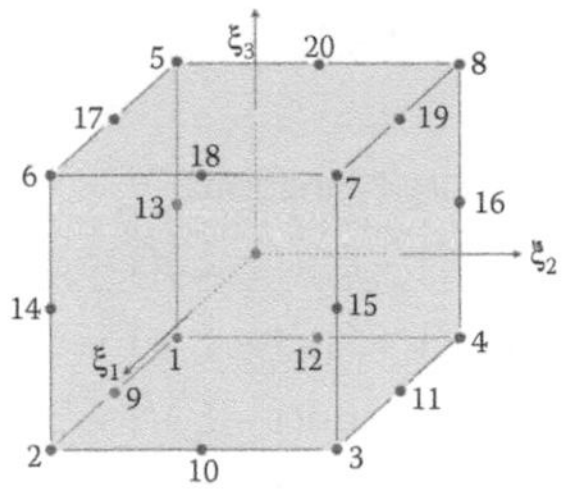

$n_e = 20$

Node id	Coordinates	Shape functions
1	(−1,−1,−1)	$N_1 = 0.125(1-\xi_1)(1-\xi_2)(1-\xi_3)(-\xi_1-\xi_2-\xi_3-2)$
2	(1,−1,−1)	$N_2 = 0.125(1+\xi_1)(1-\xi_2)(1-\xi_3)(\xi_1-\xi_2-\xi_3-2)$
3	(1,1,−1)	$N_3 = 0.125(1+\xi_1)(1+\xi_2)(1-\xi_3)(\xi_1+\xi_2-\xi_3-2)$
4	(−1,1,−1)	$N_4 = 0.125(1-\xi_1)(1+\xi_2)(1-\xi_3)(-\xi_1+\xi_2-\xi_3-2)$
5	(−1,−1,1)	$N_5 = 0.125(1-\xi_1)(1-\xi_2)(1+\xi_3)(-\xi_1-\xi_2+\xi_3-2)$
6	(1,−1,1)	$N_6 = 0.125(1+\xi_1)(1-\xi_2)(1+\xi_3)(\xi_1-\xi_2+\xi_3-2)$
7	(1,1,1)	$N_7 = 0.125(1+\xi_1)(1+\xi_2)(1+\xi_3)(\xi_1+\xi_2+\xi_3-2)$
8	(−1,1,1)	$N_8 = 0.125(1-\xi_1)(1+\xi_2)(1+\xi_3)(-\xi_1+\xi_2+\xi_3-2)$
9	(0,−1,−1)	$N_9 = 0.25(1-\xi_1^2)(1-\xi_2)(1-\xi_3)$
10	(1,0,−1)	$N_{10} = 0.25(1+\xi_1)(1-\xi_2^2)(1-\xi_3)$
11	(0,1,−1)	$N_{11} = 0.25(1-\xi_1^2)(1+\xi_2)(1-\xi_3)$
12	(−1,0,−1)	$N_{12} = 0.25(1-\xi_1)(1-\xi_2^2)(1-\xi_3)$
13	(−1,−1,0)	$N_{13} = 0.25(1-\xi_1)(1-\xi_2)(1-\xi_3^2)$
14	(1,−1,0)	$N_{14} = 0.25(1+\xi_1)(1-\xi_2)(1-\xi_3^2)$
15	(1,1,0)	$N_{15} = 0.25(1+\xi_1)(1+\xi_2)(1-\xi_3^2)$
16	(−1,1,0)	$N_{16} = 0.25(1-\xi_1)(1+\xi_2)(1-\xi_3^2)$
17	(0,−1,1)	$N_{17} = 0.25(1-\xi_1^2)(1-\xi_2)(1+\xi_3)$
18	(1,0,1)	$N_{18} = 0.25(1+\xi_1)(1-\xi_2^2)(1+\xi_3)$
19	(0,1,1)	$N_{19} = 0.25(1-\xi_1^2)(1+\xi_2)(1+\xi_3)$
20	(−1,0,1)	$N_{20} = 0.25(1-\xi_1)(1-\xi_2^2)(1+\xi_3)$

Prism element

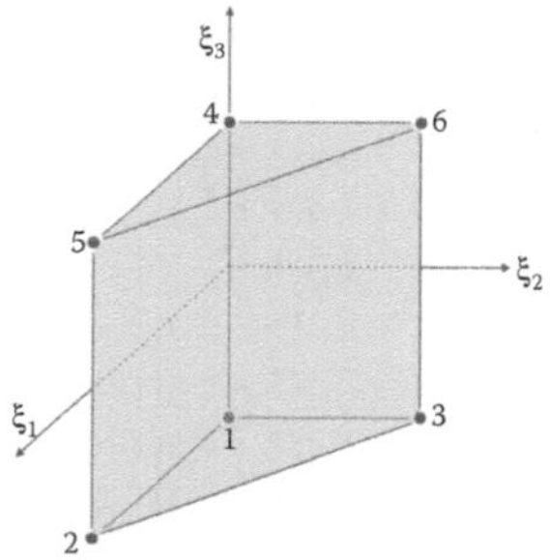

$n_e = 6$

Node id	Coordinates	Shape functions
1	(0,0,−1)	$N_1 = 0.5(1-\xi_1-\xi_2)(1-\xi_3)$
2	(1,0,−1)	$N_2 = 0.5\xi_1(1-\xi_3)$
3	(0,1,−1)	$N_3 = 0.5\xi_2(1-\xi_3)$
4	(0,0,1)	$N_4 = 0.5(1-\xi_1-\xi_2)(1+\xi_3)$
5	(1,0,1)	$N_5 = 0.5\xi_1(1+\xi_3)$
6	(0,1,1)	$N_6 = 0.5\xi_2(1+\xi_3)$

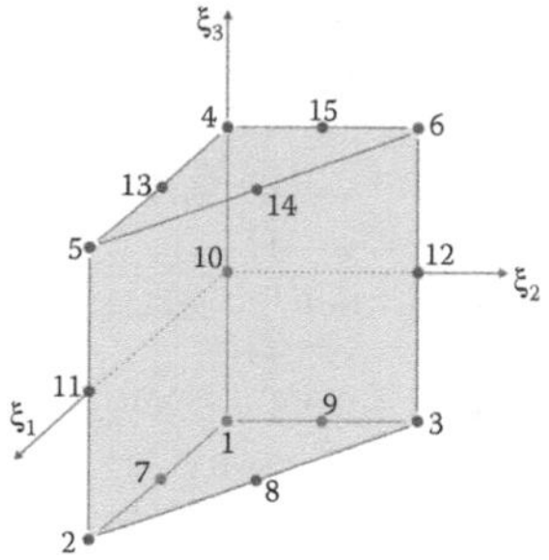

$n_e = 15$

Node id	Coordinates	Shape functions
1	(0,0,−1)	$N_1 = 0.125(1-\xi_1)(1-\xi_2)(1-\xi_3)$
2	(1,0,−1)	$N_2 = 0.125(1+\xi_1)(1-\xi_2)(1-\xi_3)$
3	(0,1,−1)	$N_3 = 0.125(1+\xi_1)(1+\xi_2)(1-\xi_3)$
4	(0,0,1)	$N_4 = 0.125(1-\xi_1)(1+\xi_2)(1-\xi_3)$
5	(1,0,1)	$N_5 = 0.125(1-\xi_1)(1-\xi_2)(1+\xi_3)$
6	(0,1,1)	$N_6 = 0.125(1+\xi_1)(1-\xi_2)(1+\xi_3)$
7	$\left(\frac{1}{2},0,-1\right)$	$N_7 = 0.5\xi_1(1-\xi_1-\xi_2)(1-\xi_3)$
8	$\left(\frac{1}{2},\frac{1}{2},-1\right)$	$N_8 = 0.5\xi_1\xi_2(1-\xi_3)$
9	$\left(0,\frac{1}{2},-1\right)$	$N_9 = 0.5\xi_2(1-\xi_1-\xi_2)(1-\xi_3)$
10	(0,0,0)	$N_{10} = (1-\xi_1-\xi_2)(1-\xi_3^2)$
11	(1,0,0)	$N_{11} = \xi_1(1-\xi_3^2)$
12	(0,1,0)	$N_{12} = \xi_2(1-\xi_3^2)$
13	$\left(\frac{1}{2},0,1\right)$	$N_{13} = 0.5\xi_1(1-\xi_1-\xi_2)(1+\xi_3)$
14	$\left(\frac{1}{2},\frac{1}{2},1\right)$	$N_{14} = 0.5\xi_1\xi_2(1+\xi_3)$
15	$\left(0,\frac{1}{2},1\right)$	$N_{15} = 0.5\xi_2(1-\xi_1-\xi_2)(1+\xi_3)$

APPENDIX 5B: CLASSIC SHAPE FUNCTIONS FOR TWO-DIMENSIONAL ELEMENTS $N_i(\xi_1,\xi_2)$

Triangular element

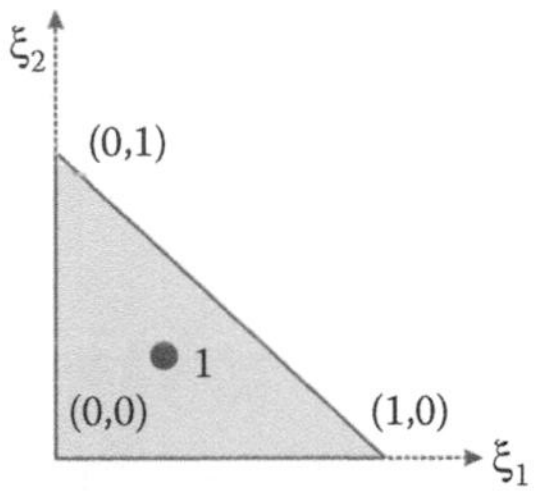

Zeroth-order interpolation: $n_e = 1$

$N_1 = 1$

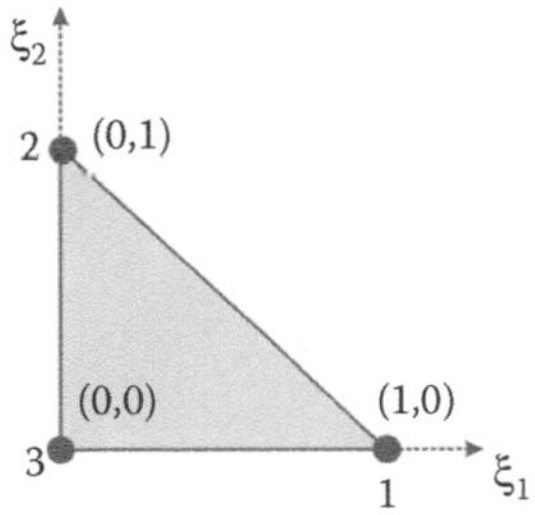

Linear interpolation $n_e = 3$

$N_1 = \xi_1$
$N_2 = \xi_2$
$N_3 = \xi_3$
$\xi_3 = 1 - \xi_1 - \xi_2$

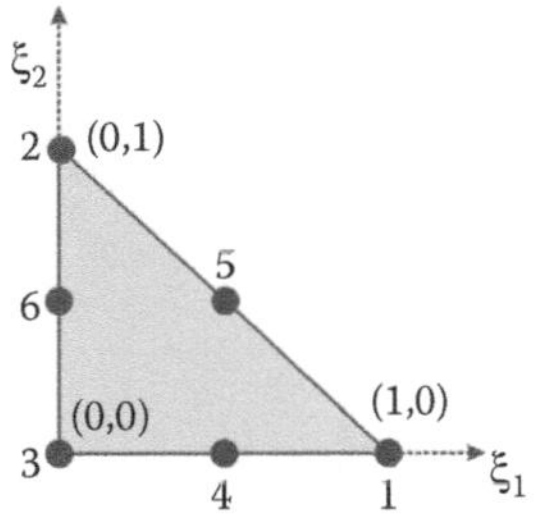

Quadratic interpolation $n_e = 6$

$N_1 = \xi_1(2\xi_1 - 1)$
$N_2 = \xi_2(2\xi_2 - 1)$
$N_3 = \xi_3(2\xi_3 - 1)$
$N_4 = 4\xi_1\xi_2$
$N_5 = 4\xi_2\xi_3$
$N_6 = 4\xi_3\xi_1$

Quadrangle element

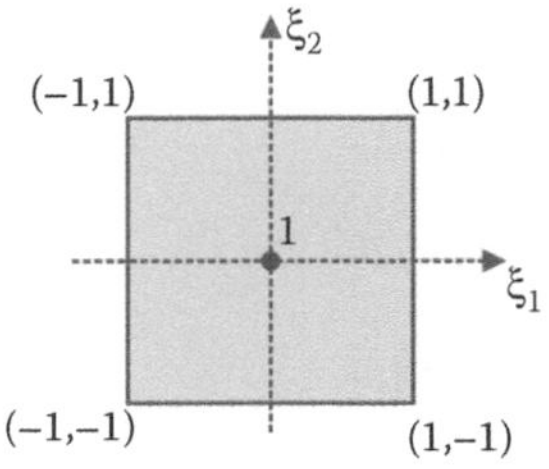

Zeroth-order interpolation: $n_e = 1$

$N_1 = 1$

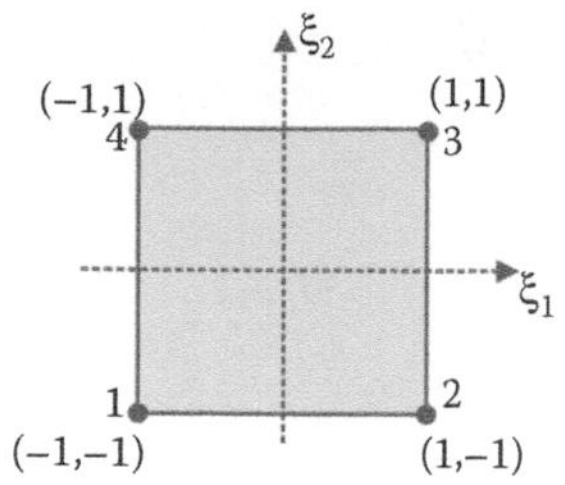

Bilinear interpolation: $n_e = 4$

$$N_1 = \frac{1}{4}(1-\xi_1)(1-\xi_2)$$
$$N_2 = \frac{1}{4}(1+\xi_1)(1-\xi_2)$$
$$N_3 = \frac{1}{4}(1+\xi_1)(1+\xi_2)$$
$$N_4 = \frac{1}{4}(1-\xi_1)(1+\xi_2)$$

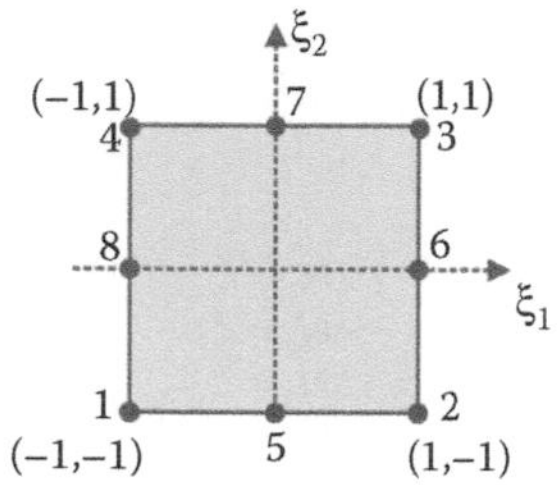

Incomplete biquadratic interpolation: $n_e = 8$

$$N_1 = -\frac{1}{4}(1-\xi_1)(1-\xi_2)(1+\xi_1+\xi_2)$$
$$N_2 = -\frac{1}{4}(1+\xi_1)(1-\xi_2)(1+\xi_1+\xi_2)$$
$$N_3 = -\frac{1}{4}(1+\xi_1)(1+\xi_2)(1+\xi_1+\xi_2)$$
$$N_4 = -\frac{1}{4}(1-\xi_1)(1+\xi_2)(1+\xi_1+\xi_2)$$
$$N_5 = \frac{1}{2}\left(1-\xi_1^2\right)(1-\xi_2)$$
$$N_6 = \frac{1}{2}(1+\xi_1)\left(1-\xi_2^2\right)$$
$$N_7 = \frac{1}{2}\left(1-\xi_1^2\right)(1+\xi_2)$$
$$N_8 = \frac{1}{2}(1-\xi_1)\left(1-\xi_2^2\right)$$

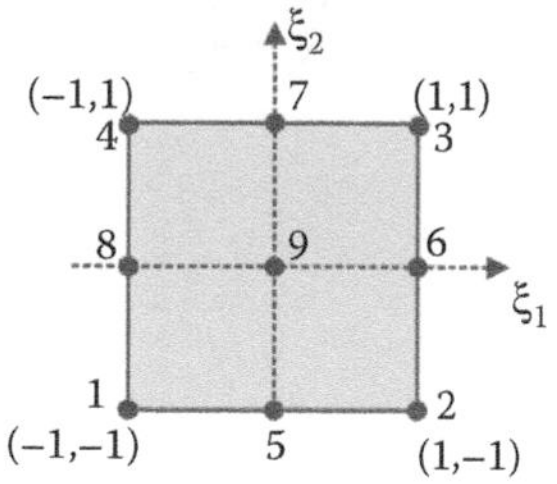

Biquadratic interpolation: $n_e = 9$

$$N_1 = \frac{1}{4}\xi_1(\xi_1-1)\xi_2(\xi_2-1)$$
$$N_2 = \frac{1}{4}\xi_1(1+\xi_1)\xi_2(\xi_2-1)$$
$$N_3 = \frac{1}{4}\xi_1(1+\xi_1)\xi_2(1+\xi_2)$$
$$N_4 = \frac{1}{4}\xi_1(\xi_1-1)\xi_2(1+\xi_2)$$
$$N_5 = \frac{1}{2}\left(1-\xi_1^2\right)\xi_2(\xi_2-1)$$
$$N_6 = \frac{1}{2}\xi_1(1+\xi_1)\left(1-\xi_2^2\right)$$
$$N_7 = \frac{1}{2}\left(1-\xi_1^2\right)\xi_2(1+\xi_2)$$
$$N_8 = \frac{1}{2}\xi_1(\xi_1-1)\left(1-\xi_2^2\right)$$
$$N_9 = \left(1-\xi_1^2\right)\left(1-\xi_2^2\right)$$

APPENDIX 5C: NUMERICAL INTEGRATION USING GAUSS-TYPE QUADRATURE RULES

Unit square elements: $\displaystyle\int_{-1}^{1}\int_{-1}^{1} f\left(x_1, x_2\right) dx_1\, dx_2$

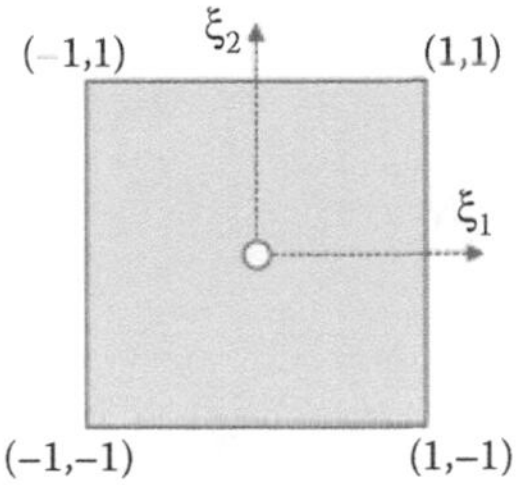

npt	id	x^g_{1i}	x^g_{2i}	W^g_i
1	1	0	0	4
4 $a = \sqrt{1/3}$	1	$-a$	$-a$	1
	2	a	$-a$	
	3	a	a	
	4	$-a$	a	
9 $a = \sqrt{3/5}$	1	$-a$	$-a$	25/81
	2	a	$-a$	
	3	a	a	
	4	$-a$	a	
	5	0	$-a$	40/81
	6	a	0	
	7	0	a	
	8	$-a$	0	
	9	0	0	64/81
16 $a = \sqrt{\dfrac{3 + 2\sqrt{6/5}}{7}}$ $b = \sqrt{\dfrac{3 - 2\sqrt{6/5}}{7}}$	1	$-a$	$-a$	0.1210029932856020
	2	$-a$	a	
	3	a	a	
	4	a	$-a$	
	5	$-b$	$-b$	0.4252933030106942
	6	$-b$	b	
	7	b	b	
	8	b	$-b$	

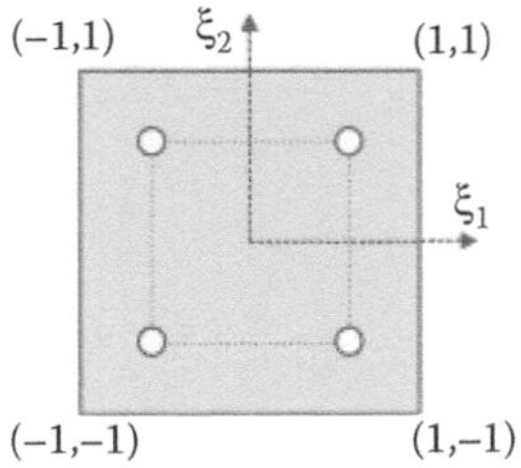

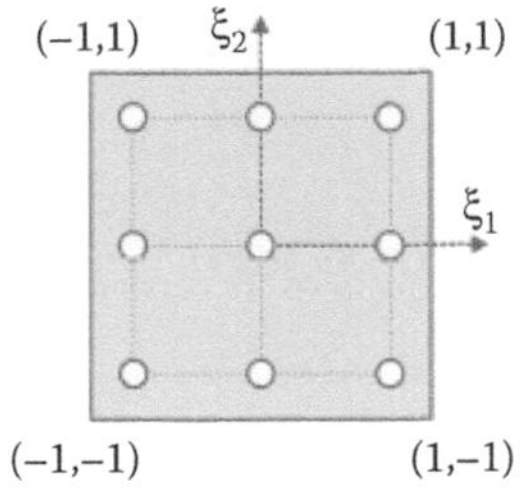

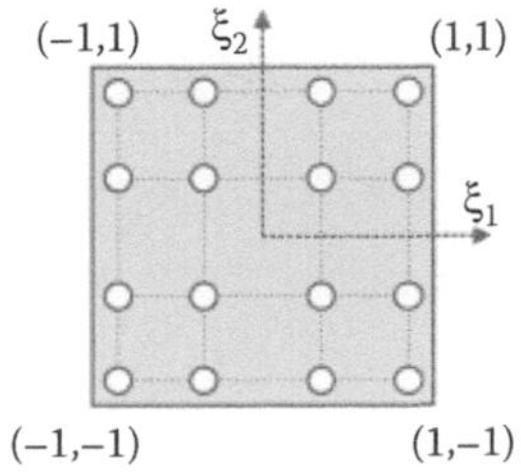

9	$-a$	$-b$	0.226851851851851 9
10	$-a$	b	
11	a	$-b$	
12	a	b	
13	$-b$	$-a$	
14	$-b$	a	
15	b	a	
16	b	$-a$	

Unit triangular elements: $\int_T f(x_1, x_2)dS = \int_0^1 \int_0^{1-x_1} f(x_1, x_2)\, dx_1 dx_2$

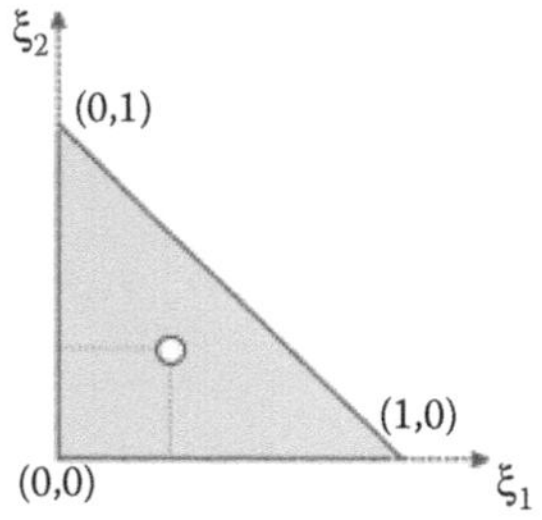

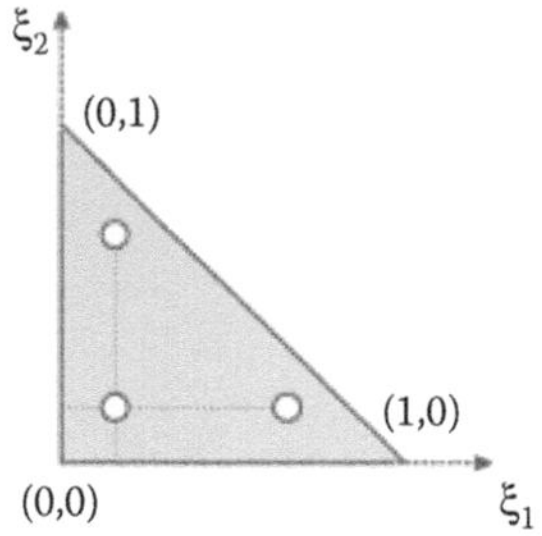

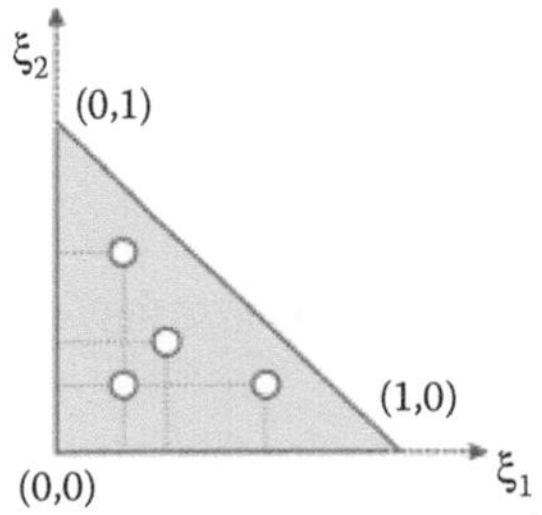

npt	*id*	x^g_{1i}	x^g_{2i}	W^g_i
1	1	1/3	1/3	0.5
3 $b = 2/3$ $a = 1/6$	1	a	a	1/6
	2	a	b	
	3	b	a	
4 $a = 1/3$ $b = 1/5$ $c = 3/5$	1	a	a	−27/96
	2	b	b	25/96
	3	b	c	
	4	c	b	

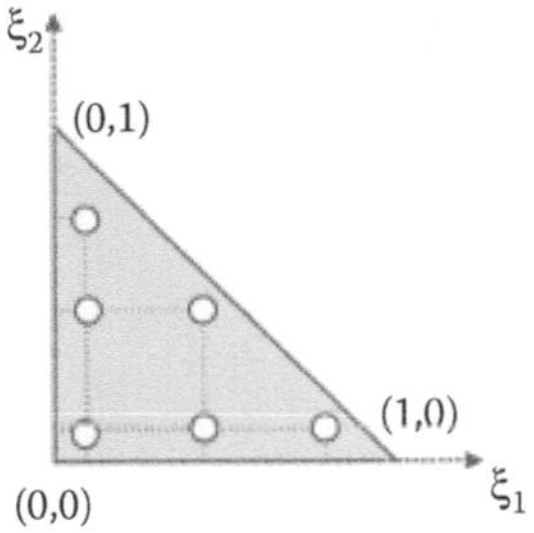

6	1	a	a	0.111690794839005
$a = 0.445948490915965$	2	c	a	
$b = 0.091576213509771$	3	a	c	
$c = 1 - 2a$	4	b	b	0.054975871827661
$d = 1 - 2b$	5	d	b	
	6	b	d	

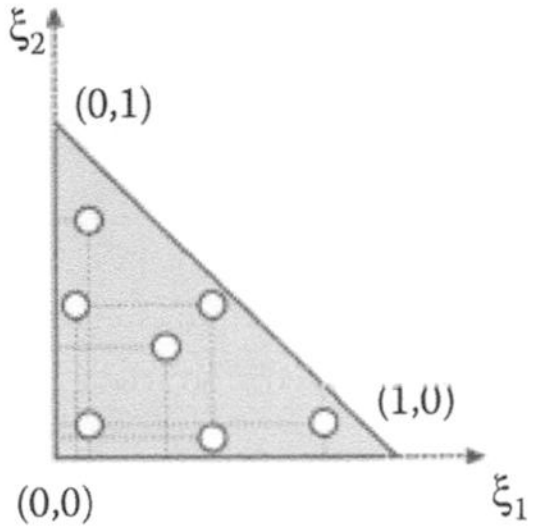

7	1	a	a	9/80
$a = 1/3$	2	b	b	$(155+\sqrt{15})/2400$
$b = \dfrac{(6+\sqrt{15})}{21}$	3	d	b	
	4	b	d	
$c = 4/7 - b$	5	c	c	$(155-\sqrt{15})/2400$
$d = 1 - 2b$	6	e	c	
$e = 1 - 2c$	7	c	e	

Unit tetrahedron elements: $\int_{\text{Tetra}} f(x_1, x_2, x_3)dV$

npt	*id*	x_{1i}^g	x_{2i}^g	x_{3i}^g	W_i^g
1	1	1/4	1/4	1/4	1/6
4	1	a	a	a	1/24
$a = \dfrac{5-\sqrt{5}}{20}$	2	a	a	b	
	3	a	b	a	
$b = \dfrac{5+3\sqrt{5}}{20}$	4	b	a	a	
5	1	a	a	a	−2/15
$a = 1/4$	2	b	b	b	3/40
$b = 1/6$	3	b	b	c	
$c = 1/2$	4	b	c	b	
	5	c	b	b	
15	1	a	a	a	8/405
$a = 1/4$	2	b	b	b	$\dfrac{2665-14\sqrt{15}}{226{,}800}$
$b = \dfrac{7+\sqrt{15}}{34}$	3	b	b	d	
	4	b	d	b	
$c = \dfrac{7-\sqrt{15}}{34}$	5	d	b	b	

$d = \frac{13 - 3\sqrt{15}}{34}$	6	c	c	c	$\frac{2665+14\sqrt{15}}{226,800}$
	7	c	c	e	
$e = \frac{13 + 3\sqrt{15}}{34}$	8	c	e	c	
	9	e	c	c	
	10	f	f	g	$\frac{5}{567}$
$f = \frac{5 - \sqrt{15}}{20}$	11	f	g	f	
	12	g	f	f	
	13	f	g	g	
$g = \frac{5 + \sqrt{15}}{20}$	14	g	f	g	
	15	g	g	f	

Unit pentahedron elements: $\int_{\text{Prism}} f(x_1, x_2, x_3)dV$

npt	*id*	x^g_{1i}	x^g_{2i}	x^g_{3i}	W^g_i
6	1	a	a	$-b$	1/6
$a = \frac{1}{2}$	2	0	a	$-b$	
	3	a	0	$-b$	
$b = \sqrt{1/3}$	4	a	a	b	
	5	0	a	b	
	6	a	0	b	
8	1	a	a	$-d$	−27/96
$a = 1/3$	2	b	c	$-d$	25/96
$b = 3/5$	3	c	b	$-d$	
$c = 1/5$	4	c	c	$-d$	
$d = \sqrt{1/3}$	5	a	a	d	−27/96
	6	b	c	d	25/96
	7	c	b	d	
	8	c	c	d	
21	1	a	a	$-f$	$b\frac{9}{80}$
$a = 1/3$					
$b = 5/9$					
$c = 8/9$	2	d	d	$-f$	$b\frac{155 + \sqrt{15}}{2400}$
	3	$1-2d$	d	$-f$	
$d = \frac{6 + \sqrt{15}}{21}$	4	d	$1-2d$	$-f$	
$e = \frac{6 - \sqrt{15}}{21}$	5	e	e	$-f$	$b\frac{155 - \sqrt{15}}{2400}$
	6	$1-2e$	e	$-f$	
$f = \sqrt{\frac{3}{5}}$	7	e	$1-2e$	$-f$	
	8	a	a	0	$c\frac{9}{80}$
	9	d	d	0	$c\frac{155 + \sqrt{15}}{2400}$

10	$1-2d$	d	0	
11	d	$1-2d$	0	
12	e	e	0	$c\dfrac{155-\sqrt{15}}{2400}$
13	$1-2e$	e	0	
14	e	$1-2e$	0	
15	a	a	f	$b\dfrac{9}{80}$
16	d	d	f	$b\dfrac{155+\sqrt{15}}{2400}$
17	$1-2d$	d	f	
18	d	$1-2d$	f	
19	e	e	f	$b\dfrac{155-\sqrt{15}}{2400}$
20	$1-2e$	e	f	
21	e	$1-2e$	f	

APPENDIX 5D: CALCULATION OF ELEMENTARY MATRICES—THREE-DIMENSIONAL WAVE PROPAGATION PROBLEM

This appendix presents Fortran subroutines to calculate elementary matrices $[Q^e]$ and $[H^e]$ for two types of linear acoustic elements: the six-node pentahedron and the eight-node hexahedron.

```
SUBROUTINE AC3PENT6(EQ,EH,XL,IPROP)
!================================================
! CODE : GAUSFEM
! ACOUSTIC LIBRARY
! ELEMENT: AC3PENT6
! FUNCTION: CALCULATION OF ELEMENTARY MATRICES OF ACOUSTIC
! AC3PENT6
! NOTE : NUMERICAL INTEGRATION USING 6 VERTICES OF THE PENTAHEDRON
! NOMENCLATURE :
! NNE : NUMBER OF NODES OF THE ELEMENT. HERE NNE=6 FOR THE
! PENTAHEDRON
! EQ ( :, :) : ELEMENTARY MATRIX [Qe] OF DIMENSION NNE X NNE
! EH ( :, :) : ELEMENTARY MATRIX [He] OF DIMENSION NNE X NNE
! XL( :, :) : COORDINATES OF THE ELEMENT NODES
! IPROP : ID OF THE ELEMENT PROPERTY
!


IMPLICIT REAL(8) (A-H,0-Z)
PARAMETER (NNE=6)
DIMENSION EH(NNE,NNE),EQ(NNE,NNE),SHPG(4,NNE),SHPL(4,NNE), &
XJ(3,3),XJTINV(3,3),XL(3,NNE),POINTS(3,NNE)
COMMON/MATERIALS/C01(32),C02(32),C03(32),G12(32),G13(32),G23(32), &
```

```
V12(32),V13(32),V23(32),RHO(32),ETA(32)
COMMON/PROPERTIES/MAT(10,10),ALPHA(10,10),H(10),H1(10,10), H2(10,10),&
NELET(10),NLAYER(10)
!DENSITY (RHO0) AND SOUND SPEED (C0) IN MEDIUM (IPROP)
IMAT=MAT(IPROP,1)
RHO0=RHO(IMAT)
C0=C01(IMAT)
! INITIALISATION OF MATRICES
EH=0.D0
EQ=0.D0
! START INTEGRATION
CALL INTEG_POINTS (POINTS,NNE) ! THE VERTICES OF THE PENTAHEDRON ARE
! THE INTEGRATION POINTS
WGT=1.D0/6.D0 ! INTEGRATION WEIGHT

INTEGRATION_LOOP: DO IN=1,NNE
! CALCULATION OF THE SHAPE FUNCTIONS AND THEIR FIRST PARTIAL
! DERIVATION AT THE INTEGRATION POINT (R,S,T)
    R=POINTS(1,IN)
    S=POINTS(2,IN)
    T=POINTS(3,IN)
    CALL SHAPEL_3D(NNE,R,S,T,SHPL)
    CALL JACOBI_3D(XJ,DET,SHPL,XL,NNE)
    CALL INVT_3X3(XJ,XJTINV)
    CALL SHAPEG_3D(XJTINV,SHPL,SHPG,NNE)
! CALCULATION OF MATRIX EH(:,:)= [He]
    WEIGHT=WGT*DET/RHO0
    DO J=1,NNE
        DO I=1,NNE
            DO K=1,3
                EH(I,J)=EH(I,J)+SHPG(K,I)*SHPG(K,J)*WEIGHT
            END DO
        END DO
    END DO
! CALCULATION OF MATRIX EQ( :, :) = [Qe]
    WEIGHT=WEIGHT/C0/C0
    DO J=1,NNE
        DO I=1,NNE
            EQ(I,J)=EQ(I,J)+SHPG(4,I)*SHPG(4,J)*WEIGHT
        END DO
    END DO
END DO INTEGRATION_LOOP
RETURN
END

SUBROUTINE AC3HEXA8(EQ,EH,XL,IPROP)
! CODE : GAUSFEM
! LIBRARY: ACOUSTIC
! ELEMENT: AC3HEXA8
```

```
! FUNCTION: CALCULATION OF ELEMENTARY MATRICES OF ACOUSTIC
! AC3HEXA8
! NOTE: NUMERICAL INTEGRATION USING A 8 GAUSS POINTS QUADRATURE
! RULE
! NOMENCLATURE :
! NNE : NUMBER OF NODES OF THE ELEMENT. HERE NNE = 8 FOR THE
! HEXAHEDRON
! EQ ( :, :) : ELEMENTARY MATRIX [Qe] OF DIMENSION NNE X NNE
! EH ( :, :) : ELEMENTARY MATRIX [He] OF DIMENSION NNE X NNE
! XL( :, :) : COORDINATES OF THE ELEMENT NODES
! IPROP : ID OF THE ELEMENT PROPERTY
!
IMPLICIT REAL(8) (A-H,O-Z)
PARAMETER (NNE = 8)
DIMENSION EH(NNE,NNE),EQ(NNE,NNE),SHPG(4,NNE),SHPL(4,NNE), &
XJ(3,3),XJTINV(3,3),XL(3,NNE),POINTS(3,NNE)
COMMON/MATERIALS/C01(32),C02(32),C03(32),G12(32),G13(32),G23(32), &
V12(32),V13(32),V23(32), RHO(32),ETA(32)
COMMON/PROPERTIES/MAT(10,10),ALPHA(10,10),H(10),H1(10,10),H2(10,10),&
NELET(10),NLAYER(10)
!DENSITY (RHO0) AND SOUND SPEED (C0) IN MEDIUM (IPROP)
IMAT = MAT(IPROP,1)
RHO0 = RHO(IMAT)
C0 = C01(IMAT)
! INITIALISATION OF MATRICES
EH = 0.D0
EQ = 0.D0
! INTEGRATION POINTS AND WEIGHTS FOR HEXAHEDRON
CALL INTEG_POINTS (POINTS,NNE)
WGT = 1d0
! START INTEGRATION:
INTEGRATION LOOP: DO IN= 1, NNE
! CALCULATION OF THE SHAPE FUNCTIONS AND THEIR FIRST PARTIAL
! DERIVATION AT THE INTEGRATION POINT (R,S,T)
    R = POINTS(1,IN)
    S = POINTS(2,IN)
    T = POINTS(3,IN)
    CALL SHAPEL_3D(NNE,R,S,T,SHPL)
    CALL JACOBI_3D(XJ,DET,SHPL,XL,NNE)
    CALL INVT_3X3(XJ,XJTINV)
    CALL SHAPEG_3D(XJTINV,SHPL,SHPG,NNE)
! CALCULATION OF MATRIX EH(:,:)= [He]
    WEIGHT = WGT*DET/RHO0
    DO J = 1,NNE
        DO I = 1,NNE
            DO K = 1,3
                EH(I,J) = EH(I,J) + SHPG(K,I)*SHPG(K,J)*WEIGHT
            END DO
        END DO
```

```
    END DO
! CALCULATION OF MATRIX EQ( :, :) = [Qe]
    WEIGHT=WEIGHT/C0/C0
    DO J=1,NNE
        DO I=1,NNE
            EQ(I,J) =EQ(I,J) +SHPG(4,I)*SHPG(4,J)*WEIGHT
        END DO
    END DO
END DO INTEGRATION_LOOP
RETURN
END

SUBROUTINE INTEG_POINTS (POINTS,NNE)
! CODE : GAUSFEM
! LIBRARY : SHAPE
! ELEMENT: PENTAHEDRON & HEXAHEDRON
! FUNCTION: RETURNS INTEGRATION POINTS FOR PARENT PENTAHEDRON AND
! HEXAHEDRON
! NOTE : INTEGRATION BASED ON VERTICES FOR PENTAHEDRON AND GAUSS
! POINTS FOR HEXAHEDRON
! INPUT :
! NNE : NUMBER OF NODES OF THE ELEMENT
! OUTPUT :
! POINTS( :, :) : COORDINATES OF INTEGRATION POINTS
!
IMPLICIT REAL(8) (A-H,O-Z)
REAL(8) :: POINTS(3,NNE)
SELECT CASE(NNE)
    CASE(6)
    ! CASE OF 6-NODE PENTAHEDRON
    POINTS(1,:)= (/0.D0, 1d0, 0.D0, 0.D0,1.D0,0.D0/)
    POINTS(2,:)= (/0.D0, 0.D0, 1d0, 0.D0,0.D0, 1d0/)
    POINTS(3,:)= (/-1d0,-1.D0,-1d0, 1d0, 1d0, 1d0/)
    CASE(8) THEN
    ! CASE OF 8-NODE HEXAHEDRON
    C= 1d0/DSQRT(3.D0)
    POINTS(1,:)= (/-C, C, C,-C,-C, C,C,-C/)
    POINTS(2,:)= (/-C,-C, C, C,-C,-C,C, C/)
    POINTS(3,:)= (/-C,-C,-C,-C, C, C,C, C/)
    CASE DEFAULT
    WRITE(*,*) 'ERROR IN SUBROUTINE VERTEX:', &
    'UNSUPPORTED ELEMENT TYPE'
END SELECT
RETURN
END SUBROUTINE INTEG_POINTS

SUBROUTINE SHAPEL_3D(NNE,R,S,T,SHPL)
! CODE : GAUSFEM
! LIBRARY: SHAPE
```

```
! ELEMENT: 3D ELEMENTS
! FUNCTION: CALCULATION OF SHAPE FUNCTIONS AND THEIR FIRST PARTIAL
! DERIVATIVES / LOCAL COORDINATES
! NOTE:
! INPUT :
! TYPE : ELEMENT TYPE
! NNE : NUMBER OF NODES OF THE ELEMENT
! NE : NUMBER OF ELEMENTS
! IE : ID OF CURRENT ELEMENT
! IEN ( :, :) : CONNECTIVITY TABLE
! EH ( :, :) : ELEMENTARY MATRIX [He] OF DIMENSION NNE X NNE
! XL( :, :) : COORDINATES OF THE ELEMENT NODES
! IPROP : ID OF THE ELEMENT PROPERTY
! R, S, T : COORDINATES OF THE INTEGRATION POINT
! OUPUT :
! SHPL( :, :) : ARRAY OF DIMENSION 4 * NNE CONTAINING THE FOLLOWING
! INFORMATION :
! SHPL(4,i) = Ni(r,s,t) for i=1,2,...NNE
! SHPL(1,i) = dNi(r,s,t) /dr for i=1,2,...NNE
! SHPL(2,i) = dNi(r,s,t) /ds for i=1,2,...NNE
! SHPL(3,i) = dNi(r,s,t) /dt for i=1,2,...NNE

!==================== =========================
IMPLICIT NONE
INTEGER :: NNE
REAL*8, DIMENSION(4,NNE) :: SHPL(4,NNE)
REAL*8 :: R,S,T
CHARACTER*10 :: SHAPE


SELECT CASE( NNE )
    CASE( 8:27)
        CALL HEXAEDRON(NNE,R,S,T,SHPL)
    CASE( 6)
        CALL PRISME(NNE,R,S,T,SHPL)
    CASE DEFAULT
        PRINT *, 'NNE=',NNE
        PRINT *, 'ERROR IN ROUTINE SHAPEL_3D: ELEMENT UNKNOWN'
END SELECT
RETURN
END SUBROUTINE SHAPEL_3D

SUBROUTINE PRISM(NNE,R,S,T, SHPL)
! CODE : GAUSFEM
! LIBRARY : SHAPE
! ELEMENT: LINEAR PENTAHEDRON
! FUNCTION: CALCULATION OF SHAPE FUNCTIONS AND THEIR FIRST PARTIAL
! DERIVATIVES / LOCAL COORDINATES
!NOTE:
! INPUT :
```

```
! NNE : NUMBER OF NODES OF THE ELEMENT (HERE=6)
! R, S, T : COORDINATES OF THE INTEGRATION POINT
! OUTPUT :
! SHPL( :, :) : ARRAY OF DIMENSION 4 * NNE CONTAINING THE FOLLOWING
! INFORMATION :
! SHPL(4,i) = Ni(r,s,t) for i=1,2,...NNE
! SHPL(1,i) = dNi(r,s,t) /dr for i=1,2,...NNE
! SHPL(2,i) = dNi(r,s,t) /ds for i=1,2,...NNE
! SHPL(3,i) = dNi(r,s,t) /dt for i=1,2,...NNE
!===============================================================

IMPLICIT NONE
integer :: NNE
real*8 :: R,S,T,A,B
real*8, dimension(4,NNE) :: SHPL
! CONSTANTS
A=0.5D0*(1d0-T)
B=0.5D0*(1d0+T)
! CALCULATION OF SHPL(4,i) = Ni(r,s,t) for i=1,2,...NNE
SHPL(4,1)=A*(1d0-R-S)
SHPL(4,2)=A*R
SHPL(4,3)=A*S
SHPL(4,4)=B*(1d0-R-S)
SHPL(4,S)=B*R
SHPL(4,6)=B*S
! CALCULATION OF SHPL(1,i) = dNi(r,s,t)/dr  for i=1,2,...NNE
SHPL(1,1)=-A
SHPL(1,2)=A
SHPL(1,3)=0.0D0
SHPL(1,4)=-B
SHPL(1,5)=B
SHPL(1,6)=0.D0
! CALCULATION OF dNi(r,s,t)/ds for i=1,2,...NNE
SHPL(2,1 )=-A
SHPL(2,2)=0.D0
SHPL(2,3)=A
SHPL(2,4 )=-B
SHPL(2,5)=0.
SHPL(2,6)=B
! CALCULATION OF dNi(r,s,t)/dt for i=1,2,...NNE
SHPL(3,1)=-0.5D0*(1d0-R-S)
SHPL(3,2)=-0.5D0*R
SHPL(3,3)=-0.5D0*S
SHPL(3,4)=0.5D00*(1d0-R-S)
SHPL(3,5)=0.5D0*R
SHPL(3,6)=0.5D0*S
RETURN
END SUBROUTINE PRISM
```

```
SUBROUTINE HEXAEDRON(NNE,R,S,T, SHPL)
! CODE : GAUSFEM
! LIBRARY : SHAPE
! ELEMENT: 8 node HEXAHEDRON
! FUNCTION: CALCULATION OF SHAPE FUNCTIONS AND THEIR FIRST PARTIAL
! DERIVATIVES / LOCAL COORDINATES
! NOTE:
! INPUT :
! NNE : NUMBER OF NODES OF THE ELEMENT (HERE= 8)
! R, S, T : COORDINDATES OF THE INTEGRATION POINT
! OUTPUT :
! SHPL( :, :) : ARRAY OF DIMENSION 4 * NNE CONTAINING THE FOLLOWING
! INFORMATION :
! SHPL(4,i) = Ni(r,s,t) for i=1,2,...NNE
! SHPL(1,i) = dNi(r,s,t) /dr for i=1,2,...NNE
! SHPL(2,i) = dNi(r,s,t) /ds for i=1,2,...NNE
! SHPL(3,i) = dNi(r,s,t) /dt for i=1,2,...NNE
IMPLICIT NONE
INTEGER :: NNE,I
REAL*8 :: R,S,T
REAL*8, DIMENSION(4,NNE) :: SHPL
! CALCULATION OF SHPL(4,i) = Ni(r,s,t) for i=1,2,...NNE
SHPL(4,1)= 0.125d0*(1d0-R)*(1d0-S)*(1d0-T)
SHPL(4,2)= 0.125d0*(1d0+R)*( 1d0-S)*(1d0-T)
SHPL(4,3)= 0.125d0*( 1d0+R)*( 1d0+S)*(1d0-T)
SHPL(4,4)= 0.125d0*(1d0-R)*(1d0+S)*(1d0-T)
SHPL(4,5)= 0.125d0*(1d0-R)*(1d0-S)*(1d0+T)
SHPL(4,6)= 0.125d0*(1d0+R)*(1d0-S)*(1d0+T)
SHPL(4,7)= 0.125d0*(1d0+R)*(1d0+S)*(1d0+T)
SHPL(4,8)= 0.125d0*(1d0-R)*(1d0+S)*(1d0+T)
! CALCULATION OF SHPL(1,i) = dNi(r,s,t)/dr  for i=1,2,...NNE
SHPL(1,1)=-0.125d0*(1d0-S)*(1d0-T)
SHPL(1,2)= 0.125d0*(1d0-S)*(1d0-T)
SHPL(1,3)= 0.125d0*(1d0+S)*(1d0-T)
SHPL(1,4)=-0.125d0*(1d0+S)*(1d0-T)
SHPL(1,5)=-0.125d0*(1.00-S)*(1d0+T)
SHPL(1,6)= 0.125d0*(1d0-S)*(1d0+T)
SHPL(1,7)= 0.125d0*(1d0+S)*(1d0+T)
SHPL(1,8)=-0.125d0*(1d0+S)*(1d0+T)
! CALCULATION OF SHPL(2,i) = dNi(r,s,t)/ds for i=1,2,...NNE
SHPL(2,1)=-0.125d0*(1d0-R)*(1d0-T)
SHPL(2,2)=-0.125d0*(1d0+R)*(1d0-T)
SHPL(2,3)= 0.125d0*(1d0+R)*(1d0-T)
SHPL(2,4)= 0.125d0*(1d0-R)*(1d0-T)
SHPL(2,5)=-0.125d0*(1d0-R)*(1d0+T)
SHPL(2,6)=-0.125d0*(1d0+R)*(1d0+T)
SHPL(2,7)= 0.125d0*(1d0+R)*(1d0+T)
SHPL(2,8)= 0.125d0*(1d0-R)*(1d0+T)
! CALCULATION OF SHPL(3,i) = dNi(r,s,t)/dt for i=1,2,...NNE
```

```
SHPL(3,1)=-0.125d0*(1d0-R)*(1d0-S)
SHPL(3,2)=-0.125d0*(1d0+R)*(1d0-S)
SHPL(3,3)=-0.125d0*(1d0+R)*(1d0+S)
SHPL(3,4)=-0.125d0*(1d0-R)*(1d0+S)
SHPL(3,5)= 0.125d0*(1d0-R)*(1d0-S)
SHPL(3,6)= 0.125d0*(1d0+R)*(1d0-S)
SHPL(3,7)= 0.125d0*(1d0+R)*(1d0+S)
SHPL(3,8)= 0.125d0*(1d0-R)*(1d0+S)

RETURN
END SUBROUTINE HEXAEDRON

SUBROUTINE JACOBI_3D(XJ,DET,SHPL,XE,NNE)
! CODE : GAUSFEM
! LIBRARY : SHAPE
! ELEMENT: 3D ELEMENT CLASS
! Function: CALCULATION OF THE JACOBIAN MATRIX
! NOTE:
! INPUT :
! SHPL( :, :) : LOCAL SHAPE FUNCTIONS AND THEIR FIRST PARTIAL
! DERIVATIVES
! LOCAL COORDINATES EVALUATED AT THE INTEGRATION POINT
! XE( :, :) : COORDINATES OF THE ELEMENT NODES
! NNE : NUMBER OF NODES OF THE ELEMENT
! OUTPUT:
! XJ : TRANSPOSE OF THE JACOBIAN MATRIX [J^e]
!=[a1,a2,a3]  with a1=dx/dxsi1, a2=dx/dxi2, a3=dx/dxsi3

! DET : DETERMINANT OF THE JACOBIAN MATRIX
!
IMPLICIT REAL*8(A-H,O-Z)
DIMENSION XJ(3,3),SHPL(4,NNE),XE(3,NNE)
XJ =0.D0
DO I=1,3
    DO IN=1,NNE
        XJ(I,1)=XJ(I,1)+SHPL(I,IN)*XE(1,IN)
        XJ(I,2)=XJ(I,2)+SHPL(I,IN)*XE(2,IN)
        XJ(I,3)=XJ(I,3)+SHPL(I,IN)*XE(3,IN)
    END DO
END DO
DET=XJ(1,1)*XJ(2,2)*XJ(3,3)+XJ(2,1)*XJ(3,2)*XJ(1,3) &
+XJ(3,1)*XJ(1,2)*XJ(2,3) - XJ(1,1)*XJ(3,2)*XJ(2,3) &
-XJ(2,1)*XJ(1,2)*XJ(3,3)- XJ(3,1)*XJ(2,2)*XJ(1,3)
RETURN
END SUBROUTINE JACOBI_3D

SUBROUTINE INVT_3x3(A,B)
!=====================================================
! CODE : GAUSFEM
! LIBRARY: UTIL
```

```
! ELEMENT: 3D ELEMENT CLASS
! Function: CALCULATION OF THE INVERSE OF THE TRANSPOSE OF A MATRIX
! 3X3 [B]=inv([A]^T)
! NOTE:
! INPUT:
! A(:,:): SQUARE MATRIX OF DIMENSIONS 3X3
! OUTPUT:
! B(:,:) : SQUARE MATRIX INVERSE OF THE TRANSPOSE OF [B]=inv([A]^T)
! ==========================================================
IMPLICIT REAL*8(A-H,O-Z)
DIMENSION A(3,3),B(3,3)
DET=A(1,1)*( A(2,2)*A(3,3)-A(3,2)*A(2,3)) &
+ A(2,1)*( A(3,2)*A(1,3) - A(1,2)*A(3,3)) &
+ A(3,1)*( A(1,2)*A(2,3) - A(2,2)*A(1,3))
RMIN=DMIN1(DABS(A(1,1)),DABS(A(2,2)),DABS(A(3,3)))
RMIN=(1D-10)*RMIN
IF (DABS(DET) .LE. RMIN) THEN
    WRITE(*,*) 'Determinant=',DET
    STOP 'Matrix A^{T} singular'
ELSE
    B(1,1)=( A(2,2)*A(3,3)-A(3,2)*A(2,3) )/DET
    B(1,2)=( A(2,3)*A(3,1)- A(2,1)*A(3,3) )/DET
    B(1,3)=( A(2,1)*A(3,2)-A(2,2)*A(3,1) )/DET
    B(2,1)=( A(3,2)*A(1,3)-A(1,2)*A(3,3) )/DET
    B(2,2)=( A(1,1)*A(3,3)-A(1,3)*A(3,1) )/DET
    B(2,3)=( A(1,2)*A(3,1) - A(1,1)*A(3,2) )/DET
    B(3,1)=( A(1,2)*A(2,3)-A(2,2)*A(1,3) )/DET
    B(3,2)=( A(1,3)*A(2,1) - A(1,1)*A(2,3) )/DET
    B(3,3)=( A(1,1)*A(2,2)- A(1,2)*A(2,1) )/DET
END IF
RETURN
END SUBROUTINE INVT_3x3

SUBROUTINE SHAPEG_3D(XJtinv,SHP1,SHPG,NNE)
! CODE : GAUSFEM
! LIBRARY: SHAPE
! ELEMENT: 3D ELEMENT CLASS
! Function: CALCULATION OF THE SHAPE FUNCTIONS AND THEIR FIRST PARTIAL
! DERIVATIVE / PHYSICAL COORDINATES
! NOTE:
! INPUT :
! SHP1( :, :): SHAPE FUNCTIONS AND THEIR FIRST PARTIAL
! DERIVATIVES / LOCAL COORDINATES
! DET : DETERMINANT OF THE JACOBIAN MATRIX
! XJTINV : INVERSE DE JACOBIAN MATRIX
! IE : CURRENT ELEMENT
! OUTPUT :
! SHPG( :, :) : ARRAY OF DIMENSION 4 * NNE CONTAINING THE FOLLOWING
! INFORMATION :
```

```
! SHPG(4,i) = Ni(r,s,t) for i=1,2,...NNE
! SHPG(1,i) = SHPL(1,i) = dNi(r,s,t)/dr  for i=1,2,...NNE
! SHPG(2,i) = dNi(r,s,t) /ds for i=1,2,...NNE
! SHPG(3,i) = dNi(r,s,t) /dt for i=1,2,...NNE
!=================================================
IMPLICIT REAL*8(A-H,O-Z)
DIMENSION SHP1(4,NNE),SHPG(4,NNE),XJtinv(3,3)
DO J=1,NNE
    DO I=1,3
        SHPG(I,J)=SHP1(1,J)*XJtinv(1,I)+SHP1(2,J)*XJtinv(2,I) &
        +SHP1(3,J)*XJtinv(3,I)
    END DO
    SHPG(4,J)=SHP1(4,J)
END DO
RETURN
END SUBROUTINE SHAPEG_3D
```

APPENDIX 5E: ASSEMBLING

In practice, the assembling procedure is intimately linked to (1) the storage format of global sparse matrices (compressed sparse row [CSR], compressed sparse column [CSC], coordinate [COO], diagonal storage, skyline storage, and block-compressed sparse row [BSR]) and (2) the technique to account for boundary conditions.

This appendix contains Fortran-assembling subroutines in the case where matrices are stored in full format. These procedures automatically account for zero-constrained variables by simply not assembling the associated entries in the elementary matrices and vectors.

```
SUBROUTINE CALCLM(NDOF, NNT, NNE,NE,NEE,NEQ,IEN,IDEST,LM)
! FUNCTION: CALCULATION OF DEGREES OF FREEDOM LOCALIZATION
! MATRIX
! LM(:,:)
! NOTE:
! INPUT :
! NDOF : NUMBER OF DEGREES OF FREEDOM PER NODE
! NNT : TOTAL NUMBER OF NODES OF THE MESH
! NNE : NUMBER OF NODES OF THE ELEMENT
! NE : TOTAL NUMBER OF ELEMENTS OF THE MESH
! NEE : NUMBER OF EQUATIONS OF THE ELEMENT (= NUMBER OF DOF
! OF THE ELEMENT)
! NEQ : NUMBER OF EQUATIONS OF THE PROBLEM (=DIMENSION OF
! ASSEMBLED MATRIX)
! IEN(:,:) : ARRAY CONTAINING THE CONNECTIVITY OF THE
! ELEMENTS
```

```
! IDEST(:,:): ARRAY CONTAINING ON INPUT THE ZERO-CONSTRAINED
! DOF
! FOR DOF "IDOF" OF NODE "IN" WE HAVE:
! IDEST(IDOF,IN) >= 0 FOR AN ACTIVE DOF
! IDEST(IDOF,IN) < 0 FOR A ZERO-CONSTRAINED DOF
! OUTPUT :
! IDEST(:,:): ARRAY CONTAINING THE POSITIONS OF THE ACTIVE
! DOF
! FOR DOF "IDOF" OF NODE "IN" WE HAVE:
! - IDEST(IDOF,IN) = POSITION IN THE VECTOR CONTAINING ALL
! THE DOFS OF
! ACTIVE DOF "IDOF" OF NODE "IN" FOR AN ACTIVE DOF
! - IDEST(IDOF,IN) = 0 FOR A ZERO-CONSTRAINED DOF
! LM(:,:) : LOCALIZATION MATRIX OF THE ELEMENTS LOCAL DOFS
! LM(IEE,IE) = POSITION IN THE VECTOR CONTAINING ALL THE DOFS
! OF LOCAL
! EQUATION "IEE" OF ELEMENT "IE"
IMPLICIT NONE
INTEGER :: NDOF, NNE, NNT, NE, NEE, NEQ, &
IE,IN,INODE,IDOF,IEE,ID
INTEGER :: IEN(NNE,NE), LM(NEE,NE), IDEST(NDOF,NNT)
! BUILDING OF THE LOCALIZATION MATRIX
ID = 0;
DO INODE = 1, NNT
     DO IDOF = 1, NDOF
          IF (IDEST(IDOF,INODE) == 0 ) THEN
                 ID = ID+1;
                 IDEST(IDOF,INODE) = ID
          ELSE
                 IDEST(IDOF,INODE) = 0
          END IF
     END DO
END DO
NEQ = ID  ! THE NUMBER OF EQUATIONS OF THE PROBLEM
LM = 0
DO IE= 1, NE
     DO IN= 1, NNE
          INODE = IEN(IN,IE)
               DO IDOF = 1, NDOF
                    IEE = NDOF*(IN-1) + IDOF
                    IF (IDEST(IDOF,INODE) >= 0 ) THEN
                          LM(IEE,IE) = ID
                    END IF
          END DO
     END DO
END DO
RETURN
END SUBROUTINE CALCLM
```

```
SUBROUTINE ASSEMB_MAT(IE,NEE,NEQ,LM,EK,GK)
! FUNCTION: ASSEMBLING OF AN ELEMENTARY MATRIX IN A GLOBAL
! MATRIX
! (FULL STORAGE)
! NOTE:
! INPUT :
! IE : ID OF THE ELEMENT TO BE ASSEMBLED
! NNE : NUMBER OF NODES OF THE ELEMENT
! NEQ : NUMBER OF EQUATIONS OF THE PROBLEM (=DIMENSION OF
! ASSEMBLED
! MATRIX)
! LM(:, :) : LOCALIZATION MATRIX OF THE ELEMENTS LOCAL DOFS
! LM(IEE,IE) = POSITION IN THE VECTOR CONTAINING ALL THE DOFS
! OF LOCAL
! EQUATION "IEE" OF ELEMENT "IE"
! EK(:,:) : REAL ELEMENTARY MATRIX
! OUTPUT:
! GK(:,:) : REAL ASSEMBLED MATRIX
IMPLICIT NONE
INTEGER :: NEE, IE,NEQ,I,J,IG,JG
INTEGER :: LM(NEE, *)
REAL*8 :: GK(NEQ, NEQ),EK(NEE,NEE)
! ASSEMBLING
DO I = 1, NEE
       IG = LM(I,IE)
       IF (IG == 0) CYCLE
       DO J = 1,NEE
              JG = LM(J,IE)
              IF (JG ==0) CYCLE
              GK(IG,JG) = GK(IG,JG) + EK(I,J)
       END DO
END DO
RETURN
END SUBROUTINE ASSEMB_MAT


SUBROUTINE ASSEMB_VEC(IE,NEE,LM,EF,GF)
! FUNCTION: ASSEMBLE AN ELEMENTARY VECTOR IN A GLOBAL VECTOR
! NOTE:
! INPUT :
! IE : ID OF THE ELEMENT TO BE ASSEMBLED
! NNE : NUMBER OF NODES OF THE ELEMENT
! NEQ : NUMBER OF EQUATIONS OF THE PROBLEM (=DIMENSION OF
! THE ASSEMBLED MATRIX)
! LM(:, :) : LOCALIZATION MATRIX OF THE DEGREES OF FREEDOM
! OF THE ELEMENTS
! LM(IEE,IE) = POSITION IN THE GLOBAL VECTOR OF THE LOCAL
! EQUATION "IEE" OF ELEMENT "IE"
! EF(:,:) : REAL ELEMENTARY VECTOR
```

```
! OUTPUT :
! GF(:,:) : REAL ASSEMBLED VECTOR


IMPLICIT NONE
INTEGER :: IE, NEE,I,  IG, LM(NEE,*)
REAL*8 :: GF(*),EF(*)
! ASSEMBLING
DO I=1, NEE
       IG=LM(I,IE)
       IF (IG == 0) CYCLE
       GF(IG)=GF(IG)+EF(I)
END DO
RETURN
END SUBROUTINE ASSEMB_VEC
```

REFERENCES

Allard, J. F. and N. Atalla. 2009. *Propagation of Sound in Porous Media, Modelling Sound Absorbing Materials, 2nd Ed.* 2nd ed. Chichester, UK: Wiley-Blackwell.

Bathe, K.-J. 1996. *Finite Element Procedures*. Englewood Cliffs, NJ, USA: Prentice-Hall.

Batoz, J.-L. and G. Dhatt. 1990. *Modélisation des structures par éléments finis*. Vol 1. Sainte-Foy, Quebec, Canada: Presses Université Laval.

CockBurn, J. A. and A. C. Jolly. 1968. *Structural–Acoustic Response, Noise Transmission Losses and Interior Noise Levels of an Aircraft Fuselage Excited by Random Pressure Fields*. Technical report AFFDL-TR-68-2. Huntsville, AL, USA: Defense Technical Information Center.

Cook, R. D., D. S. Malkus, M. E. Plesha, and R. J. Witt. 2002. *Concepts and Applications of Finite Element Analysis*. 4th ed. Hoboken, NJ, USA: Wiley & Sons.

Corcos, G. M. 1963. Resolution of pressure in turbulence. *Journal of the Acoustical Society of America* 35 (2): 192–99.

Coyette, J. P., Y. Detandt, G. Lielens, and B. Van den Nieuwenhof. 2008. *Vibro-Acoustic Simulation of Mechanical Components Subjected to Distributed Pressure Excitations*. Paris: Société Française d'acoustique.

Coyette, J. P. and K. Meerbergen. 2008. An efficient computational procedure for random vibro-acoustic simulations. *Journal of Sound and Vibration* 310: 448–58.

Craveur, J. C. 1997. *Modélisation des structures: calcul par éléments finis*. Paris, France: Masson.

Dazel, O., F. Sgard, F. X. Becot, and N. Atalla. 2008. Expressions of dissipated powers and stored energies in poroelastic media modeled by {u,U} and {u,P} formulations. *Journal of the Acoustical Society of America* 123 (4): 2054–63.

Efimtsov, B. M. 1982. Characteristics of the field of turbulent wall pressure fluctuations at large. Reynolds numbers. *Soviet Physics Acoustics* 28 (4): 289–92.

Géradin, M. and D. Rixen. 1997. *Mechanical Vibrations: Theory and Application to Structural Dynamics*. Chichester, UK: John Wiley.

Hughes, T. J. R. 2000. *The Finite Element Method: Linear Static and Dynamic Finite Element Analysis*. Mineola, NY, USA: Dover Publications.

Imbert, J. F. 1991. *Analyse des structures par éléments finis*. Toulouse, France: Editions Cépaduès.

Petyt, M. 1990. *Introduction to Finite Element Vibration Analysis*. 1st ed. Cambridge, UK: Cambridge University Press.

Pierce, A. D. 1989. *Acoustics, an Introduction to Its Physical Principles and Applications*. New York, USA: McGraw-Hill.

Reddy, J. N. 1993. *An Introduction to the Finite Element Method*. New York, USA: Mc Graw-Hill.

Shames, I. H. and C. L. Dym. 1995. *Energy and Finite Element Methods in Structural Mechanics*. New Delhi, India: New Age International.

Witting, L. E. and A. K. Sinha. 1975. Simulation of multicorrelated random processes using FFT algorithm. *Journal of the Acoustical Society of America* 58 (3): 630–34.

Zienkiewicz, O. C. and R. L. Taylor. 2000. *The Finite Element Method: The Basis*. 5th ed. Vol. 1. 3 vols. Woburn, MA, USA: Butterworth-Heinemann.

Zienkiewicz, O. C. and R. L. Taylor. 2005. *The Finite Element Method Set*. Burlington, MA, USA: Elsevier.

Chapter 6

Interior structural acoustic coupling

6.1 INTRODUCTION

Often, we have to study problems where the structure interacts with a surrounding fluid. The importance of this interaction depends on the fluid nature and on the structure of interest. A simple criterion to characterize the amount of coupling for a plate-like structure is given by (Lesueur 1988)

$$\beta_c = \frac{\rho_0 c_0}{\rho_s h_s \omega_c} \tag{6.1}$$

with

- h_s the characteristic thickness of the structure
- ρ_s the density of the structure
- ρ_0 the density of the fluid
- c_0 the sound speed in the fluid
- ω_c the characteristic frequency of the structure (critical frequency in the case of a plate).

For $\beta_c \ll 1$, the coupling is considered as weak. This is generally the case of a structure that is coupled to a light fluid. In this configuration, a sequential approach is sufficient to characterize the vibroacoustic behavior of the system. It consists of two steps:

- Calculation of the structural vibratory response by neglecting the coupling with the fluid (Chapter 5).
- Calculation of the acoustic response using the structural displacements as boundary conditions (Chapters 5 and 7).

For $\beta_c > 1$ or ≈ 1, the coupling is considered as strong. This is the case when the structure radiates in a heavy fluid. In this configuration, the fluid modifies the vibratory and acoustic response of the structure. To adequately model such situations, it is necessary to account for the fluid loading. In other words, we must solve simultaneously the coupled equations governing the wave propagation in the fluid and in the structure. This is the objective of this chapter and Chapter 8.

It is important to note that Equation 6.1 is not general. Other parameters such as the geometry, the excitation, and the dynamic characteristics of the fluid may influence the strength of the fluid–structure coupling. For example, in the case of a thin plate coupled to a shallow fluid cavity (light fluid) the coupling cannot be neglected. Another example is the vibroacoustic response of a thin plate in a light fluid with a mean flow (Sgard 1995).

6.2 THE DIFFERENT TYPES OF FLUID–STRUCTURE INTERACTION PROBLEMS

Fluid–structure coupled problems can be classified into three main categories (see Figure 6.1):

- *Interior problems*: In this case, the fluid domain is bounded. The formulations used to solve the coupled problem are generally based on

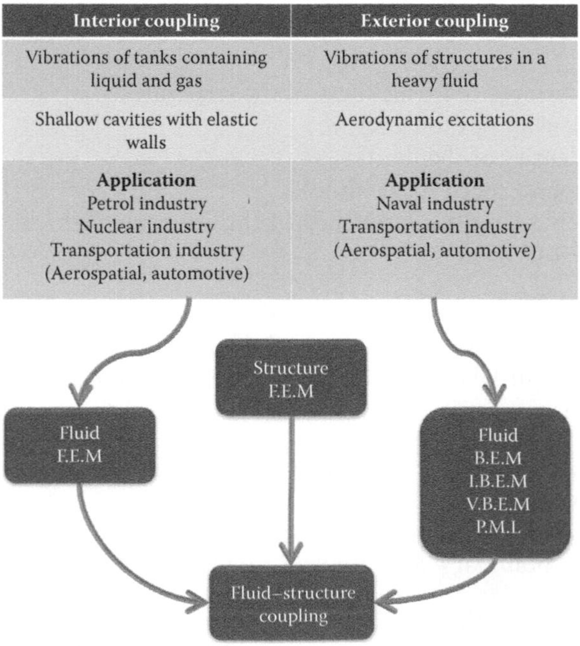

Figure 6.1 Examples of coupled problems.

the finite element method for the fluid and the structure. This is the subject of the present chapter.

- *Exterior problems*: In this case, the fluid domain is unbounded. The formulations used to solve the coupled problem are based on a mixed finite element method for the structure and a boundary element method for the fluid. These formulations are discussed in Chapter 8.
- *Interior/exterior problems*: In this case, the structure couples two fluid domains. The interior domain is bounded and the exterior domain is unbounded (example of the sound radiation of a structure coupled to an internal cavity and radiating in a heavy fluid). A combination of the finite element method for the structure and the interior fluid domain and boundary elements for the exterior fluid can be used. The numerical techniques associated to this class of problems are discussed in Chapter 8.

6.3 CLASSIC FORMULATIONS

6.3.1 $(\hat{\underline{u}}, \hat{p})$ formulation

6.3.1.1 Governing equations

The geometry of the problem is depicted in Figure 6.2.

The structure Ω_s is supposed to be linearly elastic, and the internal fluid Ω_f is assumed to be homogeneous, perfect, and at rest. Using the notations of Chapter 2, the equations for the structural displacement and the acoustic pressure in the fluid can be written as

$$\underline{\nabla} \cdot \underline{\underline{\hat{\sigma}}} + \rho_s \omega^2 \underline{\hat{u}} = 0 \quad \text{in } \Omega_s \tag{6.2a}$$

$$\underline{\underline{\hat{\sigma}}} \cdot \underline{n}^s = \underline{\hat{F}} \quad \text{over } \partial\Omega_{s,F} \tag{6.2b}$$

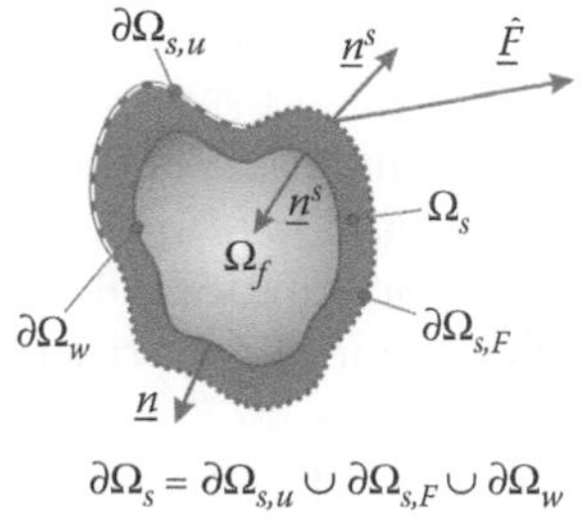

Figure 6.2 Fluid–structure coupled problem.

$$\underline{\underline{\hat{\sigma}}} \cdot \underline{n}^s = \hat{p}\underline{n} \quad \text{over } \partial\Omega_w \tag{6.2c}$$

$$\underline{\hat{u}} = \underline{\bar{u}} \quad \text{over } \partial\Omega_{s,u} \tag{6.2d}$$

$$\nabla^2\hat{p} + k^2\hat{p} = 0 \quad \text{in } \Omega_f \tag{6.2e}$$

$$\frac{\partial\hat{p}}{\partial n} = \rho_0\omega^2\underline{\hat{u}} \cdot \underline{n} \quad \text{over } \partial\Omega_w \tag{6.2f}$$

where $\underline{\hat{u}}$ is the structural displacement field, $\underline{\underline{\hat{\sigma}}}$ is the structural stress tensor, ρ_s is the structure density, $\hat{p}$ is the acoustic pressure in the fluid, ρ_0 is the fluid density, $\underline{n}^s$ and $\underline{n}$ are the outward normal to the structure and to the fluid, respectively, and $k = \omega/c_0$ is the wavenumber in Ω_f:

- Equation 6.2a represents the linear elasto-dynamic equation.
- Equation 6.2b corresponds to the external force per unit area applied on the structure.
- Equation 6.2c corresponds to the force per unit area acting on the structure induced by the internal fluid on the wetted surface $\partial\Omega_w$.
- Equation 6.2d corresponds to the imposed displacement vector.
- Equation 6.2e is Helmholtz equation.
- Equation 6.2f represents the linearized Euler's equation (continuity of fluid and structural normal displacements on the surface of the structure $\partial\Omega_s$).

Remark: The problem defined by Equations 6.2a through 6.2f is not well-posed (valid) in the static case ($\omega = 0$). This will be discussed in section 6.3.1.3 (see Equation 6.14).

6.3.1.2 Variational formulation of the coupled problem

Let us apply Galerkin's method (see Chapter 3). We proceed in two steps:

- Write the weak integral formulation of the equations for the structure subjected to the "pressure fluid loading".
- Write the weak integral formulation of the equations for the fluid subjected to the "displacement of the structural boundary."

By multiplying Equation 6.2a by $\delta\underline{\hat{u}}$, a regular admissible function defined in Ω_s, applying Green's formula (Appendix 3B), and accounting for Equations 6.2b and 6.2c we get

$$\int_{\Omega_s} \underline{\underline{\hat{\sigma}}}(\underline{\hat{u}}) : \underline{\underline{\hat{\varepsilon}}}(\delta\underline{\hat{u}})\, dV - \int_{\Omega_s} \rho_s \omega^2 \underline{\hat{u}} \cdot \delta\underline{\hat{u}}\, dV - \int_{\partial\Omega_w} \hat{p}\underline{n} \cdot \delta\underline{\hat{u}}\, dS = \int_{\partial\Omega_{s,F}} \delta\underline{\hat{u}} \cdot \underline{\hat{F}}\, dS \quad \forall\, \delta\underline{\hat{u}} \tag{6.3}$$

where $\underline{\hat{u}}$ satisfies the kinematic boundary condition as in Equation 6.2d and $\underline{\underline{\hat{\varepsilon}}}$ is the structural strain tensor.

This equation expresses the principle of virtual work. It represents the variational formulation of the problem of the harmonic vibratory response of the structure subjected to the fluid pressure in Ω_f.

In the second step, we multiply Equation 6.2e by $\delta\hat{p}$, a regular admissible function defined in Ω_f. We then apply Green's formula and account for Equation 6.2f to get

$$\int_{\Omega_f} \frac{1}{\rho_0} \underline{\nabla}\hat{p} \cdot \underline{\nabla}\delta\hat{p}\, dV - \omega^2 \int_{\Omega_f} \frac{1}{\rho_0 c_0^2} \hat{p}\delta\hat{p}\, dV - \omega^2 \int_{\partial\Omega_w} \underline{\hat{u}} \cdot \underline{n}\delta\hat{p}\, dS = 0 \quad \forall\, \delta\hat{p} \tag{6.4}$$

This equation expresses the variational formulation of the problem of the harmonic response of the fluid to a motion of its boundary induced by the structure.

The variational formulation of the coupled problem consists in finding $\underline{\hat{u}}$ and $\hat{p}$, regular admissible functions, satisfying Equations 6.3 and 6.4 for all kinematically admissible $\delta\underline{\hat{u}}$ and $\delta\hat{p}$.

6.3.1.3 Discretization of the coupled system equations

The finite element discretization of Equations 6.3 and 6.4 involves the following matrices and vectors (see Chapter 5):

$$\int_{\Omega_s} \underline{\underline{\hat{\sigma}}}(\underline{\hat{u}}) : \underline{\underline{\hat{\varepsilon}}}(\delta\underline{\hat{u}})\, dV = \langle \delta\hat{u} \rangle [K] \{\hat{u}\} \tag{6.5}$$

$$\int_{\Omega_s} \rho_s \underline{\hat{u}} \cdot \delta\underline{\hat{u}}\, dV = \langle \delta\hat{u} \rangle [M] \{\hat{u}\} \tag{6.6}$$

$$\int_{\partial\Omega_w} \hat{p}\underline{n} \cdot \delta\underline{\hat{u}}\, dS = \langle \delta\hat{u} \rangle [C_{up}] \{\hat{p}\} \tag{6.7}$$

$$\int_{\partial\Omega_{s,F}} \delta\underline{\hat{u}} \cdot \underline{\hat{F}}\, dS = \langle \delta\hat{u} \rangle \{\hat{F}\} \tag{6.8}$$

$$\int_{\Omega_f} \frac{1}{\rho_0}(\underline{\nabla}\hat{p} \cdot \underline{\nabla}\delta\hat{p})\, dV = \langle\delta\hat{p}\rangle[H]\{\hat{p}\} \tag{6.9}$$

$$\int_{\Omega_f} \frac{1}{\rho_0 c_0^2}\hat{p}\delta\hat{p}\, dV = \langle\delta\hat{p}\rangle[Q]\{\hat{p}\} \tag{6.10}$$

$$\int_{\partial\Omega_w} \underline{\hat{u}} \cdot \underline{n}\delta\hat{p}\, dS = \langle\delta\hat{p}\rangle[C_{up}]^T\{\hat{u}\} \tag{6.11}$$

where $\{\hat{u}\}$ and $\{\hat{p}\}$ denote the nodal vector of unknowns for the structure and the fluid, respectively. Matrices $[M]$ and $[K]$ denote the mass and stiffness matrices of the structure, respectively. $\{\hat{F}\}$ is the external loading nodal vector acting on the structure. Matrices $[H]$ and $[Q]$ are related to the kinetic energy and compressional energy matrices of the fluid (see Chapter 5). Finally, matrix $[C_{up}]$ is a surface coupling matrix between the structure and the fluid. It is commonly calculated using a finite element discretization of the fluid–structure interface (see Chapter 5 for its numerical calculation).

Using Equations 6.5 through 6.11 the discretized formulation of the coupled system can be written as

$$\begin{aligned}&\langle\delta\hat{u}\rangle\Big(\big([K]-\omega^2[M]\big)\{\hat{u}\}-[C_{up}]\{\hat{p}\}-\{\hat{F}\}\Big)+\\&\langle\delta\hat{p}\rangle\Big(\big([H]-\omega^2[Q]\big)\{\hat{p}\}-\omega^2[C_{up}]^T\{\hat{u}\}\Big)=0\end{aligned} \quad \forall\langle\delta\hat{u}\rangle,\ \forall\langle\delta\hat{p}\rangle \tag{6.12}$$

Invoking the stationarity of the formulation, after imposition of the boundary conditions, leads to the coupled forced system

$$\begin{bmatrix} [K]-\omega^2[M] & -[C_{up}] \\ -\omega^2[C_{up}]^T & [H]-\omega^2[Q] \end{bmatrix} \begin{Bmatrix} \{\hat{u}\} \\ \{\hat{p}\} \end{Bmatrix} = \begin{Bmatrix} \{\hat{F}\} \\ \{0\} \end{Bmatrix} \tag{6.13}$$

Note that this equation is not valid for $(\omega = 0)$ since the constraint relating the pressure to the wall normal displacement, namely:

$$\frac{1}{\rho_0 c_0^2}\int_{\Omega_f} \hat{p}\, dV + \int_{\partial\Omega_f} \underline{\hat{u}} \cdot \underline{n}\, dS = 0 \tag{6.14}$$

was not enforced. For $\omega = 0$, Equation 6.14 leads to the static pressure $\hat{p}^s$ (given by Equation 3.33). Here we omit this constraint and only consider dynamic cases $(\omega \neq 0)$. In this case, the solution of Equations 6.2a–6.2f will automatically satisfy Equation 6.14.

Dividing the second line of Equation 6.13 by ω^2 leads to a symmetric system. The direct resolution of this linear system is often numerically expensive. In practice, we rather use modal decomposition techniques (see Sections 6.3 and 6.6). However, the associated eigenvalue problem

$$\left(\begin{bmatrix} [K] & -[C_{up}] \\ [0] & [H] \end{bmatrix} - \omega^2 \begin{bmatrix} [M] & [0] \\ [C_{up}]^T & [Q] \end{bmatrix}\right) \begin{Bmatrix} \{\hat{u}\} \\ \{\hat{p}\} \end{Bmatrix} = \begin{Bmatrix} \{0\} \\ \{0\} \end{Bmatrix} \tag{6.15}$$

is nonsymmetric, which prevents the use of efficient eigenvalues and eigenvectors calculation algorithms.

Miscellaneous matrix symmetrization procedures have been proposed in the literature. Let us introduce, for example, the one presented by Irons (1970). From the second line of Equation 6.15, it is possible to write formally

$$\{\hat{p}\} = \omega^2 [H]^{-1} \left([C_{up}]^T \{\hat{u}\} + [Q]\{\hat{p}\}\right) \tag{6.16}$$

which leads, by substituting Equation 6.16 for $\{\hat{p}\}$ in the first line of Equation 6.15 and by premultiplying the second line by $[Q][H]^{-1}$, to a symmetric equivalent eigenvalue problem

$$\left(\begin{bmatrix} [K] & [0] \\ [0] & [Q] \end{bmatrix} - \omega^2 \begin{bmatrix} [M] + [C_{up}][H]^{-1}[C_{up}]^T & [C_{up}][H]^{-1}[Q] \\ [Q][H]^{-1}[C_{up}]^T & [Q][H]^{-1}[Q] \end{bmatrix}\right) \begin{Bmatrix} \{\hat{u}\} \\ \{\hat{p}\} \end{Bmatrix} = \begin{Bmatrix} \{0\} \\ \{0\} \end{Bmatrix} \tag{6.17}$$

Thus, through rather expensive numerical operations, the symmetric band feature of the global stiffness matrix is recovered together with an eigenvalue symmetric system. However, for closed cavities, matrix $[H]$ is singular and Iron's symmetrization procedure is not directly applicable. In this case, it is necessary to partition matrix $[H]$ before proceeding to the inversion of the obtained reduced matrix of rank $(n_{eq_f} - 1)$ where n_{eq_f} is the dimension of vector $\{\hat{p}\}$. Another more efficient alternative based on the preliminary modal projection of coupled system, Equation 6.13, on the uncoupled structural and fluid modal basis is discussed in Section 6.4.

Remark: In case where the interface between the structure and the cavity is occupied by an absorbing material, the latter is replaced by a massless medium with infinitesimal thickness and a complex impedance $\hat{Z}_n(\omega)$. Equations 6.2f, 6.4 and 6.13 must then be replaced by Equations 6.18a, 6.18b and 6.18c, respectively

$$\hat{p} = i\omega \hat{Z}_n(\omega)\left(\underline{\hat{U}} - \hat{u}\right) \cdot \underline{n} \quad \text{with} \quad \frac{\partial \hat{p}}{\partial n} = \rho_0 \omega^2 \underline{\hat{U}} \cdot \underline{n} \tag{6.18a}$$

$$\int_{\Omega_f} \frac{1}{\rho_0} \underline{\nabla}\hat{p} \cdot \underline{\nabla}\delta\hat{p}\, dV - \omega^2 \int_{\Omega_f} \frac{1}{\rho_0 c_0^2} \hat{p}\delta\hat{p}\, dV + i\omega \int_{\partial\Omega_w} \frac{1}{\hat{Z}_n(\omega)} \hat{p}\delta\hat{p}\, dS$$

$$- \omega^2 \int_{\partial\Omega_w} \underline{\hat{u}} \cdot \underline{n}\, \delta\hat{p}\, dS = 0 \quad \forall\, \delta\hat{p} \tag{6.18b}$$

$$\begin{bmatrix} [K] - \omega^2 [M] & -[C_{up}] \\ -\omega^2 [C_{up}]^T & [H] + i\omega[\hat{A}_{sp}] - \omega^2 [Q] \end{bmatrix} \begin{Bmatrix} \{\hat{u}\} \\ \{\hat{p}\} \end{Bmatrix} = \begin{Bmatrix} \{\hat{F}\} \\ \{0\} \end{Bmatrix} \tag{6.18c}$$

$$\int_{\Omega_f} \frac{1}{\hat{Z}_n(\omega)} \hat{p}\delta\hat{p}\, dV = \langle \delta\hat{p} \rangle [\hat{A}_{sp}] \{\hat{p}\} \tag{6.18d}$$

6.3.2 $(\underline{\hat{u}}, \hat{\varphi})$ formulation

This formulation consists in reformulating Equation 6.2 in terms of the velocity potential variable $\hat{\varphi}$ instead of the sound pressure $\hat{p}$. Using $\hat{p} = i\omega\rho_0\hat{\varphi}$, we get the coupled system from Equations 6.3 and 6.4

$$\left(\begin{bmatrix} [K] & [0] \\ [0] & [H_\varphi] \end{bmatrix} + i\omega \begin{bmatrix} [0] & -[C_{u\varphi}] \\ [C_{u\varphi}]^T & [0] \end{bmatrix} - \omega^2 \begin{bmatrix} [M] & [0] \\ [0] & [Q_\varphi] \end{bmatrix} \right) \begin{Bmatrix} \{\hat{u}\} \\ \{\hat{\varphi}\} \end{Bmatrix} = \begin{Bmatrix} \{\hat{F}\} \\ \{0\} \end{Bmatrix} \tag{6.19}$$

where matrices $[K]$, $[M]$, and vector $\{\hat{F}\}$ are given by Equations 6.5, 6.6, and 6.8. Matrices $[H_\varphi]$, $[Q_\varphi]$, and $[C_{u\varphi}]$ are, respectively, given by

$$\int_{\Omega_f} \rho_0 (\underline{\nabla}\hat{\varphi} \cdot \underline{\nabla}\delta\hat{\varphi})\, dV = \langle \delta\hat{\varphi} \rangle [H_\varphi] \{\hat{\varphi}\} \tag{6.20}$$

$$\int_{\Omega_f} \frac{\rho_0}{c_0^2} \hat{\varphi}\delta\hat{\varphi}\, dV = \langle \delta\hat{\varphi} \rangle [Q_\varphi] \{\hat{\varphi}\} \tag{6.21}$$

$$\int_{\partial\Omega_w} \hat{\varphi}\underline{n} \cdot \delta\underline{\hat{u}}\, dS = \langle \delta\hat{u} \rangle [C_{u\varphi}] \{\hat{\varphi}\} \tag{6.22}$$

By changing the sign of the second line of Equation 6.19, we end up with a system involving three symmetric matrices. However, the resolution of this system requires the use of nonstandard numerical algorithms because of the damping matrix.

Note finally that Equation 6.19 is not applicable for $\omega = 0$. In addition, the associated spectral problem includes a parasitic solution at $\omega = 0$.

6.3.3 $(\hat{\underline{u}}, \hat{\underline{U}})$ symmetric formulation

Another symmetric formulation consists in using the fluid particle displacement denoted by $\underline{U}$ instead of the sound pressure $\hat{p}$. To establish the variational formulation of the system, it suffices to consider the fluid as a particular elastic medium the behavioral law of which is

$$\hat{p} = -\rho_0 c_0^2 \underline{\nabla} \cdot \hat{\underline{U}} \tag{6.23}$$

and its displacement field satisfies the geometric condition that expresses that a perfect fluid cannot support shear deformations

$$\underline{\nabla} \times \hat{\underline{U}} = 0 \quad \text{in } \Omega_f \tag{6.24}$$

Using Equation 6.2 we derive the following weak integral formulation of the equations governing the coupled system

$$\begin{aligned} &\int_{\Omega_s} \hat{\underline{\underline{\sigma}}}(\hat{\underline{u}}) : \hat{\underline{\underline{\varepsilon}}}(\delta\hat{\underline{u}})\, dV - \int_{\Omega_s} \rho_s \omega^2 \hat{\underline{u}} \cdot \delta\hat{\underline{u}}\, dV + \int_{\Omega_f} \rho_0 c_0^2 (\underline{\nabla} \cdot \hat{\underline{U}})(\underline{\nabla} \cdot \delta\hat{\underline{U}})\, dV \\ &\quad - \omega^2 \int_{\Omega_f} \rho_0 \hat{\underline{u}} \cdot \delta\hat{\underline{U}}\, dV = \int_{\partial\Omega_{s,F}} \delta\hat{\underline{u}} \cdot \hat{\underline{F}}\, dS \quad \forall\, (\delta\hat{\underline{u}}, \delta\hat{\underline{U}}) \end{aligned} \tag{6.25}$$

where $(\hat{\underline{u}}, \hat{\underline{U}})$ are admissible, that is, they satisfy

$$\begin{cases} \hat{\underline{u}} \cdot \underline{n} = \hat{\underline{U}} \cdot \underline{n} & \text{over } \partial\Omega_w \\ \underline{\nabla} \times \hat{\underline{U}} = 0 & \text{in } \Omega_f \end{cases} \tag{6.26}$$

The discretization of Equation 6.25 leads to a symmetric system. However, this formulation $(\hat{\underline{u}}, \hat{\underline{U}})$ requires three nodal variables to describe the fluid compared to one variable for the $(\hat{\underline{u}}, \hat{p})$ formulation. An additional drawback is the enforcement of Equation 6.24. Furthermore, in the case of an incompressible fluid, it is necessary to take into account the condition $\underline{\nabla} \cdot \hat{\underline{U}} = 0$ contrary to the $(\hat{\underline{u}}, \hat{p})$ formulation where this condition is automatically accounted for (natural condition).

6.3.4 $(\hat{\underline{u}}, \hat{p}, \hat{\psi})$ symmetric formulation

To circumvent the drawbacks of symmetric formulations $(\hat{\underline{u}}, \hat{\varphi})$ and $(\hat{\underline{u}}, \hat{\underline{U}})$, Morand and Ohayon (1995) proposed a mixed formulation using two scalar variables for the fluid: the sound pressure $\hat{p}$ and the acoustic displacement potential $\hat{\psi}$. This potential is related to the velocity potential $\hat{\varphi}$ by $\hat{\varphi} = i\omega\hat{\psi}$ and consequently it is related to $\hat{p}$ by

$$\hat{\psi} = -\frac{\hat{p}}{\rho_0 \omega^2} \tag{6.27}$$

Using Equations 6.2 and 6.27, the elasto-acoustic equations can be rewritten in terms of variables $(\hat{\underline{u}}, \hat{p}, \hat{\psi})$ as

$$\underline{\nabla} \cdot \hat{\underline{\underline{\sigma}}} + \rho_s \omega^2 \hat{\underline{u}} = 0 \quad \text{in } \Omega_s \tag{6.28a}$$

$$\hat{\underline{\underline{\sigma}}} \cdot \underline{n}^s = \hat{\underline{F}} \quad \text{over } \partial\Omega_{s,F} \tag{6.28b}$$

$$\hat{\underline{\underline{\sigma}}} \cdot \underline{n}^s = -\rho_0 \omega^2 \hat{\psi} \underline{n} \quad \text{over } \partial\Omega_w \tag{6.28c}$$

$$\hat{\underline{u}} = \bar{\underline{u}} \quad \text{over } \partial\Omega_{s,u} \tag{6.28d}$$

$$-\rho_0 \nabla^2 \hat{\psi} + \frac{1}{c_0^2} \hat{p} = 0 \quad \text{in } \Omega_f \tag{6.28e}$$

$$\frac{\hat{p}}{\rho_0 c_0^2} = -\frac{\omega^2}{c_0^2} \hat{\psi} \quad \text{in } \Omega_f \tag{6.28f}$$

$$\frac{\partial \hat{\psi}}{\partial n} = -\hat{\underline{u}} \cdot \underline{n} \quad \text{over } \partial\Omega_w \tag{6.28g}$$

The Galerkin's weak formulation of Equations 6.28a through 6.28c leads to

$$\int_{\Omega_s} \hat{\underline{\underline{\sigma}}}(\hat{\underline{u}}) : \hat{\underline{\underline{\varepsilon}}}(\delta\hat{\underline{u}})\, dV - \int_{\Omega_s} \rho_s \omega^2 \hat{\underline{u}} \cdot \delta\hat{\underline{u}}\, dV + \omega^2 \int_{\partial\Omega_w} \rho_0 \hat{\psi} \underline{n} \cdot \delta\hat{\underline{u}}\, dS = \int_{\partial\Omega_{s,F}} \delta\hat{\underline{u}} \cdot \hat{\underline{F}}\, dS \tag{6.29}$$

for all admissible functions $(\hat{\underline{u}}, \delta\hat{\underline{u}})$. The Galerkin's weak formulation of Equations 6.28e and 6.28f leads to

$$\int_{\Omega_f} \rho_0 \underline{\nabla}\hat{\psi} \cdot \underline{\nabla}\delta\hat{\psi}\, dV + \int_{\Omega_f} \frac{1}{c_0^2} \hat{p}\delta\hat{\psi}\, dV + \int_{\partial\Omega_w} \rho_0 (\hat{\underline{u}} \cdot \underline{n}) \delta\hat{\psi}\, dS = 0 \tag{6.30}$$

for all admissible functions $(\hat{\psi}, \delta\hat{\psi})$. Finally, the Galerkin's weak formulation of Equation 6.28g is equivalent to the identity

$$\int_{\Omega_f} \frac{1}{\rho_0 c_0^2} \hat{p}\delta\hat{p}\, dV + \omega^2 \int_{\Omega_f} \frac{1}{c_0^2} \hat{\psi}\delta\hat{p}\, dV = 0 \tag{6.31}$$

for all admissible functions $(\hat{p}, \delta\hat{p})$.

The finite element discretization of Equations 6.29 through 6.31 combined with the imposition of the kinematic boundary conditions and the invocation of the stationarity of the functionals leads to the following matrix system:

$$\left(\begin{bmatrix} [K] & [0] & [0] \\ [0] & [Q] & [0] \\ [0] & [0] & [0] \end{bmatrix} - \omega^2 \begin{bmatrix} [M] & [0] & [C_{u\psi}] \\ [0] & [0] & [C_{p\psi}] \\ [C_{u\psi}]^T & [C_{p\psi}]^T & [H_1] \end{bmatrix} \right) \begin{Bmatrix} \{\hat{u}\} \\ \{\hat{p}\} \\ \{\hat{\psi}\} \end{Bmatrix} = \begin{Bmatrix} \{\hat{F}\} \\ \{0\} \\ \{0\} \end{Bmatrix} \tag{6.32}$$

where matrices $[M]$, $[K]$, $[H_1] = \rho_0^2[H]$, $[Q]$, and vector $\{\hat{F}\}$ are given by Equations 6.5 through 6.10. Coupling matrices $[C_{u\psi}]$ and $[C_{p\psi}]$ are given, respectively, by

$$-\left[\int_{\partial\Omega_w} \rho_0(\underline{\hat{u}} \cdot \underline{n})\delta\hat{\psi}\, dS + \int_{\partial\Omega_w} \rho_0\, (\delta\underline{\hat{u}} \cdot \underline{n})\hat{\psi}\, dS \right] = \langle \delta\hat{u} \rangle [C_{u\psi}]\{\hat{\psi}\} + \langle \delta\hat{\psi} \rangle [C_{u\psi}]^T \{\hat{u}\} \tag{6.33}$$

$$-\left[\int_{\Omega_f} \frac{1}{c_0^2} \hat{\psi}\delta\hat{p}\, dV + \int_{\Omega_f} \frac{1}{c_0^2} \hat{p}\delta\hat{\psi}\, dV \right] = \langle \delta\hat{p} \rangle [C_{p\psi}]\{\hat{\psi}\} + \langle \delta\hat{\psi} \rangle [C_{p\psi}]^T \{\hat{p}\} \tag{6.34}$$

Note that the formulation $(\underline{\hat{u}}, \hat{p}, \hat{\psi})$ leads to a symmetric system. The associated eigenvalue problem is of mass coupling type. Thus, this formulation is more interesting than symmetric formulations $(\underline{\hat{u}}, \hat{\varphi})$ and $(\underline{\hat{u}}, \underline{\hat{U}})$ because it allows for the use of efficient algorithms for calculating eigenvalue and eigenvectors. In addition, compared to the formulation $(\underline{\hat{u}}, \hat{p})$, it avoids the computationally expensive operations required by Irons's symmetrization procedure. However, in practice, formulation $(\underline{\hat{u}}, \hat{p})$ combined with the symmetric modal decomposition procedure described in Section 6.6 is used.

6.4 CALCULATION OF THE FORCED RESPONSE: MODAL EXPANSION USING UNCOUPLED MODES

We consider here the resolution of the formulation $(\underline{\hat{u}}, \hat{p})$ coupled system given by Equation 6.13. The dimension n_{eq} of this system is the sum of the active number of degrees of freedom of the structure $n_{eq,s}$ and that of the fluid $n_{eq,f}$. Although only one degree of freedom per node (the sound pressure) is used for the fluid, $n_{eq,f}$ can be important because the discretization of a fluid volume is performed. When we are interested in calculating the spectra of various vibroacoustic indicators over a large frequency band, the direct resolution of Equation 6.13 can be very costly or even impossible due to memory limitations. A simple approach based on the decomposition over the uncoupled modal basis of the structure and the fluid allows us to circumvent this issue. It consists of two steps:

- Calculation of uncoupled structural and fluid modal basis.
- Projection of the coupled system on these modal basis.

6.4.1 Calculation of uncoupled modal basis

We proceed in two steps:

- Calculation of the "*in vacuo*" normalized modal basis of the structure $([\Omega_s^2],[\Phi_s])$ truncated at order $n_{m,s}(n_{m,s} \ll n_{eq,s})$

$$\{\hat{u}\} = [\Phi_s]\{\hat{u}_m\}; \quad [\Phi_s]^T[K][\Phi_s] = [\Omega_s^2]; \quad [\Phi_s]^T[M][\Phi_s] = [I_{n_{m,s}}] \tag{6.35}$$

where $\{\hat{u}_m\}$ is the generalized displacement vector (modal coordinates vector) and

$$[\Phi_s] = [\Phi_{s,1} \quad \Phi_{s,2} \quad \cdots \quad \Phi_{s,n_{m,s}}] \tag{6.36}$$

is the matrix containing the first $n_{m,s}$ structural eigenvectors. $[\Phi_s]$ is a rectangular matrix of dimension $(n_{eq,s} \times n_{m,s})$ and

$$[\Omega_s^2] = \begin{bmatrix} \omega_{s,1}^2 & & & \\ & \omega_{s,2}^2 & & \\ & & \ddots & \\ & & & \omega_{s,n_{m,s}}^2 \end{bmatrix} \tag{6.37}$$

is a diagonal matrix of dimension $(n_{m,s} \times n_{m,s})$ containing the first $n_{m,s}$ structural natural frequencies. Finally $\left[I_{n_{m,s}}\right]$ denotes the identity matrix of size $\left(n_{m,s} \times n_{m,s}\right)$.

- Calculation of the rigid-walled cavity normalized modal basis $([\Omega_f^2],[\Phi_f])$ truncated at order $n_{m,f}$ $(n_{m,f} \ll n_{eq,f})$

$$\{\hat{p}\} = [\Phi_f]\{\hat{p}_m\}; \quad [\Phi_f]^T[H][\Phi_f] = [\Omega_f^2]; \quad [\Phi_f]^T[Q][\Phi_f] = [I_{n_{m,f}}] \tag{6.38}$$

where $\{\hat{p}_m\}$ is the generalized pressure vector (modal coordinates vector) and

$$[\Phi_f] = [\Phi_{f,1} \quad \Phi_{f,2} \quad \cdots \quad \Phi_{f,n_{m,f}}] \tag{6.39}$$

is the matrix containing the first $n_{m,f}$ eigenvectors of the fluid cavity. $[\Phi_f]$ is a rectangular matrix of dimension $(n_{eq,f} \times n_{m,f})$ and

$$[\Omega_f^2] = \begin{pmatrix} \omega_{f,1}^2 & & & \\ & \omega_{f,2}^2 & & \\ & & \ddots & \\ & & & \omega_{f,nmf}^2 \end{pmatrix} \tag{6.40}$$

is a diagonal matrix of dimension $(n_{m,f} \times n_{m,f})$ containing the first $n_{m,f}$ natural frequencies of the fluid cavity. Note that in the case of a closed rigid-walled cavity, we have $\omega_{f,1} = 0$. It is obvious that for $\omega_{f,1} = 0$ the pressure is constant; this represents a trivial solution but of no value since it doesn't satisfy the constraint $\int_{\Omega f} \hat{p} dV = 0$ (obtained from Equation 6.14 with $\underline{\hat{u}} \cdot \underline{n} = 0$). However, all the solution for $\omega_{f,i} \neq 0$ automatically verify this constraint.

6.4.2 Projection of the coupled system

Let us project the first line of Equation 6.13 on the truncated *in vacuo* modal basis of the structure $([\Omega_s^2],[\Phi_s])$ and the second line of Equation 6.13 divided by ω^2 on the truncated fluid modal basis $([\Omega_f^2],[\Phi_f])$. We get

$$\begin{pmatrix} [\Omega_s^2] - \omega^2[I_{n_{m,s}}] & -[C_{up,m}] \\ -[C_{up,m}]^T & \dfrac{1}{\omega^2}[\Omega_f^2] - [I_{n_{m,f}}] \end{pmatrix} \begin{Bmatrix} \{\hat{u}_m\} \\ \{\hat{p}_m\} \end{Bmatrix} = \begin{Bmatrix} \{\hat{F}_m\} \\ \{0\} \end{Bmatrix} \tag{6.41}$$

with

$$\{\hat{F}_m\} = [\Phi_s]^T\{\hat{F}\} \tag{6.42}$$

the generalized loading vector of the structure (vector of dimension $n_{m,s}$) and

$$[C_{up,m}] = [\Phi_s]^T[C_{up}][\Phi_f] \tag{6.43}$$

the coupling matrix projected onto the uncoupled structure/fluid modal basis (matrix of dimension $(n_{m,s} \times n_{m,f})$). The relative amplitudes of the matrix coefficients give us information about the importance of the inter-modal coupling between the structural modes and the modes of the fluid cavity.

Equation 6.41 represents a symmetric system of small size (dimension $(n_{m,s} + n_{m,f}) \times (n_{m,s} + n_{m,f})$). It can be solved using a classic algorithm such as Gauss elimination.

In the case of a modal structural damping model for the structure $\{\eta_{s,1}, \eta_{s,2}, \ldots, \eta_{s,n_{m,s}}\}$ and for the fluid $\{\eta_{f,1}, \eta_{f,2}, \ldots, \eta_{f,n_{m,f}}\}$, Equation 6.41 can be written as

$$\begin{pmatrix} [\hat{\Omega}_s^2] - \omega^2[I_{n_{m,s}}] & -[C_{up,m}] \\ -[C_{up,m}]^T & \dfrac{1}{\omega^2}[\Omega_f^2] - [\hat{I}_{n_{m,f}}] \end{pmatrix} \begin{Bmatrix} \{\hat{u}_m\} \\ \{\hat{p}_m\} \end{Bmatrix} = \begin{Bmatrix} \{\hat{F}_m\} \\ \{0\} \end{Bmatrix} \tag{6.44}$$

with

$$[\hat{\Omega}_s^2] = \begin{pmatrix} \omega_{s,1}^2(1 + i\eta_{s,1}) & & & \\ & \omega_{s,2}^2(1 + i\eta_{s,2}) & & \\ & & \ddots & \\ & & & \omega_{s,n_{m,s}}^2(1 + j\eta_{s,n_{m,s}}) \end{pmatrix} \tag{6.45}$$

and

$$[\hat{I}_{n_{m,f}}] = \begin{pmatrix} (1 - i\eta_{f,1}) & & & \\ & (1 - i\eta_{f,2}) & & \\ & & \ddots & \\ & & & (1 - i\eta_{f,n_{m,f}}) \end{pmatrix} \tag{6.46}$$

6.4.3 Pseudo-static corrections

We have seen in Section 6.4.2 that the structural and fluid modal basis are commonly truncated to a finite number of modes. This introduces errors in the solution that, in some specific cases, can be very important (see Section 6.6). In order to minimize the truncation error, the contribution of the dropped modes should be taken into account (Tournour and Atalla 2000). This can be done using an approximation of this contribution, called pseudo-static correction. Separating the dropped modes ($[\Phi_{s,d}]$ and $[\Phi_{f,d})$ from the kept modes ($[\Phi_s]$ and $[\Phi_f]$), we can write

$$\begin{aligned}\{\hat{u}\} &= [\Phi_s]\{\hat{u}_m\} + \{\hat{u}_d\} \\ &= [\Phi_s]\{\hat{u}_m\} + [\Phi_{s,d}]\{\hat{u}_{m,d}\}\end{aligned} \tag{6.47}$$

and

$$\begin{aligned}\{\hat{p}\} &= [\Phi_f]\{\hat{p}_m\} + \{\hat{p}_d\} \\ &= [\Phi_f]\{\hat{p}_m\} + [\Phi_{f,d}]\{\hat{p}_{m,d}\}\end{aligned} \tag{6.48}$$

Consider Equation 6.13 where we introduce an acoustic source nodal vector on the right-hand side denoted by $\{\hat{f}\}$. We also account for structural damping in the stiffness matrix $[\hat{K}] = [K(1 + i\eta_s)]$ and in the kinetic energy matrix $[\hat{Q}] = (1 + i\eta_f)^{-1}[Q]$. We end with

$$\begin{bmatrix} [\hat{K}] - \omega^2[M] & -[C_{up}] \\ -\omega^2[C_{up}]^T & [H] - \omega^2[\hat{Q}] \end{bmatrix} \begin{Bmatrix} \{\hat{u}\} \\ \{\hat{p}\} \end{Bmatrix} = \begin{Bmatrix} \{\hat{F}\} \\ \{\hat{f}\} \end{Bmatrix} \tag{6.49}$$

Substituting Equations 6.47 and 6.48 in Equation 6.49 and using the modes orthogonality properties, we have the following system for the dropped modes:

$$\begin{aligned}&\left[\left[\hat{\Omega}_{s,d}^2\right] - \omega^2\left[I_{n_{m,s},d}\right]\right]\left\{\hat{u}_{m,d}\right\} = \left[\Phi_{s,d}\right]^T\left(\left\{\hat{F}\right\} + \left[C_{up}\right]\left\{\hat{p}\right\}\right) \\ &\left(\left[\Omega_{f,d}^2\right] - \omega^2\left[\hat{I}_{n_{m,f},d}\right]\right)\left\{\hat{p}_{m,d}\right\} = \left[\Phi_{f,d}\right]^T\left(\left\{\hat{f}\right\} + \omega^2\left[C_{up}\right]^T\left\{\hat{u}\right\}\right)\end{aligned} \tag{6.50}$$

The contribution of the dropped modes at a frequency ω can be approximated by a pseudo-static term since their eigenfrequencies are much higher than ω. In other words, the following approximation is used:

$$\begin{aligned}&[\hat{\Omega}_{s,d}^2] - \omega^2[I_{n_{m,s},d}] \approx [\hat{\Omega}_{s,d}^2] \\ &[\Omega_{f,d}^2] - \omega^2[\hat{I}_{n_{m,f},d}] \approx [\Omega_{f,d}^2]\end{aligned} \tag{6.51}$$

and consequently Equation 6.50 becomes

$$\begin{aligned}
\left[\hat{\Omega}^2_{s,d}\right]\left\{\hat{u}_{m,d}\right\} &\approx \left[\Phi_{s,d}\right]^T\left(\left\{\hat{F}\right\} + \left[C_{up}\right]\left\{\hat{p}\right\}\right)\\
\left[\Omega^2_{f,d}\right]\left\{\hat{p}_{m,d}\right\} &\approx \left[\Phi_{f,d}\right]^T\left(\left\{\hat{f}\right\} + \omega^2\left[C_{up}\right]^T\left\{\hat{u}\right\}\right)
\end{aligned} \tag{6.52}$$

This means that the kinetic energy of the dropped displacement modes and the compressibility energy of the dropped pressure modes can be neglected. In terms of nodal variables, the pseudo-static displacements and pseudo-static pressure are then the solutions of

$$\begin{aligned}
[\hat{K}]\{\hat{u}^{(s)}\} &= \{\hat{F}\} + [C_{up}]\{\hat{p}\}\\
[H]\{\hat{p}^{(s)}\} &= \{\hat{f}\} + \omega^2[C_{up}]^T\{\hat{u}\}
\end{aligned} \tag{6.53}$$

Assuming that there are no rigid body modes, matrices $[\hat{K}]$ and $[H]$ are positive definite and

$$\begin{aligned}
\{\hat{u}^{(s)}\} &= [\hat{K}]^{-1}\left[\{\hat{F}\} + [C_{up}]\{\hat{p}\}\right]\\
\{\hat{p}^{(s)}\} &= [H]^{-1}\left[\{\hat{f}\} + \omega^2[C_{up}]^T\{\hat{u}\}\right]
\end{aligned} \tag{6.54}$$

Similarly to Equations 6.47 and 6.48, $\{\hat{u}^{(s)}\}$ and $\{\hat{p}^{(s)}\}$ can be expanded in terms of kept modes and dropped modes, namely

$$\begin{aligned}
\{\hat{u}^{(s)}\} &= \{\hat{u}^{(s)}_k\} + \{\hat{u}^{(s)}_d\}\\
&= [\Phi_s][\hat{\Omega}^2_s]^{-1}[\Phi_s]^T\left[\{\hat{F}\} + [C_{up}]\{\hat{p}\}\right] + \{\hat{u}^{(s)}_d\}
\end{aligned} \tag{6.55}$$

$$\begin{aligned}
\{\hat{p}^{(s)}\} &= \{\hat{p}^{(s)}_k\} + \{\hat{p}^{(s)}_d\}\\
&= [\Phi_f][\Omega^2_f]^{-1}[\Phi_f]^T\left[\{\hat{f}\} + \omega^2[C_{up}]^T\{\hat{u}\}\right] + \{\hat{p}^{(s)}_d\}
\end{aligned} \tag{6.56}$$

where we have used the fact that $[\hat{K}]^{-1} = [\Phi_s][\hat{\Omega}^2_s]^{-1}[\Phi_s]^T$ and $[H]^{-1} = [\Phi_f][\Omega^2_f]^{-1}[\Phi_f]^T$.

Finally, $\{\hat{u}^{(s)}_d\}$ and $\{\hat{p}^{(s)}_d\}$ are obtained from Equations 6.54 through 6.56

$$\begin{aligned}
\{\hat{u}^{(s)}_d\} &= \left[[\hat{K}]^{-1} - [\Phi_s][\hat{\Omega}^2_s]^{-1}[\Phi_s]^T\right]\left[\{\hat{F}\} + [C_{up}]\{\hat{p}\}\right]\\
\{\hat{p}^{(s)}_d\} &= \left[[H]^{-1} - [\Phi_f][\Omega^2_f]^{-1}[\Phi_f]^T\right]\left[\{\hat{f}\} + \omega^2[C_{up}]^T\{\hat{u}\}\right]
\end{aligned} \tag{6.57}$$

Using only the kept modes to approximate the displacement and the pressure on the right-hand side of Equation 6.57, the contributions of the dropped modes can be approximated by

$$\begin{aligned}
\{\hat{u}_d\} &\approx \{\hat{u}_d^{(s)}\} = \left[[\hat{K}]^{-1} - [\Phi_s][\hat{\Omega}_s^2]^{-1}[\Phi_s]^T\right]\left[\{\hat{F}\} + [C_{up}][\Phi_f]\{\hat{p}_m\}\right] \\
\{\hat{p}_d\} &\approx \{\hat{p}_d^{(s)}\} = \left[[H]^{-1} - [\Phi_f][\Omega_f^2]^{-1}[\Phi_f]^T\right]\left[\{\hat{f}\} + \omega^2[C_{up}]^T[\Phi_s]\{\hat{u}_m\}\right]
\end{aligned} \tag{6.58}$$

Therefore,

$$\begin{aligned}
\{\hat{u}\} &= [\Phi_s]\{\hat{u}_m\} + [\hat{K}_{res}]^{-1}\{\hat{F}\} + [\hat{G}_{sm}]^T\{\hat{p}_m\} \\
\{\hat{p}\} &= [\Phi_f]\{\hat{p}_m\} + [H_{res}]^{-1}\{\omega^2\hat{f}\} + \omega^2[G_{fm}]^T\{\hat{u}_m\}
\end{aligned} \tag{6.59}$$

with

$$\begin{aligned}
[\hat{K}_{res}]^{-1} &= [\hat{K}]^{-1} - [\Phi_s][\hat{\Omega}_s^2]^{-1}[\Phi_s]^T \\
[H_{res}]^{-1} &= [H]^{-1} - [\Phi_f][\Omega_f^2]^{-1}[\Phi_f]^T \\
[\hat{G}_{sm}] &= [\Phi_f]^T[C_{up}]^T[\hat{K}_{res}]^{-1} \\
[G_{fm}] &= [\Phi_s]^T[C_{up}][H_{res}]^{-1}
\end{aligned} \tag{6.60}$$

where $[\hat{K}_{res}]^{-1}$ and $[H_{res}]^{-1}$ are called structural and fluid residual flexibility matrix, respectively. In the absence of the fluid (in-vacuo), we recover the classic pseudo-static correction for the structure (Equation 5.148). Inserting Equation 6.59 in Equation 6.49 and using the modes orthogonality properties, we obtain the new coupled system

$$\begin{bmatrix}
[\hat{\Omega}_s^2] - \omega^2[I_{n_{m,s}}] - \omega^2[M_{am}] & -[C_{up,m}] \\
-[C_{up,m}]^T & \frac{1}{\omega^2}[\Omega_f^2] - [\hat{I}_{n_{m,f}}] - [\hat{Q}_{am}]
\end{bmatrix}
\begin{Bmatrix} \{\hat{u}_m\} \\ \{\hat{p}_m\} \end{Bmatrix}
= \begin{Bmatrix} [\Phi_s]^T\{\hat{F}\} + [G_{fm}]\{\hat{f}\} \\ \frac{1}{\omega^2}[\Phi_f]^T\{\hat{f}\} + [\hat{G}_{sm}]\{\hat{F}\} \end{Bmatrix} \tag{6.61}$$

where

$$\begin{aligned}
[C_{up,m}] &= [\Phi_s]^T[C_{up}][\Phi_f] \\
[M_{am}] &= [\Phi_s]^T[C_{up}][H_{res}]^{-1}[C_{up}]^T[\Phi_s] \\
[\hat{Q}_{am}] &= [\Phi_f]^T[C_{up}]^T[\hat{K}_{res}]^{-1}[C_{up}][\Phi_f]
\end{aligned} \tag{6.62}$$

In general, matrix $[H]$ is not positive definite (there exists a rigid cavity mode) and therefore cannot be inverted. The residual flexibility matrix $[H_{res}]^{-1}$ must be modified to eliminate the presence of the rigid cavity mode as

$$[H_{res}]^{-1} = [A_f]^T[H_{con}]^{-1}[A_f] - [\Phi_{f,e}][\Omega^2_{f,e}]^{-1}[\Phi_{f,e}]^T \tag{6.63}$$

where subscript "e" refers to elastic modes, $[H_{con}]^{-1}$ is the flexibility matrix of the cavity that has been made isostatic by constraining one degree of freedom* and $[A_f]$ is a filter matrix used to eliminate the rigid cavity mode $\{\Phi_{f,r}\}$

$$[A_f] = [\hat{I}_{nm,f}] - [Q]\{\Phi_{f,r}\}\{\Phi_{f,r}\}^T \tag{6.64}$$

Substituting Equation 6.63 in the second line of Equation 6.62, a new expression of $[M_{am}]$ is obtained

$$[M_{am}] = [D_{fm}]^T[H_{con}]^{-1}[D_{fm}] - [C^{f,e}_{up,m}][\Omega^2_{f,e}]^{-1}[C^{f,e}_{up,m}]^T \tag{6.65}$$

with

$$\begin{aligned}[C^{f,e}_{up,m}] &= [\Phi_s]^T[C_{up}][\Phi_{f,e}] \\ [D_{fm}] &= [A_f][C_{up}]^T[\Phi_s]\end{aligned} \tag{6.66}$$

If the structure has rigid body modes, matrix $[\hat{K}]$ is not definite positive and cannot be inverted. The residual flexibility matrix $[\hat{K}_{res}]^{-1}$ is modified in a similar way as was done above for matrix $[H_{res}]$, namely

$$[\hat{K}_{res}]^{-1} = [A_s]^T[\hat{K}_{con}]^{-1}[A_s] - [\Phi_{s,e}][\hat{\Omega}^2_{s,e}]^{-1}[\Phi_{s,e}]^T \tag{6.67}$$

where

$$[A_s] = [I_{nm,s}] - [M][\Phi_{s,r}][\Phi_{s,r}]^T \tag{6.68}$$

where $[\Phi_{s,r}]$ is the matrix containing the structural rigid body modes. Moreover, subscript "e" refers to elastic modes. $[\hat{K}_{con}]^{-1}$ is the flexibility matrix of the structure that has been made isostatic by constraining a number of degrees of freedom corresponding to the number of rigid body modes.†

* A zero is put in the matrix where the degree of freedom is constrained.

† Zeroes are put in the matrix where the degrees of freedom are constrained.

Substituting Equation 6.67 in the third line of Equation 6.62, a new expression of $[\hat{Q}_{am}]$ is obtained

$$[\hat{Q}_{am}] = [D_{sm}]^T[\hat{K}_{con}]^{-1}[D_{sm}] - [C_{up,m}^{s,e}]^T[\hat{\Omega}_{s,e}^2]^{-1}[C_{up,m}^{s,e}] \tag{6.69}$$

with

$$\begin{aligned}[C_{up,m}^{s,e}] &= [\Phi_{s,e}]^T[C_{up}][\Phi_f] \\ [D_{sm}] &= [A_s][C_{up}][\Phi_f]\end{aligned} \tag{6.70}$$

6.4.4 Calculation of vibroacoustic indicators

Once vectors $\{\hat{u}_m\}$ and $\{\hat{p}_m\}$ have been determined, vibroacoustic indicators can be calculated (see Chapter 5):

- Kinetic energy of the structure:

$$E_c = \frac{1}{4}\omega^2\langle\hat{u}\rangle[M]\{\hat{u}^*\} = \frac{1}{4}\omega^2\langle\hat{u}_m\rangle\{\hat{u}_m^*\} \tag{6.71}$$

- Mean square pressure in the fluid cavity:

$$\begin{aligned}\langle p^2\rangle &= \frac{1}{2\Omega_f}(\langle\hat{p}^*\rangle[C_{pp}]\{\hat{p}\}) = \frac{1}{2\Omega_f}\langle\hat{p}_m\rangle[C_{pp,m}]\{\hat{p}_m^*\} \\ &= \frac{1}{2\Omega_f}\rho_0 c_0^2\langle\hat{p}\rangle[Q]\{\hat{p}^*\} = \frac{1}{2\Omega_f}\rho_0 c_0^2\langle\hat{p}_m\rangle\{\hat{p}_m^*\}\end{aligned} \tag{6.72}$$

where $[C_{pp}]$ has been defined in Chapter 5 and

$$[C_{pp,m}] = [\Phi_f]^T[C_{pp}][\Phi_f] \tag{6.73}$$

- Power exchanged between the structure and the fluid cavity (radiated power):

$$\Pi_{exch} = \frac{\omega}{2}\Im(\langle\hat{u}^*\rangle[C_{up}]\{\hat{p}\}) = \frac{\omega}{2}\Im(\langle\hat{u}_m^*\rangle[C_{up,m}]\{\hat{p}_m\}) \tag{6.74}$$

where symbol $\Im$ denotes the imaginary part of a complex number.

Remark: The main problem of the decomposition of the fields over uncoupled modal basis is the choice of the number of structural modes ($n_{m,s}$) and fluid modes ($n_{m,f}$) to be kept in the modal expansions. This choice depends on the nature of the intermodal coupling. Owing to the complexity of the intermodal coupling, no general criterion exists to select the right number of modes. In practice, the structural (respectively fluid) modal basis is truncated up to an order that corresponds to one and a half (respectively twice) the maximum frequency of the calculation spectrum. However, this criterion is not general and must be used with caution (see Section 6.7).

6.5 CALCULATION OF THE ADDED MASS

In this section, we limit ourselves to the case of an incompressible fluid. We show that the coupling effect amounts to an added mass effect on the structure. Starting from the fact that the compression energy matrix $[Q]$ is zero, the equations of the coupled system defined in Equation 6.13 can be written as

$$\begin{bmatrix} [K]-\omega^2[M] & -[C_{up}] \\ -\omega^2[C_{up}]^T & [H] \end{bmatrix} \begin{Bmatrix} \{\hat{u}\} \\ \{\hat{p}\} \end{Bmatrix} = \begin{Bmatrix} \{\hat{F}\} \\ \{0\} \end{Bmatrix} \tag{6.75}$$

Let us assume that $[H]$ is invertible.* From the second line of Equation 6.75, we obtain

$$\{\hat{p}\} = \omega^2[H]^{-1}[C_{up}]^T\{\hat{u}\} \tag{6.76}$$

Substituting Equation 6.76 in the first line of Equation 6.75, we get

$$[[K]-\omega^2([M]+[M_a])]\{\hat{u}\} = \{\hat{F}\} \tag{6.77}$$

where $[M_a]$ is the mass matrix added by the presence of the fluid

$$[M_a] = [C_{up}][H]^{-1}[C_{up}]^T \tag{6.78}$$

It is a square, symmetric, positive definite matrix. It is proportional to the fluid density.

Solving the homogeneous system associated with Equation 6.77 leads to the coupled structural modes, which include the added mass effect. And solving Equation 6.77 using a direct or a modal decomposition method yields the displacements of the structure.

* This is, for example, the case of a liquid-filled tank that has a pressure release surface.

Remarks:

1. In practice, we project the added mass matrix on the "*in-vacuo*" structural modal basis ($([\Omega_s^2],[\Phi_s])$) to rewrite

$$[[\Omega_s^2] - \omega^2([I_{n_{m,s}}] + [M_{a,m}])]\{\hat{u}_m\} = \{\hat{F}_m\} \tag{6.79}$$

with $\left[M_{a,m}\right] = [\Phi_s]^T [M_a][\Phi_s]$.

The system described in Equation 6.79 is full, symmetric, and of small size and can be solved with efficient algorithms.

2. In the case of a rigid-walled closed cavity, matrix $[H]$ is not invertible and we must perform a partitioning of $[H]$ before calculating the added mass matrix (Morand and Ohayon 1995). Alternatively, an approach similar to the one used to derive Equation 6.63 can be used.

6.6 CALCULATION OF THE FORCED RESPONSE USING COUPLED MODES

Here we present a method for calculating the forced response using coupled modes obtained from the uncoupled real modes of the structure and the fluid (Ohayon 2011). Again, let us consider the general case of a compressible fluid. Let us start from the coupled system*

$$\begin{bmatrix} [K] - \omega^2[M] & -[C_{up}] \\ -[C_{up}]^T & \dfrac{[H]}{\omega^2} - [Q] \end{bmatrix} \begin{Bmatrix} \{\hat{u}\} \\ \{\hat{p}\} \end{Bmatrix} = \begin{Bmatrix} \{\hat{F}\} \\ \dfrac{1}{\omega^2}\{\hat{f}\} \end{Bmatrix} \tag{6.80}$$

First, the rigid cavity modes are calculated (order $n_{m,f}$) and the zero frequency eigenvector discarded. Let us denote the "elastic" modes eigenfrequencies and associated eigenvectors matrix by $[\Omega_{f,e}^2] = \mathrm{diag}\{\omega_{f,2}^2\ \omega_{f,3}^2 \ldots \omega_{f,n_{m,f}}^2\}$ and $[\Phi_{f,e}]$, respectively. Let $\hat{p}_s$ denote the static (constant) pressure given by Equation 6.14 when $\omega = 0$. Then, the pressure can be written as

$$\{\hat{p}\} = \{\hat{p}_s\} + \{\hat{p}_e\} \tag{6.81}$$

* To make the presentation more general, we add a right-hand side $\{\hat{f}\}$ (source term) to the equations governing the acoustic pressure inside the cavity. The generalized source vector is denoted as $\{\hat{f}_m\}$.

where $\{\hat{p}_s\}$ and $\{\hat{p}_e\}$ represent the contribution of the static pressure and the contribution of the elastic cavity modes to the pressure nodal vector $\{\hat{p}\}$, respectively. We also have

$$\{\hat{f}\} = \{\hat{f}_s\} + \{\hat{f}_e\} \tag{6.82}$$

where $\{\hat{f}_s\}$ and $\{\hat{f}_e\}$ denote the static and elastic acoustical load vector, respectively. We have

$$\begin{aligned}&\{\hat{p}_e\} = [\Phi_{f,e}]\{\hat{p}_{m,e}\}; [\Phi_{f,e}]^T[H][\Phi_{f,e}] = [\Omega^2_{f,e}];\\&[\Phi_{f,e}]^T[Q][\Phi_{f,e}] = [I_{n_{m,f}-1}]; \{\hat{f}_{e,m,e}\} = [\Phi_{f,e}]^T\{\hat{f}_e\}\end{aligned} \tag{6.83}$$

The coupling term of the structure functional (Equation 6.7) becomes

$$\int_{\partial\Omega_w} \hat{p}\underline{n}\cdot\delta\hat{\underline{u}}\,dS = -\frac{\rho_0 c_0^2}{\Omega_f}\left(\int_{\partial\Omega_w} \underline{n}\cdot\hat{\underline{u}}\,dS\right)\left(\int_{\partial\Omega_w} \underline{n}\cdot\delta\hat{\underline{u}}\,dS\right) + \int_{\partial\Omega_w} \hat{p}_e\underline{n}\cdot\delta\hat{\underline{u}}\,dS \tag{6.84}$$

Its FE approximation leads to

$$\int_{\partial\Omega_w} \hat{p}\underline{n}\cdot\delta\hat{\underline{u}}\,dS = -\langle\delta\hat{u}\rangle[K_s]\{\hat{u}\} + \langle\delta\hat{u}\rangle[C_{up}]\{\hat{p}_e\} \tag{6.85}$$

with

$$\frac{\rho_0 c_0^2}{\Omega_f}\left(\int_{\partial\Omega_w} \underline{n}\cdot\hat{\underline{u}}\,dS\right)\left(\int_{\partial\Omega_w} \underline{n}\cdot\delta\hat{\underline{u}}\,dS\right) = \langle\delta\hat{u}\rangle[K_s]\{\hat{u}\} \tag{6.86}$$

The equations of the coupled problem in Equation 6.15 reads

$$\left(\begin{bmatrix}[K]+[K^s] & -[C_{up}]\\ [0] & [H]\end{bmatrix} - \omega^2\begin{bmatrix}[M] & [0]\\ [C_{up}]^T & [Q]\end{bmatrix}\right)\begin{Bmatrix}\{\hat{u}\}\\ \{\hat{p}_e\}\end{Bmatrix} = \begin{Bmatrix}\{\hat{F}\}\\ \{\hat{f}_e\}\end{Bmatrix} \tag{6.87}$$

Using the cavity modal coordinates (Equation 6.83), the second line of Equation 6.87 leads to

$$-\omega^2[\Phi_{f,e}]^T[C_{up}]^T\{\hat{u}\} + ([\Omega^2_{f,e}] - \omega^2[I_{n_{m,f}-1}])\{\hat{p}_{m,e}\} = \{\hat{f}_{e,m,e}\} \tag{6.88}$$

Using $[\Omega_f^2]^{-1} = [\Omega_f^{-2}]$ and the notation $[C_{up,f,e}] = [C_{up}][\Phi_{f,e}]$, the latter can be written as

$$\{\hat{p}_{m,e}\} - \omega^2[\Omega_{f,e}^{-2}][C_{up,f,e}]^T\{\hat{u}\} - \omega^2[\Omega_{f,e}^{-2}]\{\hat{p}_{m,e}\} = [\Omega_{f,e}^{-2}]\{\hat{f}_{e,m,e}\} \tag{6.89}$$

In consequence, the coupled system can be written as

$$\left(\begin{bmatrix}[K]+[K^s] & -[C_{up,f,e}] \\ [0] & [I_{n_{m,f}-1}]\end{bmatrix} - \omega^2\begin{bmatrix}[M] & [0] \\ [\Omega_{f,e}^{-2}][C_{up,f,e}]^T & [\Omega_{f,e}^{-2}]\end{bmatrix}\right)\begin{Bmatrix}\{\hat{u}\} \\ \{\hat{p}_{m,e}\}\end{Bmatrix}$$
$$= \begin{Bmatrix}\{\hat{F}\} \\ [\Omega_{f,e}^{-2}]\{\hat{f}_{e,m,e}\}\end{Bmatrix} \tag{6.90}$$

Using Equation 6.89 the system can be rendered symmetric

$$\left(\begin{bmatrix}[K]+[K_s] & [0] \\ [0] & [I_{n_{m,f}-1}]\end{bmatrix} - \omega^2\begin{bmatrix}[M]+[M_e] & [C_{up,f,e}][\Omega_{f,e}^{-2}] \\ [\Omega_{f,e}^{-2}][C_{up,f,e}]^T & [\Omega_{f,e}^{-2}]\end{bmatrix}\right)\begin{Bmatrix}\{\hat{u}\} \\ \{\hat{p}_{m,e}\}\end{Bmatrix}$$
$$= \begin{Bmatrix}\{\hat{F}\} + [C_{up,f,e}][\Omega_{f,e}^{-2}]\{\hat{f}_{e,m,e}\} \\ [\Omega_{f,e}^{-2}]\{\hat{f}_{e,m,e}\}\end{Bmatrix} \tag{6.91}$$

with

$$[M_e] = [C_{up,e}][\Omega_{f,e}^{-2}][C_{up,e}]^T \tag{6.92}$$

The associated eigenvalue problem is thus symmetric and can be used to calculate the coupled modes

$$\left(\begin{bmatrix}[K]+[K_s] & 0 \\ [0] & [I_{n_{m,f}-1}]\end{bmatrix} = \omega^2\begin{bmatrix}[M]+[M_e] & [C_{up,f,e}][\Omega_{f,e}^{-2}] \\ [\Omega_{f,e}^{-2}][C_{up,f,e}]^T & [\Omega_{f,e}^{-2}]\end{bmatrix}\right)\begin{Bmatrix}\{\hat{u}\} \\ \{\hat{p}_{m,e}\}\end{Bmatrix} \tag{6.93}$$

A further reduction of the model can be obtained by solving the following eigenvalue problem:

$$([K]+[K_s])\{\hat{u}\} = \omega^2([M]+[M_e])\{\hat{u}\} \tag{6.94}$$

By denoting $\left[\Omega^2_{s,c}\right] = diag\left\{\omega^2_{s,c1}\ \omega^2_{s,c2}\ \dots\ \omega^2_{s,cn_{m,s}}\right\}$ and $[\Phi_{s,c}]$ the associated eigenfrequencies and eigenvectors (order $n_{m,s}$), respectively, the reduced coupled system reads as

$$\left(\begin{bmatrix} [\Omega^2_{s,c}] & [0] \\ [0] & \left[I_{n_{m,f}-1}\right] \end{bmatrix} - \omega^2 \begin{bmatrix} [I_{n_{m,s}}] & [\tilde{C}_{up}] \\ [\tilde{C}_{up}]^T & [\Omega^{-2}_{f,e}] \end{bmatrix}\right) \begin{Bmatrix} \{\hat{u}_m\} \\ \{\hat{p}_{m,e}\} \end{Bmatrix}$$
$$= \begin{Bmatrix} [\Phi_{s,c}]^T\{\hat{F}\} + [\Phi_{s,c}]^T[C_{up,f,e}][\Omega^{-2}_{f,e}]\{\hat{f}_{e,m,e}\} \\ [\Omega^{-2}_{f,e}]\{\hat{f}_{e,m,e}\} \end{Bmatrix} \tag{6.95}$$

with

$$\left[\tilde{C}_{up}\right] = \left[\Phi_{s,c}\right]^T\left[C_{up,f,e}\right]\left[\Omega^{-2}_{f,e}\right] \tag{6.96}$$

The associated eigenvalue problem is again symmetric and can be used to calculate the coupled modes. Let $[\Omega^2_{sf}]$ and $[\Phi_{sf}]$ be the matrices containing the coupled eigenfrequencies and the associated eigenvectors,* respectively. The forced response can then be written as

$$\begin{Bmatrix} \{\hat{u}_m\} \\ \{\hat{p}_{m,e}\} \end{Bmatrix} = \left[[\Omega^2_{sf}] - \omega^2[I_{n_{m,sf}}]\right]^{-1}[\Phi_{sf}]^T \begin{Bmatrix} [\Phi_{s,c}]^T\{\hat{F}\} + [\Phi_{s,c}]^T[C_{up,f,e}][\Omega^{-2}_{f,e}]\{\hat{f}_{e,m,e}\} \\ [\Omega^{-2}_{f,e}]\{\hat{f}_{e,m,e}\} \end{Bmatrix} \tag{6.97}$$

The system is of size $(n_{m,sf} - 1) \times (n_{m,sf} - 1)$, where $n_{m,sf} = n_{m,s} + n_{m,f}$. The pressure modal vector is recovered using Equations 6.81 and 6.83. The vibroacoustic indicators are given by Equations 6.71, 6.72, and 6.74.

Note that the method presented above relies on the prior determination of the uncoupled structural and fluid modal basis. Thus, its convergence depends on the number of modes that are kept in the structure ($n_{m,s}$) and fluid ($n_{m,f}$) modal expansions. This choice depends on the intermodal coupling. Let us recall that because of the complexity of this intermodal coupling, no general criterion exists to select these numbers. In practice, the criterion described in the last remark of Section 6.4.4 is used. However, the first example of the next section demonstrates that this criterion must be used with caution.

* The system given by Equation 6.94 being of small size, we do not need to truncate its modal basis.

6.7 EXAMPLES OF APPLICATIONS

6.7.1 Vibroacoustic response of a rectangular cavity backed by a panel

In order to exhibit the limitations of using uncoupled modal basis let us consider the following example. It consists of a simply supported aluminum plate backed by a rigid-walled acoustic cavity (Figure 6.3). The lateral dimensions of the plate are $a = 0.35$ m, $b = 0.29$ m and its thickness is 1.5 mm. The lateral dimensions of the cavity are the same as the plate and its depth is $h_c = 14$ cm. The material properties are the following: $E = 7.2 \times 10^{10}$ Nm^{-2}, $\rho_s = 2700$ kg m^{-3}, $\nu = 0.3$, and $\eta_s = 0.01$. The fluid properties are $\rho_0 = 1.21$ kg m^{-3}, $c_0 = 340$ ms^{-1}, and $\eta_f = 0.01$. A harmonic unit point force is applied normally to the plate at point ($x_{0,1} = 0.078$ m, $x_{0,2} = 0.039$ m). The plate mesh consists of (19 × 16) nodes and (19 × 16 × 4) nodes for the acoustic cavity. Figures 6.4 and 6.5 compare the mean square pressure inside the cavity for two different calculations: (i) direct resolution of the coupled system in terms of nodal values and (ii) resolution of the coupled system projected onto uncoupled structural and fluid modal basis (10 modes are used for both the structure and the cavity). Figure 6.4 taken from Tournour (1999) shows the case where the cavity is filled with air (weak coupling), whereas Figure 6.5 taken from Tournour and Atalla (2000) presents the case where the cavity is filled with water (strong coupling).

These figures show that in the case of air (weak coupling), the use of uncoupled or coupled modal basis works well. However, in the case of water (strong coupling), a large number of structure and cavity modes must be used. Even with a large number of uncoupled modes (50 plate modes and 50 cavity were used in Figure 6.5), the solution

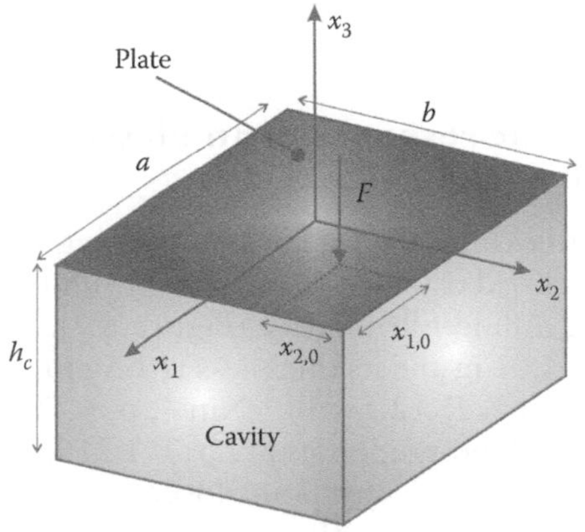

Figure 6.3 Elastic plate coupled to a rigid walled cavity.

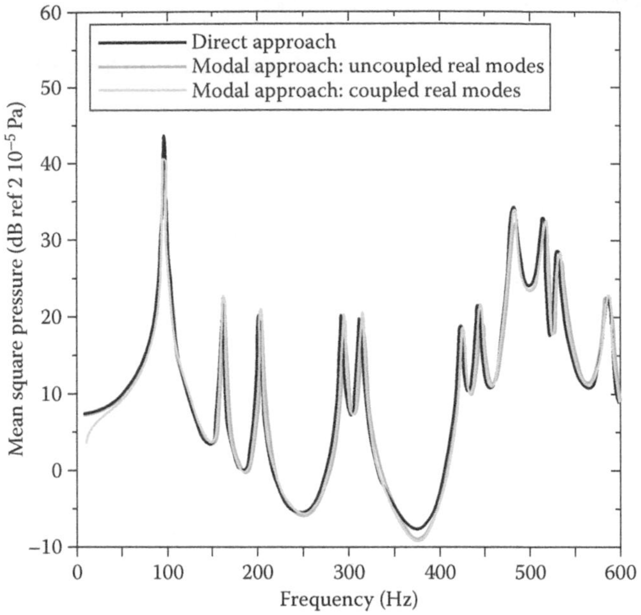

Figure 6.4 Mean square pressure in the cavity (case of air filling the cavity). Comparison of the direct approach and the modal technique using uncoupled and coupled modes. (Reproduced by permission taken from Tournour, M. 1999. *Modélisation numérique par éléments finis et éléments finis de frontière du comportement vibroacoustique de structures complexes assemblées et couplées à une cavité*. PhD Thesis, Sherbrooke, Canada: Université de Sherbrooke.)

is not converged. An excellent approximation can be obtained by adding pseudo-static corrections of Section 6.4.3 (Figure 6.6 taken from Tournour and Atalla 2000).

6.7.2 Vibroacoustic response of an elastic cylinder filled with a fluid

Another example to exhibit the limitations of using an uncoupled modal basis consists of a closed clamped steel elastic cylindrical shell with rigid end caps enclosing a fluid-filled cavity (Figure 6.7). The dimensions are $a = 0.18256$ m, $L = 1.01$ m and its thickness is 1.5 mm. The material properties are the following: $E = 2.1 \times 10^{11}$ Nm^{-2}, $\rho_s = 7800$ kg m^{-3}, $\nu = 0.3$, and $\eta_s = 0.0006$. The fluid properties are $\rho_0 = 1.21$ kg m^{-3}, $c_0 = 340$ ms^{-1}, and $\eta_f = 0.0005$. A harmonic unit point force is applied normally to the shell at a distance of $3L/16$ from the end cap. The shell is clamped on its edges. The cylinder is discretized using 28 4-node quadrilateral (quad-4)

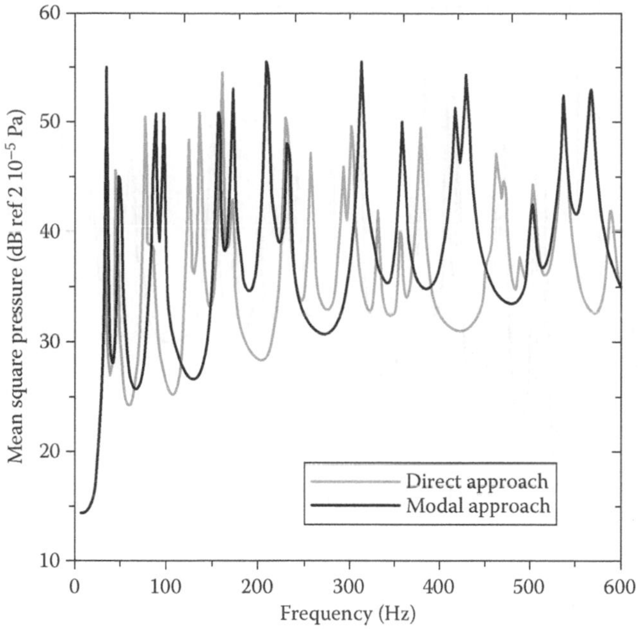

Figure 6.5 Mean square pressure in the cavity (case of water filling the cavity). Comparison of the direct approach and the modal technique. (Reproduced by permission of ASA taken from Tournour, M. and N. Atalla. Pseudo-static corrections for the forced vibroacoustic response of a structure-cavity system. *Journal of the Acoustical Society of America* 107: 2379–86. Copyright 2000, American Institute of Physics.)

Mindlin elements along the perimeter and 16 elements along the length. The cavity is discretized using 6-node wedge elements with 28 nodes along the perimeter, 17 nodes along the length, and 15 nodes along the diameter.

Figures 6.8 and 6.9 taken from Tournour and Atalla (2000) compare the mean square velocity of the cylindrical shell and the mean square pressure inside the cavity, respectively, for a (i) direct resolution of the coupled system in terms of nodal values and (ii) resolution of the coupled system projected onto uncoupled structural and fluid modal basis (using 42 modes for the shell and 80 modes in the cavity). Figure 6.8 shows that the modal approach is able to capture the structural displacement field even though the resonance peaks are overdamped. Figure 6.9 clearly shows that the modal approach fails in predicting the acoustic pressure response since certain uncoupled structural and cavity modes verify $[\Phi_s]^T [C_{up}] [\Phi_f] = 0$ and thus the contribution of several cavity modes is missing. On the contrary, the addition of the pseudo-static corrections of Section

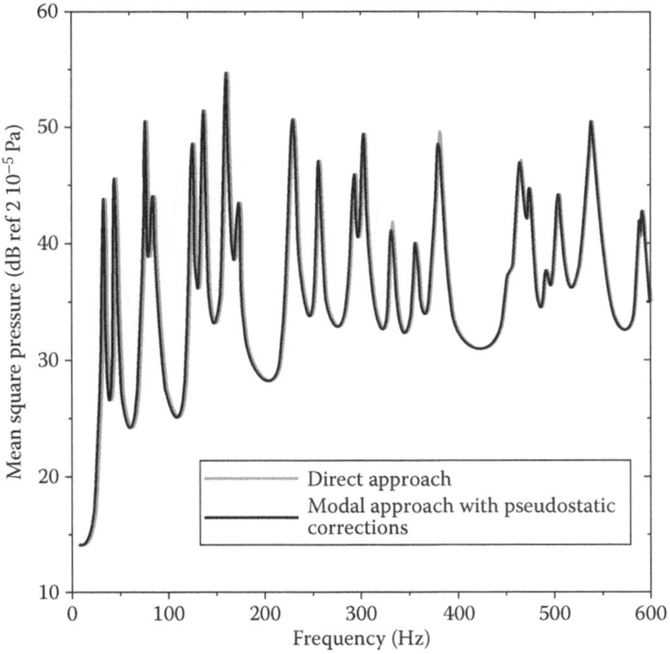

Figure 6.6 Mean square pressure in the cavity (case of water filling the cavity). Comparison of the direct approach and the modal technique including the pseudo-static correction. (Reproduced by permission of ASA taken from Tournour, M. and N. Atalla. Pseudo-static corrections for the forced vibroacoustic response of a structure-cavity system. *Journal of the Acoustical Society of America* 107: 2379–86. Copyright 2000, American Institute of Physics.)

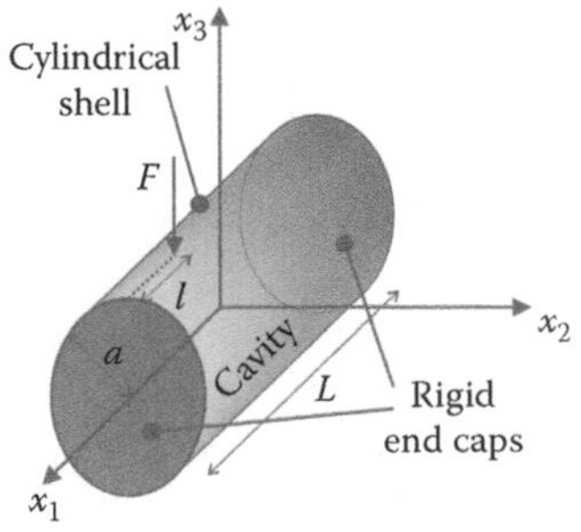

Figure 6.7 Rigid end caps closed elastic cylinder filled with a fluid and excited by a point force.

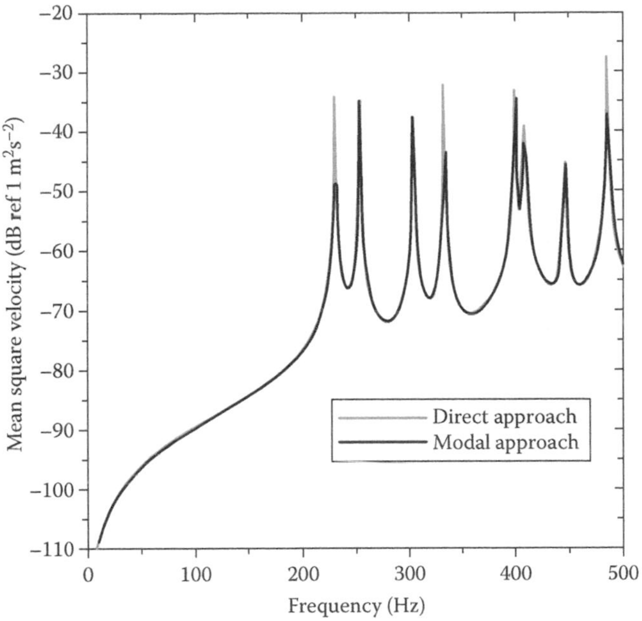

Figure 6.8 Mean square velocity of the cylindrical shell filled with air: comparison of the direct approach and the modal technique. (Reproduced by permission of ASA taken from Tournour, M. and N. Atalla. Pseudo-static corrections for the forced vibroacoustic response of a structure-cavity system. *Journal of the Acoustical Society of America* 107: 2379–86. Copyright 2000, American Institute of Physics.)

6.4.3 solves the problem (see Figures 6.10 and 6.11 taken from Tournour and Atalla 2000).

6.7.3 Vibroacoustic response of a plate with attached sound package radiating in a rectangular cavity

This example deals with the vibration response of a clamped plate coupled to an acoustic cavity. The plate is 2 cm thick, made from aluminum and measures 0.4 m × 0.85 m. It is excited by a point force and radiates into a hard-walled rectangular cavity (0.4 m × 0.85 m × 0.5 m). A sound package is attached to the plate. Two sound packages are studied. The first is a 4-cm thick melamine foam (properties in Table 5.3) and the second is a 2-cm thick melamine with an added heavy layer (mass-spring trim). The structural loss factor of the plate and the cavity are 0.02 and 0.05, respectively.

The generic FE model of this problem reads

$$\begin{bmatrix} [K] - \omega^2[M] & -[C_{up}] \\ -\omega^2[C_{up}]^T & [H] - \omega^2[Q] \end{bmatrix} \begin{Bmatrix} \{\hat{u}\} \\ \{\hat{p}\} \end{Bmatrix} = \begin{Bmatrix} \{\hat{F}\} \\ \{0\} \end{Bmatrix} - \begin{Bmatrix} \{\hat{T}\} \\ \omega^2\{\hat{U}_n\} \end{Bmatrix} \tag{6.98}$$

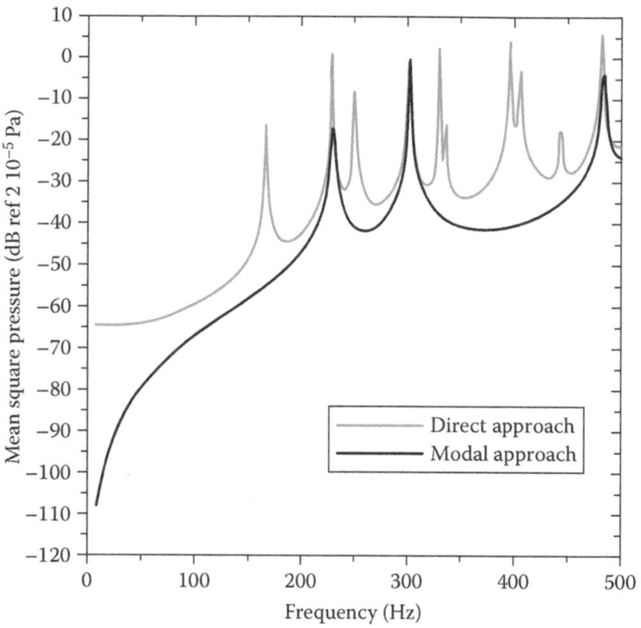

Figure 6.9 Mean square pressure in the cylindrical cavity filled with air: comparison of the direct approach and the modal technique. (Reproduced by permission of ASA taken from Tournour, M. and N. Atalla. Pseudo-static corrections for the forced vibroacoustic response of a structure-cavity system. *Journal of the Acoustical Society of America* 107: 2379–86. Copyright 2000, American Institute of Physics.)

where $\{\hat{T}\}$ and $\{\hat{U}_n\}$ represent the reaction force and normal particle displacement nodal vectors, due to the sound package, as seen from the structure and the fluid in the cavity, respectively. They can be written in terms of the displacement of the structure and the pressure in the cavity nodal vectors using the generic form

$$\begin{Bmatrix} \{\hat{T}\} \\ \omega^2\{\hat{U}_n\} \end{Bmatrix} = \begin{bmatrix} [\hat{K}_{sp}(\omega)] & -[\hat{C}_{sp}(\omega)] \\ -\omega^2[\hat{C}_{sp}(\omega)]^T & i\omega[\hat{A}_{sp}(\omega)] \end{bmatrix} \begin{Bmatrix} \{\hat{u}\} \\ \{\hat{p}\} \end{Bmatrix} \tag{6.99}$$

Here $[\hat{K}_{sp}]$ denotes the dynamic stiffness matrix added by the sound package, $[\hat{A}_{sp}]$ the dynamic admittance matrix added to the cavity by the sound package, and $[\hat{C}_{sp}]$ is an interface coupling matrix due to the sound package. In this example, these matrices are obtained using the FE approximation of the displacement–pressure formulation of Chapter 3. The details of this approximation can be found in Allard and Atalla (2009).

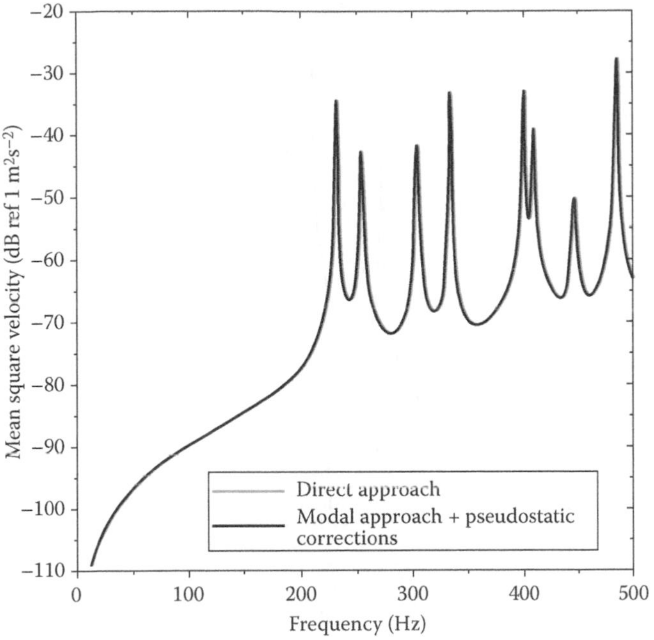

Figure 6.10 Mean square velocity of the cylindrical shell filled with air: comparison of the direct approach and the modal technique including the pseudo-static correction. (Reproduced by permission of ASA taken from Tournour, M. and N. Atalla. Pseudo-static corrections for the forced vibroacoustic response of a structure-cavity system. *Journal of the Acoustical Society of America* 107: 2379–86. Copyright 2000, American Institute of Physics.)

Another widely used approximation for this problem uses the impedance condition given by Equation 6.18. Using this equation, the coupling term seen from the plate side of the interface (recall that the sound package is assumed thickness-less and mass-less) remains unchanged

$$-\int_{\partial\Omega_w} \hat{p}\underline{n} \cdot \delta\underline{\hat{u}}\, dS = \langle \delta\hat{u} \rangle \underbrace{[C_{up}]}_{[\hat{C}_{sp}]} \{\hat{p}\} \tag{6.100}$$

However, on the cavity side of the interface we obtain

$$\begin{aligned} -\omega^2 \int_{\partial\Omega_w} \underline{\hat{U}} \cdot \underline{n}\delta\hat{p}\, dS &= -\omega^2 \int_{\partial\Omega_w} \underline{\hat{u}} \cdot \underline{n}\delta\hat{p}\, dS + i\omega \int_{\partial\Omega_w} \frac{\hat{p}}{\hat{Z}_n(\omega)} \delta\hat{p}\, dS \\ &= -\omega^2 \langle \delta\hat{p} \rangle \underbrace{[C_{up}]^T}_{[\hat{C}_{sp}]^T} \{\hat{u}\} + i\omega \langle \delta\hat{p} \rangle \underbrace{\left(\frac{1}{\hat{Z}_n(\omega)} [C_{pp}^{(2)}] \right)}_{[\hat{A}_{sp}]} \{\hat{p}\} \end{aligned} \tag{6.101}$$

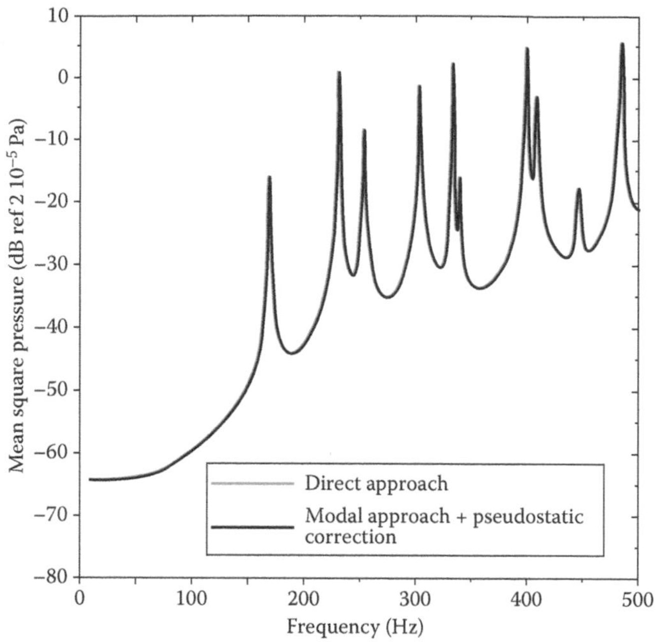

Figure 6.11 Mean square pressure in the cylindrical cavity filled with air: comparison of the direct approach and the modal technique including the pseudo-static correction. (Reproduced by permission of ASA taken from Tournour, M. and N. Atalla. Pseudo-static corrections for the forced vibroacoustic response of a structure-cavity system. *Journal of the Acoustical Society of America* 107: 2379–86. Copyright 2000, American Institute of Physics.)

The corresponding coupled matrix system is given in Equation 6.18c.

A more accurate impedance model, but still assuming thickness-less and locally reacting sound package, can be obtained using the transfer matrix model (TMM) of the sound package (Allard and Atalla 2009). Using the notation of Figure 6.12 for the normal component of the traction vector ($\hat{\underline{\underline{\sigma}}} \cdot \underline{n}$) and the normal displacement ($\hat{\underline{u}} \cdot \underline{n}$) at the structure side (denoted by subscript A) and the fluid side (denoted by subscript B), we obtain

$$\begin{Bmatrix} \hat{\sigma}_A \\ \hat{u}_B \end{Bmatrix} = \begin{bmatrix} \hat{d}_{AA}(k_t,\omega) & \hat{d}_{AB}(k_t,\omega) \\ \hat{d}_{BA}(k_t,\omega) & \hat{d}_{BB}(k_t,\omega) \end{bmatrix} \begin{Bmatrix} \hat{u}_A \\ \hat{\sigma}_B \end{Bmatrix} \tag{6.102}$$

For a given sound package (assumed to be a multilayer with infinite extent), coefficients $\hat{d}_{ij}(k_t,\omega)$ $(i, j = A \text{ or } B)$ are obtained using the TMM

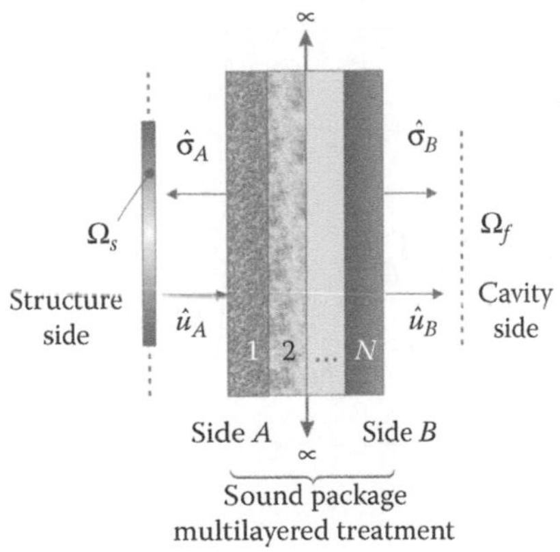

Figure 6.12 Transfer matrix model of the sound package.

at each frequency for a selected trace wavenumber k_t. Here k_t is selected equal to zero (normal incidence). Using Equation 6.102 the coupling terms read

$$\int_{\partial\Omega_w} (\hat{\underline{\underline{\sigma}}} \cdot \underline{n}) \cdot \delta\hat{\underline{u}}\, dS = \int_{\partial\Omega_w} \hat{d}_{AA}(\hat{\underline{u}} \cdot \underline{n})\, \underline{n} \cdot \delta\hat{\underline{u}}\, dS + \int_{\partial\Omega_w} \hat{d}_{AB}(\hat{p} \cdot \underline{n})\, \underline{n} \cdot \delta\hat{\underline{u}}\, dS$$
$$= \langle\delta\hat{u}\rangle[\hat{K}_{sp}]\{\hat{u}\} + \langle\delta\hat{u}\rangle[\hat{C}_{sp}]\{\hat{p}\} \qquad (6.103)$$

and

$$-\omega^2 \int_{\partial\Omega_w} \hat{\underline{U}} \cdot \underline{n}\delta\hat{p}\, dS = -\omega^2 \int_{\partial\Omega_w} \hat{d}_{BA}(\hat{\underline{u}} \cdot \underline{n})\delta\hat{p}\, dS - \omega^2 \int_{\partial\Omega_w} \hat{d}_{BB}\hat{p}\delta\hat{p}\, dS$$
$$= -\omega^2 \langle\delta\hat{p}\rangle[\hat{C}_{sp}]^T\{\hat{u}\} + i\omega\langle\delta\hat{p}\rangle[\hat{A}_{sp}]\{\hat{p}\} \qquad (6.104)$$

Figure 6.13 shows the comparison between TMM-based impedance model and full FE for a 4-cm thick melamine foam. In the FE calculation, the plate is meshed using 35×75 quad-4 elements, the cavity using $35 \times 75 \times 36$ hexa-8 elements, and the melamine, considered as limp, using $35 \times 75 \times 5$ hexa-8 elements. Overall, the approximate impedance model leads to acceptable results compared to the full FE model. Better comparison is seen for the quadratic velocity compared to the space-averaged quadratic pressure. At higher frequencies, and as expected, the comparison is excellent for both indicators. For this light sound package, the locally reacting assumption of the impedance method seems to be acceptable. Figure 6.14 shows the comparison when a 1.2 kg m^{-2} heavy layer is added to a

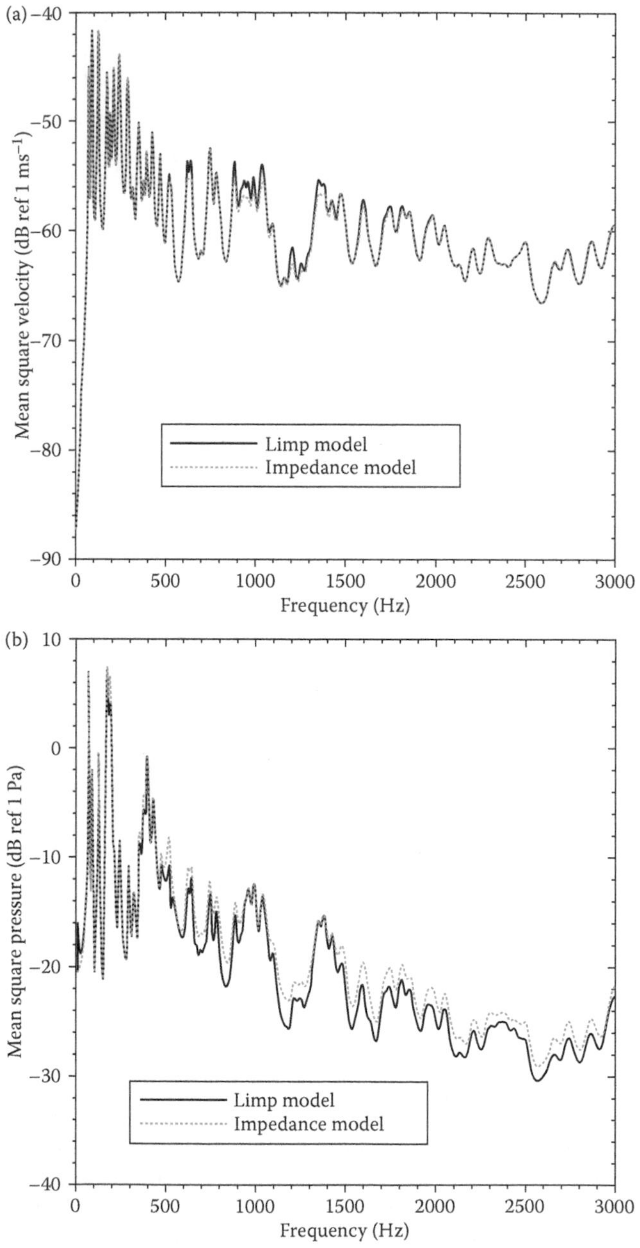

Figure 6.13 Comparison of the exact and locally reacting (impedance) model for a light sound package attached to a plate radiating in a cavity. (a) Panel mean square velocity and (b) cavity mean square pressure.

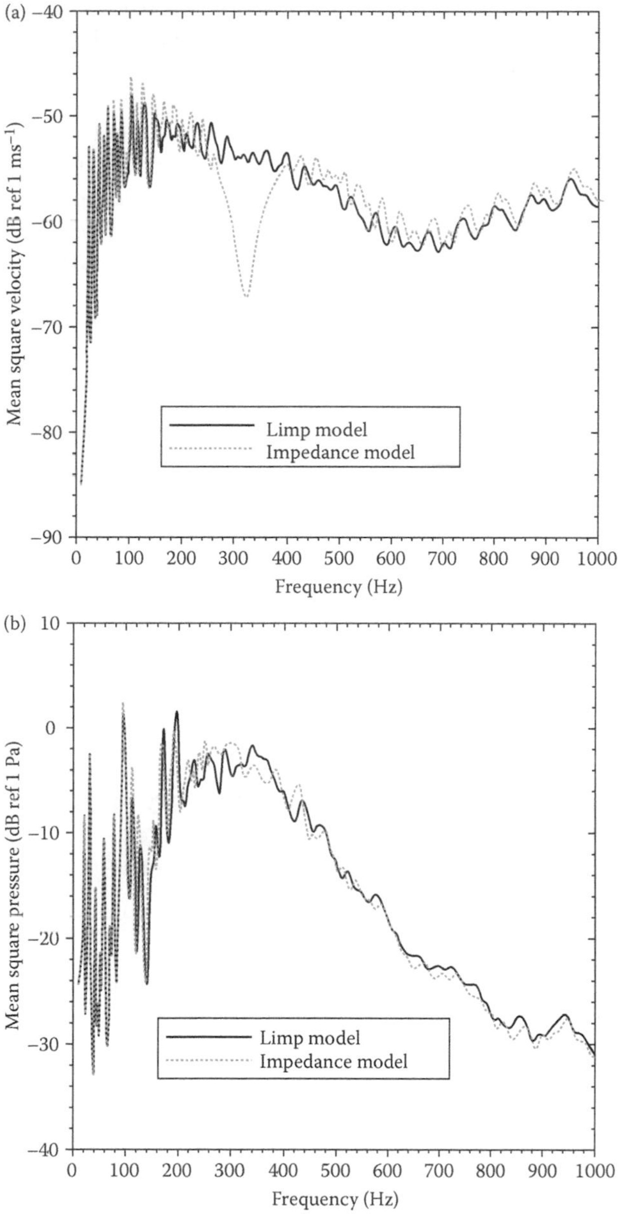

Figure 6.14 Comparison of the exact and locally reacting (impedance) model for a mass–spring type sound package attached to a plate radiating in a cavity. (a) Panel mean square velocity and (b) cavity mean square pressure.

2-cm thick melamine. In this case, the lateral dimensions of the 2-cm thick plate are 0.8 m × 1.7 m (meshed using 43 × 91 quad-4 elements), the cavity dimensions are 0.8 m × 1.7 m × 1 m (meshed using 43 × 91 × 27 hexa-8 elements); the melamine is meshed using 43 × 91 × 5 hexa-8 elements and the heavy layer using 43 × 91 × 1 hexa-8 limp solid elements. It can be clearly seen that the locally reacting method fails. In particular, it is not able to capture the vibration response of the panel around the mass–spring resonance of the system. Note that a better approximation using the TMM uses the Green's function methodology. Details on the latter can be found in Alimonti et al. (2014).

6.8 CONCLUSION

In this chapter, we have presented the coupled interior acoustic problem. Various formulations were introduced with an emphasis on the classic displacement–pressure formulation. The implementation of the latter using the uncoupled modes of the structure and the cavity was discussed in detail since it remains the most widely used in commercial codes. Mandatory corrections and transformations in the presence of strong coupling were also discussed. In the next two chapters, the exterior problem will be studied. The radiation and the scattering problems are discussed in Chapter 7. Chapter 8 deals with the exterior coupled problem together with the interior/exterior coupled problem.

APPENDIX 6A: EXAMPLE OF A COUPLED FLUID–STRUCTURE PROBLEM—PISTON–SPRING SYSTEM RADIATING IN A GAS-FILLED TUBE

6A.1 Acoustic pressure inside the tube

Consider the problem of a rigid piston of mass m_{piston} connected to a massless spring of stiffness k_{piston} and radiating in a closed tube filled with a gas (Figure 6.15). The piston is excited by a point force $\hat{F}$. We are interested in

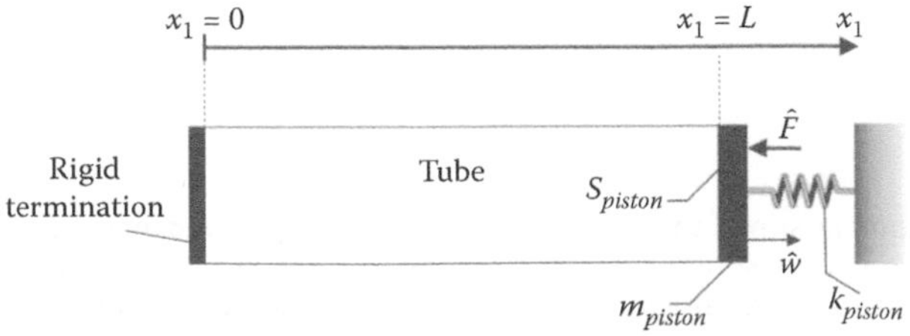

Figure 6.15 Piston–spring system attached to a gas filled tube.

solving the harmonic one-dimensional problem (frequency below the cut-off frequency of the tube). The acoustic pressure inside the tube is solution of Helmholtz equation

$$\nabla^2\hat{p} + k^2\hat{p} = 0 \tag{6.105}$$

subjected to the two following boundary conditions:

$$\frac{d\hat{p}}{dx_1}(0) = 0 \tag{6.106}$$

$$\frac{d\hat{p}}{dx_1}(L) = \rho_0\omega^2\hat{w} \tag{6.107}$$

where $\hat{w}$ is the piston displacement along the x_1-axis. The first two equations lead to

$$\hat{p} = \hat{A}\cos(kx_1) \tag{6.108}$$

The third equation allows us to relate the acoustic pressure and the piston displacement

$$\frac{d\hat{p}}{dx_1}(L) = -\hat{A}k\sin kL = \rho_0\omega^2\hat{w} \Rightarrow \hat{A} = -\frac{\rho_0\omega^2}{k\sin kL}\hat{w} \tag{6.109}$$

Finally,

$$\hat{p} = -\frac{\rho_0\omega^2}{k\sin kL}\hat{w}\cos(kx_1) \tag{6.110}$$

We end up with the specific acoustic impedance seen by the cavity at $x_1 = L$

$$\hat{Z}_{cav} = \frac{\hat{p}(L)}{i\omega\hat{w}} = \frac{i\rho_0\omega}{k\tan kL} \tag{6.111}$$

6A.2 Motion of the piston

The equation of motion of the piston–spring system is given by

$$(k_{piston} - m_{piston}\omega^2)\hat{w} = S_{piston}\hat{p}(L) + \hat{F} \tag{6.112}$$

Using Equations 6.112 and 6.111, we get

$$\hat{w} = \frac{\hat{F}}{(k_{piston} - m_{piston}\omega^2) - i\omega\hat{Z}_{cav}S_{piston}} \tag{6.113}$$

Remark: At low frequency, we can use the approximation

$$\frac{1}{kL\tan kL} \approx \frac{1}{k^2L^2} - \frac{1}{3} + O(kL)^2 \tag{6.114}$$

which gives

$$-i\omega\hat{Z}_{cav} = -iL\frac{i\rho_0\omega^2}{kL\tan kL} \cong \rho_0\omega L\left(\frac{1}{k^2L^2} - \frac{1}{3}\right) = \underbrace{\frac{\rho_0 c_0^2}{L}}_{\substack{\text{Stiffness added}\\ \text{by the cavity}}} - \underbrace{\omega^2\frac{1}{3}\rho_0 L}_{\substack{\text{Mass added}\\ \text{by the cavity}}} \tag{6.115}$$

Thus, the piston displacement is given by

$$\hat{w} = \frac{\hat{F}}{(k_{piston} + (\rho_0 c_0^2 S_{piston}/L)) - \omega^2(m_{piston} + (1/3)\rho_0 L S_{piston})} \tag{6.116}$$

6A.3 Mean square pressure in the tube

The mean square pressure in the tube is given by

$$\begin{aligned}
\langle \hat{p}^2 \rangle &= \frac{1}{2S_{piston}L}\int_0^L |\hat{A}\cos(kx_1)|^2 dx_1 \\
&= \frac{|\hat{A}|^2}{2S_{piston}L}\int_0^L \frac{1+\cos(2kx_1)}{2} dx_1 \\
&= \frac{|\hat{A}|^2}{2S_{piston}L}\left(\frac{L}{2} + \frac{\sin(2kL)}{4k}\right) \\
&= \frac{|\hat{A}|^2}{4S_{piston}}\left(1 + \frac{\sin(2kL)}{2kL}\right)
\end{aligned} \tag{6.117}$$

with $\hat{A} = -(\rho_0\omega^2/k\sin kL)\hat{w}$

This expression assumes that the damping is zero. If this is not the case, we are better off calculating the mean square velocity using an average over N points whose abscissa are such that $0 \le x_{1,i} \le L$

$$\langle p^2 \rangle = \frac{|\hat{A}|^2}{2N} \sum_{i=1}^{N} \cos^2(kx_{1,i}) \tag{6.118}$$

Remark: We can check that the low-frequency limit of the acoustic pressure is given by

$$\begin{aligned} p^s &= \lim_{\omega \to 0} \hat{p} = \hat{A} = -\lim_{\omega \to 0} \frac{\rho_0 \omega^2}{k \sin kL} \hat{w} \\ &= -\frac{\rho_0 \omega^2}{k(kL)} (\lim_{\omega \to 0} \hat{w}) \\ &= -\frac{\rho_0 c_0^2}{L} \hat{w}(0) \end{aligned} \tag{6.119}$$

The limit proved in the general three-dimensional case (Equation 3.33) is recovered

$$p^s = -\frac{\rho_0 c_0^2}{\Omega_f} \int_{\partial \Omega_w} (\hat{\underline{u}} \cdot \underline{n}) \, dS \tag{6.120}$$

where Ω_f is the cavity volume and $\partial\Omega_w$ the vibrating surface.

REFERENCES

Alimonti, L., N. Atalla, A. Berry, and F. Sgard. 2014. Assessment of a hybrid finite element-transfer matrix model for flat structures with homogeneous acoustic treatments. *Journal of the Acoustical Society of America* 135 (5): 2694–705.

Allard, J.F. and N. Atalla. 2009. *Propagation of Sound in Porous Media, Modelling Sound Absorbing Materials*. 2nd ed. Chichester, UK: Wiley-Blackwell.

Irons, B.M. 1970. Role of part inversion in fluid-structure problems with mixed variables. *AIAA Journal* 7: 568.

Lesueur, C. 1988. *Rayonnement Acoustique Des Structures Vibroacoustique, Interactions Fluide-Structure*. 1st ed. Direction des études et recherches d'Electricité de France (EDF). Paris: Eyrolles.

Morand, H.J.-P. and R. Ohayon. 1995. *Fluid Structure Interaction*. Chichester, UK: Wiley-Blackwell.

Ohayon, R. 2011. Fluid-structure interaction. In *Proceedings of the 8th International Conference on Structural Dynamics, EURODYN*. Leuven, Belgium, pp. 53–59.

Sgard, F. 1995. Etude numérique du comportement vibro-acoustique d'un système plaque-cavité rayonnant dans un écoulement uniforme, pour différents types d'excitation. PhD Thesis, Sherbrooke, QC, Canada: Université de Sherbrooke.

Tournour, M. 1999. Modélisation numérique par éléments finis et éléments finis de frontière du comportement vibroacoustique de structures complexes assemblées et couplées à une cavité. PhD Thesis, Sherbrooke, QC, Canada: Université de Sherbrooke.

Tournour, M. and N. Atalla. 2000. Pseudo-static corrections for the forced vibroacoustic response of a structure-cavity system. *Journal of the Acoustical Society of America* 107: 2379–86.

Chapter 7

Solving structural acoustics and vibration problems using the boundary element method

7.1 INTEGRAL FORMULATION FOR HELMHOLTZ EQUATION

In this section, the integral equations of problems associated with Helmholtz equation are established. These integral equations constitute the basis of the boundary element method (BEM).

7.1.1 Interior and exterior problems

Consider a tri-dimensional closed volume Ω_- with boundary $\partial\Omega$ (see Figure 7.1). The boundary $\partial\Omega$ separates an interior region (volume Ω_-) from an unbounded exterior region Ω_+. This exterior region can be seen as externally bounded by a fictitious outer sphere Γ_R whose radius R tends toward infinity (Γ_∞). $\underline{n}$ and $\underline{n}_\infty$ denote the outward normal to Ω_- and the outward normal to Ω_+ on Γ_∞, respectively. It is common to introduce the concepts of interior and exterior problems (see Figure 7.2). For the interior problem, one is interested in the sound pressure field inside Ω_-* whereas for the exterior problem, one is interested in the sound pressure field in the exterior region Ω_+† (outside Ω_-).

As we shall see, the sound pressure field in Ω_- and Ω_+ can be related to its value and its normal derivative along the surface $\partial\Omega$. Assuming that there is a volume source distribution of strength $\hat{Q}$ in Ω_-, the sound pressure field in Ω_- satisfies Helmholtz equation

$$\begin{cases} \nabla^2\hat{p} + k^2\hat{p} = -\hat{Q}(\underline{y}) \\ \text{Boundary conditions} \end{cases} \tag{7.1}$$

---

* A practical application would be the determination of the sound field inside a room due to an internal acoustical source or due to the vibration of the walls.

† An example of exterior problem is the case of the sound radiation from a vibrating surface in an anechoic room.

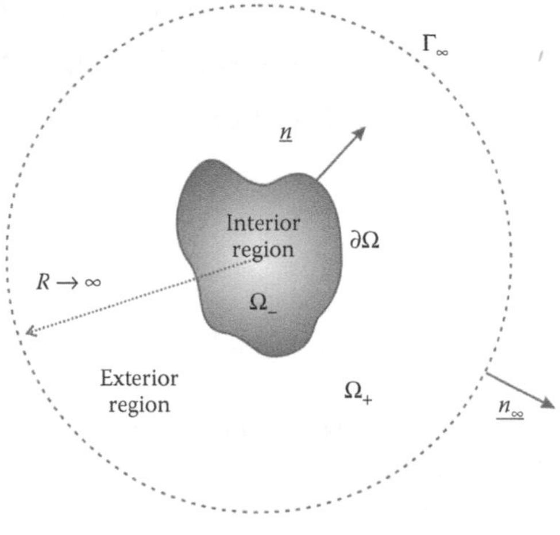

Figure 7.1 Description of the problem.

where k is the wavenumber ω/c_0, and $\hat{p}$ is the complex pressure amplitude at point $\underline{y}$ (see Section 2.2). The same equation holds if a volume source distribution is present in Ω_+. If there is no source distribution in the domain then the sound pressure field satisfies the homogeneous Helmholtz equation. The boundary conditions apply on the boundaries of Ω_- ($\partial\Omega$) or Ω_+ ($\partial\Omega \cup \Gamma_\infty$) according to the problem of interest. Equation 7.1 is also called a boundary value problem.

Before deriving the integral equations associated with the interior and exterior problems, let us introduce the elementary solutions of Helmholtz equation. This concept is very useful for solving boundary value problems. The principle is to solve for the response of the system to a unit point source (e.g., Equation 7.2). The solution is commonly called "fundamental solution" or "Green's function." We will use the latter designation in this book.

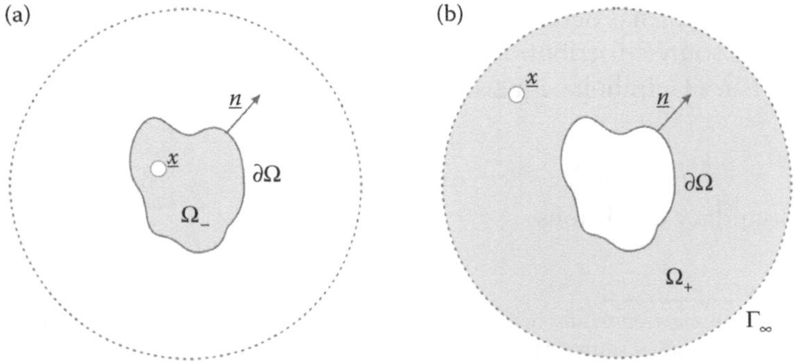

Figure 7.2 Interior problem (a); exterior problem (b).

Using the Green's function, the solution of the initial problem is written in term of an integral representation (convolution integral). Instead of solving differential equations in a given volume subjected to boundary conditions, this technique directly expresses the solution in terms of surface integrals involving its value on the boundaries of the volume and its normal derivative along the boundaries.

7.1.2 Elementary solution

The elementary or fundamental solution also called Green's function and denoted as $\hat{G}(\underline{x}, \underline{y})$ represents the sound field at a point $\underline{y}$ due to a unit point source located at a point $\underline{x}$ and satisfying all or certain homogeneous boundary conditions (Dirichlet, Newmann, Robin) of the problem. It is important to note that Green's function can be chosen according to the problem of interest. In practice, this choice involves a Green's function that can be expressed analytically and that allows for simplification of the equations. This will become clearer as we progress more into the chapter.

The Green's function satisfies the inhomogeneous Helmholtz equation*

$$\begin{cases} \nabla_y^2 \hat{G} + k^2 \hat{G} = -\delta(\underline{x} - \underline{y}) \\ \text{Boundary conditions} \end{cases} \tag{7.2}$$

The unit point source is represented here as the Dirac delta function† $\delta(\underline{x} - \underline{y})$.

A very commonly used Green's function is the one associated with a three dimensional isotropic infinite medium also called free-space Green's function denoted by $\hat{G}_\infty(\underline{x}, \underline{y})$.‡ It is given by (Wright 2005)

$$\hat{G}_\infty(\underline{x}, \underline{y}) = \frac{\exp(-ikr)}{4\pi r} \tag{7.3}$$

This is the simplest choice for $\hat{G}$. $\hat{G}$ can also be chosen so that $\hat{G}$ or its normal derivative vanishes on the volume boundary.

The gradient§ of the free-space Green's function is expressed as

$$\underline{\nabla}_y \hat{G}_\infty(\underline{x}, \underline{y}) = -\frac{1}{4\pi r^2}(1 + ikr)\exp(-ikr)\underline{\nabla}_y r \tag{7.4}$$

* It is important to indicate with respect to which variable $\underline{x}$ (the source) or $\underline{y}$ (the field point) the derivative is performed since $\hat{G}(\underline{x}, \underline{y})$ depends on two space variables.

† The Dirac delta function is defined such that $\lim_{\underline{y} \to \underline{x}} \delta(\underline{x} - \underline{y}) = +\infty$,, $\delta(\underline{x} - \underline{y}) = 0$ otherwise and $\int_\Omega f(\underline{y})\delta(\underline{x} - \underline{y})dV_y = \alpha(\underline{x})f(\underline{x})$ with $\alpha(\underline{x}) = \begin{cases} 1 & \text{if } \underline{x} \in \Omega \\ 0 & \text{otherwise} \end{cases}$.

‡ $\hat{G}_\infty(\underline{x}, \underline{y})$ is singular for $\underline{x} = \underline{y}$. $\hat{G}_\infty(\underline{x}, \underline{y})$ is symmetric, that is, $\hat{G}_\infty(\underline{x}, \underline{y}) = \hat{G}_\infty(\underline{y}, \underline{x})$.

§ Note that $\underline{\nabla}_y r = (\underline{y} - \underline{x})/r$ and $\underline{\nabla}_x r = (\underline{x} - \underline{y})/r = -\underline{\nabla}_y r$

where $r = |\underline{x} - \underline{y}| = \sqrt{(x_i - y_i)(x_i - y_i)}$ is the Euclidian distance between points $\underline{x}$ and $\underline{y}$ written using Einstein summation notation.

The exponential term* exp $(-ikr)$ can be expanded in terms of a series absolutely convergent for all values of r

$$\exp(-ikr) = \sum_{n=0}^{+\infty} \frac{(-ikr)^n}{n!} \tag{7.5}$$

Therefore, it is straightforward to show that

$$\begin{cases} \hat{G}(\underline{x},\underline{y}) - G_0(\underline{x},\underline{y}) = \dfrac{-ik}{4\pi} \displaystyle\sum_{n=0}^{+\infty} \frac{(-ikr)^n}{(n+1)!} \\ \underline{\nabla}_y \hat{G}(\underline{x},\underline{y}) - \underline{\nabla}_y G_0(\underline{x},\underline{y}) = \dfrac{-k^2}{4\pi} \displaystyle\sum_{n=0}^{+\infty} \frac{n+1}{(n+2)!}(-ikr)^n \underline{\nabla}_y r \end{cases} \tag{7.6}$$

Here

$$G_0(\underline{x},\underline{y}) = \frac{1}{4\pi r} \tag{7.7}$$

where $G_0(\underline{x},\underline{y})$ is the elementary solution of Laplace's equation $\nabla_y^2 G_0 = -\delta(\underline{x} - \underline{y})$.

Equation 7.6 is used to prove the equality of singular terms in dynamic and static elementary solutions (see Appendix 7A)

$$\begin{matrix} (kr) \to 0 \\ \forall\omega > 0 \end{matrix} \begin{cases} \hat{G}(\underline{x},\underline{y}) - \hat{G}_0(\underline{x},\underline{y}) = O(1) \\ \underline{\nabla}_y \hat{G}(\underline{x},\underline{y}) - \underline{\nabla}_y \hat{G}_0(\underline{x},\underline{y}) = O(1) \end{cases} \tag{7.8}$$

where big O notation refers to asymptotic notation. Also, it can be shown that

$$\frac{\partial^2 \hat{G}(\underline{x},\underline{y})}{\partial n_x \partial n_y} - \frac{\partial^2 G_0(\underline{x},\underline{y})}{\partial n_x \partial n_y} = O\left(\frac{1}{r}\right) \quad (kr) \to 0; \forall\omega > 0 \tag{7.9}$$

Note that other solutions of Equation 7.2 are available depending on the dimension, geometry, and boundary conditions of the problem (one-dimensional, two-dimensional, semi-infinite fluid, rigid cavity, etc.).

* Some papers and books devoted to acoustics choose a $\exp(-i\omega t)$ temporal dependency. In this case, the minus sign in the exponential term exp $(-ikr)$ should be inverted.

In the following sections, we will look for the solution of Equation 7.1 for the interior and exterior problem. For each problem, this solutions is sought in terms of the values of the pressure and its normal derivative along the boundary $\partial\Omega$.

7.2 DIRECT INTEGRAL FORMULATION FOR THE INTERIOR PROBLEM

In this section, we establish boundary integral equations for the interior problem based on the pressure field and its normal gradient. Consider the interior problem (see Figure 7.2a).

7.2.1 Integral equation for a field point

Multiplying Equation 7.1 by elementary solution $\hat{G}(\underline{x}, y)$ and Equation 7.2 by the field $\hat{p}$ and integrating both over Ω_- leads to

$$
\begin{cases}
\displaystyle\int_{\Omega_-} [\hat{G}(\underline{x}, \underline{y})\nabla_{\underline{y}}^2 p(\underline{y}) + k^2\hat{G}(\underline{x}, \underline{y})\hat{p}(\underline{y})] \, dV_y = -\int_{\Omega_-} \hat{Q}(\underline{y})\hat{G}(\underline{x}, \underline{y})dV_y \\[4mm]
\displaystyle\int_{\Omega_-} [\hat{p}(\underline{y})\nabla_{\underline{y}}^2\hat{G}(\underline{x}, \underline{y}) + k^2\hat{G}(\underline{x}, \underline{y})\hat{p}(\underline{y})] \, dV_y + \int_{\Omega_-} \hat{p}(\underline{y}) \, \delta(\underline{x} - \underline{y})dV_y = 0
\end{cases}
\tag{7.10}
$$

Subtracting these two equations leads to the identity

$$
\int_{\Omega_-} \hat{p}(\underline{y}) \, \delta(\underline{x} - \underline{y})dV_y = \int_{\Omega_-} [\hat{G}(\underline{x}, \underline{y})\nabla_{\underline{y}}^2\hat{p}(\underline{y}) - \hat{p}(\underline{y})\nabla_{\underline{y}}^2\hat{G}(\underline{x}, \underline{y})] \, dV_y
$$

$$
+ \int_{\Omega_-} \hat{Q}(\underline{y})\hat{G}(\underline{x}, \underline{y})dV_y
\tag{7.11}
$$

Using the second Green's identity (see Appendix 3B), Equation 7.11 becomes

$$
\int_{\Omega_-} [\hat{G}(\underline{x}, \underline{y})\nabla_{\underline{y}}^2\hat{p}(\underline{y}) - \hat{p}(\underline{y})\nabla_{\underline{y}}^2\hat{G}(\underline{x}, \underline{y})]dV_y
$$

$$
= \int_{\partial\Omega} \left[\hat{G}(\underline{x}, \underline{y})\frac{\partial\hat{p}(\underline{y})}{\partial n_y} - \hat{p}(\underline{y})\frac{\partial\hat{G}(\underline{x}, \underline{y})}{\partial n_y} \right]dS_y
\tag{7.12}
$$

where $\underline{n}_y$ is the outward normal to Ω_- at point $\underline{y}$ on $\partial\Omega$ and by definition $\partial/\partial n_y(\cdot) = \underline{\nabla}_y(\cdot) \cdot \underline{n}_y$.

Using the property of Dirac delta function (see footnote * page 236), Equations 7.11 and 7.12 are rewritten as

$$\alpha(\underline{x})\hat{p}(\underline{x}) = \int_{\partial\Omega}\left[\hat{G}(\underline{x},\underline{y})\frac{\partial\hat{p}(\underline{y})}{\partial n_y} - \hat{p}(\underline{y})\frac{\partial\hat{G}(\underline{x},\underline{y})}{\partial n_y}\right] dS_y + \int_{\Omega_-}\hat{Q}(\underline{y})\hat{G}(\underline{x},\underline{y})dV_y \tag{7.13}$$

This equation is called an integral representation of field $\hat{p}$, which is also known as Kirchhoff Helmholtz equation. This equation is valid for all points, which do not belong to $\partial\Omega$. $\hat{G}$ can be chosen as the free-space Green's function defined previously. Alternatively, it can be selected such that $\hat{G}$ or its normal derivative vanishes on $\partial\Omega$. Then one of the two terms of the integrand of Equation 7.13 would vanish and $\hat{p}(\underline{x})$ could be calculated only from $\hat{p}$ or $\partial\hat{p}(\underline{y})/\partial n_y$, respectively.

Remarks:

1. If there is no volume source distribution in Ω_-, then, for $\underline{x} \notin \partial\Omega$, Equation 7.13 becomes

$$\alpha(\underline{x})\hat{p}(\underline{x}) = \int_{\partial\Omega}\left[\hat{G}(\underline{x},\underline{y})\frac{\partial\hat{p}(\underline{y})}{\partial n_y} - \hat{p}(\underline{y})\frac{\partial\hat{G}(\underline{x},\underline{y})}{\partial n_y}\right] dS_y \tag{7.14}$$

2. For all $\underline{x} \notin \partial\Omega$ integrals in Equation 7.13 are regular. The integral representation of the gradient of $\hat{p}$, which is related to the particle acoustic velocity can thus be obtained by taking the derivative inside the integral of the right-hand side gradient of Equation 7.13

$$\begin{aligned}\alpha(\underline{x})\underline{\nabla}_x\hat{p}(\underline{x}) = &\int_{\partial\Omega}\left[\underline{\nabla}_x\hat{G}(\underline{x},\underline{y})\frac{\partial\hat{p}(\underline{y})}{\partial n_y} - \hat{p}(\underline{y})\underline{\nabla}_x\left(\frac{\partial\hat{G}(\underline{x},\underline{y})}{\partial n_y}\right)\right] dS_y \\ &+\int_{\Omega^-}\hat{Q}(\underline{y})\underline{\nabla}_x\hat{G}(\underline{x},\underline{y})dV_y \quad \forall\underline{x} \notin \partial\Omega\end{aligned} \tag{7.15}$$

3. Let us consider Laplace's equation $\nabla^2\hat{p}_0 = 0$ in Ω_- with G_0 the free-field elementary solution. Laplace's equation corresponds to the static case of Helmholtz equation $\omega = 0$ and therefore Equation 7.13 (without source distribution) allows for the integral representation of $\hat{p}_0$

$$\alpha(\underline{x})\hat{p}_0(\underline{x}) = \int_{\partial\Omega}\left[G_0(\underline{x},\underline{y})\frac{\partial\hat{p}_0(\underline{y})}{\partial n_y} - \hat{p}_0(\underline{y})\frac{\partial G_0(\underline{x},\underline{y})}{\partial n_y}\right]dS_y \tag{7.16}$$

The function $\hat{p}_0(\underline{y}) = 1$ is a solution of Laplace's equation. Its substitution in Equation 7.16 leads to the following identity:*

$$\alpha(\underline{x}) = -\int_{\partial\Omega}\frac{\partial G_0(\underline{x},\underline{y})}{\partial n_y}dS_y = \begin{cases}1 & \text{if } \underline{x} \in \Omega_- \\ 0 & \text{otherwise}\end{cases} \tag{7.17}$$

Equations 7.13 and 7.15 relate the values of the pressure and its gradient at a point in space $\underline{x} \notin \partial\Omega$ to the pressure $\hat{p}$ on the surface and its normal derivative $(\partial\hat{p}/\partial n_y)$ along the surface. However, these surface values have to be known. The boundary conditions on surface $\partial\Omega$ provide either the pressure or its normal derivative, or a combination of both. To solve for the remaining boundary unknowns, the boundary conditions are expressed in terms of Equation 7.13 or 7.15, evaluated at a point $\underline{x} \in \partial\Omega$. This requires a special mathematical treatment† of these integral equations, which is detailed thereafter.

Let us choose $\hat{G}(\underline{x},\underline{y}) = \hat{G}_\infty(\underline{x},\underline{y})$. $\hat{G}_\infty(\underline{x},\underline{y})$ and its normal derivative $\partial\hat{G}_\infty(\underline{x},\underline{y})/\partial n_y$ have a $1/r$ weak singularity‡ and an apparently $1/r^2$ strong singularity, respectively. In fact, one can show (see Appendix 7A) that the integral involving $\partial\hat{G}_\infty(\underline{x},\underline{y})/\partial n_y$ is also weakly singular.

* The same result can be obtained by integrating $\nabla_y^2 G_0 = -\delta(\underline{x}-\underline{y})$ over Ω_-, noting that $\nabla^2 G_0 = \nabla\cdot\underline{\nabla}G_0$ and using the divergence formula (see Appendix 3B) $\int_{\Omega_-}\nabla_y\cdot\underline{\nabla}_y G_0\, dV_y = \int_{\partial\Omega}\underline{\nabla}_y G_0\cdot\underline{n}_y dS_y = -\int_{\Omega_-}\delta(\underline{x}-\underline{y})dV_y = -\alpha(x)$.

† We cannot always interchange the order of some mathematical operations like taking the limit, integrating, or taking derivative of expressions. We have to be very careful here because the integrands contain singular terms and the order in which these operators are applied is very important.

‡ Consider a general function $f(\underline{x},\underline{y})$ with $\underline{x}$ representing the source point and $\underline{y}$ the field point. Let $I(\underline{x}) = \int_{\partial\Omega} f(\underline{x},\underline{y})dS_y$. If $f(\underline{x},\underline{y})$ is bounded everywhere for variable $\underline{y}$, the integral is said to be regular. If $f(\underline{x},\underline{y})$ becomes infinite at some point, $I(\underline{x})$ is said to be singular. Consider now $f(\underline{x},\underline{y}) = (\kappa(\underline{x},\underline{y}))/(r^\beta(\underline{x},\underline{y}))$ where $\kappa(\underline{x},\underline{y})$ is bounded everywhere and $r = |\underline{x}-\underline{y}| = \sqrt{(x_i - y_i)(x_i - y_i)}$. The level of singularity of $I(\underline{x})$ is defined as

$\beta = 0$: regular
$0 < \beta < 2$: weakly singular
$\beta = 2$: strongly singular
$2 < \beta \le 3$: hyper-singular
$\beta > 3$: super-singular

Regular and weakly singular boundary integrals are finite for a finite region problem. Strongly, hyper-singular, and super-singular boundary integrals are finite only if $\kappa(\underline{x},\underline{y})$ satisfies certain conditions.

7.2.2 Standard boundary integral equation (SBIE)

Let us first consider the evaluation of Equation 7.13 on $\partial\Omega$. $\underline{x}$ is a fixed point of $\partial\Omega$. Let ϑ_ε be an external small volume of arbitrary shape S_ε^+ surrounding point $\underline{x}$. Let us define the volume $\Omega_\varepsilon = \Omega_- \cup \vartheta_\varepsilon$. The boundary of Ω_ε is then, $\partial\Omega_\varepsilon = (\partial\Omega - e_\varepsilon) \cup S_\varepsilon^+$ where e_ε is the subregion of $\partial\Omega$ containing the singularity $e_\varepsilon = \partial\Omega \cap \vartheta_\varepsilon$ and $S_\varepsilon^+ = \partial\vartheta_\varepsilon - e_\varepsilon$. It is usual to take ϑ_ε as the external part of a sphere centered at $\underline{x}$ with radius ε (see Figure 7.3). Note that the shape of S_ε^+ and e_ε must be preserved throughout the limiting process.

To establish the boundary integral formulation, we take the limit when $\varepsilon \to 0$ of Equation 7.13 written for domain Ω_ε. Since $\underline{x}$ is inside Ω_ε, we have

$$\begin{aligned}\hat{p}(\underline{x}) &= \lim_{\varepsilon\to 0} \int_{(\partial\Omega - e_\varepsilon)\cup S_\varepsilon^+} \left[\hat{G}(\underline{x},\underline{y})\frac{\partial \hat{p}}{\partial n_y} - \hat{p}(\underline{y})\frac{\partial \hat{G}(\underline{x},\underline{y})}{\partial n_y} \right] dS_y \\ &\quad + \lim_{\varepsilon\to 0}\int_{\Omega_\varepsilon} \hat{G}(\underline{x},\underline{y})\hat{Q}(\underline{y})dV_y\end{aligned} \tag{7.18}$$

The volume integral involving $\hat{G}(\underline{x},\underline{y})$ $(= O(1/r))$ is weakly singular, therefore

$$\lim_{\varepsilon\to 0}\int_{\Omega_\varepsilon} \hat{G}(\underline{x},\underline{y})\hat{Q}(\underline{y})dV_y = \int_{\Omega_-} \hat{G}(\underline{x},\underline{y})\hat{Q}(\underline{y})dV_y \tag{7.19}$$

In addition,

$$\begin{aligned}&\lim_{\varepsilon\to 0} \int_{(\partial\Omega - e_\varepsilon)} \left[\hat{G}(\underline{x},\underline{y})\frac{\partial \hat{p}}{\partial n_y} - \hat{p}(\underline{y})\frac{\partial \hat{G}(\underline{x},\underline{y})}{\partial n_y} \right] dS_y \\ &= \int_{\partial\Omega^*} \left[\hat{G}(\underline{x},\underline{y})\frac{\partial \hat{p}}{\partial n_y} - \hat{p}(\underline{y})\frac{\partial \hat{G}(\underline{x},\underline{y})}{\partial n_y} \right] dS_y\end{aligned} \tag{7.20}$$

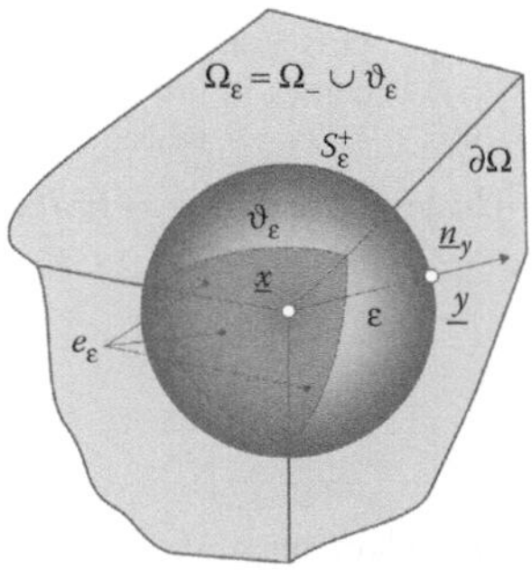

Figure 7.3 Singular point $\underline{x}$ on the boundary $\partial\Omega$.

where $\int_{\partial\Omega^*}(\cdot)dS_y$ means that the integral should be taken in the Cauchy principal value sense. Actually, we will see that it is not needed in expression (7.20) since the integrals are convergent when $\varepsilon \to 0$.

Finally, Equation 7.18 can be written as

$$\hat{p}(\underline{x}) = \int_{\partial\Omega^*}\left[\hat{G}(\underline{x},\underline{y})\frac{\partial\hat{p}}{\partial n_y} - \hat{p}(\underline{y})\frac{\partial\hat{G}(\underline{x},\underline{y})}{\partial n_y}\right]dS_y + \int_{\Omega_-}\hat{G}(\underline{x},\underline{y})\hat{Q}(\underline{y})dV_y$$
$$+ \lim_{\varepsilon\to 0}\int_{S_\varepsilon^+}\hat{G}(\underline{x},\underline{y})\frac{\partial\hat{p}}{\partial n_y}dS_y - \lim_{\varepsilon\to 0}\int_{S_\varepsilon^+}\hat{p}(\underline{y})\frac{\partial\hat{G}(\underline{x},\underline{y})}{\partial n_y}dS_y \tag{7.21}$$

Let S_ε^+ be a portion of a sphere of radius ε centered at the singular point $\underline{x}(x_1,x_2,x_3)$ then a point $\underline{y}(y_1,y_2,y_3)$ on the sphere can be located using spherical coordinates

$$\begin{cases} y_1 - x_1 = \varepsilon\sin\theta\cos\varphi \\ y_2 - x_2 = \varepsilon\sin\theta\sin\varphi \\ y_3 - x_3 = \varepsilon\cos\theta \end{cases} \tag{7.22}$$

with $\varphi \in [0,2\pi]$ and $\theta \in [\theta_{min}, \theta_{max}]$. Then

$$\frac{\partial G_0(\underline{x},\underline{y})}{\partial n_y} = \frac{\partial}{\partial\varepsilon}\left(\frac{1}{4\pi\varepsilon}\right)\underline{\nabla}_y\varepsilon\cdot\underline{n_y} = \frac{\partial}{\partial\varepsilon}\left(\frac{1}{4\pi\varepsilon}\right)\underline{\nabla}_y\varepsilon\cdot\underline{e_\varepsilon} = -\frac{1}{4\pi\varepsilon^2}\frac{\underline{y}-\underline{x}}{\varepsilon}\cdot\underline{e_\varepsilon}$$
$$= -\frac{1}{4\pi\varepsilon^2}\frac{\varepsilon}{\varepsilon}\underline{e_\varepsilon}\cdot\underline{e_\varepsilon} = -\frac{1}{4\pi\varepsilon^2} \tag{7.23}$$

Note that when $\varepsilon \to 0$, $\hat{G}(\underline{x},\underline{y}) = \exp(-ik\varepsilon)/4\pi\varepsilon \approx G_0(\underline{x},\underline{y}) = 1/4\pi\varepsilon$; $\hat{p}(\underline{y}) \approx \hat{p}(\underline{x})$; $\partial\hat{p}/\partial n_y \approx \partial\hat{p}/\partial n_x$; $\partial\hat{G}(\underline{x},\underline{y})/\partial n_y \approx \partial G_0(\underline{x},\underline{y})/\partial n_y = -1/4\pi\varepsilon^2$ therefore

$$\lim_{\varepsilon\to 0}\int_{S_\varepsilon^+}\hat{G}(\underline{x},\underline{y})(\partial\hat{p}/\partial n_y)dS_y = (\partial\hat{p}/\partial n_x)\lim_{\varepsilon\to 0}\int_0^{2\pi}\int_{\theta_{min}}^{\theta_{max}}(1/4\pi\varepsilon)\,\varepsilon^2\sin\theta d\theta d\varphi = 0$$

and thus

$$\hat{p}(\underline{x}) = \int_{\partial\Omega^*}\left[\hat{G}(\underline{x},\underline{y})\frac{\partial\hat{p}}{\partial n_y} - \hat{p}(\underline{y})\frac{\partial\hat{G}(\underline{x},\underline{y})}{\partial n_y}\right]dS_y + \int_{\Omega_-}\hat{G}(\underline{x},\underline{y})\hat{Q}(\underline{y})dV_y$$
$$- \lim_{\varepsilon\to 0}\hat{p}(\underline{x})\int_{S_\varepsilon^+}\frac{\partial\hat{G}_0(\underline{x},\underline{y})}{\partial n_y}dS_y \tag{7.24}$$

which can be rewritten as

$$C^-(\underline{x})\hat{p}(\underline{x}) = \int_{\partial\Omega^*}\left[\hat{G}(\underline{x},\underline{y})\frac{\partial\hat{p}}{\partial n_y} - \hat{p}(\underline{y})\frac{\partial\hat{G}(\underline{x},\underline{y})}{\partial n_y}\right]dS_y + \int_{\Omega_-}\hat{G}(\underline{x},\underline{y})\hat{Q}(\underline{y})dV_y \quad (7.25)$$

where $C^-(\underline{x})$ is called a free coefficient and is given by

$$C^-(\underline{x}) = 1 + \lim_{\varepsilon\to 0}\int_{S_\varepsilon^+}\frac{\partial G_0(\underline{x},\underline{y})}{\partial n_y}dS_y \quad (7.26)$$

There remains to calculate $C^-(\underline{x})$. In the case where $\partial\Omega$ is a regular surface, S_ε^+ can be chosen as a half-sphere of radius ε centered at point $\underline{x}$ and Equation 7.26 can be rewritten as

$$C^-(\underline{x}) = 1 + \lim_{\varepsilon\to 0}\int_0^{2\pi}\int_0^{\pi/2} -\frac{1}{4\pi\varepsilon^2}\varepsilon^2\sin\theta d\theta d\varphi = \frac{1}{2} \quad (7.27)$$

Therefore, for a smooth surface $C^-(\underline{x}) = 1/2$.

In the general case (e.g., surface with corners, edges, etc.), Equation 7.26 can be rewritten as

$$\begin{aligned} C^-(\underline{x}) &= -\int_{\partial\Omega_\varepsilon}\frac{\partial G_0}{\partial n_y}dS_y + \lim_{\varepsilon\to 0}\int_{S_\varepsilon^+}\frac{\partial G_0}{\partial n_y}dS_y \\ &= -\left(\int_{\partial\Omega^*}\frac{\partial G_0}{\partial n_y}dS_y + \lim_{\varepsilon\to 0}\int_{S_\varepsilon^+}\frac{\partial G_0}{\partial n_y}dS_y\right) + \lim_{\varepsilon\to 0}\int_{S_\varepsilon^+}\frac{\partial G_0}{\partial n_y}dS_y \end{aligned} \quad (7.28)$$

where we have used Equation 7.17 and the fact that $\underline{x}$ is inside V_ε. Hence, in the general case

$$C^-(\underline{x}) = -\int_{\partial\Omega^*}\frac{\partial G_0}{\partial n_y}dS_y \quad (7.29)$$

In other words, the free coefficient can be calculated from the surface integral of the normal derivative of G_0 over boundary $\partial\Omega^*$. This may be much more convenient than using formula (7.26). Note that for a closed smooth surface, $-\int_{\partial\Omega^*}(\partial G_0/\partial n_y)dS_y = 1/2$. Note also that $\underline{n}_y$ denotes the outward normal of the interior problem.

In summary, the integral formulation for the interior problem, which is valid for all points of Ω including its boundary $\partial\Omega$, is given by*

$$C^{-}(\underline{x})\hat{p}(\underline{x}) = \int_{\partial\Omega^*}\left[\hat{G}(\underline{x},\underline{y})\frac{\partial\hat{p}(\underline{y})}{\partial n_y} - \hat{p}(\underline{y})\frac{\partial\hat{G}(\underline{x},\underline{y})}{\partial n_y}\right]dS_y + \int_{\Omega_-}\hat{Q}(\underline{y})\,\hat{G}(\underline{x},\underline{y})\,dV_y \tag{7.30}$$

with

$$\begin{cases} C^{-}(\underline{x}) = 1 & \text{if } \underline{x} \text{ is inside } \Omega_- \\ C^{-}(\underline{x}) = 0 & \text{if } \underline{x} \text{ is outside } \Omega_- \\ C^{-}(\underline{x}) = -\displaystyle\int_{\partial\Omega^*}\frac{\partial G_0(\underline{x},\underline{y})dS_y}{\partial n_y} & \text{if } \underline{x} \in \partial\Omega \end{cases} \tag{7.31}$$

Remarks: In Equation 7.30, the integral should be *a priori* taken in the principal value sense. Actually, it is not needed since both improper integrals converge (the singularity at $\underline{x}$ is integrable). The proof that it is so is given in Appendix 7B.

7.2.3 Regularized boundary integral equation (RBIE)

An alternate form of Equation 7.30 defined for all point $\underline{x}$ belonging to the definition domain of the elementary solution of interest can be obtained. This form is called the regularized form of the boundary integral equation. For a point $\underline{x} \in \partial\Omega$, Equation 7.30 can be rewritten as

* In many books, this equation is derived starting with a point $\underline{x}$ outside Ω_ε instead of inside Ω_ε. In this case, we start with Equation 7.13 in which $\alpha(\underline{x}) = 0$ and the boundary is deformed such that point $\underline{x}$ remains outside Ω_ε. The same procedure as above is followed and we obtain for the free-term coefficient $\lim_{\varepsilon\to 0}\int_{S_\varepsilon^-}\partial G_0(\underline{x},\underline{y})/\partial n_y dS_y$ where S_ε^- is the part of a sphere centered at $\underline{x}$ of radius ε and protruding inside Ω_ε and n_y is the normal to S_ε^- which points toward $\underline{x}$. In Appendix 7A, it is shown that $\lim_{\varepsilon\to 0}\int_{S_\varepsilon^-}\partial G_0(\underline{x},\underline{y})/\partial n_y dS_y = 1 + \lim_{\varepsilon\to 0}\int_{S_\varepsilon^+}\partial G_0(\underline{x},\underline{y})/\partial n_y dS_y$ which is equal to $C^-(\underline{x})$ and we obtain the same equation as Equation 7.30.

$$-\left(\int_{\partial\Omega} \frac{\partial G_0(\underline{x},\underline{y})}{\partial n_y} dS_y\right)\hat{p}(\underline{x}) = \int_{\partial\Omega}\left[\hat{G}(\underline{x},\underline{y})\frac{\partial \hat{p}(\underline{y})}{\partial n_y} - \hat{p}(\underline{y})\frac{\partial \hat{G}(\underline{x},\underline{y})}{\partial n_y}\right] dS_y + \int_{\Omega_-} \hat{Q}(\underline{y})\hat{G}(\underline{x},\underline{y}) dV_y \tag{7.32}$$

Equation 7.32 can be rewritten in turn

$$\int_{\partial\Omega}\left[\hat{p}(\underline{y})\frac{\partial \hat{G}(\underline{x},\underline{y})}{\partial n_y} - \frac{\partial G_0(\underline{x},\underline{y})}{\partial n_y}\hat{p}(\underline{x})\right] dS_y - \int_{\partial\Omega} \hat{G}(\underline{x},\underline{y})\frac{\partial \hat{p}(\underline{y})}{\partial n_y} dS_y = \int_{\Omega_-} \hat{Q}(\underline{y})\, \hat{G}(\underline{x},\underline{y}) dV_y \tag{7.33}$$

Equation 7.33 can be rewritten equivalently as

$$\int_{\partial\Omega} \hat{p}(\underline{y})\left(\frac{\partial \hat{G}(\underline{x},\underline{y})}{\partial n_y} - \frac{\partial G_0(\underline{x},\underline{y})}{\partial n_y}\right) dS_y + \int_{\partial\Omega} (\hat{p}(\underline{y}) - \hat{p}(\underline{x}))\frac{\partial G_0(\underline{x},\underline{y})}{\partial n_y} dS_y = \int_{\partial\Omega} \hat{G}(\underline{x},\underline{y})\frac{\partial \hat{p}(\underline{y})}{\partial n_y} dS_y + \int_{\Omega_-} \hat{Q}(\underline{y})\, \hat{G}(\underline{x},\underline{y}) dV_y \tag{7.34}$$

In the neighborhood of $\underline{x}$, $(\partial\hat{G}/\partial n_y - \partial G_0/\partial n_y)$ has a weak singularity (see Equation 7.8) and the first integral on the left-hand side of Equation 7.34 is convergent. Regarding the second integral, assuming that $\hat{p}$ is continuous in the Hölder sense of exponent α_h^*, then in the neighborhood of $\underline{x}$, $|(\hat{p}(\underline{y}) - \hat{p}(\underline{x}))\partial G_0/\partial n_y| \le C_1\ |\underline{y} - \underline{x}|^{\alpha_h - 2}$ which ensures that this integral converges.

All the integrals in Equation 7.33 are regular and can be computed efficiently using a polar coordinate transformation of the surface element as will be shown in Section 7.6.4.2. Equation 7.33 is valid for any point $\underline{x}$ inside or outside Ω_- and over $\partial\Omega$.

* A function f is said to satisfy a Hölder condition of exponent α_h if $\exists C_1 > 0; \exists \alpha_h \in\ ]0;1]$ such that $|f(\underline{y}) - f(\underline{x})| \le C_1\ |\underline{y} - \underline{x}|^{\alpha_h}$ (f is said to belong to C^{0,α_h}). If its derivative also satisfies a Hölder condition, then f is said to belong to C^{1,α_h}.

7.3 DIRECT INTEGRAL FORMULATION FOR THE EXTERIOR PROBLEM

7.3.1 Integral equation for a field point

The fundamental interest of the BEM is the treatment of unbounded domains, namely the exterior problem. Let us consider a domain Ω_- bounded by boundary $\partial\Omega$ of inward normal $\underline{\bar{n}}$. Assume, without loss of generality, that there is no source distribution inside Ω_- and $\Omega_+ = \Re^3 - \Omega_-$ (i.e., $\hat{\underline{Q}} = 0$) to simplify the derivation. Let $\underline{x}$ be a point of Ω_+ bounded by $\partial\Omega$ and the fictitious outer sphere Γ_∞ (see Section 7.1).

Point $\underline{x}$ being interior to Ω_+, the integral representation (7.14) is valid so that

$$\alpha(\underline{x})\hat{p}(\underline{x}) = \int_{\partial\Omega\cup\Gamma_\infty} \left[\hat{G}(\underline{x},\underline{y})\frac{\partial\hat{p}(\underline{y})}{\partial\bar{n}_y} - \hat{p}(\underline{y})\frac{\partial\hat{G}(\underline{x},\underline{y})}{\partial\bar{n}_y} \right] dS_y \tag{7.35}$$

Note that

$$\begin{aligned}
&\int_{\Gamma_\infty} \left[\hat{G}(\underline{x},\underline{y})\frac{\partial\hat{p}(\underline{y})}{\partial\bar{n}_y} - \hat{p}(\underline{y})\frac{\partial\hat{G}(\underline{x},\underline{y})}{\partial\bar{n}_y} \right] dS_y \\
&= \lim_{R\to\infty} \int_{\Gamma_R} \left[\hat{G}(\underline{x},\underline{y})\frac{\partial\hat{p}(\underline{y})}{\partial\bar{n}_y} - \hat{p}(\underline{y})\frac{\partial\hat{G}(\underline{x},\underline{y})}{\partial\bar{n}_y} \right] dS_y = 0
\end{aligned} \tag{7.36}$$

since both $\hat{p}$ and the elementary solution $\hat{G}(\underline{x},\underline{y})$ satisfy the Sommerfeld radiation condition at infinity, namely

$$\lim_{r\to\infty} r\left(\frac{\partial\hat{p}}{\partial r} + ik\hat{p} \right) = 0 \tag{7.37}$$

Thus, the integral formulation of the exterior problem reduces to

$$\alpha(\underline{x})\hat{p}(\underline{x}) = \int_{\partial\Omega} \left[\hat{G}(\underline{x},\underline{y})\frac{\partial\hat{p}(\underline{y})}{\partial\bar{n}_y} - \hat{p}(\underline{y})\frac{\partial\hat{G}(\underline{x},\underline{y})}{\partial\bar{n}_y} \right] dS_y \tag{7.38}$$

Let $\underline{n}_y = -\underline{\bar{n}}_y$, the outward normal to Ω_-, we then obtain for all $\underline{x}$ exterior to Ω_-

$$\alpha(\underline{x})\hat{p}(\underline{x}) = \int_{\partial\Omega} \left[\hat{p}(\underline{y})\frac{\partial\hat{G}(\underline{x},\underline{y})}{\partial n_y} - \hat{G}(\underline{x},\underline{y})\frac{\partial\hat{p}(\underline{y})}{\partial n_y} \right] dS_y \tag{7.39}$$

where

$$\alpha(\underline{x}) = \begin{cases} 1 \text{ if } \underline{x} \in \Omega_+ \\ 0 \text{ if } \underline{x} \in \Omega_- \end{cases}$$

7.3.2 SBIE

Following the same approach as for the interior problem, it can be shown that for all points $\underline{x} \in \Re^3$, the pressure satisfies the following boundary integral equation:

$$C^+(\underline{x})\hat{p}(\underline{x}) = \int_{\partial\Omega} \left[\hat{p}(\underline{y}) \frac{\partial \hat{G}(\underline{x},\underline{y})}{\partial n_y} - \hat{G}(\underline{x},\underline{y}) \frac{\partial \hat{p}(\underline{y})}{\partial n_y} \right] dS_y \tag{7.40}$$

where

$$\begin{cases} C^+(\underline{x}) = 1 & \text{if } \underline{x} \text{ is outside } \Omega_- \\ C^+(\underline{x}) = 0 & \text{if } \underline{x} \text{ is inside } \Omega_- \\ C^+(\underline{x}) = 1 + \displaystyle\int_{\partial\Omega} \frac{\partial G_0(\underline{x},\underline{y})}{\partial n_y} dS_y & \text{if } \underline{x} \in \partial\Omega \end{cases} \tag{7.41}$$

Remarks:

1. The proof of the third equation of Equation (7.41) is provided in Appendix 7A. Also, note that $\underline{n}_y$ is the normal pointing into the exterior domain.
2. Using Equations 7.31 and 7.41, we note that

$$C^+(\underline{x}) + C^-(\underline{x}) = 1 \tag{7.42}$$

3. A direct consequence of the previous equation is that $C^+(\underline{x}) = 1/2$ for a smooth surface.
4. The regularized form of Equation 7.40 writes for all $\underline{x}$

$$\hat{p}(\underline{x}) = \int_{\partial\Omega} \left[\hat{p}(\underline{y}) \frac{\partial \hat{G}(\underline{x},\underline{y})}{\partial n_y} - \frac{\partial G_0(\underline{x},\underline{y})}{\partial n_y} \hat{p}(\underline{x}) \right] dS_y - \int_{\partial\Omega} \hat{G}(\underline{x},\underline{y}) \frac{\partial \hat{p}(\underline{y})}{\partial n_y} dS_y \tag{7.43}$$

5. Let us consider the particular case of a planar structure embedded in an infinite rigid baffle. Assume that the baffle is in the plane $x_3 = 0$.

Let Ω_+ and Ω_- be the domains corresponding to $x_3 > 0$ and $x_3 < 0$, respectively. Let $\underline{x}$ be a point in Ω_+ and $\underline{x}'$ its image with respect to the baffle. We are interested in the exterior problem ($x_3 > 0$). In this case $C^+(\underline{x}) = 1$ and $C^+(\underline{x}') = 0$ since $\underline{x}'$ is in the interior domain ($x_3' < 0$).

$$\begin{aligned} \hat{p}(\underline{x}) &= \int_{\partial\Omega} \left[\hat{p}(\underline{y}) \frac{\partial \hat{G}(\underline{x},\underline{y})}{\partial n_y} - \hat{G}(\underline{x},\underline{y}) \frac{\partial \hat{p}(\underline{y})}{\partial n_y} \right] dS_y \\ 0 &= \int_{\partial\Omega} \left[\hat{p}(\underline{y}) \frac{\partial \hat{G}(\underline{x}',\underline{y})}{\partial n_y} - \hat{G}(\underline{x}',\underline{y}) \frac{\partial \hat{p}(\underline{y})}{\partial n_y} \right] dS_y \end{aligned} \tag{7.44}$$

Summing both terms in Equation 7.44 leads to

$$\hat{p}(\underline{x}) = \int_{\partial\Omega} \left[\hat{p}(\underline{y}) \left(\frac{\partial \hat{G}(\underline{x},\underline{y})}{\partial n_y} + \frac{\partial \hat{G}(\underline{x}',\underline{y})}{\partial n_y} \right) - (\hat{G}(\underline{x},\underline{y}) + \hat{G}(\underline{x}',\underline{y})) \frac{\partial \hat{p}(\underline{y})}{\partial n_y} \right] dS_y \tag{7.45}$$

To simplify Equation 7.45, we can use the relationship $\underline{x}' = \underline{x} - 2(\underline{x} \cdot \underline{n}_B)\underline{n}_B$, where $\underline{n}_B$ is the normal to the baffle pointing into Ω_+.

Since $\underline{n}_B = \underline{n}_y$, we also have $(\underline{y} - \underline{x}') \cdot \underline{n}_y = (\underline{y} - \underline{x} + 2(\underline{x} \cdot \underline{n}_y)\underline{n}_y) \cdot \underline{n}_y = (\underline{y} + \underline{x}) \cdot \underline{n}_y$.

For a point $\underline{y}$ on the structure, we have $\underline{y} \cdot \underline{n}_y = y_3 = 0$. In addition, from Equation 7.4, we have

$$\begin{aligned} \frac{\partial \hat{G}(x,\underline{y})}{\partial n_y} + \frac{\partial \hat{G}(\underline{x}',\underline{y})}{\partial n_y} &= \underline{\nabla}_y \hat{G}(\underline{x},\underline{y}) \cdot \underline{n}_y + \underline{\nabla}_y \hat{G}(\underline{x}',\underline{y}) \cdot \underline{n}_y \\ &= -\frac{1}{4\pi r^2}(1 + ikr)\exp(-ikr)\frac{\underline{y} - \underline{x}}{r} \cdot \underline{n}_y \\ &\quad - \frac{1}{4\pi r'^2}(1 + ikr')\exp(-ikr')\frac{\underline{y} - \underline{x}'}{r'} \cdot \underline{n}_y \\ &= -\frac{1}{2\pi r^2}(1 + ikr)\exp(-ikr)\frac{\underline{y}}{r} \cdot \underline{n}_y = 0 \end{aligned} \tag{7.46}$$

Moreover, $\hat{G}(\underline{x},\underline{y}) = \hat{G}(\underline{x}',\underline{y})$, therefore Equation 7.45 reduces to

$$\hat{p}(\underline{x}) = -2\int_{\partial\Omega} \hat{G}(\underline{x},\underline{y}) \frac{\partial \hat{p}(\underline{y})}{\partial n_y} dS_y \tag{7.47}$$

This is *Rayleigh's integral*. Note that this integral can only be applied to planar baffled structures.

6. Other examples of applications of the boundary integral equation are given in Section 7.13.

7.4 DIRECT INTEGRAL FORMULATION FOR THE SCATTERING PROBLEM

Let us consider some acoustic source located outside a body occupying the volume Ω_- of boundary $\partial\Omega$. We are interested in the calculation of the acoustic field $\hat{p}$ resulting from the scattering by the body of the incident pressure field. The incident pressure field $\hat{p}_{inc}$ is the field due to the acoustic source when there is no foreign body inside the surface $\partial\Omega$.

Let us denote $\hat{p}_{sc}$ the scattered field* due to the presence of Ω_-. We write

$$\hat{p} = \hat{p}_{sc} + \hat{p}_{inc} \tag{7.48}$$

The scattered field $\hat{p}_{sc}$ is radiated toward infinity and must satisfy the Sommerfeld radiation condition equation (7.37) and consequently it must satisfy the integral representation of the exterior problem given by Equation 7.40, namely

$$C^+(\underline{x})\hat{p}_{sc}(\underline{x}) = \int_{\partial\Omega}\left[\hat{p}_{sc}(\underline{y})\frac{\partial\hat{G}(\underline{x},\underline{y})}{\partial n_y} - \hat{G}(\underline{x},\underline{y})\frac{\partial\hat{p}_{sc}(\underline{y})}{\partial n_y}\right]dS_y \tag{7.49}$$

Furthermore, the incident pressure field exists in the volume Ω_- enclosed by $\partial\Omega$. It satisfies the Helmholtz equation within Ω_- and therefore satisfies the integral representation of the interior problem equation (7.30). We have

$$C^-(\underline{x})\hat{p}_{inc}(\underline{x}) = \int_{\partial\Omega}\left[\hat{G}(\underline{x},\underline{y})\frac{\partial\hat{p}_{inc}(\underline{y})}{\partial n_y} - \hat{p}_{inc}(\underline{y})\frac{\partial\hat{G}(\underline{x},\underline{y})}{\partial n_y}\right]dS_y \tag{7.50}$$

Subtracting Equations 7.49 and 7.50 and using Equation 7.42 leads to the integral representation of the total pressure field $\hat{p}$

* This field can be considered as the superposition of the field scattered by the body occupying Ω_- as if it were rigid acoustically (blocked pressure) and the radiated field induced by the vibration of the boundary $\partial\Omega$.

$$C^{+}(\underline{x})\hat{p}(\underline{x}) = \int_{\partial\Omega}\left[\hat{p}(\underline{y})\frac{\partial \hat{G}(\underline{x},\underline{y})}{\partial n_y} - \hat{G}(\underline{x},\underline{y})\frac{\partial \hat{p}(\underline{y})}{\partial n_y}\right]dS_y + \hat{p}_{inc}(\underline{x}) \tag{7.51}$$

The regularized form for all $\underline{x} \in \partial\Omega$ is given by

$$\hat{p}(\underline{x}) = \int_{\partial\Omega}\left[\hat{p}(\underline{y})\frac{\partial \hat{G}(\underline{x},\underline{y})}{\partial n_y} - \frac{\partial G_0(\underline{x},\underline{y})}{\partial n_y}\hat{p}(\underline{x})\right]dS_y - \int_{\partial\Omega}\hat{G}(\underline{x},\underline{y})\frac{\partial \hat{p}(\underline{y})}{\partial n_y}\, dS_y + \hat{p}_{inc}(\underline{x}) \tag{7.52}$$

Equations 7.51 and 7.52 allow one to calculate the total pressure field resulting from the scattering of the incident pressure field by the body occupying the volume Ω_-.

Remarks:

Equation 7.51 can be derived directly from Equation 7.40 with its source term in volume Ω_+

$$C^{+}(\underline{x})\hat{p}(\underline{x}) = \int_{\partial\Omega}\left[\hat{p}(\underline{y})\frac{\partial \hat{G}(\underline{x},\underline{y})}{\partial n_y} - \hat{G}(\underline{x},\underline{y})\frac{\partial \hat{p}(\underline{y})}{\partial n_y}\right]dS_y + \int_{\Omega^+}\hat{\underline{Q}}(\underline{y})\,\hat{G}(\underline{x},\underline{y})\,dV_y \tag{7.53}$$

The source term $\hat{p}_{inc}(\underline{x})$ is given by $\int_{\Omega^+}\hat{\underline{Q}}(y)\hat{G}(\underline{x},\underline{y})dV_y$. In the case of monopole (point source) located at point $\underline{x}_0$, the source density $\hat{\underline{Q}}(\underline{y})$ can be written as

$$\hat{\underline{Q}}(\underline{y}) = i\omega\rho_0\hat{\underline{Q}}_s\delta(\underline{y} - \underline{x}_0) \tag{7.54}$$

where $\hat{\underline{Q}}_s$ represents the source strength or the volume velocity in m^3s^{-1}. The total pressure field is then given by*

* In the case of a dipole located at position $\underline{x}_0$ with moment amplitude vector $\hat{\underline{D}}$, the source term reads $\int_\Omega \hat{\underline{D}} \cdot \nabla_{x_0}(\delta(\underline{y} - \underline{x}_0))\hat{G}(\underline{x},\underline{y})\,dV_y$ which transforms into $-\int_\Omega \delta(\underline{y} - \underline{x}_0)\hat{\underline{D}} \cdot \nabla_{x_0}\hat{G}(\underline{x},\underline{y})dV_y = \int_\Omega \delta(\underline{y} - \underline{x}_0)\,\hat{\underline{D}} \cdot \nabla_y\hat{G}(\underline{x},\underline{y})\,dV_y = \hat{\underline{D}} \cdot \nabla_{x_0}\,\hat{G}(\underline{x},\underline{x}_0)$. When the dipole is modeled using two out of phase identical monopoles separated by a small distance given by vector $\underline{d}$, the dipole moment vector $\hat{\underline{D}}$ is related to the monopoles' source strength $\hat{\underline{Q}}_s$ by $\hat{\underline{D}} = -\hat{\underline{Q}}\underline{d} = (-i\omega\rho_0\hat{\underline{Q}}_s)\underline{d}$. Similar equations can be written for a quadrupole-type point source.

$$C^+(\underline{x})\hat{p}(\underline{x}) = \int_{\partial\Omega} \left[\hat{p}(\underline{y}) \frac{\partial \hat{G}(\underline{x},\underline{y})}{\partial n_y} - \hat{G}(\underline{x},\underline{y}) \frac{\partial \hat{p}(\underline{y})}{\partial n_y} \right] dS_y + i\omega\rho_0 \hat{Q}_s \hat{G}(\underline{x},\underline{x}_0) \quad (7.55)$$

In the particular case of the scattering by a rigid body, Equation 7.53 reduces to

$$C^+(\underline{x})\hat{p}(\underline{x}) = \int_{\partial\Omega} \hat{p}(\underline{y}) \frac{\partial \hat{G}(\underline{x},\underline{y})}{\partial n_y} dS_y + \int_{\Omega^+} \hat{Q}(\underline{y})\, \hat{G}(\underline{x},\underline{y}) dV_y \quad (7.56)$$

7.5 INDIRECT INTEGRAL FORMULATIONS

It is possible to derive integral representation using the superposition method or Huygens principle. The latter states that the field radiated or scattered by a body can be constructed from a distribution of monopole and dipole sources. This construction is direct for classic boundary conditions:

- Dirichlet problem (single-layer potential formulation): The field $\hat{p}$ being known on the boundary $\partial\Omega$, the flux $\partial\hat{p}/\partial n$ is sought. $\hat{p}$ can be constructed from a distribution of monopole sources

$$\hat{p}(\underline{x}) = \int_{\partial\Omega} \hat{\sigma}(\underline{y}) \hat{G}(\underline{x},\underline{y})\, dS_y \quad \underline{x} \notin \partial\Omega \quad (7.57)$$

  This identity represents the integral formulation of $\hat{p}$ by a single-layer potential of density $\hat{\sigma}$.
- Neumann problem (double-layer potential formulation): The flux of the field $\hat{p}$, $\partial\hat{p}/\partial n$ is known on the boundary $\partial\Omega$, and $\hat{p}$ is sought. $\hat{p}$ can then be reconstructed from a distribution of dipole sources

$$\hat{p}(\underline{x}) = \int_{\partial\Omega} \hat{\mu}(\underline{y}) \frac{\partial \hat{G}(\underline{x},\underline{y})}{\partial n_y} dS_y \quad \underline{x} \notin \partial\Omega \quad (7.58)$$

  This identity represents the integral formulation of $\hat{p}$ by a double-layer potential of density $\hat{\mu}$.
- General problem (single and double layer potential formulation): In the general case, the field $\hat{p}$ can be reconstructed from a distribution of both monopole and dipole sources

$$\hat{p}(\underline{x}) = \int_{\partial\Omega} \left[\hat{\mu}(\underline{y}) \frac{\partial \hat{G}(\underline{x}, \underline{y})}{\partial n_y} - \hat{\sigma}(\underline{y})\hat{G}(\underline{x}, \underline{y}) \right] dS_y \quad \underline{x} \notin \partial\Omega \tag{7.59}$$

Remarks:

1. The previous integral representations are referred to as indirect formulations because the calculation of the field $\hat{p}$ requires the determination of intermediate unknowns namely the single- and double-layer potential densities $\hat{\sigma}$ and $\hat{\mu}$.
2. The previous integral representations can be directly obtained from the results of Sections 7.2 through 7.4.

Indeed, let Ω_- be a bounded domain of boundary $\partial\Omega$ and outward normal $\underline{n}$. Each side of $\partial\Omega$ is indicated by a "+" or a "−" depending on whether we are on the side of the normal (exterior of Ω_-) or on the other side (interior of Ω_-) (see Figure 7.4).

Let $\hat{p}^+$ and $\partial\hat{p}^+/\partial n$ be the values of the field $\hat{p}$ and its flux $\partial\hat{p}/\partial n$ on the positive side of $\partial\Omega$. Similarly, let $\hat{p}^-$ and $\partial\hat{p}^-/\partial n$ be the values of the field $\hat{p}$ and its flux $\partial\hat{p}/\partial n$ on the negative side of $\partial\Omega$.

For all $\underline{x} \in \Re^3$, the integral representations of the interior problem equation (7.30) and the exterior problem equation (7.40) allows us to write

$$C^+(\underline{x})\hat{p}^+(\underline{x}) = \int_{\partial\Omega} \left[\hat{p}^+(\underline{y}) \frac{\partial \hat{G}(\underline{x}, \underline{y})}{\partial n_y} - \hat{G}(\underline{x}, \underline{y}) \frac{\partial \hat{p}^+(\underline{y})}{\partial n_y} \right] dS_y \tag{7.60}$$

$$C^-(\underline{x})\hat{p}^-(\underline{x}) = \int_{\partial\Omega} \left[\hat{G}(\underline{x}, \underline{y}) \frac{\partial \hat{p}^-(\underline{y})}{\partial n_y} - \hat{p}^-(\underline{y}) \frac{\partial \hat{G}(\underline{x}, \underline{y})}{\partial n_y} \right] dS_y \tag{7.61}$$

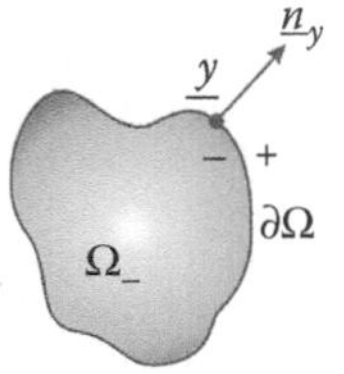

Figure 7.4 Convention of sign for the indirect formulation.

where $\hat{p}^+(\underline{x})$ (respectively, $\hat{p}^-(\underline{x})$) is the field created at point $\underline{x}$ by the field $[\hat{p}^+, \partial\hat{p}^+/\partial n]$ (respectively, $[\hat{p}^-, \partial\hat{p}^-/\partial n]$) over $\partial\Omega$.

Summing the two previous equations leads to

$$C^-(\underline{x})\hat{p}^-(\underline{x}) + C^+(\underline{x})\hat{p}^+(\underline{x}) = \int_{\partial\Omega} \left[\hat{\mu}(\underline{y}) \frac{\partial \hat{G}(\underline{x}, \underline{y})}{\partial n_y} - \hat{\sigma}(\underline{y})\, \hat{G}(\underline{x}, \underline{y}) \right] dS_y \quad \forall \underline{x} \in \mathbb{R}^3 \tag{7.62}$$

where

$$\hat{\sigma} = \frac{\partial \hat{p}^+}{\partial n_y} - \frac{\partial \hat{p}^-}{\partial n_y} = \left[\frac{\partial \hat{p}}{\partial n_y} \right] \tag{7.63}$$

is the jump of flux $\partial\hat{p}/\partial n$ through $\partial\Omega$ and

$$\hat{\mu} = \hat{p}^+ - \hat{p}^- = [\hat{p}] \tag{7.64}$$

is the jump of pressure $\hat{p}$ through $\partial\Omega$.

We thus obtain the integral representations Equations 7.57 through 7.59. Let us consider some examples.

For the exterior problem, we have for a point $\underline{x} \in \Omega_+, C^+ = 1; C^- = 0$ and consequently Equation 7.62 can be written as

$$\hat{p}^+(\underline{x}) = \int_{\partial\Omega} \left[\hat{\mu}(\underline{y}) \frac{\partial \hat{G}(\underline{x}, \underline{y})}{\partial n_y} - \hat{\sigma}(\underline{y})\, \hat{G}(\underline{x}, \underline{y}) \right] dS_y \tag{7.65}$$

Let us now assume that the boundary $\partial\Omega$ is thin* (thickness infinitely small). In the case of Dirichlet problem, $\hat{p}$ is given over $\partial\Omega$, which implies $\hat{\mu} = 0$ and consequently Equation 7.65 reduces to Equation 7.57.

In the case of Neumann problem, the flux $\partial\hat{p}/\partial n$ is known on the boundary which implies $\hat{\sigma} = 0$ and consequently Equation 7.65 reduces to Equation 7.58.

Similar results can be obtained for the interior problem.

* The interpretation of densities $\hat{\sigma}$ and $\hat{\mu}$ in terms of pressure gradient jump and pressure jump only holds if the boundary is thin, that is, we can define at a given point on the surface a pressure gradient and a pressure on each side of the surface. If the domain is enclosed by a thick boundary, we can still use densities $\hat{\sigma}$ and $\hat{\mu}$ but they cannot be interpreted in terms of jumps across the boundary.

7.5.1 SBIE

To calculate densities $\hat{\sigma}$ and $\hat{\mu}$, we must take the limit of $\hat{p}$ and $\partial\hat{p}/\partial n$ on the boundary before applying the boundary conditions.

We start from Equation 7.62, which is valid at all points of $\mathbb{R}^3$. Using the identity equation (7.42), we have

$$\hat{p}^+(\underline{x}) = C^-(\underline{x})\hat{\mu}(\underline{x}) + \int_{\partial\Omega}\left[\hat{\mu}(\underline{y})\frac{\partial\hat{G}(\underline{x},\underline{y})}{\partial n_y} - \hat{\sigma}(\underline{y})\hat{G}(\underline{x},\underline{y})\right]dS_y \quad \forall\underline{x}\in\partial\Omega \tag{7.66}$$

For a smooth surface,

$$\hat{p}^+(\underline{x}) = \frac{\hat{\mu}(\underline{x})}{2} + \int_{\partial\Omega}\left[\hat{\mu}(\underline{y})\frac{\partial\hat{G}(\underline{x},\underline{y})}{\partial n_y} - \hat{\sigma}(\underline{y})\,\hat{G}(\underline{x},\underline{y})\right]dS_y \quad \forall\underline{x}\in\partial\Omega \tag{7.67}$$

In addition, $\hat{p}^-(\underline{x}) = \hat{p}^+(\underline{x}) - \hat{\mu}(\underline{x})$ so that

$$\hat{p}^-(\underline{x}) = -C^+(\underline{x})\hat{\mu}(\underline{x}) + \int_{\partial\Omega}\left[\hat{\mu}(\underline{y})\frac{\partial\hat{G}(\underline{x},\underline{y})}{\partial n_y} - \hat{\sigma}(\underline{y})\hat{G}(\underline{x},\underline{y})\right]dS_y \quad \forall\underline{x}\in\partial\Omega \tag{7.68}$$

For a smooth surface,

$$\hat{p}^-(\underline{x}) = -\frac{\hat{\mu}(\underline{x})}{2} + \int_{\partial\Omega}\left[\hat{\mu}(\underline{y})\frac{\partial\hat{G}(\underline{x},\underline{y})}{\partial n_y} - \hat{\sigma}(\underline{y})\hat{G}(\underline{x},\underline{y})\right]dS_y \quad \forall\underline{x}\in\partial\Omega \tag{7.69}$$

Furthermore,

$$\frac{\partial\hat{p}^+}{\partial n}(\underline{x}) = C^+(\underline{x})\hat{\sigma}(\underline{x}) - \int_{\partial\Omega}\left[\hat{\sigma}(\underline{y})\frac{\partial\hat{G}(\underline{x},\underline{y})}{\partial n_x}\right]dS_y + FP\int_{\partial\Omega}\left[\hat{\mu}(\underline{y})\frac{\partial^2\hat{G}(\underline{x},\underline{y})}{\partial n_x\partial n_y}\right]dS_y \tag{7.70}$$

and since $\frac{\partial \hat{p}^-}{\partial n}(\underline{x}) = \frac{\partial \hat{p}^+}{\partial n}(\underline{x}) - \hat{\sigma}(\underline{x})$, we get

$$\frac{\partial \hat{p}^-}{\partial n}(\underline{x}) = -C^-(\underline{x})\hat{\sigma}(\underline{x}) - \int_{\partial\Omega}\left[\hat{\sigma}(\underline{y})\frac{\partial \hat{G}(\underline{x},\underline{y})}{\partial n_x}\right]dS_y + FP\int_{\partial\Omega}\left[\hat{\mu}(\underline{y})\frac{\partial^2 \hat{G}(\underline{x},\underline{y})}{\partial n_x \partial n_y}\right]dS_y \tag{7.71}$$

where FP denotes the Hadamard finite part of the integral which is defined by

$$\begin{aligned} FP\int_{\partial\Omega}\left[\hat{\mu}(\underline{y})\frac{\partial^2 \hat{G}(\underline{x},\underline{y})}{\partial n_x \partial n_y}\right]dS_y &= \lim_{\underline{x}\to\partial\Omega}\frac{\partial}{\partial n_x}\int_{\partial\Omega}\hat{\mu}(\underline{y})\frac{\partial \hat{G}(\underline{x},\underline{y})}{\partial n_y}dS_y \\ &= \lim_{\underline{x}\to\partial\Omega}\underline{n}_\zeta \cdot \underline{\nabla}_x \int_{\partial\Omega}\hat{\mu}(\underline{y})\frac{\partial \hat{G}(\underline{x},\underline{y})}{\partial n_y}dS_y \end{aligned} \tag{7.72}$$

where $\underline{n}_\zeta$ is the normal to the boundary $\partial\Omega$ at point $\underline{\zeta}$, which is the limit of point $\underline{x}$ when $\underline{x} \to \partial\Omega$.

Remarks:

The integral $\int_S \hat{\mu}(y)(\partial^2\hat{G}(\underline{x},\underline{y})/\partial n_x \partial n_y)dS_y$ is hyper-singular. It does not exist for $\underline{x} \in \partial\Omega$. Actually, the order of differentiation and integration on the left-hand side of Equation 7.72 cannot be strictly interchanged since the kernel would be nonintegrable. However, this change of order can be done if the resulting hyper-singular integral is interpreted in the sense of Hadamard finite part. The change of order allows one to obtain an integral operator with the kernel $\partial^2\hat{G}(\underline{x},\underline{y})/\partial n_x \partial n_y$, which is much more suitable for numerical calculations. This numerical calculation is, however, tricky and requires special mathematical treatment on the integral operator before using a numerical quadrature. Several techniques allow for its computation (Guiggiani et al. 1992; Harris 1992; Silva 1994; Bonnet 1995; Harris et al. 2006). The reader can refer to Section 7.10 for examples of such techniques. One particular regularization formula based on the transformation proposed by Maue (1949) and Stallybrass (1967), which is used in the variational formulation of boundary integral equations (VBIE) (see Section 7.7.1) is given in Hamdi (1988) and Pierce (1993)

$$\begin{aligned} FP\int_{\partial\Omega}\left[\hat{\mu}(\underline{y})\frac{\partial^2 \hat{G}(\underline{x},\underline{y})}{\partial n_x \partial n_y}\right]dS_y = \int_{\partial\Omega}&[k^2\underline{n}_x \cdot \underline{n}_y\hat{G}(\underline{x},\underline{y})\hat{\mu}(\underline{y}) \\ &+ [\underline{n}_x \times \underline{\nabla}_x\hat{G}(\underline{x},\underline{y})]\cdot[\underline{n}_y \times \underline{\nabla}_y\hat{\mu}(\underline{y})]]\,dS_y \end{aligned} \tag{7.73}$$

where $\partial\Omega$ is supposed to be a closed surface. In this formula, $\underline{x}$ is on the boundary. Note that the operator $[\underline{n}_x \times \underline{\nabla}_x]$ only involves the tangential derivative of $\hat{\mu}$ along the surface at point $\underline{x}$. Indeed, if we denote by $\underline{\nabla}_{x,S}\hat{\mu}$ the tangential gradient of $\hat{\mu}$ given by $\underline{\nabla}_{x,S}\,\hat{\mu}(\underline{x}) = \underline{\nabla}_x\,\hat{\mu}(\underline{x}) - \underline{n}_x(\partial\hat{\mu}(\underline{x})/\partial n_x)$, we see that $\underline{n}_x \times \underline{\nabla}_x\hat{\mu}(\underline{x}) = \underline{n}_x \times \underline{\nabla}_{x,S}\hat{\mu}(\underline{x})$. Appendix 7G provides a demonstration of this formula. Other regularization methods are discussed in Sections 7.10.2 and 7.10.3.

7.5.2 RBIE

Replacing $C^-(\underline{x})$ by Equation 7.31 in Equation 7.66 leads to

$$\hat{p}^+(\underline{x}) = \int_{\partial\Omega}\left[\hat{\mu}(\underline{y})\frac{\partial\hat{G}(\underline{x},\underline{y})}{\partial n_y} - \frac{\partial G_0(\underline{x},\underline{y})}{\partial n_y}\hat{\mu}(\underline{x})\right]dS_y - \int_{\partial\Omega}\hat{\sigma}(\underline{y})\hat{G}(\underline{x},\underline{y})\,dS_y \qquad (7.74)$$

Finally, using the method of Section 7.5.1 it can be shown that the flux $\partial\hat{p}^+/\partial n$ on $\partial\Omega$ is given by

$$\frac{\partial\hat{p}^+}{\partial n}(\underline{x}) = \hat{\sigma}(\underline{x}) - \int_{\partial\Omega}\left[\hat{\sigma}(\underline{y})\frac{\partial\hat{G}(\underline{x},\underline{y})}{\partial n_x} - \frac{\partial G_0(\underline{x},\underline{y})}{\partial n_x}\hat{\sigma}(\underline{x})\right]dS_y + FP\int_{\partial\Omega}\left[\hat{\mu}(\underline{y})\frac{\partial^2\hat{G}(\underline{x},\underline{y})}{\partial n_x\partial n_y}\right]dS_y \qquad (7.75)$$

7.6 NUMERICAL IMPLEMENTATION: COLLOCATION METHOD

7.6.1 The principle: Case of the zeroth-order interpolation

The discretization by boundary elements results from a direct transposition of the classic approach used for the discretization by finite elements (see Chapter 6).

To simplify the presentation, we consider the numerical implementation of the RBIE for the exterior problem (7.43)

$$\hat{p}(\underline{x}) = \int_{\partial\Omega}\left[\hat{p}(\underline{y})\frac{\partial\hat{G}(\underline{x},\underline{y})}{\partial n_y} - \frac{\partial G_0(\underline{x},\underline{y})}{\partial n_y}\hat{p}(\underline{x})\right]dS_y - \int_{\partial\Omega}\hat{G}(\underline{x},\underline{y})\frac{\partial\hat{p}(\underline{y})}{\partial n_y}\,dS_y \qquad (7.76)$$

The advantage of the regularized form is that the numerical treatments are more tractable. This equation is rewritten as

$$\hat{p}(\underline{x}) = \int_{\partial\Omega} \left[\frac{\partial \hat{G}(\underline{x}, \underline{y})}{\partial n_y} - \frac{\partial G_0(\underline{x}, \underline{y})}{\partial n_y} \right] \hat{p}(\underline{y})\, dS_y - \int_{\partial\Omega} [\hat{p}(\underline{x}) - \hat{p}(\underline{y})] \frac{\partial G_0(\underline{x}, \underline{y})}{\partial n_y}\, dS_y - \int_{\partial\Omega} \hat{G}(\underline{x}, \underline{y}) \frac{\partial \hat{p}(\underline{y})}{\partial n_y}\, dS_y \tag{7.77}$$

Note that even if the integrals are weakly singular, a special numerical treatment is required because Gaussian quadratures formulas, which are commonly used to evaluate integrals numerically, are only applicable to regular integrands.

We start from Equation 7.77 to describe the principle of the boundary element discretization.

7.6.1.1 Representation and discretization of the geometry

The boundary $\partial\Omega$ is discretized into N_e elements: $\partial\Omega = \cup_{e=1}^{N_e} \partial\Omega^e$. For three-dimensional problems, these boundary elements are two-dimensional and can be either planar or curved.

These elements satisfy the same properties as finite elements (see Chapter 6). Equation 7.77 can be rewritten as

$$\hat{p}(\underline{x}) = \sum_{e=1}^{N_e} \int_{\partial\Omega_e} \left[\frac{\partial \hat{G}(\underline{x}, \underline{y})}{\partial n_y^e} - \frac{\partial G_0(\underline{x}, \underline{y})}{\partial n_y^e} \right] \hat{p}(\underline{y})\, dS_y - \sum_{e=1}^{N_e} \int_{\partial\Omega_e} [\hat{p}(\underline{x}) - \hat{p}(\underline{y})] \frac{\partial G_0(\underline{x}, \underline{y})}{\partial n_y^e}\, dS_y - \sum_{e=1}^{N_e} \int_{\partial\Omega_e} \hat{G}(\underline{x}, \underline{y}) \frac{\partial \hat{p}(\underline{y})}{\partial n_y^e}\, dS_y \tag{7.78}$$

A point $\underline{y} = \begin{Bmatrix} y_1 \\ y_2 \\ y_3 \end{Bmatrix}$ pertaining to an element $\partial\Omega^e$ is assumed to be described by a suitable piecewise parametric representation of intrinsic* coordinates $\underline{\eta} = (\eta_1, \eta_2)$

* Also called natural or local coordinates.

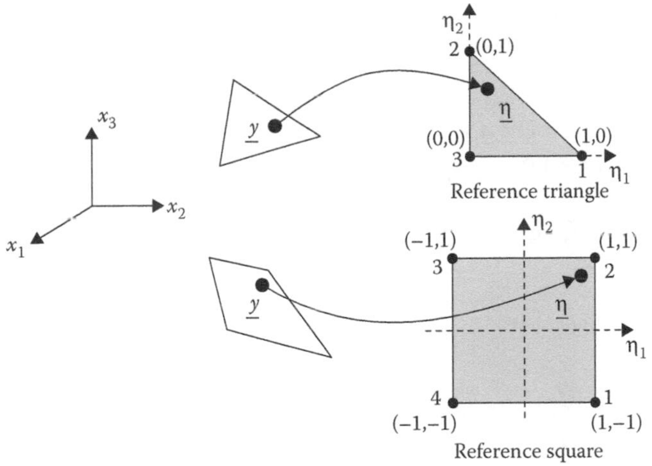

Figure 7.5 Mapping from the physical space to the reference space.

$$\begin{aligned} y_i(\underline{\eta}) &= \langle N^e(\underline{\eta})\rangle \{x_i^e\} \\ &= N_j^e(\underline{\eta}) x_{i,j}^e \end{aligned} \tag{7.79}$$

where $\langle N^e\rangle = \langle N_1^e,\ldots,N_{N_{ne}}^e\rangle$ is the vector of local shape functions of the element, N_{ne} is the number of nodes of the element. $\{x_i^e\}$ is the vector of nodal coordinates of each node of element $e(1$ to $N_{ne})$ in direction i. This representation is a one-to-one mapping between the physical space and the local element space $\underline{\eta}$ (see Figure 7.5).

The mapping can be written explicitly as

$$\begin{cases} y_1 = \langle N^e(\eta_1,\eta_2)\rangle\{x_1^e\} \\ y_2 = \langle N^e(\eta_1,\eta_2)\rangle\{x_2^e\} \\ y_3 = \langle N^e(\eta_1,\eta_2)\rangle\{x_3^e\} \end{cases} \tag{7.80}$$

The expressions of the local shape functions for classic elements (triangles and/or quadrangles) are given in Appendix 5B.

7.6.1.2 Calculation of the normal vector and gradient transformations

Usual theory of surfaces formulas allow one to carry out this calculation (see also Chapter 6).

Let $\underline{a}_1^e = (\partial y/\partial \xi_1)$ and $\underline{a}_2^e = (\partial \underline{y}/\partial \xi_2)$ be the vectors defining the local tangent plane to $\partial\Omega^e$ at point $\underline{y}$. We have

$$a^e_{i,\alpha}(\underline{\eta}) = \left\langle \frac{\partial N^e}{\partial \eta_\alpha} \right\rangle \{x^e_i\} \quad i = 1,2,3; \alpha = 1,2 \tag{7.81}$$

The surface element dS_y and the unit normal vector $\underline{n}^e_y$ at point $\underline{y}$ of element $\partial\Omega^e$ are given by

$$dS_y = \left|j^e\right| \, d\xi_1 d\xi_2 \tag{7.82}$$

with $\left|j^e\right| = \left|\underline{a}^e_1 \times \underline{a}^e_2\right|$ and

$$\underline{n}^e_y = \frac{\underline{a}^e_1 \times \underline{a}^e_2}{\left|\underline{a}^e_1 \times \underline{a}^e_2\right|} = \frac{1}{\left|j^e\right|}\left(\underline{a}^e_1 \times \underline{a}^e_2\right) \tag{7.83}$$

We have

$$\left|j^e\right| = \sqrt{(g^e_1)^2 + (g^e_2)^2 + (g^e_3)^2} \tag{7.84}$$

with

$$\underline{n}^e_y = \frac{1}{\left|j^e\right|}\begin{Bmatrix} g^e_1 \\ g^e_2 \\ g^e_3 \end{Bmatrix} \tag{7.85}$$

where

$$\begin{cases} g^e_1(\eta_1, \eta_2) = \left[\dfrac{\partial y_2}{\partial \eta_1}\dfrac{\partial y_3}{\partial \eta_2} - \dfrac{\partial y_3}{\partial \eta_1}\dfrac{\partial y_2}{\partial \eta_2}\right] \\ g^e_2(\eta_1, \eta_2) = \left[\dfrac{\partial y_3}{\partial \eta_1}\dfrac{\partial y_1}{\partial \eta_2} - \dfrac{\partial y_1}{\partial \eta_1}\dfrac{\partial y_3}{\partial \eta_2}\right] \\ g^e_3(\eta_1, \eta_2) = \left[\dfrac{\partial y_1}{\partial \eta_1}\dfrac{\partial y_2}{\partial \eta_2} - \dfrac{\partial y_2}{\partial \eta_1}\dfrac{\partial y_1}{\partial \eta_2}\right] \end{cases} \tag{7.86}$$

7.6.1.3 Discretization of unknowns

The field $\hat{p}$ and its flux $\hat{q} = (\partial\hat{p}/\partial n_y)$ at a point $\underline{y}$ of element $\partial\Omega^e$ are expressed in terms of nodal values at N_{ie} points of the element $\partial\Omega^e$, which are called interpolation nodes

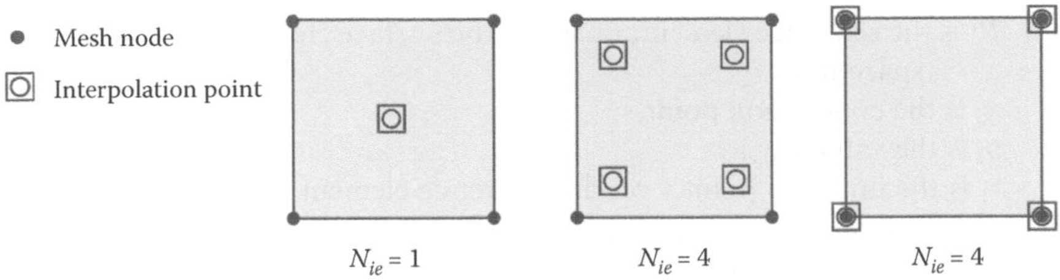

Figure 7.6 Examples of mesh and interpolation nodes for a quadrilateral element.

$$\hat{p}(\underline{y}) = \langle \overline{N}^e(\eta_1, \eta_2) \rangle \{\hat{p}^e\} \tag{7.87}$$

$$\hat{q}(\underline{y}) = \langle \overline{N}^e(\eta_1, \eta_2) \rangle \{\hat{q}^e\} \tag{7.88}$$

where $\{\hat{p}^e\}$ and $\{\hat{q}^e\}$ are the vector of nodal values of $(\hat{p}, \hat{q})$ at the interpolation nodes (Figure 7.6).

Note that the interpolation nodes must be chosen so that the appropriate continuity conditions for the solution are satisfied. In particular, the evaluation of hyper-singular integrals (see Section 7.10.2) requires the solution to be sufficiently smooth to exist. C^0 (piecewise constant interpolation functions) or discontinuous boundary elements must then be used.

7.6.2 Construction of the discretized problem: Collocation method

For a well-posed problem, at each point of $\partial\Omega$, half of the variables $(\hat{p}, \hat{q})$ are known. Thus, for N interpolation nodes, we have a set of N unknowns. To determine these unknowns, we enforce integral equation (7.78) to be satisfied at exactly the N points of $\partial\Omega$. This is the principle of the collocation method. In practice, we carry out the collocation at all the interpolation nodes. The discretized boundary integral evaluated at a collocation point $\underline{x}_i$, can be written as

$$\begin{aligned}\hat{p}_i = &-\sum_{e=1}^{N_e} \int_{R^e} \left[\frac{\partial \hat{G}(\underline{x}_i, \underline{y}(\underline{\eta}))}{\partial n_y^e} - \frac{\partial G_0(\underline{x}_i, \underline{y}(\underline{\eta}))}{\partial n_y^e} \right] \overline{N}_j^e(\underline{\eta})\, |j^e(\underline{\eta})| d\eta_1 d\eta_2\, \hat{p}_j^e \\ &+ \sum_{e=1}^{N_e} \int_{R^e} \left[\frac{\partial G_0(\underline{x}_i, \underline{y}(\underline{\eta}))}{\partial n_y^e} [\overline{N}_j^e(\underline{\xi}) - \overline{N}_j^e(\underline{\eta})] \right] |j^e(\underline{\eta})|\, d\eta_1 d\eta_2\, \hat{p}_j^e \\ &- \sum_{e=1}^{N_e} \int_{R^e} \hat{G}(\underline{x}_i, \underline{y}(\underline{\eta}))\, \overline{N}_j^e(\underline{\eta})\, |j^e(\underline{\eta})|\, d\eta_1 d\eta_2\, \hat{q}_j^e \qquad \forall i \in [1, N]\end{aligned} \tag{7.89}$$

where

R^e is the reference element, image of the surface element $\partial\Omega^e$ in the local space η_1,η_2.

$\underline{x}_i$ is the collocation point.

p_i is the value $\hat{p}(\underline{x}_i)$.

$\underline{\eta}$ is the image of point $\underline{y}$ on the reference element R^e.

$(\hat{p}_j^e, \hat{q}_j^e)$ are the values of $\hat{p}$ and $\hat{q} = (\partial\hat{p}/\partial n)$ at the interpolation point j of element $\partial\Omega^e$.

Recall that in Equation 7.89, Einstein summation notation is used. The Green's function together with its normal derivative can be expressed in terms of local coordinates

$$\hat{G}(\underline{x}_i, \underline{y}) = \frac{\exp\left(-ik\left|\underline{x}_i - \underline{y}\right|\right)}{4\pi\left|\underline{x}_i - \underline{y}\right|} = \frac{\exp(-ikr)}{4\pi r} \tag{7.90}$$

with

$$\begin{aligned} r &= \sqrt{(x_{i,1} - y_1)^2 + (x_{i,2} - y_2)^2 + (x_{i,3} - y_3)^2} \\ &= \sqrt{(x_{i,1} - N_j(\eta)\, x_{1,j}^e)^2 + (x_{i,2} - N_j(\eta)\, x_{2,j}^e)^2 + (x_{i,3} - N_j(\eta)\, x_{3,j}^e)^2} \end{aligned} \tag{7.91}$$

where the sum over index j is carried out from 1 to N_{ie}.

In addition,

$$\frac{\partial\hat{G}}{\partial n_y^e}(\underline{x}_i, \underline{y}) = \frac{\partial\hat{G}}{\partial r}\frac{(\underline{y} - \underline{x}_i)}{r} \cdot \underline{n}_y^e = \frac{-(1 + ikr)}{4\pi r^2}\frac{(\underline{y} - \underline{x}_i)}{r} \cdot \underline{n}_y^e \exp(-ikr) \tag{7.92}$$

where $\underline{n}_y^e$ is given by Equation 7.83. Similarly,

$$\frac{\partial G_0}{\partial n_y^e}(\underline{x}_i, \underline{y}) = \frac{\partial G_0}{\partial r}\frac{(\underline{y} - \underline{x}_i)}{r} \cdot \underline{n}_y^e = -\frac{1}{4\pi r^2}\frac{(\underline{y} - \underline{x}_i)}{r} \cdot \underline{n}_y^e \tag{7.93}$$

Equation 7.89 can be written for each of the N collocation points, which leads to the following linear system

$$[\hat{\mathcal{H}}]\{\hat{p}\} = [\hat{\mathcal{G}}]\{\hat{q}\} \tag{7.94}$$

where $\{\hat{p}\}$ and $\{\hat{q}\}$ are vectors of dimension N containing the values of field $\hat{p}$ and its flux $\hat{q} = \partial\hat{p}/\partial n$ at the N interpolation nodes. The matrices

$[\hat{\mathcal{H}}]$ and $[\hat{\mathcal{G}}]$ are square of dimension $N \times N$. They are full, complex valued, and frequency-dependent. This represents a main drawback compared to the finite element method where matrices are sparse, symmetric, and the frequency dependency can be taken out of the elementary matrices (case of the acoustic interior problem). For a spectral calculation, matrices $[\hat{\mathcal{H}}]$ and $[\hat{\mathcal{G}}]$ must be calculated at each frequency. When dealing with multifrequency analysis, special interpolation techniques allow for reducing the number of frequency points at which the solution is computed (see Section 7.11). Furthermore, because the Green's function is a function of $1/r$, a special numerical treatment is necessary when the collocation point $\underline{x}_i$ coincides with the interpolation node $\underline{y}_j$. This is the subject of Section 7.6.4.

7.6.3 Imposition of boundary conditions and resolution of the discretized system

The components of $\{\hat{p}\}$ and $\{\hat{q}\}$ which are constrained by the boundary conditions, are now substituted in Equation 7.94. Let us assume that N_p values of $\hat{p}$ are unknown over a region $\partial\Omega_p$ of $\partial\Omega$ and N_q values of $\hat{q} = \partial\hat{p}/\partial n$ are unknown over a region $\partial\Omega_q = \partial\Omega - \partial\Omega_p$. Let us rearrange the $N = N_p + N_q$ unknowns in a vector $\langle\hat{x}\rangle = \langle\hat{p}_1 \cdots \hat{p}_{N_p}, \hat{q}_1 \cdots \hat{q}_{N_q}\rangle$, which leads to the square linear system

$$[\hat{\mathfrak{B}}(\omega)]\{\hat{x}\} = \{\hat{f}(\omega)\} \tag{7.95}$$

where matrix $[\hat{\mathfrak{B}}(\omega)]$ is full, square, complex-valued, nonsymmetric, and frequency-dependent. This system can be solved using classic algorithms such as Gauss elimination.

7.6.4 Numerical quadrature

The numerical evaluation of integrals appearing in Equation 7.89 requires the calculation of three kinds of elementary integrals

$$\begin{cases} \hat{I}_{1,ij} = \displaystyle\int_{R^e} \left[\frac{\partial \hat{G}(\underline{x}_i, \underline{y}(\underline{\eta}))}{\partial n_y} - \frac{\partial G_0(\underline{x}_i, \underline{y}(\underline{\eta}))}{\partial n_y} \right] \overline{N}_j(\underline{\eta}) \left| j(\underline{\eta}) \right| d\eta_1 d\eta_2 \\ \hat{I}_{2,ij} = \displaystyle\int_{R^e} \frac{\partial G_0(\underline{x}_i, \underline{y}(\underline{\eta}))}{\partial n_y} [\overline{N}_j(\underline{\xi}) - \overline{N}_j(\underline{\eta})] \left| j(\underline{\eta}) \right| d\eta_1 d\eta_2 \\ \hat{I}_{3,ij} = \displaystyle\int_{R^e} \hat{G}(\underline{x}_i, \underline{y}(\underline{\eta}))\, \overline{N}_j(\underline{\eta}) \left| j(\underline{\eta}) \right| d\eta_1 d\eta_2 \end{cases} \tag{7.96}$$

where we have removed the superscript "e" in all the expressions to simplify the notation. It is implicit that the shape functions, the Jacobian, and the normal are related to the element $\partial\Omega^e$. Two cases must be considered:

i. The collocation point $\underline{x}_i$ does not belong to element $\partial\Omega^e$.
ii. The collocation point $\underline{x}_i$ does belong to element $\partial\Omega^e$.

7.6.4.1 Calculation of regular elementary integrals

For $\underline{x}_i \notin \partial\Omega^e$, the integrals $\hat{I}_{1,ij}$, $\hat{I}_{2,ij}$, and $\hat{I}_{3,ij}$ are regular. The numerical calculation is based, like the finite element method (FEM), on the use of Gauss quadrature formulas (see Chapter 5 and Appendix 5C). Let us recall here the examples of square and triangular elements.

For the square, the integration usually relies on tensor products of the one-dimensional Gauss quadrature formula

$$\int_{-1}^{1}\int_{-1}^{1} g(\eta_1,\eta_2)d\eta_1 d\eta_2 \approx \sum_{k=1}^{N_G}\sum_{l=1}^{N_G} w_k w_l g(\eta_{1,k},\eta_{2,l}) \tag{7.97}$$

where $\eta_{1,k}$ and w_k are the one-dimensional Gauss points and weights of order N_G on the interval [−1; 1].

For the triangle, we use the following change of variables:

$$\eta_1 = \frac{(1+u)(1-v)}{4}, \quad \eta_2 = \frac{(1+u)(1+v)}{4} \tag{7.98}$$

This change of variable maps the reference triangle into the reference square (see Figure 7.7)

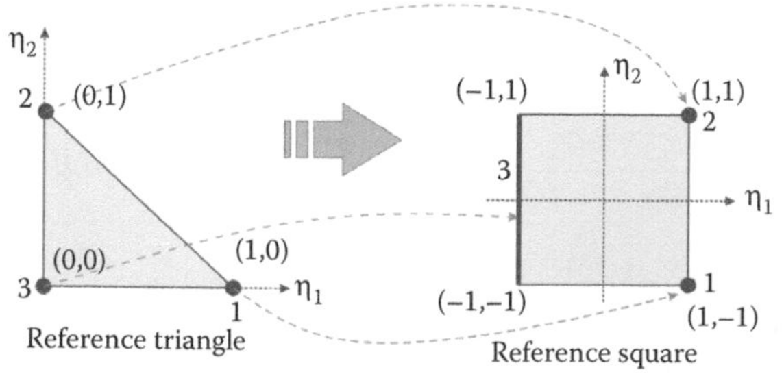

Figure 7.7 Mapping of a reference triangle into a reference square.

$$\int_0^1\int_0^{1-\eta_2} g(\eta_1,\eta_2)d\eta_1 d\eta_2 = \frac{1}{8}\int_{-1}^{1}\int_{-1}^{1} \hat{g}(u,v)\,(1+u)\,du dv$$
$$= \frac{1}{8}\sum_{k=1}^{N_G}\sum_{l=1}^{N_G} w_k w_l\, \hat{g}(u_k,v_l)\,(1+u_k) \tag{7.99}$$

Regarding the numerical implementation, this trick has the advantage of using a single Gauss quadrature scheme for both quadrilateral and triangular elements. Note that there exist other integration formulas adapted to the triangle (Hammer et al. 1956; Abramowitz and Stegun 1965; Lyness and Jespersen 1975; Dunavant 1985) (see also Appendix 5C).

7.6.4.2 Calculation of singular integrals

For $\underline{x}_i \notin \partial\Omega^e$, the integrals $\hat{I}_{1,ij}$, $\hat{I}_{2,ij}$, and $\hat{I}_{3,ij}$ are either regular or weakly singular because of the terms $|\underline{y}-\underline{x}|^{-1}$ and $|\underline{y}-\underline{x}|^{-2}$ that appear in elementary solutions $\hat{G}(\underline{x},\underline{y})$, $\partial\hat{G}(\underline{x},\underline{y})/\partial n_y$, and $\partial G_0(\underline{x},\underline{y})/\partial n_y$.

$\hat{I}_{1,ij}$ is regular and a simple Gauss quadrature is sufficient (see Equation 7.6 for the proof of the regularity of $\hat{I}_{1,ij}$). We have shown in Section 7.2.2 that $\hat{I}_{2,ij}$ and $\hat{I}_{3,ij}$ are weakly singular and to evaluate them, a suitable numerical technique must be chosen. This technique is based on a polar coordinates transformation. To give the main idea of the method, let us consider the square reference element.

Let $\underline{\xi}$ be the point of R^e image of a given $\underline{x}_i \in \partial\Omega^e$. The square element R^e is subdivided into four triangles of common vertex $\underline{\xi}$ (Figure 7.8).

Then, $R^e = \cup_{k=1}^{4} T_k$.

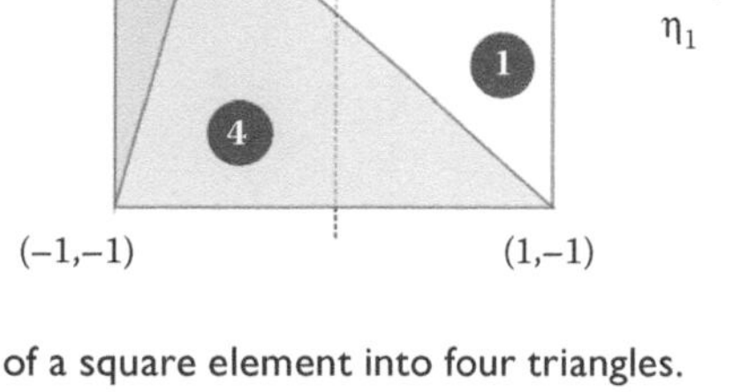

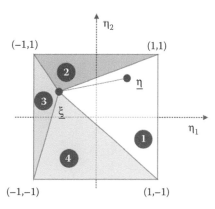

Figure 7.8 Subdivision of a square element into four triangles.

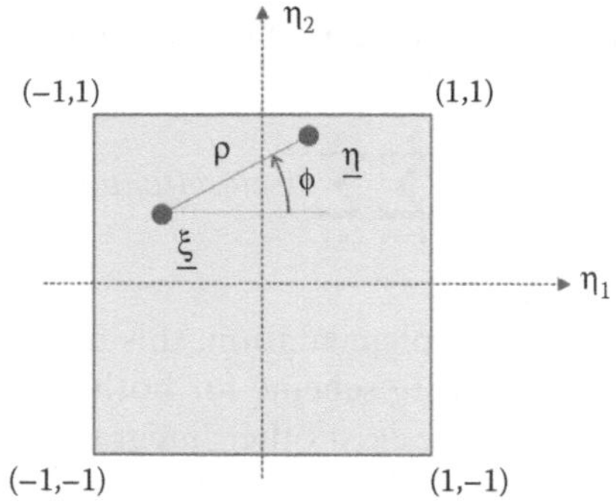

Figure 7.9 Use of polar coordinates system to evaluate the singular integrals.

Let us introduce the polar coordinate system (ρ,ϕ) centered on $\underline{\xi}$ (Figure 7.9)

$$\begin{cases} \eta_1 = \xi_1 + \rho\cos\phi \\ \eta_2 = \xi_2 + \rho\sin\phi \\ d\eta_1 d\eta_2 = \rho\, d\rho d\phi \end{cases} \tag{7.100}$$

In the case where point $\underline{\xi}$ corresponds to a vertex, the square reference element is split into two triangles only and the procedure is identical. The integrals $\hat{I}_{2,ij}$ and $\hat{I}_{3,ij}$ can be written as follows:

$$\begin{cases} \hat{I}_{2,ij} = \sum_{k=1}^{4} \int_{T_k} \frac{\partial G_0}{\partial n_y} [N_j(\underline{\xi}) - N_j(\rho,\phi,\underline{\xi})] \, |j(\rho,\phi,\underline{\xi})| \rho d\rho d\phi \\ \hat{I}_{3,ij} = \sum_{k=1}^{4} \int_{T_k} \hat{G}(\underline{\xi},\underline{\eta}) N_j(\rho,\phi,\underline{\xi}) \, |j(\rho,\phi,\underline{\xi})| \, \rho d\rho d\phi \end{cases} \tag{7.101}$$

where the integral over each triangle T_k takes the form (Figure 7.10)

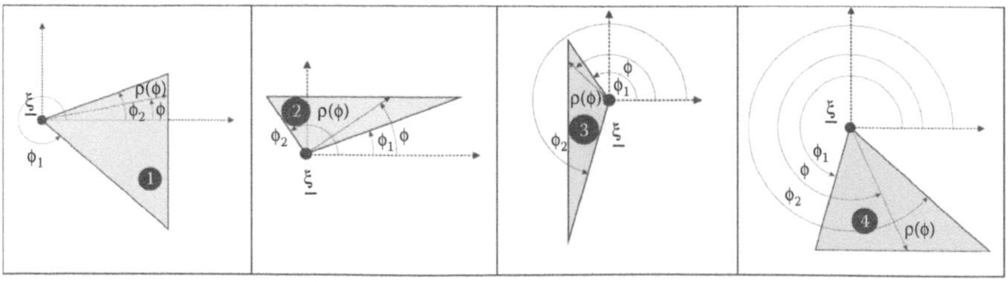

Figure 7.10 Integration over the four triangles.

$$\int_{T_k} g(\rho,\phi)\,\rho d\rho d\phi = \int_{\phi_1}^{\phi_2}\int_{0}^{\rho(\phi)} g(\rho,\phi)\,\rho d\rho d\phi \tag{7.102}$$

The shape functions being polynomials, it can be shown that (see e.g., Bonnet 1995)

$$\begin{cases} \bar{N}_j(\underline{\xi}) - \bar{N}_j(\underline{\eta}) = \rho\,\tilde{\bar{N}}_j(\rho,\phi,\underline{\xi}) \\ r = |\underline{y} - \underline{x}| = \rho\,\tilde{r}(\rho,\phi,\underline{\xi}) \\ \hat{G}(\underline{x},\underline{y}) = \dfrac{1}{\rho}\tilde{\hat{G}}(\rho,\phi,\underline{\xi}) \\ \dfrac{\partial G_0(\underline{x},\underline{y})}{\partial n_y} = \dfrac{1}{\rho^2}\tilde{H}_0(\rho,\phi,\underline{\xi}) \end{cases} \tag{7.103}$$

where $\tilde{\bar{N}}_j(\rho,\phi,\underline{\xi})$ are reduced shape functions for the solution (see Appendix 7C for some examples), $\tilde{\hat{G}}$ and $\tilde{H}_0$ are nonsingular when $\rho = 0$. $\tilde{\hat{G}}$ and $\tilde{H}_0$ are given by

$$\tilde{\hat{G}}(\rho,\phi,\underline{\xi}) = \frac{\exp(-ik\rho\tilde{r})}{4\pi\tilde{r}} \tag{7.104}$$

$$\tilde{H}_0(\rho,\phi,\underline{\xi}) = -\frac{\tilde{r}_j}{4\pi\tilde{r}^3}n_{y_j} \tag{7.105}$$

$$\tilde{r}(\underline{\xi},\rho,\alpha) = [\tilde{r}_1^2 + \tilde{r}_2^2 + \tilde{r}_3^2]^{\frac{1}{2}} \tag{7.106}$$

with $\tilde{r}_j$ defined by

$$\begin{aligned} y_j - x_j &= (N_k(\xi_1 + \rho\cos\phi, \xi_2 + \rho\sin\phi) - N_k(\xi_1,\xi_2))x_{j,k} \\ &= \rho\tilde{N}_k(\rho,\phi,\underline{\xi})x_{j,k} = \rho\tilde{r}_j \end{aligned} \tag{7.107}$$

where $\tilde{N}_k(\rho,\phi,\underline{\xi})$ are the reduced shape functions for the geometry.

These expressions allow us to rewrite integrals $\hat{I}_{2,ij}$ and $\hat{I}_{3,ij}$ in the regular form

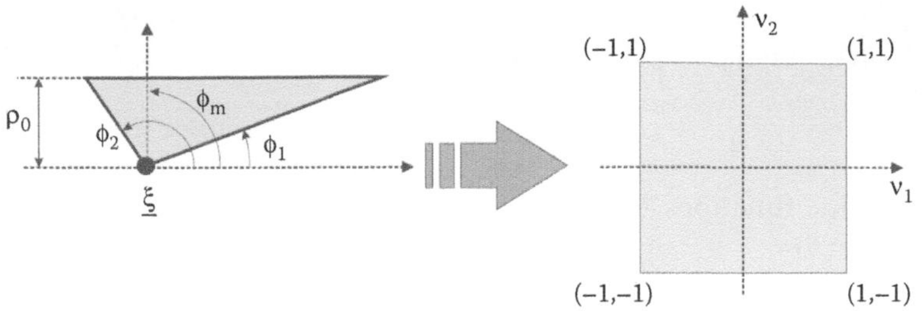

Figure 7.11 Mapping of a triangle into the square reference element using change of variable given by Equation 7.109.

$$\begin{cases} \hat{I}_{2,ij} = \sum_{k=1}^{4} \int_{T_k} \tilde{H}_0(\rho,\phi,\underline{\xi})\, \tilde{\tilde{N}}_j(\rho,\phi,\underline{\xi})\, |j(\rho,\phi,\underline{\xi})|\, d\rho d\phi \\ \hat{I}_{3,ij} = \sum_{k=1}^{4} \int_{T_k} \tilde{\hat{G}}(\rho,\phi,\underline{\xi})\, N_j(\underline{\eta})\, |j(\rho,\phi,\underline{\xi})|\, d\rho d\phi \end{cases} \tag{7.108}$$

Finally, for each triangle T_k, the following change of variable maps a triangle into the square reference element (Figure 7.11).

$$\begin{cases} \rho = \dfrac{1}{2} \dfrac{\rho_0^k}{\cos(\phi - \phi_m^k)}(1 + v_1) \\ \phi = \dfrac{(\phi_2^k - \phi_1^k)}{2} v_2 + \dfrac{(\phi_2^k + \phi_1^k)}{2} \end{cases} \tag{7.109}$$

The integration over the square reference element is carried out using a Gauss integration scheme.

Note that if triangular reference elements are used (the mapping which maps a triangle into a square element described in Equation 7.98 is not used), the above procedure can be applied. If point $\underline{\xi}$ is inside the element, then the triangle is split into three triangles. If it is on a vertex, the triangle itself is used.

7.7 VARIATIONAL FORMULATION OF INTEGRAL EQUATIONS

7.7.1 Problem formulation

In this section, a variational formulation is associated with the acoustic radiation problem (exterior problem). This formulation, thanks to double surface integrals, allows for a regularization of singularities in classic integral equations. This simplifies the numerical implementation and leads, after a boundary element discretization, to a symmetric linear system, which is suitable for the use of efficient resolution algorithms.

In order to simplify the presentation of the variational formulation, we are going to focus on the indirect formulation of the radiation problem of a structure. Let Ω be the external domain of a vibrating structure whose boundary is denoted by $\partial\Omega$ (Figure 7.12). $\underline{n}$ is the normal to $\partial\Omega$ pointing into Ω. The acoustic radiation problem is governed by the following system of equations:

$$\begin{cases} \nabla^2 \hat{p} + k^2 \hat{p} = 0 \quad \text{in} \quad \Omega \\ \hat{p}\big|_{\partial\Omega_1} = \bar{p} \\ \dfrac{\partial \hat{p}}{\partial n_y}\bigg|_{\partial\Omega_2} = \rho_0 \omega^2 \hat{u}_n \\ \lim\limits_{r\to\infty} \; r\left(\dfrac{\partial \hat{p}}{\partial r} + jk\hat{p}\right) = 0 \end{cases} \tag{7.110}$$

where $k = \omega/c_0$ is the wavenumber, c_0 and ρ_0 are the sound speed and the density of medium Ω, respectively. $\hat{u}_n$ denotes the normal displacement of the vibrating structure. Let $\hat{\mu} = \hat{p}^+ - \hat{p}^-$ be the pressure jump across $\partial\Omega$ and $\hat{\sigma} = \partial\hat{p}^+/\partial n_y - \partial\hat{p}^-/\partial n_y$ the normal pressure gradient jump across $\partial\Omega$. The indirect integral representation of Equation 7.110 can be written as

$$\hat{p}^+(\underline{x}) = \int_{\partial\Omega} \left[\hat{\mu}(\underline{y}) \frac{\partial \hat{G}(\underline{x},\underline{y})}{\partial n_y} - \hat{\sigma}(\underline{y}) \hat{G}(\underline{x},\underline{y}) \right] dS_y \quad \forall \underline{x} \notin \partial\Omega \tag{7.111}$$

The boundary conditions in Equation 7.110 become (we assume that $\hat{p}$ and $\hat{w}_n$ are continuous across $\partial\Omega_1$ and $\partial\Omega_2$, respectively)

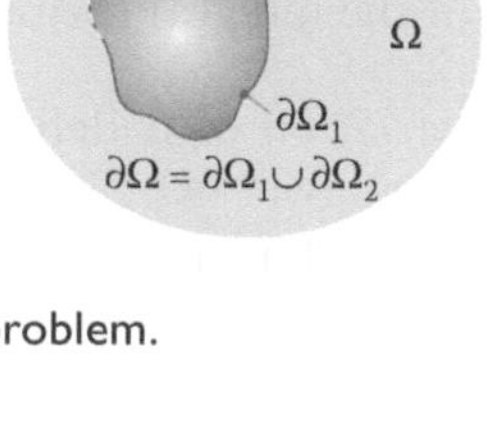

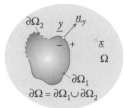

Figure 7.12 Exterior radiation problem.

$$\begin{cases} \hat{\mu} = 0 & \text{along } \partial\Omega_1 \\ \hat{\sigma} = 0 & \text{along } \partial\Omega_2 \end{cases} \tag{7.112}$$

Equation 7.111 can be rewritten as

$$\hat{p}^+(\underline{x}) = \int_{\partial\Omega_2} \hat{\mu}(\underline{y}) \frac{\partial \hat{G}(\underline{x},\underline{y})}{\partial n_y}\, dS_y - \int_{\partial\Omega_1} \hat{\sigma}(\underline{y}) \hat{G}(\underline{x},\underline{y})\, dS_y \quad \forall \underline{x} \notin \partial\Omega \tag{7.113}$$

Let us apply the boundary conditions by making $\underline{x}$ tend toward $\partial\Omega_1$ and then toward $\partial\Omega_2$. This leads to:

- $\underline{x} \in \partial\Omega_1$

$$\bar{p}(\underline{x}) = \int_{\partial\Omega_2} \hat{\mu}(\underline{y}) \frac{\partial \hat{G}(\underline{x},\underline{y})}{\partial n_y}\, dS_y - \int_{\partial\Omega_1} \hat{\sigma}(\underline{y}) \hat{G}(\underline{x},\underline{y})\, dS_y \tag{7.114}$$

- $\underline{x} \in \partial\Omega_2$

$$\rho_0\omega^2 \hat{u}_n = FP \int_{\partial\Omega_2} \hat{\mu}(\underline{y}) \frac{\partial^2 \hat{G}(\underline{x},\underline{y})}{\partial n_x \partial n_y}\, dS_y - \int_{\partial\Omega_1} \hat{\sigma}(\underline{y}) \frac{\partial \hat{G}(\underline{x},\underline{y})}{\partial n_x}\, dS_y \tag{7.115}$$

The discretization of this system using the collocation method requires the evaluation of the Hadamard finite part. This is a complex task as discussed in Section 7.5. It will be discussed in Sections 7.10.2 and 7.10.3 dealing with the uniqueness problem. In addition, the matrix system which results from the collocation method is full, complex-valued, and nonsymmetric.

The evaluation of Hadamard finite part can be circumvented by associating a variational form to the system of equations (7.110) (Hamdi 1982, 1988; Pierce 1993). This approach allows us to

- Regularize hyper-singular integrals
- End up with a symmetric matrix system after discretization
- Improve the accuracy of the integral formulation

Let $\delta\hat{\mu}$ be an admissible variation of $\hat{\mu}$ defined over $\partial\Omega_2$ and $\delta\hat{\sigma}$ be an admissible variation of $\hat{\sigma}$ defined over $\partial\Omega_1$. Let us multiply Equation 7.114 (respectively, Equation 7.115) by $\delta\hat{\sigma}$ (respectively, $\delta\hat{\mu}$) and integrate over $\partial\Omega_1$ (respectively, $\partial\Omega_2$). We end up with

$$\begin{cases} \int\limits_{\partial\Omega_1} \delta\hat{\sigma}(\underline{x})\overline{p}(\underline{x})dS_x = \int\limits_{\partial\Omega_1}\int\limits_{\partial\Omega_2} \delta\hat{\sigma}(\underline{x})\,\hat{\mu}(\underline{y})\dfrac{\partial\hat{G}(\underline{x},\underline{y})}{\partial n_y}dS_y dS_x \\ \qquad\qquad - \int\limits_{\partial\Omega_1}\int\limits_{\partial\Omega_1} \delta\hat{\sigma}(\underline{x})\hat{\sigma}(\underline{y})\hat{G}(\underline{x},\underline{y})\,dS_y dS_x \\ \rho_0\omega^2 \int\limits_{\partial\Omega_2} \delta\hat{\mu}(\underline{x})\hat{u}_n(\underline{x})dS_x = \int\limits_{\partial\Omega_2}\int\limits_{\partial\Omega_2} \delta\hat{\mu}(\underline{x})\hat{\mu}(\underline{y})\dfrac{\partial^2\hat{G}(\underline{x},\underline{y})}{\partial n_x \partial n_y}dS_y dS_x \\ \qquad\qquad - \int\limits_{\partial\Omega_2}\int\limits_{\partial\Omega_1} \delta\hat{\mu}(\underline{x})\hat{\sigma}(\underline{y})\dfrac{\partial\hat{G}(\underline{x},\underline{y})}{\partial n_x}dS_y dS_x \end{cases} \tag{7.116}$$

The double integral $\int_{\partial\Omega_2}\int_{\partial\Omega_2}(\partial^2\hat{G}/\partial n_x\partial n_y)\delta\hat{\mu}(\underline{x})\hat{\mu}(\underline{y})dS_y dS_x$ being regular, the Hadamard finite part is not necessary anymore. Indeed, the integral can be regularized using Equation 7.73 (Hamdi 1982; Pierce 1993)

$$\begin{aligned} &\int\limits_{\partial\Omega_2}\int\limits_{\partial\Omega_2} \delta\hat{\mu}(\underline{x})\frac{\partial^2\hat{G}(\underline{x},\underline{y})}{\partial n_x\partial n_y}\hat{\mu}(\underline{y})dS_x dS_y \\ &= \int\limits_{\partial\Omega_2}\int\limits_{\partial\Omega_2} k^2(\underline{n}_x\cdot\underline{n}_y)\hat{G}(\underline{x},\underline{y})\delta\hat{\mu}(\underline{x})\hat{\mu}(\underline{y})dS_x dS_y \\ &- \int\limits_{\partial\Omega_2}\int\limits_{\partial\Omega_2} [(\underline{n}_x\times\underline{\nabla}_x\delta\hat{\mu}(\underline{x}))\cdot(\underline{n}_y\times\underline{\nabla}_y\hat{\mu}(\underline{y}))]\,\hat{G}(\underline{x},\underline{y})\,dS_x dS_y \end{aligned} \tag{7.117}$$

Following Galerkin's approach, the system given by Equation 7.116 is equivalent to the variational statement

$$\delta J(\hat{\sigma},\hat{\mu}) = 0 \quad \forall\hat{\sigma}\in\partial\Omega_1 \quad \text{and} \quad \forall\hat{\mu}\in\partial\Omega_2 \tag{7.118}$$

where

$$\begin{aligned} J(\hat{\sigma},\hat{\mu}) = &\frac{1}{2}\int\limits_{\partial\Omega_1}\int\limits_{\partial\Omega_1} \hat{\sigma}(\underline{x})\hat{\sigma}(\underline{y})\hat{G}(\underline{x},\underline{y})dS_x dS_y + \frac{1}{2}\int\limits_{\partial\Omega_2}\int\limits_{\partial\Omega_2} \hat{\mu}(\underline{x})\hat{\mu}(\underline{y})\frac{\partial^2\hat{G}(\underline{x},\underline{y})}{\partial n_x\partial n_y}dS_x dS_y \\ &- \int\limits_{\partial\Omega_1}\int\limits_{\partial\Omega_2} \hat{\sigma}(\underline{x})\hat{\mu}(\underline{y})\frac{\partial\hat{G}(\underline{x},\underline{y})}{\partial n_y}dS_x dS_y + \int\limits_{\partial\Omega_1} \hat{\sigma}(\underline{x})\overline{p}(\underline{x})\,dS_x \\ &- \rho_0\omega^2\int\limits_{\partial\Omega_2} \hat{\mu}(\underline{x})\hat{u}_n(\underline{x})dS_x \end{aligned} \tag{7.119}$$

All integrals appearing in $J(\hat{\sigma},\hat{\mu})$ are regular. In addition, the functional $J(\hat{\sigma},\hat{\mu})$ is symmetric thus leading to a symmetric matrix system after discretization using boundary elements.

Remarks:

1. For Neumann problem, the functional reduces to

$$J(\hat{\mu}) = \frac{1}{2}\int_{\partial\Omega}\int_{\partial\Omega} \hat{\mu}(\underline{x})\,\hat{\mu}(\underline{y})\frac{\partial^2 \hat{G}(\underline{x},\underline{y})}{\partial n_x \partial n_y} dS_x dS_y - \rho_0\omega^2 \int_{\partial\Omega} \hat{\mu}(\underline{x})\hat{u}_n(\underline{x}) dS_x \tag{7.120}$$

2. For Newmann problem with condition $\hat{u}_n(\underline{x}) = C$ where C is a constant, namely the set of points of $\partial\Omega$ vibrate in phase, it can be shown that the variational formulation of the problem is equivalent to minimizing the acoustic power Π_{rad} radiated by the structure (Pierce 1993).
3. It is interesting to note that the variational formulation guarantees that the boundary conditions are satisfied at all points of $\partial\Omega$, contrary to the collocation method where the boundary conditions are only satisfied at the collocation nodes. Consequently, the variational formulation is numerically, "more accurate" than classic integral formulations.
4. In the case of the scattering by a rigid body of the field created by a source of volume density $\hat{Q}$ occupying the volume Ω_S, the variational formulation reads

$$J(\hat{\mu}) = \frac{1}{2}\int_{\partial\Omega}\int_{\partial\Omega} \hat{\mu}(\underline{x})\hat{\mu}(\underline{y})\frac{\partial^2 \hat{G}(\underline{x},\underline{y})}{\partial n_x \partial n_y} dS_x dS_y + \int_{\partial\Omega}\int_{\Omega_S} \hat{\mu}(\underline{x})\hat{Q}(\underline{y})\frac{\partial \hat{G}(\underline{x},\underline{y})}{\partial n_x} dS_x d\Omega_y \tag{7.121}$$

For a point source (monopole) located at $\underline{x}_0$, the above equation reduces to

$$J(\hat{\mu}) = \frac{1}{2}\int_{\partial\Omega}\int_{\partial\Omega} \hat{\mu}(\underline{x})\hat{\mu}(\underline{y})\frac{\partial^2 \hat{G}(\underline{x},\underline{y})}{\partial n_x \partial n_y} dS_x dS_y + i\rho_0\omega\hat{Q}_s \int_{\partial\Omega} \hat{\mu}(\underline{x})\frac{\partial \hat{G}(\underline{x},\underline{x}_0)}{\partial n_x} dS_x \tag{7.122}$$

For a plane wave excitation $\hat{p}_{inc}$, it takes the form

$$J(\hat{\mu}) = \frac{1}{2}\int_{\partial\Omega}\int_{\partial\Omega} \hat{\mu}(\underline{x})\hat{\mu}(\underline{y})\frac{\partial^2 \hat{G}(\underline{x},\underline{y})}{\partial n_x \partial n_y}\, dS_x dS_y + \int_{\partial\Omega} \hat{\mu}(\underline{x})\frac{\partial \hat{p}_{inc}(\underline{x})}{\partial n_x}\, dS_x \tag{7.123}$$

7.7.2 Numerical implementation

The numerical implementation of the VBIE is similar to that of the finite element method and differs little from the one discussed in Section 7.6.2 regarding the collocation method. The boundary $\partial\Omega = \partial\Omega_1 \cup \partial\Omega_2$ is subdivided into $N_e = N_{e,1} + N_{e,2}$ boundary elements $\partial\Omega = \bigcup_{e=1}^{N_e} \partial\Omega^e$; $\partial\Omega_1 = \bigcup_{e=1}^{N_{e,1}} \partial\Omega_1^e$; and $\partial\Omega_2 = \bigcup_{e=1}^{N_{e,2}} \partial\Omega_2^e$.

A point $\underline{x} = \begin{Bmatrix} x_1 \\ x_2 \\ x_3 \end{Bmatrix}$ (a point $\underline{y} = \begin{Bmatrix} y_1 \\ y_2 \\ y_3 \end{Bmatrix}$) belonging to an element $\partial\Omega^e$ (and an element $\partial\Omega^{e'}$, respectively) are mapped in the reference space using Equation 7.79: $x_i(\underline{\xi}) = \langle N^e(\underline{\xi})\rangle\{x_i^e\}$, $y_i(\underline{\eta}) = \langle N^{e'}(\underline{\eta})\rangle\{x_i^{e'}\}$ $i = 1,2,3$.

Using an isoparametric representation ($\langle \bar{N}^e\rangle = \langle N^e\rangle$), the variables $\hat{\sigma}$ and $\hat{\mu}$ are written for $\underline{x} \in \partial\Omega^e$ and $\underline{y} \in \partial\Omega^{e'}$

$$\begin{aligned}
\hat{\sigma}(\underline{x}) &= \langle N^e(\xi_1,\xi_2)\rangle\{\hat{\sigma}^e\} \\
\hat{\mu}(\underline{x}) &= \langle N^e(\xi_1,\xi_2)\rangle\{\hat{\mu}^e\} \\
\hat{\sigma}(\underline{y}) &= \langle N^{e'}(\eta_1,\eta_2)\rangle\{\hat{\sigma}^{e'}\} \\
\hat{\mu}(\underline{y}) &= \langle N^{e'}(\eta_1,\eta_2)\rangle\{\hat{\mu}^{e'}\}
\end{aligned} \tag{7.124}$$

where $\{\hat{\sigma}^e\}$, $\{\hat{\mu}^e\}$ ($\{\hat{\sigma}^{e'}\}$, $\{\hat{\mu}^{e'}\}$ respectively) are the nodal values vectors of $\hat{\sigma}$ and $\hat{\mu}$ over the element $\partial\Omega^e$ ($\partial\Omega^{e'}$, respectively).

Consequently, the discretized form of Equation 7.119 is written as

$$\begin{aligned}
J(\hat{\sigma},\hat{\mu}) &= \frac{1}{2}\Bigg\{\sum_{e=1}^{N_{e1}}\sum_{e'=1}^{N_{e2}}\langle\hat{\sigma}^e\rangle\Bigg[\int_{R^e}\int_{R^{e'}}\{N^e(\xi_1,\xi_2)\}\hat{G}(\underline{x}(\xi_1,\xi_2),\underline{y}(\eta_1,\eta_2))\langle N^{e'}(\eta_1,\eta_2)\rangle \\
&\quad \times |j^e||j^{e'}|d\xi_1 d\xi_2 d\eta_1 d\eta_2\Bigg]\{\hat{\sigma}^{e'}\} \\
&\quad + \sum_{e=1}^{N_{e1}}\sum_{e'=1}^{N_{e2}}\langle\hat{\mu}^e\rangle\Bigg[\int_{R^e}\int_{R^{e'}}\{N^e(\xi_1,\xi_2)\}\frac{\partial^2\hat{G}(\underline{x}(\xi_1,\xi_2),\underline{y}(\eta_1,\eta_2))}{\partial n_x\partial n_y}\langle N^{e'}(\eta_1,\eta_2)\rangle \\
&\quad \times |j^e||j^{e'}|d\xi_1 d\xi_2 d\eta_1 d\eta_2\Bigg]\{\hat{\mu}^{e'}\}
\end{aligned}$$

$$
\begin{aligned}
&-2\sum_{e=1}^{N_{e1}}\sum_{e'=1}^{N_{e2}}\langle\hat{\sigma}^e\rangle\left[\int_{R^e}\int_{R^{e'}}\{N^e(\xi_1,\xi_2)\}\frac{\partial\hat{G}(\underline{x}(\xi_1,\xi_2),\underline{y}(\eta_1,\eta_2))}{\partial n_y}\langle N^{e'}(\eta_1,\eta_2)\rangle\right.\\
&\left.\times|j^e||j^{e'}|\,d\xi_1 d\xi_2 d\eta_1 d\eta_2\right]\{\hat{\mu}^{e'}\}\}\\
&-\sum_{e=1}^{N_{e1}}\langle\hat{\sigma}^e\rangle\left[\int_{R^e}\{N^e(\xi_1,\xi_2)\}\langle N^e(\xi_1,\xi_2)\rangle\,|j^e|\,d\xi_1 d\xi_2\right]\{\bar{p}^e\}\\
&+\sum_{e=1}^{N_{e2}}\langle\hat{\mu}^e\rangle\left[\int_{R^e}\{N^e(\xi_1,\xi_2)\}\langle N^e(\xi_1,\xi_2)\rangle\,|j^e|\,d\xi_1 d\xi_2\right]\{\rho_0\omega^2\hat{u}_n^e\}
\end{aligned}
\tag{7.125}
$$

where R^e ($R^{e'}$, respectively) the reference element associated with $\partial\Omega^e$ ($\partial\Omega^{e'}$, respectively).

Note that formula (7.117) must be used to evaluate $\int_{\partial\Omega_2}\int_{\partial\Omega_2}\hat{\mu}(\underline{x})\hat{\mu}(\underline{y})\,\partial^2\hat{G}(\underline{x},\underline{y})/\partial n_x\partial n_y\,dS_x dS_y$. Appendix 7D gives details about the discretization of this term together with the calculation of normals $\underline{n}_x$ and $\underline{n}_y$ and Jacobian matrices j^e and $j^{e'}$ (see also Chapter 5 about finite elements).

After assembling and imposition of boundary conditions, the discretized functional can be written as

$$
\begin{aligned}
J(\hat{\sigma},\hat{\mu}) &= \frac{1}{2}\langle\hat{\sigma}\rangle[\hat{\mathcal{A}}_{\sigma\sigma}(\omega)]\{\hat{\sigma}\} + \frac{1}{2}\langle\hat{\mu}\rangle[\hat{\mathcal{D}}(\omega)]\{\hat{\mu}\} - \langle\hat{\sigma}\rangle[\hat{\mathcal{C}}_{\sigma\mu}(\omega)]\{\hat{\mu}\}\\
&\quad -\langle\hat{\sigma}\rangle\{\hat{f}_\sigma\} - \langle\hat{\mu}\rangle\{\hat{f}_\mu\}
\end{aligned}
\tag{7.126}
$$

The stationarity of this functional with respect to $\hat{\sigma}$ and $\hat{\mu}$ leads to

$$
\begin{bmatrix}[\hat{\mathcal{A}}_{\sigma\sigma}(\omega)] & -[\hat{\mathcal{C}}_{\sigma\mu}(\omega)]\\ -[\hat{\mathcal{C}}_{\sigma\mu}(\omega)]^T & [\hat{\mathcal{D}}(\omega)]\end{bmatrix}\begin{Bmatrix}\{\hat{\sigma}\}\\ \{\hat{\mu}\}\end{Bmatrix} = \begin{Bmatrix}\{\hat{f}_\sigma\}\\ \{\hat{f}_\mu\}\end{Bmatrix}
\tag{7.127}
$$

Note that the system is full, complex-valued, frequency-dependent but symmetric.

7.7.3 Calculation of elementary integrals

The numerical implementation of the VBIE is considered to be its main drawback (Atalla and Bernhard 1994). Indeed, this formulation involves double surface integrals with $1/r$, $1/r^2$, and $1/r^3$ singularities. The relationship given by Equation 7.117 allows us to remove the $1/r^3$ singularity.

The calculation of integrals in Equation 7.119 is similar to that in the collocation method (see Section 7.6.4). Two kinds of integrals need to be evaluated:

- The cross-influence integrals between two different elements. These integrals are regular and computed numerically using a Gauss integration scheme.
- The self-influence integrals of an element ($\underline{x} \in \partial\Omega^e$ and $\underline{y} \in \partial\Omega^e$). In this case, a special procedure must be used because of the presence of the term $|\underline{y} - \underline{x}|^{-1}$. Similarily to the collocation method, a polar coordinates transformation can be used (Hamdi 1982; Silva 1994; Sgard 1995). For quadrangle elements (4 to 9 nodes), an efficient algorithm that accounts explicitly for the $|\underline{y} - \underline{x}|^{-1}$ singularity has been developed by Wang and Atalla (1997).

Note that special techniques exist to accelerate the computation of the VBIE. For example, the use of the multiple multipole expansion approach is discussed in Tournour and Atalla (1998).

7.8 STRUCTURES IN PRESENCE OF RIGID BAFFLES

An important configuration occurring in vibroacoustic problems is the case of a structure embedded in or above a rigid infinite baffle (e.g., sound transmission loss through panels, interaction of a sound field with an object above a rigid plane, etc.)

Let us consider the case of the direct formulation for the exterior problem to illustrate the technique to account for a rigid baffle. Figure 7.13

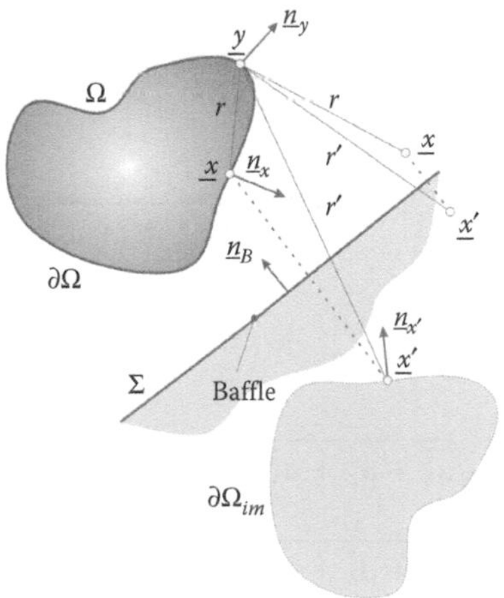

Figure 7.13 Baffled problem.

describes schematically the methodology. The field created in the half-space above the baffle is the same as that generated by the radiating body Ω and its image through the baffle, without any baffle. This is done through a modification of the Green's function. The acoustic pressure at a point $\underline{x}$ in Ω in the presence of a rigid baffle is given by

$$\hat{p}(\underline{x}) = \int_{\partial\Omega} \hat{p}(\underline{y}) \frac{\partial \hat{G}_b(\underline{x},\underline{y})}{\partial n_y} dS_y - \int_{\partial\Omega} \hat{G}_b(\underline{x},\underline{y}) \frac{\partial \hat{p}}{\partial n_y}(\underline{y}) dS_y \tag{7.128}$$

with

$$\hat{G}_b(\underline{x},\underline{y}) = \underbrace{\frac{\exp(-ikr)}{4\pi r}}_{\hat{G}_\infty(\underline{x},\underline{y})} + \underbrace{\frac{\exp(-ikr')}{4\pi r'}}_{\hat{G}_{im}(\underline{x},\underline{y})} \tag{7.129}$$

where $r = \sqrt{(x_1 - y_1)^2 + (x_2 - y_2)^2 + (x_3 - y_3)^2}$ and $r' = \sqrt{(x_1' - y_1)^2 + (x_2' - y_2)^2 + (x_3' - y_3)^2}$. In addition

$$\begin{aligned} x_1' &= x_1 - 2d_B(\underline{x})n_{B_1} \\ x_2' &= x_2 - 2d_B(\underline{x})n_{B_2} \\ x_3' &= x_3 - 2d_B(\underline{x})n_{B_3} \end{aligned} \tag{7.130}$$

The point $\underline{x}'$ is the image of point $\underline{x}$ with respect to the baffle (Figure 7.13). $d_B(\underline{x})$ is the distance of point $\underline{x}$ to the baffle and $\underline{n}_B$ is the unit normal vector to the baffle pointing into Ω. If the equation of the baffle plane is given by $a_1y_1 + a_2y_2 + a_3y_3 + a_4 = 0$, then we have

$$\begin{aligned} d_B(\underline{x}) &= \frac{a_1x_1 + a_2x_2 + a_3x_3 + a_4}{\sqrt{a_1^2 + a_2^2 + a_3^2}} \\ &= n_{B_1}x_1 + n_{B_2}x_2 + n_{B_3}x_3 + \frac{a_4}{\sqrt{a_1^2 + a_2^2 + a_3^2}} \end{aligned} \tag{7.131}$$

where we have used the fact that

$$\underline{n}_B = \begin{Bmatrix} n_{B_1} \\ n_{B_2} \\ n_{B_3} \end{Bmatrix} = \begin{Bmatrix} a_1 \big/ \sqrt{a_1^2 + a_2^2 + a_3^2} \\ a_2 \big/ \sqrt{a_1^2 + a_2^2 + a_3^2} \\ a_3 \big/ \sqrt{a_1^2 + a_2^2 + a_3^2} \end{Bmatrix} \tag{7.132}$$

Note that

$$\hat{G}_{im}(\underline{x},\underline{y}) = \hat{G}_{\infty}(\underline{x}',\underline{y}) \tag{7.133}$$

The Jacobian matrix of the mapping $\underline{x} \rightarrow \underline{x}'$ defined by Equation 7.130 is given by

$$[J] = \begin{bmatrix} \frac{\partial x_1'}{\partial x_1} & \frac{\partial x_2'}{\partial x_1} & \frac{\partial x_3'}{\partial x_1} \\ \frac{\partial x_1'}{\partial x_2} & \frac{\partial x_2'}{\partial x_2} & \frac{\partial x_3'}{\partial x_2} \\ \frac{\partial x_1'}{\partial x_3} & \frac{\partial x_2'}{\partial x_3} & \frac{\partial x_3'}{\partial x_3} \end{bmatrix} = \begin{bmatrix} 1-2n_{B_1}^2 & -2n_{B_1}n_{B_2} & -2n_{B_1}n_{B_3} \\ -2n_{B_1}n_{B_2} & 1-2n_{B_2}^2 & -2n_{B_2}n_{B_3} \\ -2n_{B_1}n_{B_3} & -2n_{B_2}n_{B_3} & 1-2n_{B_3}^2 \end{bmatrix} \tag{7.134}$$

The determinant of $[J]$ is equal to $1-2(n_{B_1}^2+n_{B_2}^2+n_{B_3}^2) = 1-2\|\underline{n}_B\|^2 = -1$. Moreover, $[J]^{-1} = [J]$. In addition, we have

$$\underline{\nabla}_x(\cdot) = \begin{Bmatrix} \frac{\partial(\cdot)}{\partial x_1} \\ \frac{\partial(\cdot)}{\partial x_2} \\ \frac{\partial(\cdot)}{\partial x_3} \end{Bmatrix} = [J]\begin{Bmatrix} \frac{\partial(\cdot)}{\partial x_1'} \\ \frac{\partial(\cdot)}{\partial x_2'} \\ \frac{\partial(\cdot)}{\partial x_3'} \end{Bmatrix} = [J]\underline{\nabla}_{x'}(\cdot) \tag{7.135}$$

We also have the relationship

$$\underline{n}_{x'} = \underline{n}_x - 2(\underline{n}_x\cdot\underline{n}_B)\underline{n}_B = [J]\underline{n}_x \tag{7.136}$$

Thus,

$$\begin{aligned} \frac{\partial}{\partial n_x} &= \underline{n}_x\cdot\underline{\nabla}_x(\cdot) = ([J]^{-1}\underline{n}_{x'})^T[J]\underline{\nabla}_{x'}(\cdot) \\ &= \underline{n}_{x'}^T[J]^{-T}[J]\underline{\nabla}_{x'}(\cdot) \underbrace{=}_{[J]^{-1}\text{ is symmetric}} \underline{n}_{x'}^T\underline{\nabla}_{x'}(\cdot) = \underline{n}_{x'}\cdot\underline{\nabla}_{x'}(\cdot) \\ &= \frac{\partial}{\partial n_{x'}} \end{aligned} \tag{7.137}$$

The surface element dS_x is related to the surface element $dS_{x'}$ by

$$dS_{x'} = \det[J]dS_x = -dS_x \tag{7.138}$$

For a point $\underline{x}$ belonging to $\partial\Omega$ supposed to be regular, we have

$$\frac{\hat{p}(\underline{x})}{2} = \int_{\partial\Omega} \hat{p}(\underline{y}) \frac{\partial \hat{G}_\infty(\underline{x}, \underline{y})}{\partial n_y} dS_y + \int_{\partial\Omega} \hat{p}(\underline{y}) \frac{\partial \hat{G}_{im}(\underline{x}, \underline{y})}{\partial n_y} dS_y$$
$$- \int_{\partial\Omega} \hat{G}_\infty(\underline{x}, \underline{y}) \frac{\partial \hat{p}}{\partial n_y}(\underline{y}) dS_y - \int_{\partial\Omega} \hat{G}_{im}(\underline{x}, \underline{y}) \frac{\partial \hat{p}}{\partial n_y}(\underline{y}) dS_y \quad (7.139)$$

In addition,

$$\frac{1}{2}\frac{\partial \hat{p}(\underline{x})}{\partial n_x} = FP\int_{\partial\Omega} \hat{p}(\underline{y}) \frac{\partial^2 \hat{G}_\infty(\underline{x}, \underline{y})}{\partial n_x \partial n_y} dS_y + \int_{\partial\Omega} \hat{p}(\underline{y}) \frac{\partial^2 \hat{G}_{im}(\underline{x}, \underline{y})}{\partial n_x \partial n_y} dS_y$$
$$- \int_{\partial\Omega} \frac{\partial \hat{G}_\infty(\underline{x}, \underline{y})}{\partial n_x} \frac{\partial \hat{p}}{\partial n_y}(\underline{y}) dS_y - \int_{\partial\Omega} \frac{\partial \hat{G}_{im}(\underline{x}, \underline{y})}{\partial n_x} \frac{\partial \hat{p}}{\partial n_y}(\underline{y}) dS_y \quad (7.140)$$

For example, consider the baffled exterior Neumann problem

$$\begin{cases} \nabla^2 \hat{p} + k^2 \hat{p} = 0 \quad \text{in} \quad \Omega \\ \left.\dfrac{\partial \hat{p}}{\partial n}\right|_{\partial\Omega} = \rho_0 \omega^2 \hat{u}_n \\ \left.\dfrac{\partial \hat{p}}{\partial n}\right|_{\Sigma} = 0 \\ \lim\limits_{r \to \infty} \quad r\left(\dfrac{\partial \hat{p}}{\partial r} + ik\hat{p}\right) = 0 \end{cases} \quad (7.141)$$

The associated VBIE can be written as

$$\frac{1}{2}\int_{\partial\Omega} \rho_0 \omega^2 \hat{u}_n(\underline{x}) \delta\hat{p}(\underline{x}) dS_x = \int_{\partial\Omega}\int_{\partial\Omega} \hat{p}(\underline{y}) \frac{\partial^2 \hat{G}_\infty(\underline{x}, \underline{y})}{\partial n_x \partial n_y} \delta\hat{p}(\underline{x}) dS_y dS_x$$
$$+ \int_{\partial\Omega}\int_{\partial\Omega} \hat{p}(\underline{y}) \frac{\partial^2 \hat{G}_{im}(\underline{x}, \underline{y})}{\partial n_x \partial n_y} \delta\hat{p}(\underline{x}) dS_y dS_x$$
$$- \int_{\partial\Omega}\int_{\partial\Omega} \frac{\partial \hat{G}_\infty(\underline{x}, \underline{y})}{\partial n_x} \frac{\partial \hat{p}}{\partial n_y}(\underline{y}) \delta\hat{p}(\underline{x}) dS_y dS_x$$
$$- \int_{\partial\Omega}\int_{\partial\Omega} \frac{\partial \hat{G}_{im}(\underline{x}, \underline{y})}{\partial n_x} \frac{\partial \hat{p}}{\partial n_y}(\underline{y}) \delta\hat{p}(\underline{x}) dS_y dS_x \quad (7.142)$$

Equivalently, using Equation 7.133

$$\begin{aligned}\frac{1}{2}\int_{\partial\Omega}\rho_0\omega^2\hat{u}_n(\underline{x})\delta\hat{p}(\underline{x})dS_x &= \int_{\partial\Omega}\int_{\partial\Omega}\hat{p}(\underline{y})\frac{\partial^2\hat{G}_\infty(\underline{x},\underline{y})}{\partial n_x\partial n_y}\delta\hat{p}(\underline{x})dS_y dS_x \\ &- \int_{\partial\Omega_{im}}\int_{\partial\Omega}\hat{p}(\underline{y})\frac{\partial^2\hat{G}_\infty(\underline{x}',\underline{y})}{\partial n_{x'}\partial n_y}\delta\hat{p}(\underline{x}')dS_y dS_{x'} \\ &- \int_{\partial\Omega}\int_{\partial\Omega}\rho_0\omega^2\hat{u}_n(\underline{y})\frac{\partial\hat{G}_\infty(\underline{x},\underline{y})}{\partial n_x}\delta\hat{p}(\underline{x})dS_y dS_x \\ &+ \int_{\partial\Omega_{im}}\int_{\partial\Omega}\rho_0\omega^2\hat{u}_n(\underline{y})\frac{\partial\hat{G}_\infty(\underline{x}',\underline{y})}{\partial n_{x'}}\delta\hat{p}(\underline{x}')dS_y dS_{x'} \qquad (7.143)\end{aligned}$$

where $\partial\Omega_{im}$ is the surface image of $\partial\Omega$ through the baffle (see Figure 7.13). Equation 7.143 will be used in Chapter 8 to construct the variational formulation associated with some baffled coupled fluid–structure problems.

Remarks: When deriving the variational formulation of the direct boundary integral equation, Equations 7.139 and 7.140 are used since the point on the surface is a Gauss integration point, which is internal to boundary elements that are smooth surfaces. If the collocation method is to be used and the pressure has to be evaluated at a node where the boundary is not smooth, Equation 7.128 can be rewritten as

$$C^+(\underline{x})\hat{p}(\underline{x}) = \int_{\partial\Omega}\hat{p}(\underline{y})\frac{\partial\hat{G}_b(\underline{x},\underline{y})}{\partial n_y}dS_y - \int_{\partial\Omega}\hat{G}_b(\underline{x},\underline{y})\frac{\partial\hat{p}}{\partial n_y}(\underline{y})dS_y \qquad (7.144)$$

where the calculation of $C^+(\underline{x})$ requires a special treatment, depending on the location of point $\underline{x}$ with respect to the baffle (Seybert and Wu 1989).

For instance, consider the exterior problem where the boundary $\partial\Omega \cup \partial\Omega_c$ enclosing the volume Ω^- is immersed in a semi-infinite external fluid Ω^+ (see Figure 7.14). $\partial\Omega$ is in contact with the semi-infinite external fluid Ω^+. $\partial\Omega_c$ lies in the plane of the infinite rigid baffle Σ and is not in contact with the acoustic domain Ω^+.

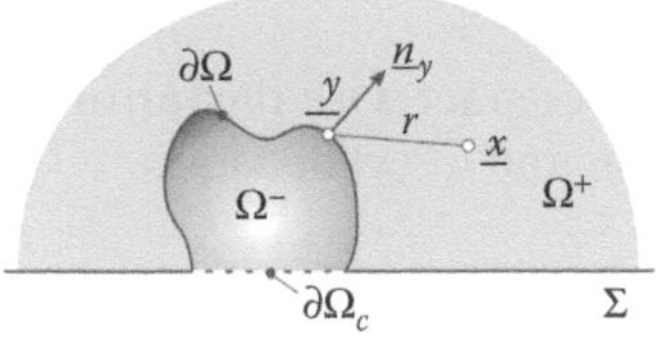

Figure 7.14 Body sitting on an infinite rigid baffle Σ.

If $\underline{x}$ is on $\partial\Omega$ but not on the infinite rigid plane Σ, it can be shown that (see Appendix 7E)

$$C^{+}(\underline{x}) = 1 + \int_{\partial\Omega+\partial\Omega_c} \frac{\partial G_0(\underline{x},\underline{y})}{\partial n_y} dS_y \tag{7.145}$$

If $\underline{x}$ is on the intersection of $\partial\Omega$ and the infinite rigid baffle Σ, it can be shown that (see Appendix 7E)

$$C^{+}(\underline{x}) = 1 + 2\int_{\partial\Omega+\partial\Omega_c} \frac{\partial G_0(\underline{x},\underline{y})}{\partial n_y} dS_y \tag{7.146}$$

Remarks: In case where the structure is planar, $\hat{G}_b(\underline{x},\underline{y}) = 2\hat{G}_\infty(\underline{x},\underline{y})$ and $\partial\hat{G}_b(\underline{x},\underline{y})/\partial n_y = 0$ on the plane of the baffle. In addition, $C^{+}(\underline{x}) = 1$ since $\int_{\partial\Omega+\partial\Omega_c} \partial G_0(\underline{x},\underline{y})/\partial n_y \, dS_y = 0$ and Equation 7.144 reduces to the Rayleigh integral given by Equation 7.47.

Another case where this treatment is necessary is when the indirect formulation is used. In this case,

$$\hat{p}^{+}(\underline{x}) = C^{-}(\underline{x})\hat{\mu}(\underline{x}) + \int_{\partial\Omega}\left[\hat{\mu}(\underline{y})\frac{\partial\hat{G}_b(\underline{x},\underline{y})}{\partial n_y} - \hat{\sigma}(\underline{y})\hat{G}_b(\underline{x},\underline{y})\right] dS_y \quad \forall \underline{x}\in\partial\Omega \tag{7.147}$$

$$\hat{p}^{-}(\underline{x}) = -C^{+}(\underline{x})\hat{\mu}(\underline{x}) + \int_{\partial\Omega}\left[\hat{\mu}(\underline{y})\frac{\partial\hat{G}_b(\underline{x},\underline{y})}{\partial n_y} - \hat{\sigma}(\underline{y})\hat{G}_b(\underline{x},\underline{y})\right] dS_y \quad \forall \underline{x}\in\partial\Omega \tag{7.148}$$

If $\underline{x}$ is on $\partial\Omega$ and off the infinite rigid plane Σ then $C^{+}(\underline{x})$ is given by Equation 7.145 and $C^{-}(\underline{x})$ by

$$C^{-}(\underline{x}) = -\int_{\partial\Omega+\partial\Omega_c} \frac{\partial G_0(\underline{x},\underline{y})}{\partial n_y} dS_y \tag{7.149}$$

If $\underline{x}$ is on $\partial\Omega$ and also in contact with the infinite rigid baffle Σ then $C^{+}(\underline{x})$ is given by Equation 7.146 and $C^{-}(\underline{x})$ becomes

$$C^{-}(\underline{x}) = -2\int_{\partial\Omega+\partial\Omega_c} \frac{\partial G_0(\underline{x},\underline{y})}{\partial n_y} dS_y \tag{7.150}$$

7.9 CALCULATION OF ACOUSTIC AND VIBRATORY INDICATORS

7.9.1 Surface pressure and pressure gradient

The pressure on the surface of the body can be directly obtained from the resolution of the direct integral formulation whether the collocation method or the VBIE is used since this is the unknown variable in both formulations. The normal pressure gradient is also obtained directly from the resolution of the direct integral formulation when the collocation method is used. If the direct VBIE is used, the normal pressure gradient, on the boundary where it is not prescribed (Dirichlet problem), can be included as an unknown of the problem in the variational formulation and obtained directly from the resolution of the system. If the indirect formulation is used, the pressure field (respectively, the normal pressure gradient) on either side of the boundary are obtained using Equations 7.66 and 7.68 (respectively, Equations 7.70 and 7.71). The numerical calculations are carried out according to the procedure described in Section 7.6.4 for the terms involving the Green's function together with its normal derivative. The term involving the second normal derivative of the Green's function must be taken in the Hadamard finite part sense. Note that regularized forms of the SBIE (Equations 7.33 and 7.43) are mainly used to solve for the unknown variables on the boundary.

7.9.2 Field pressure and pressure gradient

Once the values of the unknowns on the surface have been found, the pressure at a point $\underline{x} \notin \partial\Omega$ can be calculated from the appropriate boundary integral equation depending on the problem of interest and the chosen formulation. Table 7.1 summarizes the equations to be used to recover the pressure at a point inside the domain.

Note that the distance of point $\underline{x}$ (source point) to the boundary should be sufficiently large to apply these formulas. If point $\underline{x}$ is close to the

Table 7.1 Equations to be used to recover the sound pressure at a point inside a domain

		Interior problem	*Exterior problem*	*Scattering problem*
Direct integral formulation	SBIE	Equation 7.30 with $C^-(\underline{x}) = 1$	Equation 7.40 with $C^+(\underline{x}) = 1$	Equation 7.51 with $C^+(\underline{x}) = 1$
	RBIE	Equation 7.33	Equation 7.43	Equation 7.52
Indirect integral formulation	-	Equation 7.59	Equation 7.59	Equation 7.59 where the left-hand side is the scattered pressure

boundary the integrals become *nearly singular** and the conventional Gaussian quadrature becomes inefficient. Special integration techniques are required to provide accurate results. The main ones are the element subdivision technique, the analytical and semianalytical methods, the nonlinear transformation techniques, and the distance transformation techniques (see Qin et al. 2011 for a review).

To illustrate how the calculations are carried out, let us take Equation 7.40. The discretized form of Equation 7.40 can be written as

$$\hat{p}(\underline{x}) = \sum_{e=1}^{N_e} \langle \hat{p}^e \rangle \int_{R^e} \{N^e(\eta_1,\eta_2)\} \frac{\partial \hat{G}(\underline{x}, \underline{y}(\eta_1,\eta_2))}{\partial n_y} |j^e| d\eta_1 d\eta_2 \\ - \sum_{e=1}^{N_e} \left\langle \frac{\partial \hat{p}^e}{\partial n_y} \right\rangle \int_{R^e} \{N^e(\eta_1,\eta_2)\} \hat{G}(\underline{x}, \underline{y}(\eta_1,\eta_2)) |j^e| d\eta_1 d\eta_2 \tag{7.151}$$

where $\{\hat{p}^e\}$ and $\{\partial \hat{p}^e / \partial n_y\}$ are the nodal vector of the pressure and its normal derivative on the element $\partial\Omega^e$ whose associated reference element is R^e. The integrals can be calculated using a Gauss quadrature scheme since they are regular.

The pressure gradient for the interior problem (similar equation for the exterior problem) is calculated according to Equation 7.15. Again the integrals are regular and can be computed using a Gauss quadrature scheme.

7.9.3 Sound intensity

The sound intensity vector at a point $\underline{x}$ is defined as

$$\hat{\underline{I}}(\underline{x}) = \frac{1}{2} \Re(\hat{p}(\underline{x}) \hat{\underline{v}}^*(\underline{x})) = -\frac{1}{2} \Re\left(\hat{p}(\underline{x}) \frac{\underline{\nabla}^* \hat{p}(\underline{x})}{i\rho_0\omega} \right) \tag{7.152}$$

The pressure and the pressure gradient at a field point $\underline{x}$ are calculated as shown in Section 7.9.2.

7.10 UNIQUENESS PROBLEM

Although the solution of the physical exterior problem expressed by the direct and indirect boundary integral formulations of the Helmholtz equation derived previously does exist, it may not be unique. This non-uniqueness problem is not associated with the physical problem, nor with a differential

* In acoustics, nearly singular integral arises in mainly four cases: (a) the boundary of interest is thin; (b) neighboring element sizes over a surface are quite different; (c) the element shape is very irregular; (d) the field points are close to the boundary in post-processing.

equation formulation of the problem but rather with the fact that the solution is represented under an integral form and has to be related to the theory of integral equations. This sole integral equation is not sufficient to determine uniquely the solution. The problem occurs whenever the wavenumber corresponds to an eigenfrequency for the interior region where the boundary is acoustically rigid or pressure-release. In practice, this problem happens for the radiation problem (the exterior Neumann problem), the Dirichlet problem, or the scattering problem from a rigid or pressure-release surface. A very good discussion about why this problem occurs is given in Pierce (1993), p. 285–292. Several approaches have been proposed to circumvent this problem. We are going to recall the most popular ones based on the collocation method and the VBIE. Other methods to eliminate the irregular frequencies are discussed in Marburg and Wu (2008).

7.10.1 CHIEF method and its variants

The combined Helmholtz integral equation formulation (CHIEF) method was originally proposed by Shenck (1968). Since the pressure in the interior domain is not zero, his idea was to force it to be zero at a certain number of collocation points* of the interior domain for which $C^+(\underline{x}) = 0$. This leads to an overdetermined matrix system.

If the SBIE representation given by Equation 7.40 is chosen instead of the RBIE (Equation 7.76), it can be shown that matrix $[\hat{\mathcal{H}}]$ (see Equation 7.94) for the CHIEF method reads

$$[\hat{\mathcal{H}}] = \begin{bmatrix} \hat{c}_1 + \hat{h}_{11} & \hat{h}_{12} & \cdots & \hat{h}_{1N} \\ \hat{h}_{21} & \hat{c}_2 + \hat{h}_{22} & \cdots & \hat{h}_{2N} \\ \vdots & \vdots & \cdots & \vdots \\ \hat{h}_{N1} & \hat{h}_{N2} & \cdots & \hat{c}_N + \hat{h}_{NN} \\ \hat{h}_{N+1\,1} & \hat{h}_{N+1\,2} & \cdots & \hat{h}_{N+1\,N} \\ \vdots & \vdots & \cdots & \vdots \\ \hat{h}_{N+M\,1} & \hat{h}_{N+M\,2} & \cdots & \hat{h}_{N+M\,N} \end{bmatrix} \tag{7.153}$$

where

$$\hat{h}_{ij} = -\int_{R^e} \frac{\partial \hat{G}(\underline{x}_i, \underline{y}(\underline{\eta}))}{\partial n_y^e} \overline{N}_j^e(\underline{\eta}) \left| j^e(\underline{\eta}) \right| d\eta_1 d\eta_2 \quad \forall (i,j) \in [1, N+M] \times [1, N] \tag{7.154}$$

* Referred to as CHIEF points.

and

$$\hat{c}_i = 1 + \sum_{e=1}^{N} \int_{R^e} \frac{\partial G_0(\underline{x}_i, y(\underline{\eta}))}{\partial n_y^e} \left| j^e(\underline{\eta}) \right| d\eta_1 d\eta_2 \quad \forall i \in [1, N] \tag{7.155}$$

where N and M correspond to the number of collocation points and the number of constraint points which have been selected in the interior domain, respectively. Note that the matrix system to be solved is rectangular and requires special resolution techniques such as least square algorithms or singular value decomposition (see Ciskowski and Brebbia 1991).

The main drawback of this method is its sensitivity to the position of the points selected in the interior domain. If the points are on a nodal surface of an eigenmode of the interior problem, the system to be solved will be ill-conditioned.

Apart from the irregular frequencies, the accuracy of the solution is not affected significantly. Other variations of the CHIEF method (CHIEF with a square matrix, enhanced CHIEF, weighted residual CHIEF) have been proposed over the years. For an overview and detailed information about the performance of these methods, the reader can refer to Marburg and Wu (2008).

7.10.2 Burton and Miller's method

Burton and Miller's technique, following the work of Panich (Panich 1965) for the indirect formulation, is the most popular one to overcome the non-uniqueness problem. It relies on the use of a linear combination of the SBIE and its normal derivative, which leads to a hyper-singular boundary integral equation (HBIE):

$$\begin{aligned} C^+(\underline{x})\left[\hat{p}(\underline{x}) - i\alpha_{bm}\frac{\partial \hat{p}(\underline{x})}{\partial n_x}\right] &= \int_{\partial\Omega} \hat{p}(\underline{y}) \left(\frac{\partial \hat{G}(\underline{x}, \underline{y})}{\partial n_y} - i\alpha_{bm} \frac{\partial^2 \hat{G}(\underline{x}, \underline{y})}{\partial n_x \partial n_y} \right) dS_y \\ &\quad - \int_{\partial\Omega} \frac{\partial \hat{p}(\underline{y})}{\partial n_y} \left(\hat{G}(\underline{x}, \underline{y}) - i\alpha_{bm} \frac{\partial \hat{G}(\underline{x}, \underline{y})}{\partial n_x} \right) dS_y \end{aligned} \tag{7.156}$$

where α_{bm} is an arbitrary positive constant, which depends not only on the wavenumber but also on the shape of the boundary $\partial\Omega$.

It can be shown that the resulting equation is valid for all wavenumbers and can be used for both the Dirichlet problem and the Robin problem. Note that in Equation 7.156, all the kernels except the one associated with the hyper-singular integral $\partial^2\hat{G}(\underline{x},\underline{y})/\partial n_x \partial n_y$ are at worst weakly singular and can be numerically evaluated using the procedure described in Section 7.6.4. The value of α_{bm} strongly influences the conditioning of the system to be solved. Studies have shown that $\alpha_{bm} = (1/k)$ is an "almost optimal" choice independent of $\partial\Omega$, provided that the surface is not too thin or elongated. The main difficulty associated with this approach is to evaluate the hyper-singular integral.

No matter what the calculation technique is, the solution must satisfy some continuity requirements,[*,†] which are demanded by the nature of the hyper-singularity. The discretization scheme has to be chosen accordingly. Several techniques relying on the use of piecewise constant elements have been proposed. The solution then satisfies the appropriate smoothness conditions (see footnote †) since the solution is infinitely smooth over such elements and the hyper-singular integral exists. Calculating this hyper-singular integral using high-order piecewise polynomial basis functions is more tricky since the classic isoparametric surface elements are defined in terms of the nodal geometrical points. If these nodes are used as collocation points, the discretized boundary does not possess a uniquely defined normal at these points,[‡] which prevents the computation of the second normal derivative of the Green's function. In addition, the discretized solution does not satisfy the required smoothness conditions. Nonconforming elements (discontinuous elements) where the collocation points are inside the elements and different from the geometrical nodes can be used to circumvent this issue. For these elements, the variables are discontinuous across element edges but this discontinuity does not prevent the convergence of the approach. In particular, these elements make it easier to deal with edges and corners since the normal vector at a node always exists.[§] However, there are several drawbacks associated with their use among which we can cite an increased number of degrees of freedom compared to continuous elements, a dependency of the interpolation functions on the position of the node inside the element, the need for advanced numerical integration techniques to evaluate quasi-singu-

* The solution $\hat{p}$ (or the double-layer density function in the case of an indirect formulation) is assumed to be differentiable at point $\underline{x}$ with its first derivative satisfying a Hölder condition.

† These continuity requirements are also called Lyapunov–Tauber smoothness conditions (see Silva 1994) for mathematical details.

‡ Indeed the normal is not uniquely defined at nodes if flat elements are used to discretize a curved surface or if edges or vertices are present.

§ Consequently, $C^-(\underline{x}) = C^+(\underline{x}) = 1/2$ at nodes even if the boundary is not smooth.

lar integrals when integrating over neighboring elements. An alternative technique consists in taking $\alpha_{bm} = 0$ over part of the boundary where the collocation point is on an edge or a vertex thereby avoiding to compute the part related to the normal derivative of the SBIE at these points. This still results in a well-conditioned formulation (Harris 1992). To bypass the strict $C^{1,\alpha}$ continuity requirement on every collocation point, several authors (Ingber and Hickox 1992; Marburg and Amini 2005) proposed modified implementations of the Burton and Miller method using continuous high-order elements such as quadratic quadrilateral elements. In these works, the HBIE is evaluated at the internal node and the SBIE is applied at all the other nodes on the surface where only $C^{0,\alpha}$ continuity is ensured. The authors claimed that they obtained promising results.

Various regularization methods have been proposed to reduce the order of the integrand singularity (Maue 1949; Stallybrass 1967; Guiggiani et al. 1992; Harris 1992; Silva 1994). The resulting integral operators have kernel functions that are at worst weakly singular and hence are relatively straightforward to approximate by standard numerical methods.

We have already seen one regularization formula (7.73) based on Maue (1949) and Stallybrass's (1967) works. Let us recall it here

$$FP\int_{\partial\Omega} \hat{p}(\underline{y})\frac{\partial^2\hat{G}(\underline{x},\underline{y})}{\partial n_x \partial n_y}dS_y = \int_{\partial\Omega}\Big(k^2\underline{n}_x\cdot\underline{n}_y\hat{G}(\underline{x},\underline{y})\hat{p}(\underline{y}) + [\underline{n}_x\times\underline{\nabla}_x\hat{G}(\underline{x},\underline{y})]\cdot[\underline{n}_y\times\underline{\nabla}_y\hat{p}(\underline{y})]\Big)dS_y \qquad (7.157)$$

Harris (1992) provided regularization formulas for zeroth-order (piecewise constant) and higher order interpolation boundary elements. The equation used for zeroth-order interpolation is

$$FP\int_{\partial\Omega} \hat{p}(\underline{y})\frac{\partial^2\hat{G}(\underline{x},\underline{y})}{\partial n_x \partial n_y}dS_y = \int_{\partial\Omega}(\hat{p}(\underline{y})-\hat{p}(\underline{x}))\frac{\partial^2\hat{G}(\underline{x},\underline{y})}{\partial n_x \partial n_y}dS_y + \hat{p}(\underline{x})k^2\int_{\partial\Omega}\underline{n}_x\cdot\underline{n}_y\hat{G}(\underline{x},\underline{y})dS_y \qquad (7.158)$$

This expression indicates that when a collocation point belongs to the same element as the integration point then the first integral on the right-hand side of Equation 7.158 involving the hyper-singular kernel function vanishes and we are left with the calculation of a weakly singular integral.

Harris (1992) uses another equation for higher order interpolation since the first integral is not zero anymore. It reads

$$
\begin{aligned}
FP\int_{\partial\Omega} \hat{p}(\underline{y})\frac{\partial^2 \hat{G}(\underline{x},\underline{y})}{\partial n_x \partial n_y} dS_y &= \int_{\partial\Omega} \hat{p}(\underline{y})\left(\frac{\partial^2 \hat{G}(\underline{x},\underline{y})}{\partial n_x \partial n_y} - \frac{\partial^2 G_0(\underline{x},\underline{y})}{\partial n_x \partial n_y}\right) dS_y \\
&+ \int_{\partial\Omega} (\hat{p}(\underline{y}) - \hat{p}(\underline{x}) - \underline{\nabla}_x \hat{p}(\underline{x}) \cdot (\underline{y} - \underline{x}))\frac{\partial^2 G_0(\underline{x},\underline{y})}{\partial n_x \partial n_y} dS_y \\
&+ \int_{\partial\Omega} \underline{\nabla}_x \hat{p}(\underline{x}) \cdot \underline{n}_y \frac{\partial G_0(\underline{x},\underline{y})}{\partial n_x} dS_y - \frac{1}{2}\frac{\partial \hat{p}(\underline{x})}{\partial n_x}
\end{aligned}
\tag{7.159}
$$

It can be proved that all the integrals are at most weakly singular (Chen et al. 2009). The author claims that numerical results based on the use of higher order piecewise polynomials are considerably more accurate.

Another technique has been proposed by Guiggiani et al. (1992) and Silva (1994). It starts with deforming locally the boundary around the singular point as explained in Section 7.2.2. Here the point is excluded from the exterior domain Ω^+. The boundary of the deformed volume becomes $\partial\Omega_\varepsilon = (\partial\Omega - e_\varepsilon) \cup S_\varepsilon^-$ where S_ε^- is the surface of a small sphere of radius ε centered at point $\underline{x} \in \partial\Omega$ and e_ε is the subregion of $\partial\Omega$ lying inside the small sphere. The definition of the Hadamard finite part of $\int_{\partial\Omega} \hat{p}(\underline{y})\partial^2\hat{G}(\underline{x},\underline{y})/\partial n_x \partial n_y \, dS_y$ is given by

$$
FP\int_{\partial\Omega} \hat{p}(\underline{y})\frac{\partial^2 \hat{G}(\underline{x},\underline{y})}{\partial n_x \partial n_y} dS_y = \lim_{\varepsilon\to 0} \int_{\partial\Omega - e_\varepsilon \cup S_\varepsilon^-} \hat{p}(\underline{y})\frac{\partial^2 \hat{G}(\underline{x},\underline{y})}{\partial n_x \partial n_y} dS_y \tag{7.160}
$$

Equation 7.160 can be rewritten as

$$
\begin{aligned}
FP\int_{\partial\Omega} \hat{p}(\underline{y})\frac{\partial^2 \hat{G}(\underline{x},\underline{y})}{\partial n_x \partial n_y} dS_y &= \int_{\partial\Omega - \delta S} \hat{p}(\underline{y})\frac{\partial^2 \hat{G}(\underline{x},\underline{y})}{\partial n_x \partial n_y} dS_y \\
+ \lim_{\varepsilon\to 0} \int_{\delta S - e_\varepsilon} \hat{p}(\underline{y})\frac{\partial^2 \hat{G}(\underline{x},\underline{y})}{\partial n_x \partial n_y} dS_y &+ \lim_{\varepsilon\to 0} \int_{S_\varepsilon^-} \hat{p}(\underline{y})\frac{\partial^2 \hat{G}(\underline{x},\underline{y})}{\partial n_x \partial n_y} dS_y
\end{aligned}
\tag{7.161}
$$

where δS is a small subregion of $\partial\Omega$ which contains e_ε. In practice, δS is chosen as a portion of $\partial\Omega$ which surrounds the singular point in the real space (usually a boundary element). The limit process is carried out after the boundary element discretization and in the reference coordinate system.*

Using the nodal values $\langle \hat{p}_j \rangle$ of the pressure on the singular element, it can be shown that

$$FP\int_{\partial\Omega} \hat{p}(\underline{y})\frac{\partial^2 \hat{G}(\underline{x},\underline{y})}{\partial n_x \partial n_y} dS_y = \int_{\partial\Omega-\delta S} \hat{p}(\underline{y})\frac{\partial^2 \hat{G}(\underline{x},\underline{y})}{\partial n_x \partial n_y} dS_y$$
$$+ \langle \hat{p}_j \rangle \left\{ \int_0^{2\pi}\int_0^{\breve{\rho}(\phi)} \left[F^j(\rho,\phi) - \left(\frac{f_{-2}^j(\phi)}{\rho^2} + \frac{f_{-1}^j(\phi)}{\rho} \right) \right] d\rho d\phi \right.$$
$$\left. + \int_0^{2\pi} \left[f_{-1}^j(\phi) \ln \left| \frac{\breve{\rho}(\phi)}{\varpi(\phi)} \right| - f_{-2}^j(\phi) \left(\frac{\chi(\phi)}{\varpi^2(\phi)} + \frac{1}{\breve{\rho}(\phi)} \right) \right] d\phi \right\} \tag{7.162}$$

where (ρ, ϕ) corresponds to a polar coordinate system centered at the source point:

$$F^j(\rho,\theta) = \rho \frac{\partial^2 \hat{G}(\underline{\xi},\underline{\eta})}{\partial n_x \partial n_y} \bar{N}_j(\underline{\eta}) \left| j(\underline{\eta}) \right| \tag{7.163}$$

Note that to alleviate the notations, the dependency on $\underline{\xi}$ (and $\underline{\eta}$) is omitted in all the functions. The terms $f_{-1}^j(\phi)$ and $f_{-2}^j(\phi)$ represent the first two terms of the Laurent series of function:

$$f^j(\rho,\theta) = \rho \frac{n_i(\underline{\xi}) n_i(\underline{\eta})}{4\pi r^3(\underline{\xi},\underline{\eta})} \bar{N}_j(\underline{\eta}) \left| j(\underline{\eta}) \right| \tag{7.164}$$

They are given by

$$\begin{aligned} f_{-1}^j(\phi) &= \frac{1}{4\pi}\left[S_{-2}(\phi)\bar{N}_j^0 j_0 + S_{-3}(\phi)(\bar{N}_j^1 j_0 + \bar{N}_j^0 n_i^0 j_i^1(\phi)) \right] \\ f_{-2}^j(\phi) &= \frac{1}{4\pi} S_{-3}(\phi)\bar{N}_j^0 j_0 \end{aligned} \tag{7.165}$$

* The singular element δS is mapped (using mapping defined in Equation 7.95) into a regular element in the reference space (η_1, η_2) and the subregion e_ε is mapped into a subregion around the singular point $\underline{\xi}$.

where

$$S_{-3}(\phi) = \frac{1}{A^3(\phi)}; \quad S_{-2}(\phi) = -\frac{3A_i(\phi)B_i(\phi)}{A^5(\phi)}$$
$$\varpi(\phi) = \frac{1}{A(\phi)}; \quad \chi(\phi) = -\frac{A_i(\phi)B_i(\phi)}{A^4(\phi)} \tag{7.166}$$
$$A(\phi) = (A_i(\phi)A_i(\phi))^{\frac{1}{2}}$$

and $A_i(\phi)$ and $B_i(\phi)$ are defined below. Recall that Einstein summation notation is adopted in the previous equations. $\rho = \breve{\rho}(\phi)$ is the equation of the external contour of the singular element in the reference space (e.g., triangle or quadrilateral).

The values of $A_i(\phi)$, $B_i(\phi)$, $f^j_{-1}(\phi)$, and $f^j_{-2}(\phi)$ are obtained using a Taylor series around the singular point $\underline{\xi}$ of $y_i(\underline{\eta}) - x_i(\underline{\xi})$. Given that $y_i(\underline{\eta}) = N_j(\underline{\eta})x_{i,j}$ and $x_i(\underline{\xi}) = N_j(\underline{\xi})x_{i,j}$ (see Equation 7.79) where $x_{i,j}$ $i = 1,2,3$ are the nodal coordinates along direction i of each node of the singular element, N_j are the shape function on the singular element, the Taylor series around the singular point $\underline{\xi}$ of $y_i(\underline{\eta}) - x_i(\underline{\xi})$ can be written as

$$\begin{aligned} y_i(\underline{\eta}) - x_i(\underline{\xi}) &= (\eta_1 - \xi_1)\left.\frac{\partial N_j(\underline{\eta})}{\partial \eta_1}\right|_{\underline{\eta}=\underline{\xi}} x_{i,j} + (\eta_2 - \xi_2)\left.\frac{\partial N_j(\underline{\eta})}{\partial \eta_2}\right|_{\underline{\eta}=\underline{\xi}} x_{i,j} \\ &+ \frac{(\eta_1-\xi_1)^2}{2}\left.\frac{\partial^2 N_j(\underline{\eta})}{\partial \eta_1^2}\right|_{\underline{\eta}=\underline{\xi}} x_{i,j} + \frac{(\eta_1-\xi_1)(\eta_2-\xi_2)}{2}\left.\frac{\partial^2 N_j(\underline{\eta})}{\partial \eta_1 \partial \eta_2}\right|_{\underline{\eta}=\underline{\xi}} x_{i,j} \\ &+ \frac{(\eta_2-\xi_2)^2}{2}\left.\frac{\partial^2 N_j(\underline{\eta})}{\partial \eta_2^2}\right|_{\underline{\eta}=\underline{\xi}} x_{i,j} + \cdots \end{aligned} \tag{7.167}$$

which can be rewritten as

$$\begin{aligned} y_i(\underline{\eta}) - x_i(\underline{\xi}) &= \rho\left[\cos\phi\left.\frac{\partial N_j(\underline{\eta})}{\partial \eta_1}\right|_{\underline{\eta}=\underline{\xi}} + \sin\phi\left.\frac{\partial N_j(\underline{\eta})}{\partial \eta_2}\right|_{\underline{\eta}=\underline{\xi}}\right] x_{i,j} + \rho^2\left[\frac{\cos^2\phi}{2}\left.\frac{\partial^2 N_j(\underline{\eta})}{\partial \eta_1^2}\right|_{\underline{\eta}=\underline{\xi}}\right. \\ &\left. + \frac{\cos\phi\sin\phi}{2}\left.\frac{\partial^2 N_j(\underline{\eta})}{\partial \eta_1 \partial \eta_2}\right|_{\underline{\eta}=\underline{\xi}} + \frac{\sin^2\phi}{2}\left.\frac{\partial^2 N_j(\underline{\eta})}{\partial \eta_2^2}\right|_{\underline{\eta}=\underline{\xi}}\right] x_{i,j} + O(\rho^3) \\ &= \rho \widehat{N}^1_j(\phi)x_{i,j} + \rho^2 \widehat{N}^2_j(\phi)x_{i,j} + O(\rho^3) \\ &= \rho A_i(\phi) + \rho^2 B_i(\phi) + O(\rho^3) \end{aligned} \tag{7.168}$$

using the change of variable defined in Equation 7.100. $\widehat{N}_j^1(\phi)$ and $\widehat{N}_j^2(\phi)$ represent reduced shape functions of order 1 and 2. Similarly, we have

$$\bar{N}_j(\underline{\eta}) = \bar{N}_j^0 + \rho\bar{N}_j^1(\phi) + O(\rho^2) \tag{7.169}$$

with $\bar{N}_j^0 = \bar{N}_j(\underline{\xi})$ and

$$j_i(\underline{\eta}) = n_i(\underline{\eta})\left|j(\underline{\eta})\right| = j_i^0 + \rho j_i^1(\phi) + O(\rho^2) \tag{7.170}$$

with $j_i^0 = n_i(\underline{\xi})j_0 = n_i(\underline{\xi})|j(\underline{\xi})|$

All the integrals in Equation 7.162 can be evaluated using standard quadrature rules. More details regarding the derivations can be found in Silva (1994), Amini and Harris (1990), Amini and Wilton (1986).

7.10.3 Regularized Burton Miller formulation

Silva revisited a formulation originally derived by Panich (Panich 1965). To address the Neumann (normal derivative of the pressure boundary condition) and the Robin problem (impedance boundary condition), Panich proposed an indirect formulation with single-layer density $\hat{\sigma}$ and double-layer density $\hat{V}_{\sigma,0}$ as

$$\hat{p}(\underline{x}) = \hat{V}_\sigma(\underline{x}) + \alpha_{panich}\hat{W}_\sigma(\underline{x}) \tag{7.171}$$

where α_{panich}* is a complex-valued coupling parameter

$$\hat{V}_\sigma(\underline{x}) = \int_{\partial\Omega} \hat{\sigma}(\underline{y})\hat{G}(\underline{x},\underline{y})dS_y \tag{7.172}$$

$$\hat{W}_\sigma(\underline{x}) = \int_{\partial\Omega} \hat{V}_{\sigma,0}(\underline{y})\frac{\partial\hat{G}(\underline{x},\underline{y})}{\partial n_y}dS_y \tag{7.173}$$

are Helmholtz single-layer and double-layer potential, respectively. $\hat{V}_{\sigma,0}$, which is the unknown double-layer density, is given as an integral of the unknown single-layer density $\hat{\sigma}$

$$\hat{V}_{\sigma,0}(\underline{y}) = \int_{\partial\Omega} \hat{\sigma}(\underline{z})G_0(\underline{y},\underline{z})dS_z \tag{7.174}$$

* Good numerical results are obtained with a value $\alpha_{panich} = i$.

where $G_0(\underline{y},\underline{z})$ is the Green's function for Laplace's equation. In fact, $\hat{V}_{\sigma,0}$ is a harmonic single-layer potential solution of Laplace's equation.

It can be shown that for a smooth boundary the normal derivative of Equation 7.171 on the surface can be written as

$$\begin{aligned}\frac{\partial\hat{p}}{\partial n_x}(\underline{x}) &= \frac{1}{2}\hat{\sigma}(\underline{x}) + \int_{\partial\Omega}\hat{\sigma}(\underline{y})\frac{\partial\hat{G}(\underline{x},\underline{y})}{\partial n_x}dS_y \\ &+ \alpha_{panich}\left[\int_{\partial\Omega}\left(\frac{\partial^2\hat{G}(\underline{x},\underline{y})}{\partial n_x\partial n_y} - \frac{\partial^2 G_0(\underline{x},\underline{y})}{\partial n_x\partial n_y}\right)\left(\int_{\partial\Omega}\hat{\sigma}(\underline{z})G_0(\underline{y},\underline{z})dS_z\right)dS_y\right. \\ &\left. -\frac{1}{4}\hat{\sigma}(\underline{x}) + \int_{\partial\Omega}\frac{\partial G_0(\underline{x},\underline{y})}{\partial n_x}\left(\int_{\partial\Omega}\frac{\partial G_0(\underline{y},\underline{z})}{\partial n_y}\hat{\sigma}(\underline{z})dS_z\right)dS_y\right]\end{aligned} \tag{7.175}$$

Equation 7.175 referred to as improved boundary integral equation possesses a unique solution at all frequencies as proved in Silva (1994). The advantage of this formulation is that it can be used with classic isoparametric continuous elements since the smoothness requirements for the existence of the normal derivative of the double-layer potential at the collocation points are satisfied. Silva proved that a $C^{0,\alpha}$ single-layer density yields a single-layer potential $C^{1,\alpha}$ at the collocation points.

Following the general procedure described in Section 7.6, the boundary $\partial\Omega$ can be discretized using isoparametric continuous elements. This means that the shape functions used to discretize the geometry are the same as the interpolation or shape functions used to discretize the density function $\hat{\sigma}$. We end up with the following integrals to calculate

$$\left\{\begin{aligned}\hat{J}_{1,ij} &= \int_{R^e}\frac{\partial\hat{G}(\underline{x}_i,\underline{y}(\underline{\eta}))}{\partial n_y}\overline{N}_j(\underline{\eta})\left|j(\underline{\eta})\right|d\eta_1 d\eta_2 \\ \hat{J}_{2,ij} &= \int_{R^e}\left[\frac{\partial^2\hat{G}(\underline{x}_i,\underline{y}(\underline{\eta}))}{\partial n_x\partial n_y} - \frac{\partial^2 G_0(\underline{x}_i,\underline{y}(\underline{\eta}))}{\partial n_x\partial n_y}\right]\overline{N}_j(\underline{\eta})\left|j(\underline{\eta})\right|d\eta_1 d\eta_2 \\ J_{3,ij} &= \int_{R^e}\frac{\partial G_0(\underline{x}_i,\underline{y}(\underline{\eta}))}{\partial n_y}\overline{N}_j(\underline{\eta})\left|j(\underline{\eta})\right|d\eta_1 d\eta_2 \\ J_{4,ij} &= \int_{R^e}G_0(\underline{x}_i,\underline{y}(\underline{\eta}))\,\overline{N}_j(\underline{\eta})\left|j(\underline{\eta})\right|d\eta_1 d\eta_2\end{aligned}\right. \tag{7.176}$$

which can be evaluated using a standard Gauss quadrature scheme. The integration of the kernels having a singularity of order $1/r$ is handled as described in Section 7.6.4.2 using a polar coordinate transformation.

7.10.4 Dual surface (DS) method

The DS method has been used in the field of electromagnetic problems for more than a decade and adapted to acoustics more recently (Mohsen et al. 2011; Burgschweiger et al. 2013). The method consists in imposing the vanishing of the pressure field inside a closed volume of boundary $\partial\Omega$ on a virtual surface located inside the volume (see Figure 7.15). The virtual surface is obtained by shifting the original boundary along the boundary inward normal by a distance δ_{DS}, which depends on the wavelength. This is similar to CHIEF method but it makes use of a continuous surface instead of discrete points to impose the vanishing of the pressure field inside the volume.

For the scattering problem, the associated equation reads

$$C^+(\underline{x})\hat{p}(\underline{x}) = \int_{\partial\Omega}\left(\hat{p}(\underline{y})\frac{\partial\hat{G}(\underline{x},\underline{y})}{\partial n_y} - \hat{G}(\underline{x},\underline{y})\frac{\partial\hat{p}(\underline{y})}{\partial n_y}\right)dS_y + \hat{p}_{inc}(\underline{x})$$

$$+\,\alpha_{DS}\int_{\partial\Omega}\left(\hat{p}(\underline{y})\frac{\partial\hat{G}(\underline{x}_{DS},\underline{y})}{\partial n_y} - \hat{G}(\underline{x}_{DS},\underline{y})\frac{\partial\hat{p}(\underline{y})}{\partial n_y}\right)dS_y + \alpha_{DS}\hat{p}_{inc}(\underline{x}_{DS}) \qquad (7.177)$$

This equation maintains the simplicity of the original boundary integral equation and ensures the uniqueness of the solution provided that the coupling constant α_{DS} is complex-valued and the inner surface is at a distance δ_{DS} less than $\lambda/2$ where λ is the acoustic wavelength. The authors (Mohsen et al. 2011) propose a value of $\alpha_{DS} = -i$ for the hard scattering problem, a value of $\alpha_{DS} = ik$ for the soft scattering problem, and a value $\alpha_{DS} = -ika$ for

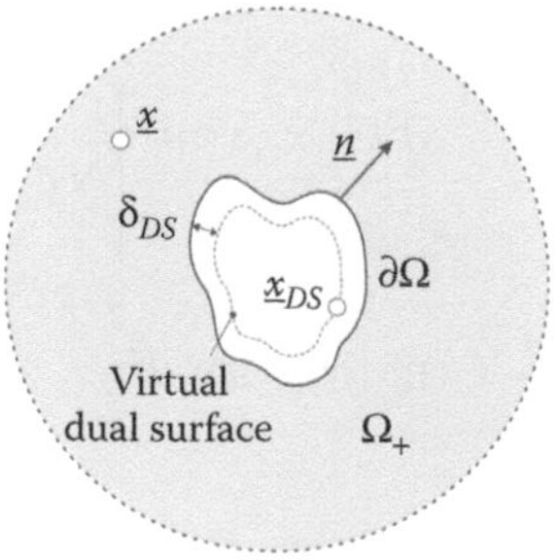

Figure 7.15 DS method.

the radiation problem of a pulsating sphere. Unlike Burton and Miller's approach, this method avoids the introduction of hyper-singular integrals. The method also solves the problem of the adequate choice of interior CHIEF points and avoids the introduction of an overdetermined system of equations. Equation 7.177 is then discretized and the integrals are calculated as described in Section 7.6.4.

7.10.5 Prolongation by continuity

This technique proposed by Quevat et al. (1988) consists in prolonging the exterior problem with an interior problem, which admits a unique solution. If the exterior problem is defined as

$$\begin{cases} \nabla^2\hat{p} + k^2\hat{p} = 0 \quad \text{in} \quad \Omega^+ \\ \hat{p}\big|_{\partial\Omega_1} = \bar{p} \\ \left.\dfrac{\partial\hat{p}}{\partial n}\right|_{\partial\Omega_2} = \bar{q} \\ \lim\limits_{r\to\infty} r\left(\dfrac{\partial\hat{p}}{\partial r} + ik\hat{p}\right) = 0 \end{cases} \tag{7.178}$$

The associated interior problem is defined as

$$\begin{cases} \nabla^2\hat{p} + k^2\hat{p} = 0 & \text{in } \Omega_- \\ \hat{p}\big|_{\partial\Omega_1} = \bar{p} & \\ \left.\dfrac{\partial\hat{p}}{\partial n}\right|_{\partial\Omega_2} = \bar{q} & \\ \dfrac{\partial\hat{p}^+}{\partial n} + ik\hat{\beta}\hat{p}^+ = 0 & \text{over } \partial\Omega_{\beta+} \\ \dfrac{\partial\hat{p}^-}{\partial n} - ik\hat{\beta}^*\hat{p}^- = 0 & \text{over } \partial\Omega_{\beta-} \end{cases} \tag{7.179}$$

where $\partial\Omega_\beta = \partial\Omega_{\beta+} \cup \partial\Omega_{\beta-}$ is an interior thin surface coated with a locally reacting material with normalized acoustic admittance $\hat{\beta}$ on the positive side and a locally reacting material with normalized acoustic admittance $\hat{\beta}^*$ on the negative side (see Figure 7.16). The problem defined by Equation 7.179 has a unique solution since the homogeneous problem (i.e., $\bar{p} = \bar{q} = 0$) only admits the trivial solution (Quevat et al. 1988). The solution $\hat{p}$ of the entire problem is sought for under the form

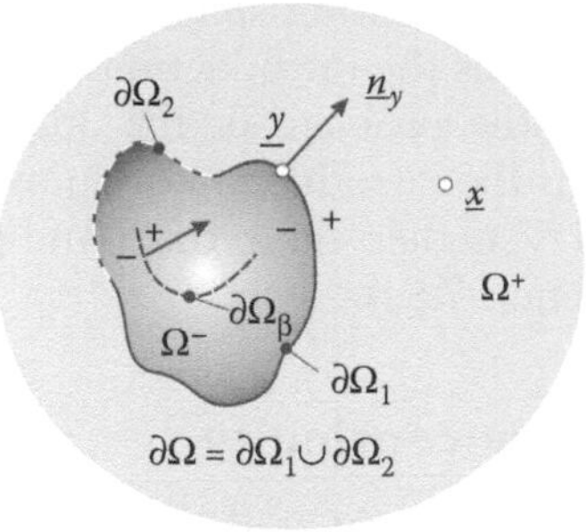

Figure 7.16 Configuration for the associated interior problem.

$$\hat{p}(\underline{x}) = \int_{\partial\Omega_2 \cup \partial\Omega_\beta} \hat{\mu}(\underline{y}) \frac{\partial \hat{G}(\underline{x},\underline{y})}{\partial n_y}\, dS_y - \int_{\partial\Omega_1 \cup \partial\Omega_\beta} \hat{\sigma}(\underline{y}) \hat{G}(\underline{x},\underline{y})\, dS_y \tag{7.180}$$

Applying the boundary conditions and following the procedure described in Section 7.7.1, we end up with the following variational statement:

$$\delta J(\hat{\sigma},\hat{\mu}) = 0 \quad \forall \hat{\sigma} \in \partial\Omega_1 \quad \text{and} \quad \forall \hat{\mu} \in \partial\Omega_2 \tag{7.181}$$

where

$$\begin{aligned}
J(\hat{\sigma},\hat{\mu}) = & \frac{1}{2} \int_{\partial\Omega_1 \cup \partial\Omega_\beta} \int_{\partial\Omega_1 \cup \partial\Omega_\beta} \hat{\sigma}(\underline{x}) \hat{G}(\underline{x},\underline{y}) \hat{\sigma}(\underline{y}) dS_x dS_y \\
& + \frac{1}{2} \int_{\partial\Omega_2 \cup \partial\Omega_\beta} \int_{\partial\Omega_2 \cup \partial\Omega_\beta} \hat{\mu}(\underline{x}) \frac{\partial^2 \hat{G}(\underline{x},\underline{y})}{\partial n_x \partial n_y} \hat{\mu}(\underline{y})\, dS_x dS_y \\
& - \int_{\partial\Omega_1 \cup \partial\Omega_\beta} \int_{\partial\Omega_2 \cup \partial\Omega_\beta} \hat{\sigma}(\underline{x}) \frac{\partial \hat{G}(\underline{x},\underline{y})}{\partial n_y} \hat{\mu}(\underline{y}) dS_x dS_y \\
& + \frac{i}{2} \Re(\hat{\beta}) \Im(\hat{\beta}) \int_{\partial\Omega_1} \hat{\sigma}(\underline{x}) \hat{\mu}(\underline{x})\, dS_x \\
& - \frac{i}{4} \frac{\Re(\hat{\beta})}{k} \int_{\partial\Omega_\beta} \hat{\sigma}^2(\underline{x})\, dS_x - \frac{i}{4} k \Re(\hat{\beta}) \left|\hat{\beta}\right|^2 \int_{\partial\Omega_\beta} \hat{\mu}^2(\underline{x})\, dS_x \\
& - \int_{\partial\Omega_1} \hat{\sigma}(\underline{x}) \bar{p}(\underline{x})\, dS_x - \int_{\partial\Omega_2} \hat{\mu}(\underline{x}) \bar{q}(\underline{x}) dS_x
\end{aligned} \tag{7.182}$$

7.10.6 Impedance coating

Another idea to avoid irregular frequencies is to use a variant of the idea described in Section 7.10.5 (Zhang et al. 2001; D'Amico et al. 2010). The impedance coating approach consists in applying unequal appropriate acoustic impedance boundary conditions on the internal and external side of the boundary instead of on a fictitious surface located inside the interior volume enclosed by $\partial\Omega$. The resonant behavior in the interior of the boundary element model is artificially damped by choosing a small impedance value on the interior side of the boundary and a high impedance value on the exterior side that corresponds to a hard wall condition. To illustrate the method, consider the case of the radiation problem. The scattering problem is tackled in the case of a sphere in Section 7.13. Let us consider the following system of equations:

$$\begin{cases} \nabla^2 \hat{p} + k^2 \hat{p} = 0 & \text{in } \Omega^+ \cup \Omega^- \\ \dfrac{\partial \hat{p}^+}{\partial n} - ik\hat{\beta}^+ \hat{p}^+ - \rho_0 \omega^2 \hat{u}_n = 0 & \text{over } \partial\Omega^+ \\ \dfrac{\partial \hat{p}^-}{\partial n} + ik\hat{\beta}^- \hat{p}^- - \rho_0 \omega^2 \hat{u}_n = 0 & \text{over } \partial\Omega^- \end{cases} \tag{7.183}$$

where $\hat{p}^+$ and $\hat{p}^-$ are the total pressure in domain Ω^+ and Ω^-, $\hat{\beta}^+$ and $\hat{\beta}^-$ are the normalized acoustic admittance applied on $\partial\Omega^+$ and $\partial\Omega^-$, respectively. $\hat{u}_n$ is the normal displacement of boundary $\partial\Omega$. We have

$$\hat{p}^{\pm}(\underline{x}) = \int_{\partial\Omega} \hat{\mu}(\underline{y}) \frac{\partial \hat{G}(\underline{x}, \underline{y})}{\partial n_y} dS_y - \int_{\partial\Omega} \hat{\sigma}(\underline{y}) \hat{G}(\underline{x}, \underline{y}) dS_y \tag{7.184}$$

where

$$\begin{aligned} \hat{\mu}(\underline{x}) &= \hat{p}^+(\underline{x}) - \hat{p}^-(\underline{x}) \\ \hat{\sigma}(\underline{x}) &= \frac{\partial \hat{p}^+(\underline{x})}{\partial n_x} - \frac{\partial \hat{p}^-(\underline{x})}{\partial n_x} \end{aligned} \tag{7.185}$$

are the double-layer and single-layer density functions. For $\underline{x} \in \partial\Omega$, we have

$$\begin{aligned} \hat{p}^+(\underline{x}) &= C^-(\underline{x})\hat{\mu}(\underline{x}) + \int_{\partial\Omega} \hat{\mu}(\underline{y}) \frac{\partial \hat{G}(\underline{x}, \underline{y})}{\partial n_y} dS_y - \int_{\partial\Omega} \hat{\sigma}(\underline{y}) \hat{G}(\underline{x}, \underline{y}) dS_y \\ \hat{p}^-(\underline{x}) &= -C^+(\underline{x})\hat{\mu}(\underline{x}) + \int_{\partial\Omega} \hat{\mu}(\underline{y}) \frac{\partial \hat{G}(\underline{x}, \underline{y})}{\partial n_y} dS_y - \int_{\partial\Omega} \hat{\sigma}(\underline{y}) \hat{G}(\underline{x}, \underline{y}) dS_y \end{aligned} \tag{7.186}$$

Using Equation 7.183 and the second line of Equation 7.185, we get

$$\hat{\sigma}(\underline{x}) = ik(\hat{\beta}^+\hat{p}^+(\underline{x}) + \hat{\beta}^-\hat{p}^-(\underline{x})) \tag{7.187}$$

That is, by using Equation 7.186

$$\begin{aligned}\hat{\sigma}(\underline{x}) &= ik(\hat{\beta}^+C^-(\underline{x}) - \hat{\beta}^-C^+(\underline{x}))\hat{\mu}(\underline{x}) \\ &\quad + ik(\hat{\beta}^+ + \hat{\beta}^-)\int_{\partial\Omega}\hat{\mu}(\underline{y})\frac{\partial\hat{G}(\underline{x},\underline{y})}{\partial n_y}dS_y - \int_{\partial\Omega}\hat{\sigma}(\underline{y})\hat{G}(\underline{x},\underline{y})dS_y\end{aligned} \tag{7.188}$$

In addition, we have

$$\begin{aligned}\frac{\partial\hat{p}^+(\underline{x})}{\partial n_x} &= C^+(\underline{x})\hat{\sigma}(\underline{x}) + FP\int_{\partial\Omega}\hat{\mu}(\underline{y})\frac{\partial^2\hat{G}(\underline{x},\underline{y})}{\partial n_x\partial n_y}dS_y - \int_{\partial\Omega}\hat{\sigma}(\underline{y})\frac{\partial\hat{G}(\underline{x},\underline{y})}{\partial n_x}dS_y \\ \frac{\partial\hat{p}^-(\underline{x})}{\partial n_x} &= -C^-(\underline{x})\hat{\sigma}(\underline{x}) + FP\int_{\partial\Omega}\hat{\mu}(\underline{y})\frac{\partial^2\hat{G}(\underline{x},\underline{y})}{\partial n_x\partial n_y}dS_y - \int_{\partial\Omega}\hat{\sigma}(\underline{y})\frac{\partial\hat{G}(\underline{x},\underline{y})}{\partial n_x}dS_y\end{aligned} \tag{7.189}$$

Using Equation 7.183 and the first line of Equation 7.185, we get

$$\hat{\mu}(\underline{x}) = \frac{1}{ik\hat{\beta}^+}\left[\frac{\partial\hat{p}^+}{\partial n_x} - \rho_0\omega^2\hat{u}_n(\underline{x})\right] + \frac{1}{ik\hat{\beta}^-}\left[\frac{\partial\hat{p}^-}{\partial n_x} - \rho_0\omega^2\hat{u}_n(\underline{x})\right] \tag{7.190}$$

Using Equation 7.189 and Equation 7.190, we get

$$\begin{aligned}\hat{\mu}(\underline{x}) &= \frac{1}{ik\hat{\beta}^+}\Bigg[C^+(\underline{x})\hat{\sigma}(\underline{x}) + FP\int_{\partial\Omega}\hat{\mu}(\underline{y})\frac{\partial^2\hat{G}(\underline{x},\underline{y})}{\partial n_x\partial n_y}dS_y \\ &\quad - \int_{\partial\Omega}\hat{\sigma}(\underline{y})\frac{\partial\hat{G}(\underline{x},\underline{y})}{\partial n_x}dS_y - \rho_0\omega^2\hat{u}_n(\underline{x})\Bigg] \\ &\quad + \frac{1}{ik\hat{\beta}^-}\Bigg[-C^-(\underline{x})\hat{\sigma}(\underline{x}) + FP\int_{\partial\Omega}\hat{\mu}(\underline{y})\frac{\partial^2\hat{G}(\underline{x},\underline{y})}{\partial n_x\partial n_y}dS_y \\ &\quad - \int_{\partial\Omega}\hat{\sigma}(\underline{y})\frac{\partial\hat{G}(\underline{x},\underline{y})}{\partial n_x}dS_y - \rho_0\omega^2\hat{u}_n(\underline{x})\Bigg]\end{aligned} \tag{7.191}$$

The associated variational formulation is given by

$$
\begin{aligned}
\delta\Pi(\hat{\mu},\hat{\sigma}) = & -\int_{\partial\Omega} \hat{\mu}(\underline{x})\delta\hat{\mu}(\underline{x})dS_x + \frac{1}{jk\hat{\beta}^+}\left[\int_{\partial\Omega}\int_{\partial\Omega} \hat{\mu}(\underline{y})\frac{\partial^2\hat{G}(\underline{x},\underline{y})}{\partial n_x \partial n_y}\delta\hat{\mu}(\underline{x})dS_y\, dS_x\right. \\
& - \int_{\partial\Omega}\int_{\partial\Omega} \hat{\sigma}(\underline{y})\frac{\partial\hat{G}(\underline{x},\underline{y})}{\partial n_x}\delta\hat{\mu}(\underline{x})dS_y dS_x + \frac{1}{2}\int_{\partial\Omega}\hat{\sigma}(\underline{x})\delta\hat{\mu}(\underline{x})dS_x \\
& \left. - \int_{\partial\Omega}\rho_0\omega^2\hat{u}_n(\underline{x})\delta\hat{\mu}(\underline{x})dS_x\right] + \frac{1}{jk\hat{\beta}^-}\left[\int_{\partial\Omega}\int_{\partial\Omega}\hat{\mu}(\underline{y})\frac{\partial^2\hat{G}(\underline{x},\underline{y})}{\partial n_x \partial n_y}\delta\hat{\mu}(\underline{x})dS_y dS_x\right. \\
& - \int_{\partial\Omega}\int_{\partial\Omega} \hat{\sigma}(\underline{y})\frac{\partial\hat{G}(\underline{x},\underline{y})}{\partial n_x}\delta\hat{\mu}(\underline{x})dS_y dS_x - \frac{1}{2}\int_{\partial\Omega}\hat{\sigma}(\underline{x})\delta\hat{\mu}(\underline{x})dS_x \\
& \left. - \int_{\partial\Omega}\rho_0\omega^2\hat{u}_n(\underline{x})\delta\hat{\mu}(\underline{x})dS_x\right] - \int_{S}\hat{\sigma}(\underline{x})\delta\hat{\sigma}(\underline{x})dS_x \\
& + ik\frac{(\hat{\beta}^+ - \hat{\beta}^-)}{2}\int_{\partial\Omega}\hat{\mu}(\underline{x})\delta\hat{\sigma}(\underline{x})dS_x \\
& + ik(\hat{\beta}^+ + \hat{\beta}^-)\left[\int_{\partial\Omega}\int_{\partial\Omega}\hat{\mu}(\underline{y})\frac{\partial\hat{G}(\underline{x},\underline{y})}{\partial n_y}\delta\hat{\sigma}(\underline{x})dS_y dS_x\right. \\
& \left. - \int_{\partial\Omega}\int_{\partial\Omega}\hat{\sigma}(\underline{y})\hat{G}(\underline{x},\underline{y})\delta\hat{\sigma}(\underline{x})dS_y dS_x\right] \forall(\delta\hat{\mu},\delta\hat{\sigma}) \text{ admissible}
\end{aligned}
\tag{7.192}
$$

This leads to the following matrix system:

$$
\begin{bmatrix}
ik(\hat{\beta}^+ + \hat{\beta}^-)[\hat{\mathcal{D}}] + k^2\hat{\beta}^+\hat{\beta}^-[C_{\mu\mu}] & -ik\left[(\hat{\beta}^+ + \hat{\beta}^-)[\hat{\mathcal{C}}_{\sigma\mu}] + (\hat{\beta}^+ - \hat{\beta}^-)\dfrac{[C_{\sigma\mu}]}{2}\right] \\
-ik\left[(\hat{\beta}^+ + \hat{\beta}^-)[\hat{\mathcal{C}}_{\sigma\mu}]^T + (\hat{\beta}^+ - \hat{\beta}^-)\dfrac{[C_{\sigma\mu}]^T}{2}\right] & [C_{\sigma\sigma}] + ik(\hat{\beta}^+ + \hat{\beta}^-)[\hat{\mathcal{M}}_1']
\end{bmatrix}
\begin{Bmatrix}\{\hat{\mu}\} \\ \{\hat{\sigma}\}\end{Bmatrix} = \begin{Bmatrix}\{\hat{S}_{r,\mu}\} \\ \{0\}\end{Bmatrix}
\tag{7.193}
$$

where

$$\begin{aligned}
&\int_{\partial\Omega}\int_{\partial\Omega} \hat{\mu}(\underline{y})\frac{\partial^2 \hat{G}(\underline{x},\underline{y})}{\partial n_x \partial n_y}\delta\hat{\mu}(\underline{x})dS_y dS_x \Rightarrow \langle\delta\hat{\mu}\rangle[\hat{\mathcal{D}}]\{\hat{\mu}\} \\
&\int_{\partial\Omega}\int_{\partial\Omega} \hat{\sigma}(\underline{y})\frac{\partial \hat{G}(\underline{x},\underline{y})}{\partial n_x}\delta\hat{\mu}(\underline{x})dS_y dS_x \Rightarrow \langle\delta\hat{\mu}\rangle[\hat{\mathcal{C}}_{\sigma\mu}]\{\hat{\sigma}\} \\
&ik(\hat{\beta}^+ + \hat{\beta}^-)\int_{\partial\Omega} \rho_0\,\omega^2 \hat{u}_n(\underline{x})\delta\hat{\mu}(\underline{x})dS_x \Rightarrow \langle\delta\hat{\mu}\rangle\{\hat{S}_{r,\mu}\} \\
&\int_{\partial\Omega}\int_{\partial\Omega} \hat{\sigma}(\underline{y})\hat{G}(\underline{x},\underline{y})\delta\hat{\sigma}(\underline{x})dS_y dS_x \Rightarrow \langle\delta\hat{\sigma}\rangle[\hat{\mathcal{M}}_1']\{\hat{\sigma}\} \\
&\int_{\partial\Omega}\int_{\partial\Omega} \hat{\mu}(\underline{y})\frac{\partial \hat{G}(\underline{x},\underline{y})}{\partial n_y}\delta\hat{\sigma}(\underline{x})dS_y dS_x \Rightarrow \langle\delta\hat{\sigma}\rangle[\hat{\mathcal{C}}_{\sigma\mu}]^T\{\hat{\mu}\} \\
&\int_{\partial\Omega} \hat{\sigma}(\underline{x})\delta\hat{\sigma}(\underline{x})dS_x \Rightarrow \langle\delta\hat{\sigma}\rangle[C_{\sigma\sigma}]\{\hat{\sigma}\} \\
&\int_{\partial\Omega} \hat{\sigma}(\underline{x})\delta\hat{\mu}(\underline{x})dS_x \Rightarrow \langle\delta\hat{\sigma}\rangle[C_{\sigma\mu}]\{\hat{\mu}\} \\
&\int_{\partial\Omega} \hat{\mu}(\underline{x})\delta\hat{\sigma}(\underline{x})dS_x \Rightarrow \langle\delta\hat{\mu}\rangle[C_{\sigma\mu}]^T\{\hat{\sigma}\} \\
&\int_{\partial\Omega} \hat{\mu}(\underline{x})\delta\hat{\mu}(\underline{x})dS_x \Rightarrow \langle\delta\hat{\mu}\rangle[C_{\mu\mu}]\{\hat{\mu}\}
\end{aligned} \tag{7.194}$$

Using the second equation of Equation 7.193, we get

$$\{\hat{\sigma}\} = \left(ik(\hat{\beta}^+ + \hat{\beta}^-)[\hat{\mathcal{M}}_1'] + [C_{\sigma\sigma}]\right)^{-1}\left((ik(\hat{\beta}^+ + \hat{\beta}^-)[\hat{\mathcal{C}}_{\sigma\mu}] + ik(\hat{\beta}^+ - \hat{\beta}^-)\frac{[C_{\sigma\mu}]}{2}\right)^T\{\hat{\mu}\} \tag{7.195}$$

and finally $\{\hat{\mu}\}$ is the solution of

$$\begin{aligned}
&\Big[ik(\hat{\beta}^+ + \hat{\beta}^-)[\hat{\mathcal{D}}] + k^2\hat{\beta}^+\hat{\beta}^-[C_{\mu\mu}] \\
&\quad - \left(ik(\hat{\beta}^+ + \hat{\beta}^-)[\hat{\mathcal{C}}_{\sigma\mu}] + ik(\hat{\beta}^+ - \hat{\beta}^-)\frac{[C_{\sigma\mu}]}{2}\right)(ik(\hat{\beta}^+ + \hat{\beta}^-)[\hat{\mathcal{M}}_1'] + [C_{\sigma\sigma}])^{-1} \\
&\quad \times \left(ik(\hat{\beta}^+ + \hat{\beta}^-)[\hat{\mathcal{C}}_{\sigma\mu}] + ik(\hat{\beta}^+ - \hat{\beta}^-)\frac{[C_{\sigma\mu}]}{2}\right)^T\Big]\{\hat{\mu}\} = \{\hat{S}_{r,\mu}\}
\end{aligned} \tag{7.196}$$

This system is symmetric and can be solved using efficient resolution algorithms. Note, however, that the inversion of a matrix is needed.

For all $\underline{x} \in \partial\Omega$ or $\underline{x} \in \Omega$ the acoustic pressure on the surface can be recovered from Equation 7.186.* This method provides very good results but requires the calculation of additional matrices that increase the computation time. Some numerical examples are presented in Section 7.13 to illustrate the performance of this approach and to discuss the values of impedance to be chosen to eliminate the irregular frequencies. In particular, the radiation problem of a pulsating sphere and the scattering by a sphere and by a half-sphere resting on a rigid wall will be studied.

7.11 SOLVING MULTIFREQUENCY PROBLEMS USING BEM

In practical problems in acoustics, the frequency response is often sought over a large frequency band. We have seen that one shortcoming of the BEM is that the matrices coming out from the integral equations discretization are frequency-dependent. These matrices require a large number of numerical integral calculations and have to be recomputed at each frequency of interest. This can be very time consuming for a large number of individual frequencies. This makes the BEM a cumbersome method to deal with multifrequency analysis of acoustical problems. Several techniques dedicated to the resolution of multifrequency problems have been proposed to circumvent this issue and improve computational efficiency. These techniques can act at two levels: (i) approximation of the matrices and (ii) approximation of the final results. Among the first category, one can cite the frequency interpolation technique (Vanhille and Lavie 1998), the Green's function interpolation procedure (Wu et al. 1993), the matrix interpolation and solution iteration process (Raveendra 1999), and the polynomial approximation of the Green's function numerator (Li 2005). In the second category, one finds mainly *the frequency interpolated transfer function* (Von Estorff 2003) and *the frequency response function approximation using Padé approximants* (Coyette et al. 1999; Rumpler and Göransson 2013). The latter deal mainly with interior problems. Advanced similar Padé-like methods can be, for instance, found in the work of Farhat et al (Hetmaniuk et al. 2012; Amsallem and Farhat 2012; Hetmaniuk et al. 2013).

In the following paragraphs, we present the main ideas of the different methods and focus on the most promising to date: the Padé approximants.

* Note that when $\underline{x} \in \partial\Omega$ is a smooth node, $\hat{p}^+$ can be recovered from the first line of Equations 7.200 and 7.202 using $\hat{p}^+(\underline{x}) = (ik\hat{\beta}^-\hat{\mu}(\underline{x}) + \hat{\sigma}(\underline{x}))/(ik(\hat{\beta}^+ + \hat{\beta}^-))$ If the node is not smooth (general case), Equation 7.201 must be used.

The so-called *frequency interpolation technique* (Vanhille and Lavie 1998) consists in eliminating the fast frequency oscillating feature of the BEM matrix coefficients induced by the large variations of the argument of the exp(–*ikr*) term in the Green's kernel. A fluctuating term* is factored out from the integrals to get slowly varying new frequency coefficients. These coefficients are evaluated at two master frequencies. Between these two frequencies, the new coefficients are interpolated linearly. Multiplication by the extracted fluctuating terms at the interpolating frequency allows for recovering the original coefficients. The frequency interpolation technique proved to save a significant amount of CPU time for problems of small to medium size. Its main drawback is that a large amount of disk space is necessary to store the matrices at the master frequencies.

The idea of the *Green's function interpolation* procedure (Wu et al. 1993) is to interpolate the Green's function itself using shape functions. The numerical integral is then carried out on a frequency independent integrand, which needs to be calculated once. For a multifrequency analysis, the BEM matrix coefficients calculation at subsequent frequencies requires only the computation of the new nodal values of the Green's function.† This algorithm uses much less space than the frequency interpolation method.

In the matrix interpolation and solution iteration process, the idea is to avoid the computation of the elementary matrices at each frequency. Thus, the system matrices are only calculated at a few predetermined master frequencies and then the matrices at other intermediate frequencies are evaluated by quadratic interpolation. The matrix solution process is made efficient by iterating the solutions using the factored form of the master frequency matrices.

In the polynomial approximation of the Green's function numerator, least-square approximating polynomials are used for the cosine and sine part of the complex exponential exp(–*ikr*). For example, Li (2005) approximates cos(*kr*) and sin(*kr*) with polynomials of the form $P_{\cos}(kr)=a_0+a_1(kr)^2+a_2(kr)^4+a_3(kr)^6$ and $P_{\sin}(kr) = a_0(kr) + a_1(kr)^3 + a_2(kr)^5 + a_3(kr)^7$, respectively, for $kr < 5$. The BEM matrix coefficients can then be written as a power series in terms of the circular frequency ω whose coefficients are frequency-independent integrals. These coefficients only need to be evaluated once for all the frequencies. At subsequent frequencies, the final global BEM matrix coefficients can be computed by a simple frequency power series whose coefficients are

* In collocation methods, this fluctuating term is of the form $\exp(-ikr_{ij})$ where r_{ij} is the distance between the interpolation node *i* and the mesh node *j*. Once this term is factored out of the integrals over the element to which node *j* belongs, the integrand of the resulting integral varies slowly.

† The approach is straightforward for the regular integral. For singular terms, the technique is applied to the difference between the free-space Green's function $\exp(-ikr)/4\pi r$ and the Green's function for Laplace's equation $1/4\pi r$. The add-back term containing $1/4\pi r$ is not frequency-dependent and needs to be calculated once.

the previously calculated frequency-independent matrix coefficients. The CPU time saving is particularly interesting for small- to medium-size multifrequency problems when the CPU time is mainly spent in the integration and assembling steps. For large-scale problems, however, the time spent in the system resolution dominates and this technique loses its advantage.

The frequency interpolated transfer function (Von Estorff 2003; Von Estorff and Zaleski 2003) was developed for radiation problems. The method consists in using an indirect BEM to calculate the acoustic transfer function (ATF) values at master frequencies. Between these frequencies, ATF values are interpolated using a spherical harmonics expansion. The coefficients of this expansion are found by imposing that the expansion satisfies two conditions. First, it must equal the ATF at the master frequencies at a field point of interest. Second, it must satisfy the velocity boundary conditions defined at selected nodes on the boundary of the vibrating object. Once the spherical harmonics expansion coefficients have been determined, the ATF can be approximated at slave (intermediate) frequencies. The number of master frequencies to be taken into account depends on the frequency range in which the ATF values shall be approximated. The authors show that in a model where the number of nodes of the acoustic mesh is much higher than the number of terms in the approximation series, a significant reduction of the overall computation time to calculate a multifrequency spectrum can be achieved.

In the frequency response function approximation using Padé approximants (Coyette et al. 1999), the acoustic response is calculated at a central frequency using a classic boundary element technique and the extension of this solution to the whole frequency range is carried out using a Padé approximation of the frequency response function. The Padé approximation consists in approximating the acoustic response $\hat{X}(f)$ (f denotes the frequency) using a ratio of polynomials (rational function) instead of a Taylor expansion. Indeed $\hat{X}(f)$ can have singular points (poles) which may induce convergence problem of the Taylor series around these points. The Padé approximation provides an excellent approximation of the FRF beyond the disk of convergence of the power series. The following description is based on Coyette et al's work (1999).

The rational fraction $\hat{P}(f)/\hat{Q}(f)$ ($\hat{P}(f) = \sum_{k=0}^{m} \hat{p}_k f^k$ and $\hat{Q}(f) = 1 + \sum_{k=1}^{n} \hat{q}_k f^k$ are polynomials of order m and n, respectively) is a Padé approximation of order $[m,n]$ to the function $\hat{X}(f)$ if the following system of equations has a unique solution:

$$\hat{Q}(f)\hat{X}(f) - \hat{P}(f) = O(f^{m+n+1})^{*} \tag{7.197}$$

[*] Formally, the power series of $\hat{Q}(f)\hat{X}(f) - \hat{P}(f)$ around a point or pole f_0 begins with the term $(f - f_0)^{n+m+1}$.

It can be shown that coefficients $\hat{q}_k$ are solution of the Hankel* system

$$\begin{bmatrix} \hat{X}_{m-n+1} & \hat{X}_{m-n+2} & \cdots & \hat{X}_{m-1} & \hat{X}_{m} \\ \hat{X}_{m-n+2} & \hat{X}_{m-n+3} & \cdots & \hat{X}_{m} & \hat{X}_{m+1} \\ \vdots & \vdots & \vdots & \vdots & \vdots \\ \hat{X}_{m-1} & \hat{X}_{m} & \cdots & \hat{X}_{m+n-3} & \hat{X}_{m+n-2} \\ \hat{X}_{m} & \hat{X}_{m+1} & \cdots & \hat{X}_{m+n-2} & \hat{X}_{m+n-1} \end{bmatrix} \begin{Bmatrix} \hat{q}_n \\ \hat{q}_{n-1} \\ \vdots \\ \hat{q}_2 \\ \hat{q}_1 \end{Bmatrix} = - \begin{Bmatrix} \hat{X}_{m+1} \\ \hat{X}_{m+2} \\ \vdots \\ \hat{X}_{m+n-1} \\ \hat{X}_{m+n} \end{Bmatrix} \tag{7.198}$$

where

$$\hat{X}_k = \frac{1}{k!} \frac{d^k \hat{X}(f)}{df^k}\bigg|_{f=f_0} \tag{7.199}$$

is the kth coefficient of the Taylor expansion for $\hat{X}(f)$. The Padé approximation requires, therefore, the knowledge of the successive derivatives of the acoustic field with respect to frequency, evaluated at frequency f_0. The coefficients $\hat{p}_k$ of $\hat{P}(f)$ are then obtained by selecting the $m+1$ first coefficients of the Taylor expansion of the product $\hat{Q}(f)\hat{X}(f)$. The convergence radius of the Padé approximation is limited to the distance between the initial frequency f_0 (computation point) and the first pole of $\hat{X}(f)$ not included in $\hat{Q}(f)$.

To illustrate how the procedure is handled numerically, let us consider the acoustic radiation problem described by the indirect boundary element method in (Coyette et al. 1999). The associated discretized variational indirect integral formulation given by Equation 7.126 reduces to

$$J(\hat{\mu}) = \frac{1}{2}\langle \hat{\mu} \rangle [\hat{\mathcal{D}}(k)]\{\hat{\mu}\} - \langle \hat{\mu} \rangle \{\hat{f}_\mu(\omega)\} \tag{7.200}$$

where $\hat{\mu}$ is the potential density and $\{\hat{f}_\mu(\omega)\}$ is the load vector which is related to the normal gradient of the pressure on the surface. The associated solution is obtained from the resolution of

$$[\hat{\mathcal{D}}(k)]\{\hat{\mu}\} = \{\hat{f}_\mu(\omega)\} \tag{7.201}$$

Equation 7.201 is first solved at frequency f_0. The corresponding acoustic pressure at a point in space is then recomputed from Equation 7.58 using the nodal vector $\{\hat{\mu}\}$ at frequency f_0. To calculate the pressure at a different

* A Hankel matrix is a square matrix with constant skew-diagonals.

frequency f, the surface potential density nodal vector $\{\hat{\mu}\}$ is calculated using Padé approximation. A given nodal value $\hat{\mu}_n$ is thus equivalent to the variable $\hat{X}(f)$ mentioned previously. The N first derivatives of $\{\hat{\mu}\}$ have then to be computed. This can be achieved by differentiating Equation 7.201 with respect to frequency f at order N. For example, the first derivative of $\{\hat{\mu}\}$ is the solution of system

$$[\hat{\mathcal{D}}]\frac{\partial}{\partial f}\{\hat{\mu}\} = \frac{\partial}{\partial f}\{\hat{f}_\mu\} - \frac{\partial}{\partial f}[\hat{\mathcal{D}}]\{\hat{\mu}\} \tag{7.202}$$

which is basically the same system as Equation 7.201, which was solved to get $\{\hat{\mu}\}$ at frequency f_0 but with a different right-hand side involving the frequency derivative of matrix $[\hat{\mathcal{D}}]$ multiplied by the vector $\{\hat{\mu}\}$ and the frequency derivative of the original right-hand side $\{\hat{f}_\mu\}$. Similarly, the second derivative of $\{\hat{\mu}\}$ is the solution of system

$$[\hat{\mathcal{D}}]\frac{\partial^2}{\partial f^2}\{\hat{\mu}\} = \frac{\partial^2}{\partial f^2}\{\hat{f}_\mu\} - 2\frac{\partial}{\partial f}[\hat{\mathcal{D}}]\frac{\partial}{\partial f}\{\hat{\mu}\} - \frac{\partial^2}{\partial f^2}[\hat{\mathcal{D}}]\{\hat{\mu}\} \tag{7.203}$$

Actually, all successive derivatives of $\{\hat{\mu}\}$ can be obtained in the same way and only matrix $[\hat{\mathcal{D}}]$ needs to be factorized. The successive derivatives of $[\hat{\mathcal{D}}]$ coefficients with respect to frequency can be computed explicitly since the frequency appears in the Green's function and as a squared term via the wavenumber k (see Equation 7.289). The successive derivatives of $\{\hat{f}_\mu\}$ with respect to frequency can be calculated analytically if the normal gradient on the surface is known analytically. In the case where the normal gradient on the surface is obtained from the results of a structural finite element calculation, $\{\hat{f}_\mu\}$ is only known for a discrete set of frequencies and its derivatives cannot be calculated accurately at the frequency chosen for the Padé approximation. A special procedure must then be applied to solve the problem (see Coyette et al. 1999 for details).

7.12 PRACTICAL CONSIDERATIONS

7.12.1 Convergence criteria

In order for the mesh to capture the acoustic response, the size of the element should be small enough compared to the wavelength. This is a consequence of Shannon's theorem (Shannon 1949). The choice of this size is commonly ruled by a convergence criterion based on the number of elements per wavelength, which depends on the desired accuracy and the type of element. If continuous linear boundary elements are used, a minimum of six elements per acoustic wavelength is used in practice. For quadratic boundary elements, the criterion can be reduced to four elements per wavelength. For further information, the reader may refer, for example, to Marburg (2008) who discusses in detail the convergence criteria for continuous and discontinuous boundary elements of various topologies in the scope of the collocation method.

7.12.2 Numerical integration

As we have seen in Sections 7.6.4 and 7.7.3, the calculation of singular integrals requires the use of appropriate semianalytical quadrature techniques based on a polar change of variable. The transformed integral can be integrated using a classic Gauss quadrature. In the special case of 4 and 8-noded quadrangle boundary elements, an efficient algorithm that accounts explicitly for the $|\underline{y} - \underline{x}|^{-1}$ singularity can be used (Wang and Atalla 1997). For 3 and 6-noded triangular boundary elements, Equation 7.109 that maps triangular into quadrangle elements can be utilized to allow for the use of the previous efficient numerical quadrature. An example of number of Gauss points and Wang and Atalla points, which can be used to achieve a good accuracy, is given in Table 7.2 (see also Appendix 7F). For regular integrals, a Gauss integration technique is used. The number of Gauss points to be selected depends on the topology of the element. It also depends on the ratio r/h where r is the distance between the collocation point and an element node and h is a characteristic size of the element, in the case of the collocation method. Three zones can be specified (near, intermediate, far) depending on the value of r/h (see Table 7.3). In the case of the VBEM, r is the distance between one Gauss point belonging to element i and another Gauss point belonging to element j. In practice, 4 points in the near and intermediate zones and 1 point in the far zone are usually sufficient but these numbers should be increased if the ratio r/h gets smaller than 1.

Table 7.2 Typical number of Gauss and Wang and Atalla points used to calculate boundary element singular integrals

Element shape	*Element topology*	*Number of Gauss points*	*Number of Wang and Atalla points*
Triangles	Linear 3 nodes	3	2
	Quadratic 6 nodes	7	4
Quadrangles	Linear 4 nodes	4	2
	Quadratic 8 nodes	9	4

Table 7.3 Typical number of Gauss points used to calculate boundary element regular integrals according to the ratio (*r/h*)

r/h	*Element topology*	*Triangles*	*Quadrangles*
$r/h < 2$	Linear	4	4
	Quadratic	7	9
$2 < r/h < 5$	Linear	4	4
	Quadratic	7	9
$r/h > 5$	Linear	1	1
	Quadratic	4	4

7.13 EXAMPLES OF APPLICATIONS

7.13.1 Acoustic field inside a cavity with locally reacting walls

Consider the case of a cavity with locally reacting walls. The cavity is filled with a fluid domain Ω of density ρ_0 and sound speed c_0 and is excited by a point source. The system of equations satisfied by the pressure field inside the cavity can be written as

$$\nabla^2\hat{p} + k^2\hat{p} = -\hat{Q}(\underline{y}) \quad \forall \underline{y} \in \Omega \tag{7.204a}$$

$$\frac{\partial \hat{p}}{\partial n_y} + ik\hat{\beta}(\underline{y}) = 0 \quad \forall \underline{y} \in \partial\Omega \tag{7.204b}$$

where $k = \omega/c_0$. The absorption of the walls is characterized by a normalized acoustic admittance $\hat{\beta}$ (inverse of the specific acoustic impedance $\hat{\beta} = (\rho_0 c_0/\hat{Z}_s)$.

The sound pressure at a point $\underline{x}$ inside the cavity is given by the integral formulation of the interior problem

$$\hat{p}(\underline{x}) = \int_{\partial\Omega}\left[\hat{G}(\underline{x},\underline{y})\frac{\partial\hat{p}(\underline{y})}{\partial n_y} - \hat{p}(\underline{y})\frac{\partial\hat{G}(\underline{x},\underline{y})}{\partial n_y}\right] dS_y + \int_{\Omega}\hat{Q}(\underline{y})\hat{G}(\underline{x},\underline{y})\, dV_y \tag{7.205}$$

If the Green's function is chosen such that $\partial\hat{G}(\underline{x},\underline{y})/\partial n_y = 0$ then

$$\hat{p}(\underline{x}) = \int_{\partial\Omega}\hat{G}(\underline{x},\underline{y})\frac{\partial\hat{p}(\underline{y})}{\partial n_y}\, dS_y + \int_{\Omega}\hat{Q}(\underline{y})\hat{G}(\underline{x},\underline{y})\, dV_y \tag{7.206}$$

Using the admittance condition Equation 7.204b, Equation 7.206 can be written as

$$\hat{p}(\underline{x}) = -ik\int_{\partial\Omega} \hat{\beta}(\underline{y})\hat{G}(\underline{x},\underline{y})\, dS_y + \int_{\Omega} \hat{Q}(\underline{y})\hat{G}(\underline{x},\underline{y})\, dV_y \tag{7.207}$$

For a point source located at $\underline{x}_0$

$$\hat{Q}(\underline{y}) = i\omega\rho_0\hat{Q}_s\delta(\underline{y} - \underline{x}_0) \tag{7.208}$$

Here, $\hat{Q}_s$ is the volume source strength in m^3s^{-1}. Consequently,

$$\hat{p}(\underline{x}) = -ik\int_{\partial\Omega} \hat{\beta}(\underline{y})\hat{G}(\underline{x},\underline{y})dS_y + i\omega\rho_0\hat{Q}_s\hat{G}(\underline{x},\underline{x}_0) \tag{7.209}$$

In the particular case of a rigid-walled cavity, $\hat{\beta} = 0$ and

$$\hat{p}(\underline{x}) = i\omega\rho_0\hat{Q}_s\hat{G}(\underline{x},\underline{x}_0) \tag{7.210}$$

It can be shown that the Green's function satisfies

$$\hat{G}(\underline{x},\underline{y}) = \sum_{p,q,r} \frac{\Psi_{pqr}(\underline{x})\Psi_{pqr}(\underline{y})}{N_{pqr}(k_{pqr}^2 - k^2)} = \sum_{p,q,r} \frac{c^2\Psi_{pqr}(\underline{x})\Psi_{pqr}(\underline{y})}{N_{pqr}(\omega_{pqr}^2 - \omega^2)} \tag{7.211}$$

where $\Psi_{pqr}(\underline{x})$ are the eigenmodes of the cavity with rigid walls, $k_{pqr} = (\omega_{pqr}/c_0)$ are the associated wavenumbers, and ω_{pqr} are the circular eigenfrequencies. In addition,

$$N_{pqr} = \int_{\Omega} \Psi_{pqr}^2(\underline{y})dV_y \tag{7.212}$$

is the norm of mode Ψ_{pqr}.

For a rectangular cavity of dimensions (a,b,h_c), we have (Figure 7.17)

$$\begin{aligned}
f_{pqr} &= \frac{c_0}{2}\sqrt{\frac{p^2}{a^2} + \frac{q^2}{b^2} + \frac{r^2}{h_c^2}} \\
\Psi_{pqr}(\underline{x}) &= \cos\left(\frac{p\pi x_1}{a}\right)\cos\left(\frac{q\pi x_2}{b}\right)\cos\left(\frac{r\pi x_3}{h_c}\right) \\
N_{pqr} &= \frac{abh_c}{8}\varepsilon_p\varepsilon_q\varepsilon_r;\ \varepsilon_i = \begin{cases} 2 & i=0 \\ 1 & i>0 \end{cases} \\
&(0 \le x_1 \le a;\quad 0 \le x_2 \le b;\quad 0 \le x_3 \le h_c)
\end{aligned} \tag{7.213}$$

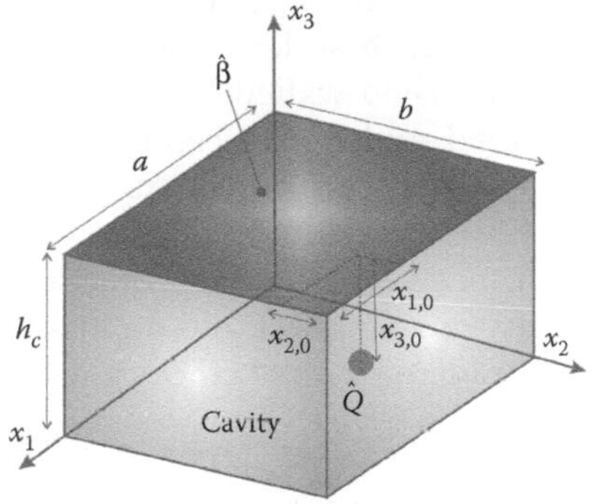

Figure 7.17 Rectangular cavity with a point source.

For a cylindrical cavity (radius a, length L) (Figure 7.18) (see, e.g., Bruneau 2006)

$$\begin{aligned}
f_{pqr} &= \frac{c_0}{2}\sqrt{\left(\frac{\mu_{pq}}{\pi a}\right)^2 + \frac{r^2}{L^2}} \\
\Psi'_{pqr}(\underline{x}) &= \alpha_{pqr} J_p\left(\frac{\mu_{pq} r}{a}\right)\cos\left(\frac{r\pi x_3}{L}\right)\begin{Bmatrix}\cos(p\theta) \\ \sin(p\theta)\end{Bmatrix} \\
\alpha_{pqr} &= \alpha_{pq}\alpha_r, \quad \alpha_{pq} = \frac{\mu_{pq}}{\sqrt{(\mu_{pq}^2 - p^2)}J_p(\mu_{pq})}, \quad \alpha_r = \sqrt{\varepsilon_r}\sqrt{\frac{1}{L}} \qquad (7.214) \\
N_{pqr} &= \pi a^2; \quad \varepsilon_i = \begin{cases} 2 & i = 0 \\ 1 & i > 0 \end{cases} \\
&(0 \le r \le a; \quad 0 \le \theta \le 2\pi; \quad 0 \le x_3 \le L)
\end{aligned}$$

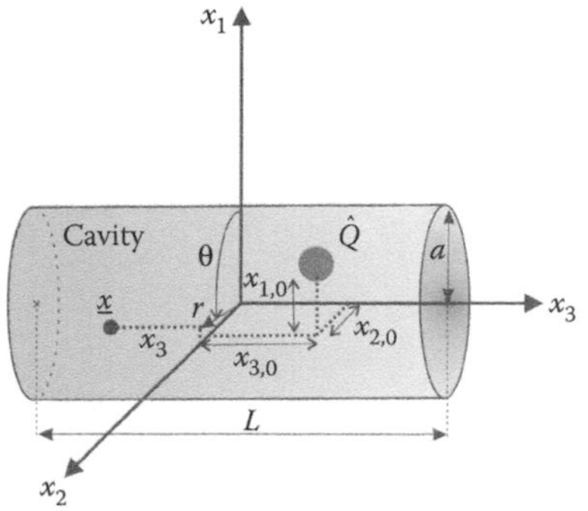

Figure 7.18 Cylindrical cavity with a point source.

Here, J_p denotes the Bessel function of the first kind of order p and μ_{pq} ($q = 1,2\ldots$) is the qth zero of the first derivative of J_p: $J'_p(\mu_{pq}) = 0$.

As an example consider a rigid-walled cylinder with one end cap being rigid and the other baffled and open (Figure 7.19). A point source is located at a point $(r_0,\theta_0,x_{3,0})$ inside the cylinder. An analytical solution to this problem can be obtained using a modal expansion method (see, e.g., the book of Bruneau (Bruneau 2006) for a presentation of acoustic waves propagation in cylindrical waveguides) and applying at the open end an impedance given by the radiation impedance of a baffled circular piston (see Appendix 7I).

$$Z_{rad} = \rho_0 c_0 \left(1 - \frac{2J_1(2ka)}{2ka} + i\frac{2S_1(2ka)}{2ka}\right) \tag{7.215}$$

where $J_1(x)$ is the Bessel function of the first kind of order 1 and $S_1(x)$ is the first-order Struve function.

Denoting by $\hat{\beta}$ the associated normalized acoustic admittance, the pressure at a point (r,θ,x_3) inside the cylinder, given by Equation 7.209, reduces to

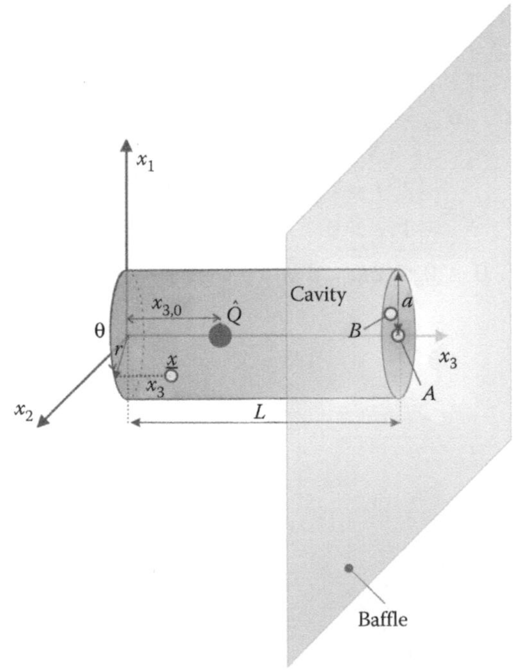

Figure 7.19 Monopole in a rigid-walled cylinder with one end cap rigid and the other baffled and open.

$$\hat{p}(r,\theta,x_3) = \sum_{pqr} \hat{P}_{pqr}(\omega)\Psi_{pqr}(r,\theta,x_3) \tag{7.216}$$

The modal contribution coefficients $\hat{P}_{pqr}$ are obtained from the solution of the following system of equations:

$$\hat{P}_{pqr}(k_{pqr}^2 - k^2 + ik\hat{\beta}(-1)^{2r}\alpha_r^2) + ik\hat{\beta}\alpha_r \sum_{l\neq w} \hat{P}_{pql}\alpha_l(-1)^{l+r} = i\rho_0\omega\hat{Q}_s \frac{\Psi_{pqr}(r_0,\theta_0,x_{3,0})}{\pi a^2} \tag{7.217}$$

In the case of a rigid-walled cylinder, the previous equation reduces to

$$\hat{P}_{pqr} = i\rho_0\omega\hat{Q}_s \frac{\Psi_{pqr}(r_0,\theta_0,x_{3,0})}{\pi a^2(k_{pqr}^2 - k^2)} \tag{7.218}$$

Figure 7.20 shows a comparison between the numerical solution and the modal solution for a cylinder of radius $a = 0.415$ m and length $L = 0.67$ m. The source is located on the cylinder axis at (0,0,0.1) (x_3 axis measured from the rigid end). The numerical solution is obtained using a FEM/VBEM approach. The cylindrical cavity is modeled using FEM while a radiation impedance is added to the outlet, assumed baffled (see Chapter 8). The mesh consists of 2379 hexa-8 fluid elements and 184 quad-4 impedance elements. For the modal approach, all cavity modes having their resonance frequency below 2 times the maximum frequency of the spectrum are kept. Cavity damping is assumed zero in both methods. The comparison is shown for two locations on the outlet: $A(0,0,0.67)$ on the x_3 axis and $B(0.3,0.15,0.67)$ away from the center. Overall, the comparison is very good.

7.13.2 Sound radiation of a monopole above ground

Let us consider a monopole with source strength $\hat{Q}_s$ located at point $\underline{x}_0$, radiating above a rigid flat ground of infinite lateral extent located in the plane $\partial\Omega$ of equation $x_3 = 0$. The sound pressure at a point $\underline{x}$ is given by

$$\hat{p}(\underline{x}) = \int_{\partial\Omega} \left[\hat{p}(\underline{y}) \frac{\partial \hat{G}(\underline{x},\underline{y})}{\partial n_y} - \hat{G}(\underline{x},\underline{y}) \frac{\partial \hat{p}(\underline{y})}{\partial n_y} \right] dS_y + i\omega\rho_0\hat{Q}_s\hat{G}(\underline{x},\underline{x}_0) \quad \forall \underline{x} \in \Omega \tag{7.219}$$

where the normal points inside Ω.

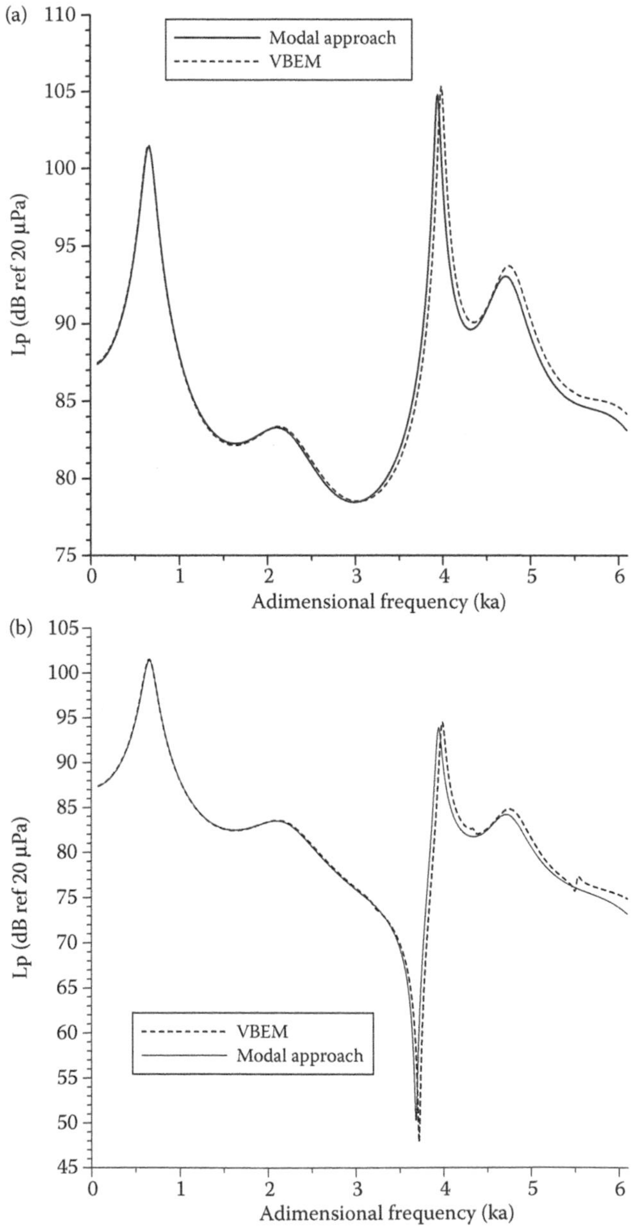

Figure 7.20 Sound pressure level at a location on the outlet of the cylinder. Comparisons between two approaches (analytical solution, VBEM). (a) Position on the axis (point A) and (b) position away from the axis (point B).

Assume that the ground is rigid. Then, $\partial \hat{p}(\underline{y})/\partial n_y = 0$ over $\partial\Omega$ so that Equation 7.219 reduces to

$$\hat{p}(\underline{x}) = \int_{\partial\Omega} \hat{p}(\underline{y}) \frac{\partial \hat{G}(\underline{x},\underline{y})}{\partial n_y} dS_y + i\omega\rho_0 \hat{Q}_s \hat{G}(\underline{x},\underline{x}_0) \quad \forall \underline{x} \in \Omega \tag{7.220}$$

Following the methodology of Section 7.8, let us consider a particular choice for $\hat{G}(\underline{x},\underline{y})$ satisfying $\partial\hat{G}(\underline{x},\underline{y})/\partial n_y = 0 \;\forall \underline{y} \in \partial\Omega$. Let us refer to this Green's function as $\hat{G}_b(\underline{x},\underline{y})$. It is then given by

$$\hat{G}_b(\underline{x},\underline{y}) = \frac{\exp(ikr)}{4\pi r} + \frac{\exp(ikr')}{4\pi r'} \tag{7.221}$$

where $r = \sqrt{(y_1 - x_1)^2 + (y_2 - x_2)^2 + (y_3 - x_3)^2}$ and $r' = \sqrt{(y_1 - x_1)^2 + (y_2 - x_2)^2 + (y_3 + x_3)^2}$, r' is the Euclidian distance between the point $\underline{y}$ and the image of point $\underline{x}$ with respect to the plane of equation $x_3 = 0$.

With this choice for $\hat{G}_b(\underline{x},\underline{y})$, the sound pressure at all points of Ω is simply given by

$$\hat{p}(\underline{x}) = i\omega\rho_0 \hat{Q}_s \hat{G}_b(\underline{x},\underline{x}_0) = i\omega\rho_0 \hat{Q}_s \left(\frac{\exp(ikr)}{4\pi r} + \frac{\exp(ikr')}{4\pi r'} \right) \tag{7.222}$$

where $r = \sqrt{(x_1 - x_{01})^2 + (x_2 - x_{02})^2 + (x_3 - x_{03})^2}$ and $r' = \sqrt{(x_1 - x_{01})^2 + (x_2 - x_{02})^2 + (x_3 + x_{03})^2}$.

Consider now an absorbing ground characterized by its normalized acoustic admittance ($\hat{\beta} = \rho_0 c_0 / \hat{Z}_s$) such that

$$\frac{\partial \hat{p}}{\partial n_y} + ik\hat{\beta}(\underline{y}) = 0 \quad \forall \underline{y} \in \partial\Omega \tag{7.223}$$

For a monopole at point $\underline{x}_0$ and with the choice of the free-field Green's function, the sound pressure field above the ground is given by

$$\hat{p}(\underline{x}) = \int_{\partial\Omega} \hat{p}(\underline{y}) \left[\frac{\partial \hat{G}(\underline{x},\underline{y})}{\partial n_y} + ik\hat{\beta}(\underline{y})\hat{G}(\underline{x},\underline{y}) \right] dS_y + i\omega\rho_0 \hat{Q}_s \hat{G}(\underline{x},\underline{x}_0) \quad \forall \underline{x} \in \Omega \tag{7.224}$$

Now, choosing a Green's function satisfying

$$\frac{\partial \hat{G}(\underline{x},\underline{y})}{\partial n_y} + ik\hat{\beta}(\underline{y})\hat{G}(\underline{x},\underline{y}) = 0 \quad \forall \underline{y} \in \partial\Omega \tag{7.225}$$

The ground normalized acoustic admittance is related to the oblique incidence reflection coefficient $\hat{\mathfrak{R}}(\theta)$. It can be shown that

$$\hat{G}(\underline{x},\underline{y}) = \frac{\exp(ikr)}{4\pi r} + \hat{\mathfrak{R}}(\theta)\frac{\exp(ikr')}{4\pi r'}; \quad \cos\theta = \frac{x_{30}}{r} \tag{7.226}$$

and the sound pressure above the ground is then given by

$$\hat{p}(\underline{x}) = i\omega\rho_0\hat{Q}_s\hat{G}(\underline{x},\underline{x}_0) = i\omega\rho_0\hat{Q}_s\left(\frac{\exp(ikr)}{4\pi r} + \hat{\mathfrak{R}}(\theta)\frac{\exp(ikr')}{4\pi r'}\right) \tag{7.227}$$

In both the hard and absorbing ground cases, the pressure field in Ω at a receiver $R(\underline{x})$ is found as the superposition of a direct field radiated by the monopole and a reflected field by the ground. The reflected field can be seen as the field generated by the source $S(\underline{x}_0)$ located at a virtual point R' image of $R(\underline{x})$ through $\partial\Omega$. Alternatively, it can be seen as the field generated by the virtual source S' image of $S(\underline{x}_0)$ through $\partial\Omega$ since the distance r' between the source $S(\underline{x}_0)$ and the image receiver R' is identical to the distance between image source S' and the receiver $R(\underline{x})$ (see Figure 7.21). In the presence of ground absorption, the reflected field is multiplied by the reflection coefficient of the ground. This is the basis of the image sources method (Figure 7.22). It consists in replacing a problem involving sound sources radiating in a domain with boundary conditions with an equivalent problem without any boundary conditions involving the original source

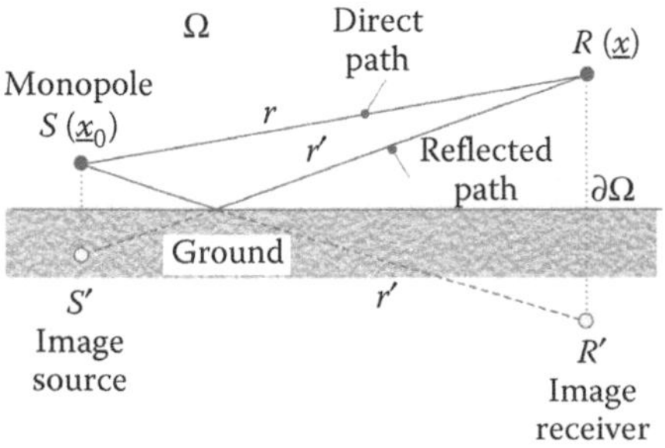

Figure 7.21 Sound radiation of a monopole above the ground.

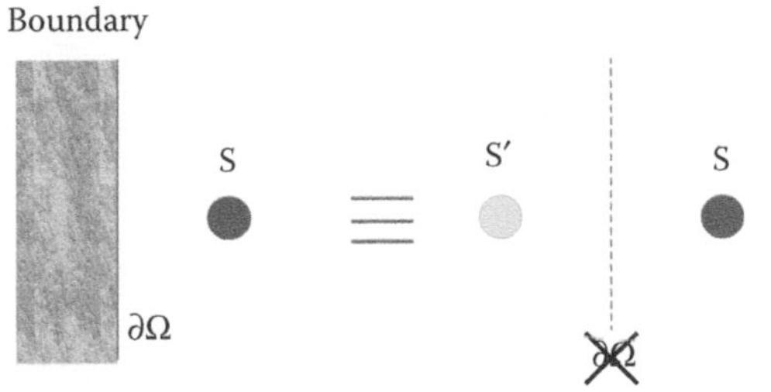

Figure 7.22 Image sources method: S is the physical source, S′ is the virtual source image of S with respect to the boundary. The problem of the source in the presence of the boundary is equivalent to that of the source S in the presence of S′ without any boundary.

and a collection of image sources. For plane boundaries, an image source is defined as the symmetric source of the original source with respect to these boundaries (see Figure 7.22). Each image source can also have its own images. Let us assume that the sound propagates as rays whose energies decrease because of geometric divergence, absorption by boundaries, and atmospheric attenuation. Then for a given observer, a ray specularly (mirror-like) reflected on a plane seems to be coming from the corresponding image source.

7.13.3 Sound radiation from a pulsating sphere

Consider a pulsating sphere of radius a vibrating with uniform normal velocity $\hat{v}_n$ and radiating in free space Ω (Figure 7.23). Its surface is referred to as $\partial\Omega$. The sound pressure $\hat{p}$ satisfies

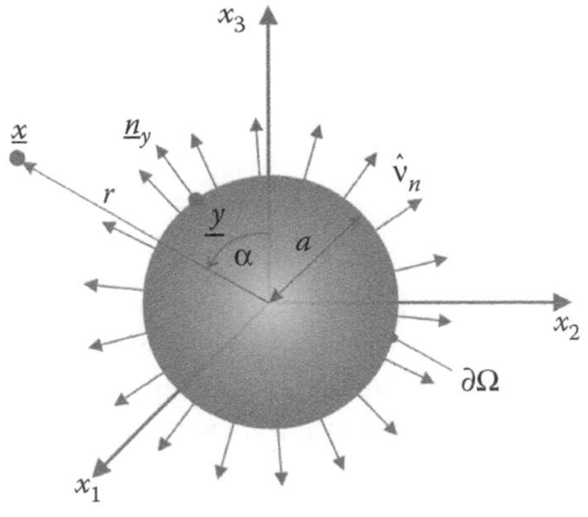

Figure 7.23 Pulsating sphere radiating in free space.

$$\begin{cases} \nabla^2\hat{p} + k^2\hat{p} = 0 & \text{in } \Omega \\ \dfrac{\partial \hat{p}}{\partial n} = -i\rho_0\omega\hat{v}_n & \text{over } \partial\Omega \end{cases} \tag{7.228}$$

This boundary value problem can be solved using a direct variational formulation of the SBIE equation (7.40), following the approach described in Section 7.7 (see also Chapter 8 for the elastic sphere case). Applying Equation 7.40 at a point on surface $\partial\Omega$,* the variation of the associated functional can be written as

$$\begin{aligned} \delta\Pi(\hat{p}) = & \int_{\partial\Omega}\int_{\partial\Omega} \hat{p}(\underline{y})\frac{\partial^2\hat{G}(\underline{x},\underline{y})}{\partial n_x \partial n_y}\delta\hat{p}(\underline{x})dS_y dS_x \\ & + \int_{\partial\Omega}\int_{\partial\Omega} i\rho_0\,\omega\hat{v}_n(\underline{y})\frac{\partial\hat{G}(\underline{x},\underline{y})}{\partial n_x}\delta\hat{p}(\underline{x})dS_y dS_x \\ & + \frac{1}{2}\int_{\partial\Omega} i\rho_0\omega\hat{v}_n(\underline{x})\delta\hat{p}(\underline{x})dS_x = 0 \end{aligned} \tag{7.229}$$

This leads to the following linear system:

$$[\hat{\mathcal{D}}]\{\hat{p}\} = -i\omega\rho_0\left([\hat{\mathcal{C}}_{up}^{(2)}] + \frac{1}{2}[\mathrm{C}_{up}^{(2)}]\right)\{\hat{v}_n\} \tag{7.230}$$

where matrices $[\hat{\mathcal{D}}]$ is given by Equation 7.194 and

$$\begin{aligned} & \int_{\partial\Omega}\int_{\partial\Omega} \hat{v}_n(\underline{y})\frac{\partial\hat{G}(\underline{x},\underline{y})}{\partial n_x}\delta\hat{p}(\underline{x})dS_y dS_x \Rightarrow \langle\delta\hat{p}\rangle[\hat{\mathcal{C}}_{up}^{(2)}]\{\hat{v}_n\} \\ & \int_{\partial\Omega} \hat{v}_n(\underline{x})\delta\hat{p}(\underline{x})dS_x \Rightarrow [\mathrm{C}_{up}^{(2)}]\{\hat{v}_n\} \end{aligned} \tag{7.231}$$

Equation 7.228 can also be solved using an indirect variational formulation as described in Section 7.7. Since we have a Newmann boundary condition, Equation 7.58 can be used. Taking the normal derivative of Equation 7.58 (in the Hadamard finite part sense) and applying the boundary condition (second line of Equation 7.228), we get

* $C^+ = 1/2$ since the point $\underline{x}$ is actually a Gauss point located inside a boundary element which has a unique normal at an interior point.

$$\delta\Pi(\hat{\mu}) = \int_{\partial\Omega}\int_{\partial\Omega} \hat{\mu}(\underline{y}) \frac{\partial^2 \hat{G}(\underline{x},\underline{y})}{\partial n_x \partial n_y} \delta\hat{\mu}(\underline{x}) dS_y dS_x + \int_{\partial\Omega} i\rho_0 \omega \hat{v}_n(\underline{x}) \delta\hat{\mu}(\underline{x}) dS_x = 0 \tag{7.232}$$

which leads to the following system:

$$[\hat{\mathcal{D}}]\{\hat{\mu}\} = -i\omega\rho_0 [C^{(2)}_{u\mu}]\{\hat{v}_n\} \tag{7.233}$$

where matrix $[\hat{\mathcal{D}}]$ is given by Equation 7.194 and

$$\int_{\partial\Omega} \hat{v}_n(\underline{x}) \delta\hat{\mu}(\underline{x}) dS_x \Rightarrow [C^{(2)}_{u\mu}]\{\hat{v}_n\} \tag{7.234}$$

The sound pressure $\hat{p}^+$ can then be recovered using

$$\hat{p}^+(\underline{x}) = C^-(\underline{x})\hat{\mu}(\underline{x}) + \int_{\partial\Omega} \hat{\mu}(\underline{y}) \frac{\partial \hat{G}(\underline{x},\underline{y})}{\partial n_y} dS(\underline{y}) \tag{7.235}$$

If point $\underline{x}$ is not on $\partial\Omega$, $C^- = 0$. If point $\underline{x}$ corresponds to a node on $\partial\Omega$, C^- needs to be computed using, for example, the third line of Equation 7.31 since the normal is not necessarily defined at a node. As discussed in Section 7.10, we are going to see that irregular frequencies appear in the sound pressure spectrum when Equation 7.230 or 7.233 is solved. To remove these numerical artifacts, the impedance coating method presented in Section 7.10.6 is applied to our problem. We then have to solve Equation 7.193. The sound pressure $\hat{p}^+$ can then be recovered using Equation 7.186 when point $\underline{x}$ is on $\partial\Omega$ and using Equation 7.184 when it is not.

In what follows, the three above-mentioned methods referred to as "direct," "indirect," and "impedance-coating" will be compared. Since an analytical solution for the sound pressure field radiated by a pulsating sphere is known (see e.g., Kinsler and Frey 2000), we are going to use this solution as a reference in our comparisons. The sound pressure radiated by a pulsating sphere only depends on the distance r and is given by

$$\hat{p}(r) = \rho_0 c_0 \hat{v}_n \frac{ika^2}{r} \frac{\exp(-ik(r-a))}{1+ika} \tag{7.236}$$

In order to compare the four approaches, we evaluate the sound pressure at one point on the boundary $\partial\Omega$ (see Figure 7.24), one point close to the boundary $\partial\Omega$ (see Figure 7.25), and one point far from the boundary $\partial\Omega$.

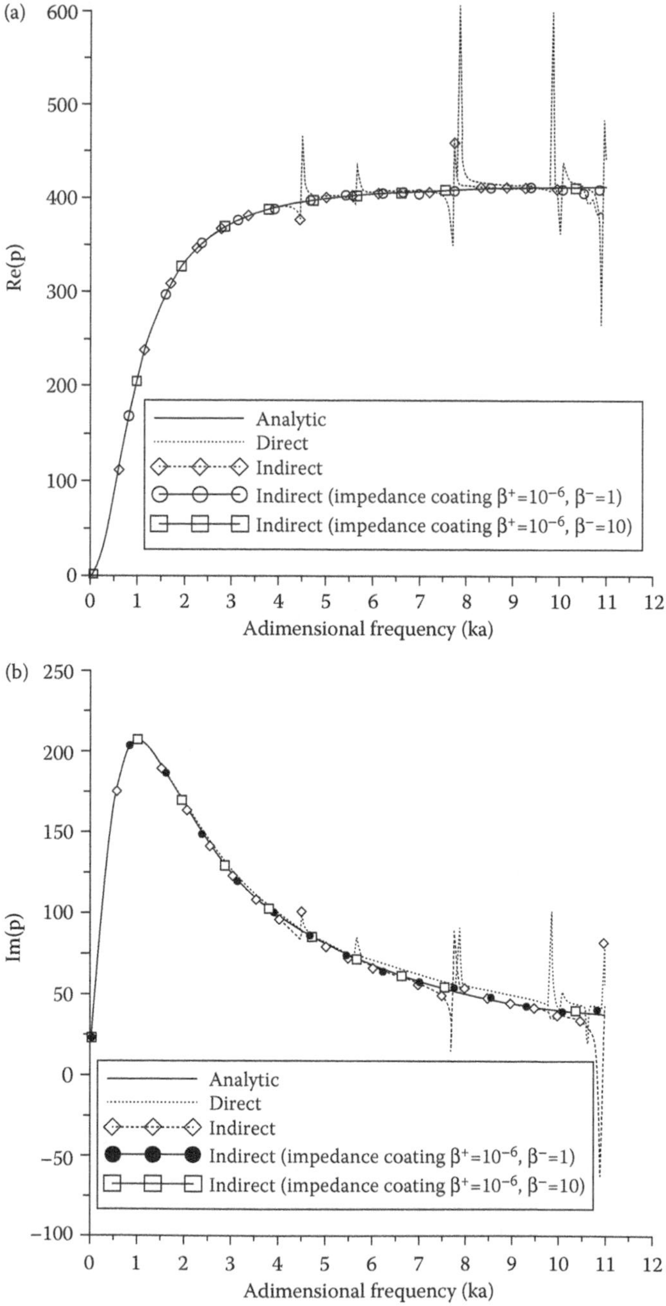

Figure 7.24 Parietal pressure radiated by a pulsating sphere radiating in free space. Comparisons between four approaches. (a) Real part and (b) imaginary part.

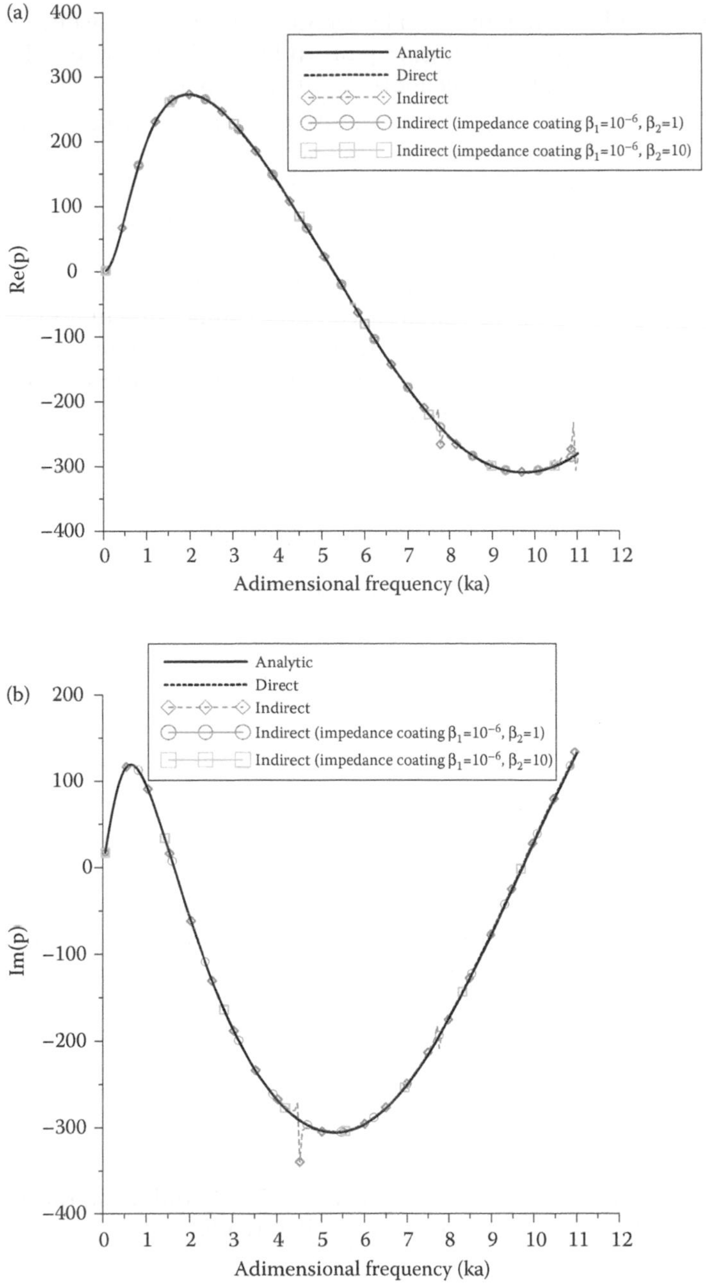

Figure 7.25 Pressure radiated at r = 0.4 m by a pulsating sphere radiating in free space. Comparisons between four approaches. (a) Real part and (b) imaginary part.

The considered sphere has a radius $a = 0.3$ m, a normal surface velocity $\hat{v}_n = 1\text{ms}^{-1}$, and radiates in air. The irregular frequencies (spikes) can be clearly seen on the parietal pressure plots in Figure 7.24. These spikes occur both for the direct and indirect approach. The use of the impedance-coating technique with values of $\hat{\beta}^+ = 10^{-6}$ and $\hat{\beta}^- = 1$ allows us to eliminate these artifacts. Actually, numerical experiments show that values of $\hat{\beta}^+ = 10^{-6}$ and $|\hat{\beta}^-| \geq 1$ yield an excellent agreement with the analytical solution.

For the sound pressure radiated in the surrounding fluid (Figure 7.25), the irregular frequencies are much less visible and only appear at a few locations for the indirect method (not for the direct method). This is due to the fact that in the direct approach, the second term appearing in Equation 7.40 dominates so that errors on the sound pressure do not influence the result.

The irregular frequencies are not visible in the plot of the sound pressure radiated in the far field (Figure 7.25) whether the direct or indirect method is used (Figure 7.26).

7.13.4 Scattering of a plane wave by a sphere

Let us consider the case of a rigid sphere of radius a excited by an incoming oblique incidence plane wave $\hat{p}_{inc}(\underline{x}) = \hat{A}\exp(-i\underline{k}.\underline{x})$. Its surface is referred to as $\partial\Omega$ (see Figure 7.27).

The total sound pressure $\hat{p} = \hat{p}_{sc} + \hat{p}_{inc}$ satisfies

$$\begin{cases} \nabla^2\hat{p} + k^2\hat{p} = 0 & \text{in } \Omega^+ \\ \dfrac{\partial\hat{p}}{\partial n} = 0 & \text{over } \partial\Omega \end{cases} \tag{7.237}$$

We are interested in calculating the pressure field scattered by the sphere.

The associated direct SBIE for the total pressure field $\hat{p}(\underline{x})$ is given by Equation 7.51, which using the boundary condition in Equation 7.237 can be rewritten as

$$C^+(\underline{x})\,\hat{p}(\underline{x}) = \int_{\partial\Omega} \hat{p}(\underline{y})\frac{\partial\hat{G}(\underline{x},\underline{y})}{\partial n_y}\,dS_y + \hat{p}_{inc}(\underline{x}) \tag{7.238}$$

Taking the normal derivative of Equation 7.238 and using the approach described in Section 7.7, the associated VBIE reads

$$\int_{\partial\Omega}\int_{\partial\Omega} \delta\hat{p}(\underline{x})\,\hat{p}(\underline{y})\,\frac{\partial^2\hat{G}(\underline{x},\underline{y})}{\partial n_x \partial n_y}\,dS_x dS_y + \int_{\partial\Omega} \delta\hat{p}(\underline{x})\,\frac{\partial\hat{p}_{inc}(\underline{x})}{\partial n_x}\,dS_x = 0 \tag{7.239}$$

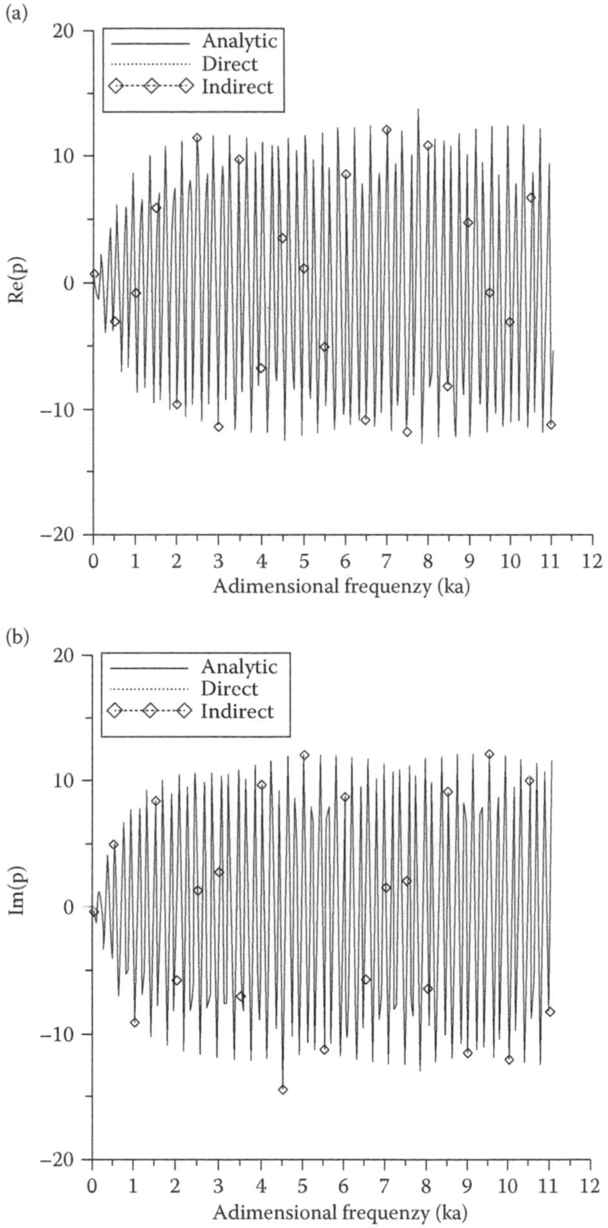

Figure 7.26 Pressure radiated at r = 10 m by a pulsating sphere radiating in free space. Comparisons between three approaches. (a) Real part and (b) imaginary part.

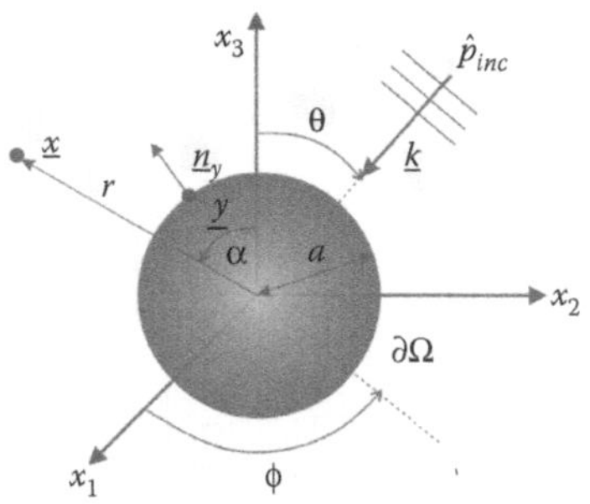

Figure 7.27 Rigid sphere excited by a plane wave.

which leads to the system

$$[\hat{\mathcal{D}}]\{\hat{p}\} = \{\hat{f}_{sc}\} \tag{7.240}$$

where matrix $[\hat{\mathcal{D}}]$ is given by Equation 7.194 and

$$-\int_{\partial\Omega} \delta\hat{p}(\underline{x}) \frac{\partial \hat{p}_{inc}(\underline{x})}{\partial n_x} dS_x \Rightarrow \langle \delta\hat{p} \rangle \{\hat{f}_{sc}\} \tag{7.241}$$

The scattered pressure field is simply $\hat{p}_{sc}(\underline{x}) = \hat{p}(\underline{x}) - \hat{p}_{inc}(\underline{x})$.

As in the previous section, we are going to see that irregular frequencies appear in the sound pressure spectrum when Equation 7.240 is solved. Alternatively to the direct variational formulation, the indirect variational formulation can be used. The same system as Equation 7.240 is obtained

$$[\hat{\mathcal{D}}]\{\hat{\mu}\} = \{\hat{f}_{sc}\} \tag{7.242}$$

The sound pressure can then be recovered from

$$C^{+}(\underline{x})\, \hat{p}(\underline{x}) = \int_{\partial\Omega} \hat{\mu}(\underline{y}) \frac{\partial \hat{G}(\underline{x}, \underline{y})}{\partial n_y} dS_y + \hat{p}_{inc}(\underline{x}) \tag{7.243}$$

The use of Equation 7.243 allows us to eliminate these numerical artifacts. Note in passing that, in this example, the coating method is found very sensitive to the selection of $\hat{\beta}^+$ and $\hat{\beta}^-$; the best solution is obtained for small values of these two constants. This emphasizes the difficulties of deriving a universal formulation for solving the non-uniqueness problem.

The numerical results can be compared with a reference solution, which can be found in Junger and Feit (1972). The analytical derivation of the solution for an elastic sphere is given in Appendix 8A. Using the notations of Figure 7.27 and assuming a temporal dependency $\exp(-i\omega t)$,* $\hat{p}_{inc}(\underline{x})$ can be expressed as $\hat{p}_i \exp(i\underline{k}\cdot\underline{x}) = \hat{p}_i \exp(ikr\cos(\pi+\alpha-\theta))$ where α is the angle between the x_3-axis and the vector $\underline{x}$. The scattered pressure at a point $\underline{x}$ is then given by

$$\hat{p}_{sc}(r,\alpha) = -\hat{P}_i \sum_n (2n+1) i^n P_n(\cos(\pi+\alpha-\theta)) \frac{j_n'(ka)}{h_n'(ka)} h_n(kr) \tag{7.244}$$

and the total pressure field can be written as

$$\begin{aligned}\hat{p}(r,\alpha) &= \hat{p}_{inc}(r,\alpha) + \hat{p}_{sc}(r,\alpha)\\ &= \hat{P}_i \sum_n (2n+1) i^n P_n(\cos(\pi+\alpha-\theta)) j_n(kr)\\ &\quad - \hat{P}_i \sum_n (2n+1) i^n P_n(\cos(\pi+\alpha-\theta)) \frac{j_n'(ka)}{h_n'(ka)} h_n(kr)\end{aligned} \tag{7.245}$$

where $P_n(\cos(\pi+\alpha-\theta))$, j_n, and h_n denote the Legendres polynomial of order n, the spherical Bessel function of order n, and the spherical Hankel function of order n, respectively. j_n' and h_n' are the first derivatives with respect to the argument of j_n and h_n, respectively.

In order to compare the three approaches (analytical, direct VBEM, and indirect VBEM), consider a sphere of radius $a = 0.3$ m excited by a plane wave propagating along the positive x_3 direction (incidence angle $\theta = \pi$). We evaluate the scattered sound pressure at three points on the sphere: $A(r=0.3,\alpha=\pi)$, $B(r=0.3,\alpha=0)$, $E(r=0.3,\alpha=\pi/2)$ and at two exterior points away from the surface $C(r=0.4,\alpha=\pi)$ and $D(r=0.4,\alpha=0)$ (see Figure 7.28).

Different boundary element meshes† are used for the calculations: 1680 4-noded quadrangles, 6512 4-noded quadrangles, and 744 8-noded quadrangles (see Figure 7.29). The first two 4-noded quadrangles meshes

* Recall that the convention in this book is $\exp(i\omega t)$. To compare the numerical results with the analytical solution the sign of the imaginary part of the latter should be inverted.

† Note that symmetries are not used in the calculation.

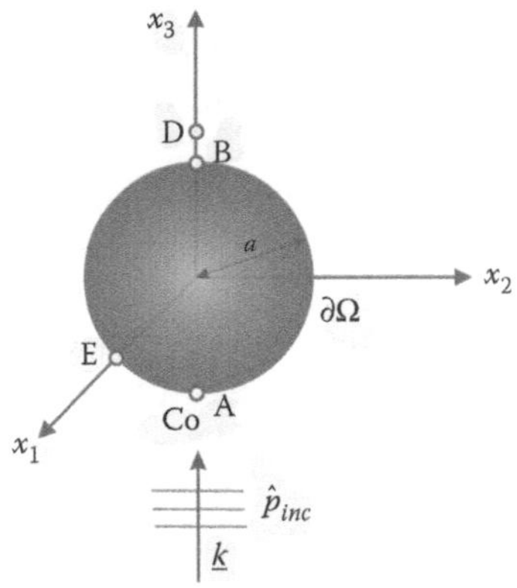

Figure 7.28 Studied configuration.

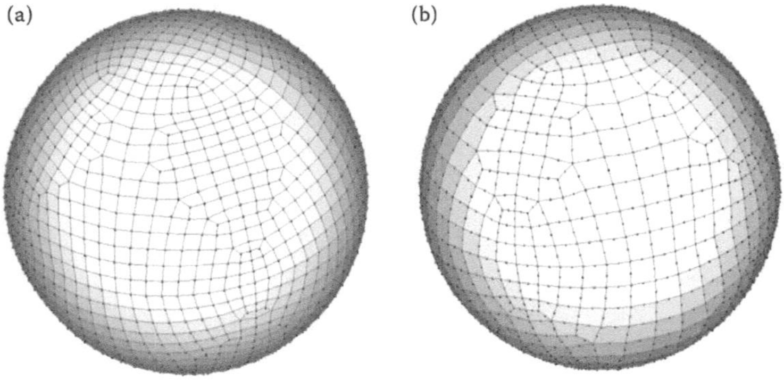

Figure 7.29 Boundary element meshes used for the calculation. (a) 1680 quad4 and (b) 744 quad8.

correspond, respectively, to 6 elements per wavelength (at the maximum frequency) and 12 elements per wavelength. For the quadratic mesh, 4 elements per wavelength are used. The results are shown in Figures 7.30 through 7.34. Irregular frequencies can be clearly seen on the parietal pressure plots when the direct VBEM is used. However, for this configuration they are not visible in the pressure plots in the external medium. We can observe that the use of the indirect VBEM allows for regularizing the problem. The spikes disappear. The above-mentioned meshing criteria lead to a very good convergence of the numerical calculation toward the analytical solution. Refined meshes and the use of quadratic elements improve the convergence as expected.

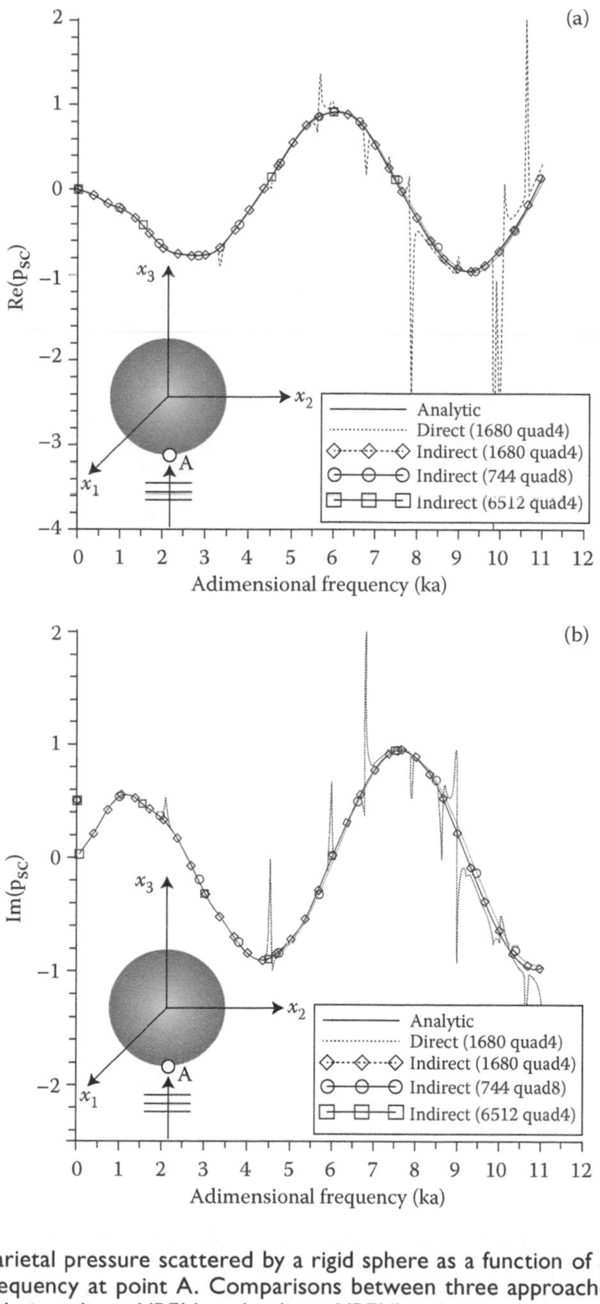

Figure 7.30 Parietal pressure scattered by a rigid sphere as a function of adimensional frequency at point A. Comparisons between three approaches (analytical solution, direct VBEM, and indirect VBEM) and two topologies of boundary elements 4-noded (quad4) and 8-noded (quad8) quadrangles. (a) Real part and (b) imaginary part.

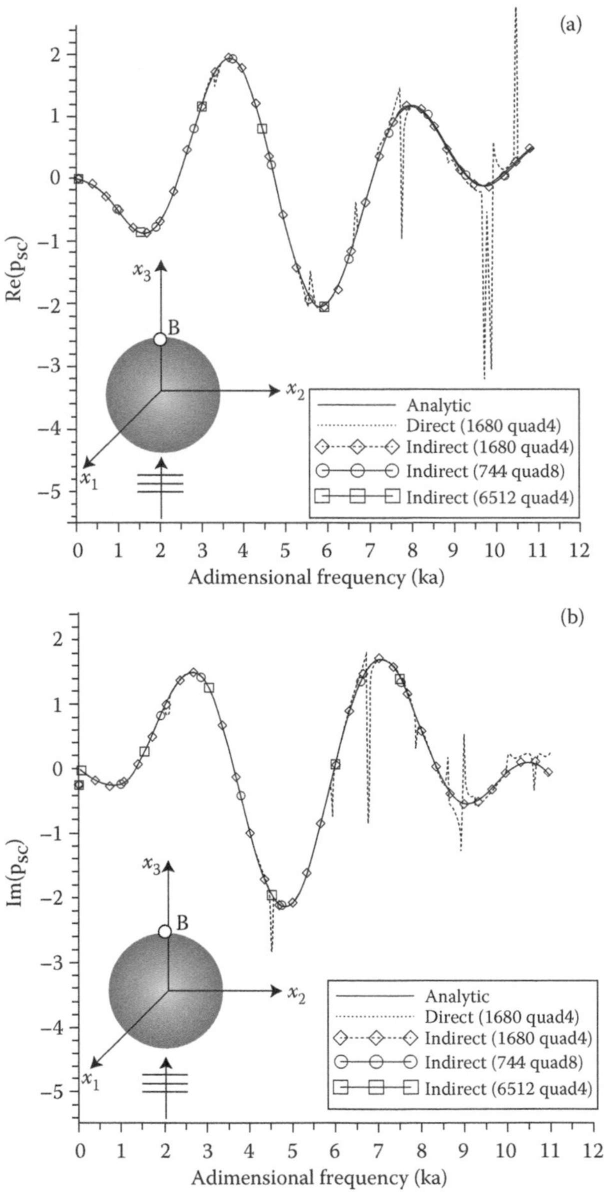

Figure 7.31 Parietal pressure scattered by a rigid sphere as a function of adimensional frequency at point B. Comparisons between three approaches (analytical solution, direct VBEM, and indirect VBEM) and two topologies of boundary elements 4-noded (quad4) and 8-noded (quad8) quadrangles. (a) Real part and (b) imaginary part.

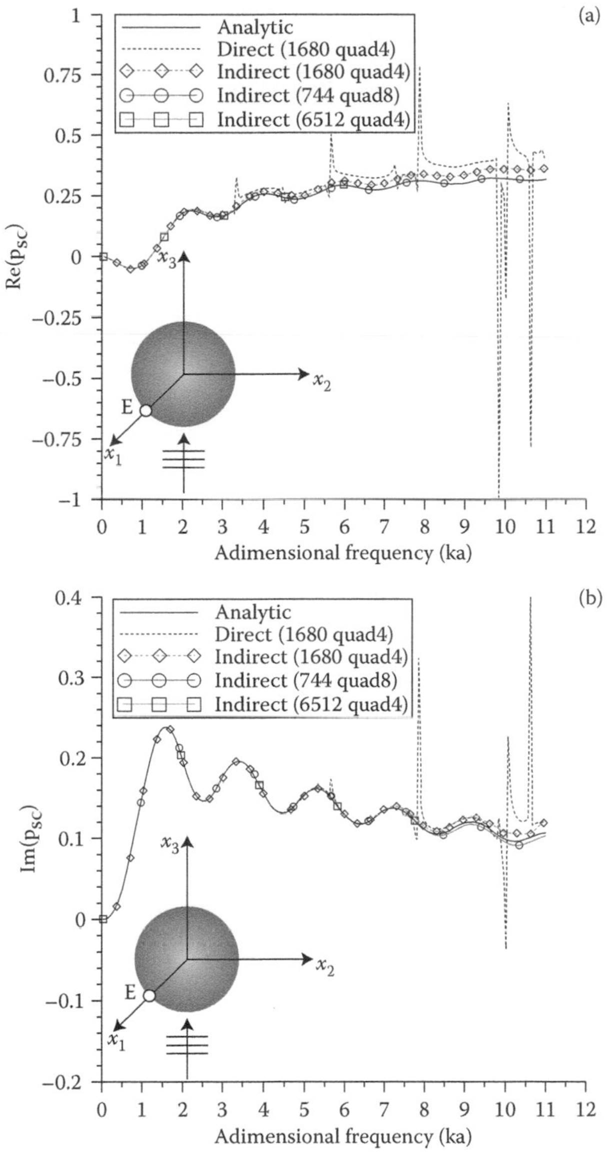

Figure 7.32 Parietal pressure scattered by a rigid sphere as a function of adimensional frequency at point E. Comparisons between three approaches (analytical solution, direct VBEM, and indirect VBEM) and two topologies of boundary elements 4-noded (quad4) and 8-noded (quad8) quadrangles. (a) Real part and (b) imaginary part.

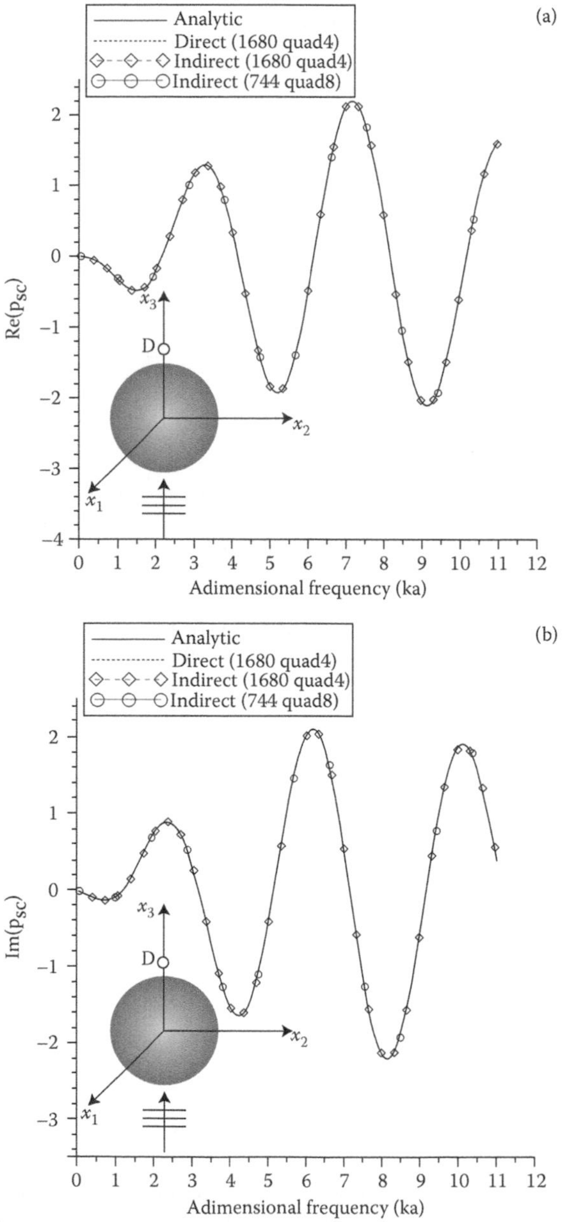

Figure 7.33 Pressure scattered by a rigid sphere as a function of adimensional frequency at point D. Comparisons between three approaches (analytical solution, direct, and indirect VBEM) and two topologies of boundary elements 4-noded (quad4) and 8-noded (quad8) quadrangles. (a) Real part and (b) imaginary part.

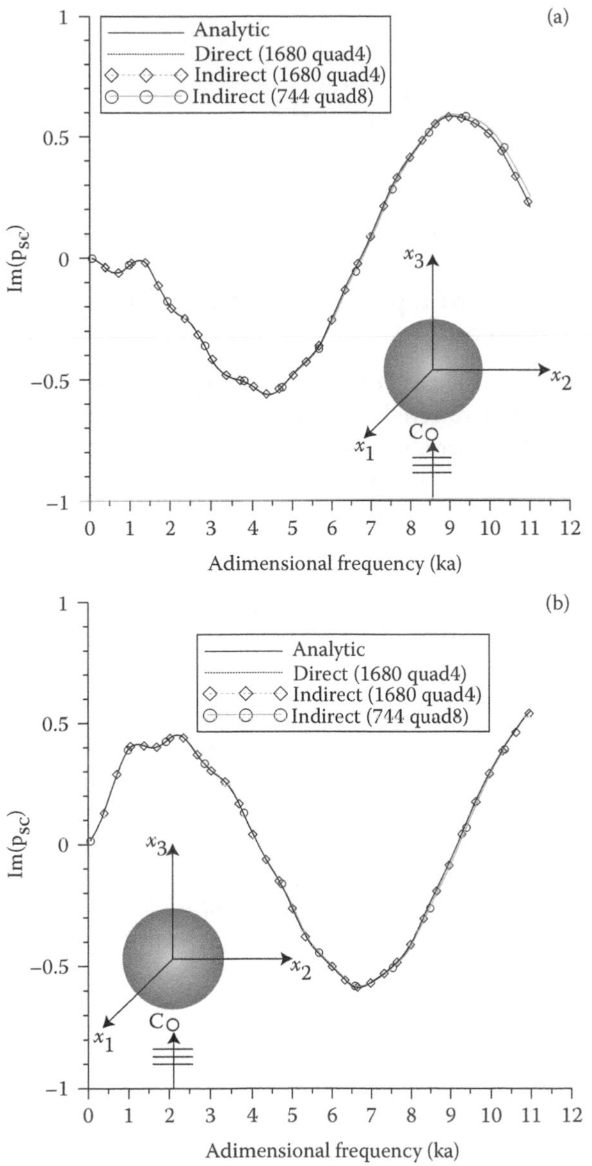

Figure 7.34 Pressure scattered by a rigid sphere as a function of adimensional frequency at point C. Comparisons between three approaches (analytical solution, direct VBEM, and indirect VBEM) and two topologies of boundary elements 4-noded (quad4) and 8-noded (quad8) quadrangles. (a) Real part and (b) imaginary part.

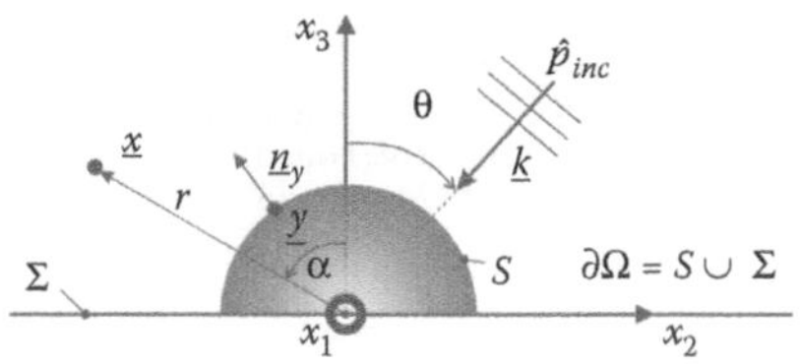

Figure 7.35 Rigid half-sphere excited by a plane wave.

7.13.5 Scattering of a plane wave by a baffled half-sphere

Let us now consider the case of a rigid half-sphere lying on a rigid infinite ground (rigid baffle) referred to as Σ (see Figure 7.35). Let S denote the surface of the half–sphere, which is excited with an incoming oblique incidence plane wave. The boundary $\partial\Omega$ is equal to $S \cup \Sigma$.

The associated boundary integral equation is given by (see Section 7.8)

$$C^{+}(\underline{x})\,\hat{p}(\underline{x}) = \int_S \hat{p}(\underline{y})\frac{\partial \hat{G}_b(\underline{x},\underline{y})}{\partial n_y}\,dS_y + \hat{p}_{inc}(\underline{x}) + \hat{p}_{ref}(\underline{x}) \tag{7.246}$$

where $\hat{G}_b(\underline{x},\underline{y})$ is the baffled Green's function and $\hat{p}_{ref}(\underline{x}) = \hat{A}\exp(-i\underline{k}^{ref}\cdot\underline{x})$ is the reflected pressure field induced by the presence of the baffle only ($\underline{k}^{ref}$ is the reflected wave vector). The associated VBIE can be written as (see Section 7.8 and Figure 7.36)

$$\int_S\int_S \delta\hat{p}(\underline{x})\,\hat{p}(\underline{y})\,\frac{\partial^2 \hat{G}_\infty(\underline{x},\underline{y})}{\partial n_x \partial n_y}\,dS_y dS_x - \int_{S_{im}}\int_S \delta\hat{p}(\underline{x})\,\hat{p}(\underline{y})\,\frac{\partial^2 \hat{G}_\infty(\underline{x},\underline{y})}{\partial n_x \partial n_y}\,dS_y dS_x$$
$$+ \int_S \delta\hat{p}(\underline{x})\left(\frac{\partial \hat{p}_{inc}(\underline{x})}{\partial n_x} + \frac{\partial \hat{p}_{ref}(\underline{x})}{\partial n_x}\right)dS_x = 0 \tag{7.247}$$

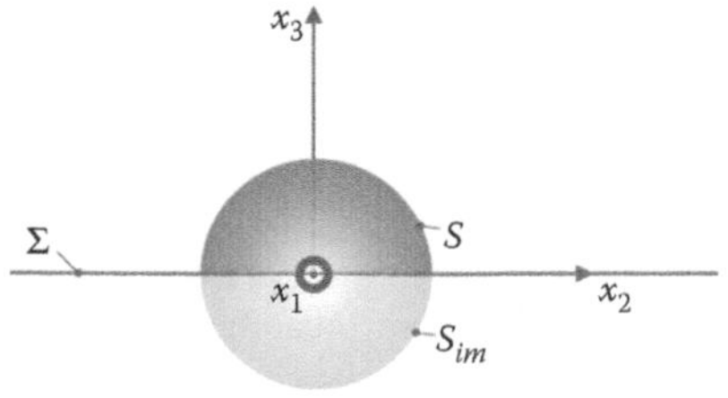

Figure 7.36 Principle of symmetry for the half-rigid sphere above a rigid plane.

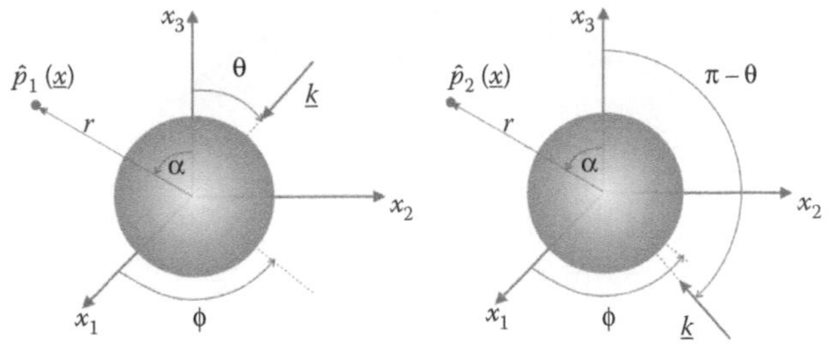

Figure 7.37 Equivalent problem.

The scattered pressure field is simply $\hat{p}_{sc}(\underline{x}) = \hat{p}(\underline{x}) - \hat{p}_{inc}(\underline{x}) - \hat{p}_{ref}(\underline{x})$. The irregular frequencies can be treated using the indirect VBEM approach like in the previous section.

Again, we are going to compare the results obtained using the direct and indirect VBEM for different meshes with the analytical solution, which can be obtained using the symmetry of the problem.* The scattered sound pressure (see Figure 7.37) is given by

$$\hat{p}_{sc,\frac{1}{2}sphere}(r,\alpha) = \hat{p}_{sc,sph}(r,\theta+\alpha) + \hat{p}_{sc,sph}(r,\pi+\alpha-\theta) \qquad (7.248)$$

In order to compare the three approaches (analytical, direct VBEM, and indirect VBEM), consider a half-sphere of radius $a = 0.3$ m excited by a

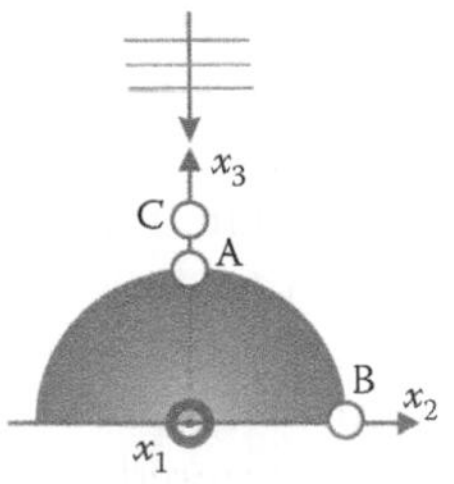

Figure 7.38 Studied configuration.

* Note that for the numerical approach, the pressure field scattered by the baffled half-rigid sphere can also be calculated as the sum of the scattered pressure fields $\hat{p}_{sc,1}(\underline{x})$ and $\hat{p}_{sc,2}(\underline{x})$ resulting from the interaction of the entire sphere and two plane waves of respective wave vectors symmetric with respect to the baffle plane (see Figure 7.37). $\hat{p}_{sc,1}(\underline{x})$ and $\hat{p}_{sc,2}(\underline{x})$ can be calculated using Equation 7.254 and $\hat{p}_{sc}(\underline{x}) = \hat{p}(\underline{x}) - \hat{p}_{inc}(\underline{x})$ where $\hat{p}_{inc}(\underline{x})$ corresponds to $\hat{p}_{inc,1}(\underline{x})$ and $\hat{p}_{inc,2}(\underline{x})$, respectively.

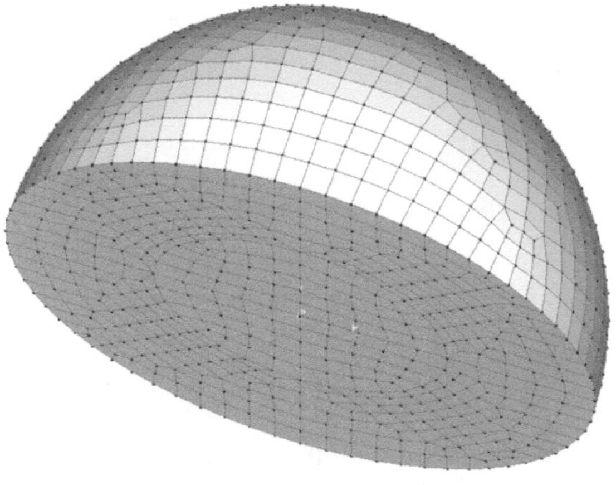

Figure 7.39 Mesh of the half-sphere.

plane wave propagating along the negative x_3 direction (incidence angle $\theta = 0$). We evaluate the scattered sound pressure at two points on the surface of the sphere: A($r = 0.3, \alpha = 0$), B($r = 0.3, \alpha = \pi/2$) and at another exterior point C($r = 0.4$, $\alpha = 0$) (see Figure 7.38).

Again different boundary element meshes are used for the calculations. Classic meshing criteria are utilized. For the direct VBEM, the half-sphere is discretized using 840 4-noded quadrangles. For the indirect VBEM 1328 4-noded quadrangles and 632 8-noded quadrangles are used. Note that when the indirect VBEM is used, the coefficient C^+ must be evaluated according to the procedure described in Section 7.8. This requires to mesh the bottom disk of the half-sphere (see Figure 7.39).

The results are shown in Figures 7.27 through 7.31. Excellent agreement is observed. However, irregular frequencies can be clearly seen on the parietal pressure plots when the direct VBEM is used (Figures 7.40 through 7.42).

Another comparison is given in Figure 7.43. It shows the polar plot of the modulus of the pressure scattered at a distance $3a$ from the center of a baffled hemisphere excited by an oblique incidence plane wave ($\theta = 45°$) for an adimensional frequency $ka = 1$. This configuration was also presented in Seybert and Wu (1989). Here we limit the comparison to the indirect VBEM and use the same mesh as previously described. A perfect match between the analytical results and the VBEM approach is observed.

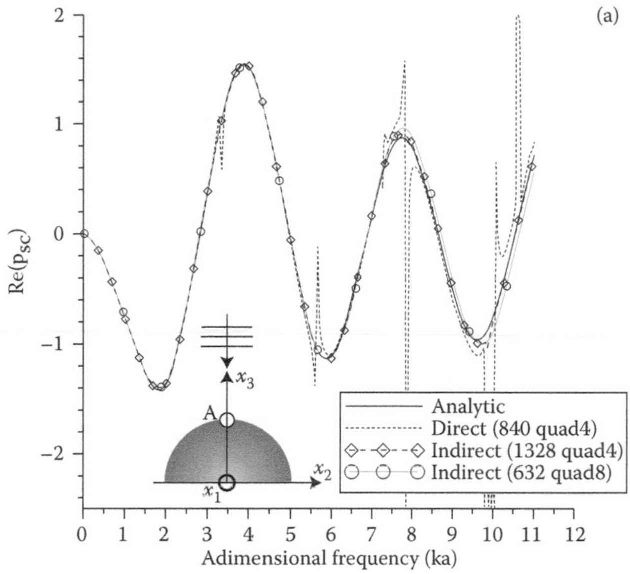

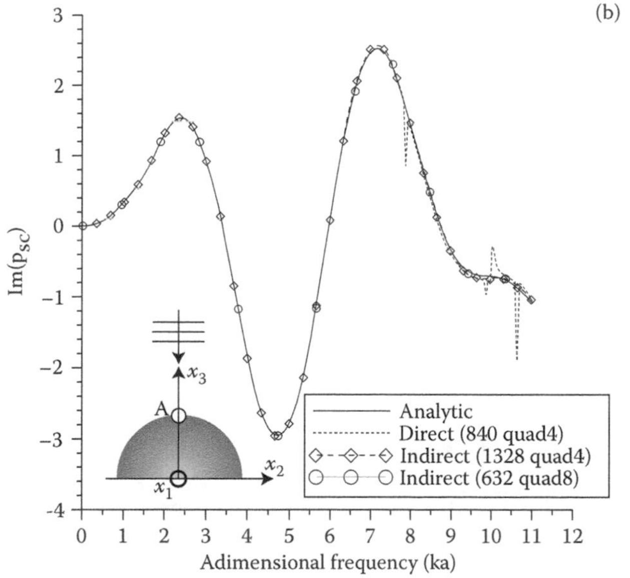

Figure 7.40 Pressure scattered by a half-rigid sphere as a function of adimensional frequency at point A. Comparisons between three approaches (analytical solution, direct VBEM, and indirect VBEM) and two topologies of boundary elements 4-noded (quad4) and 8-noded (quad8) quadrangles. (a) Real part and (b) imaginary part.

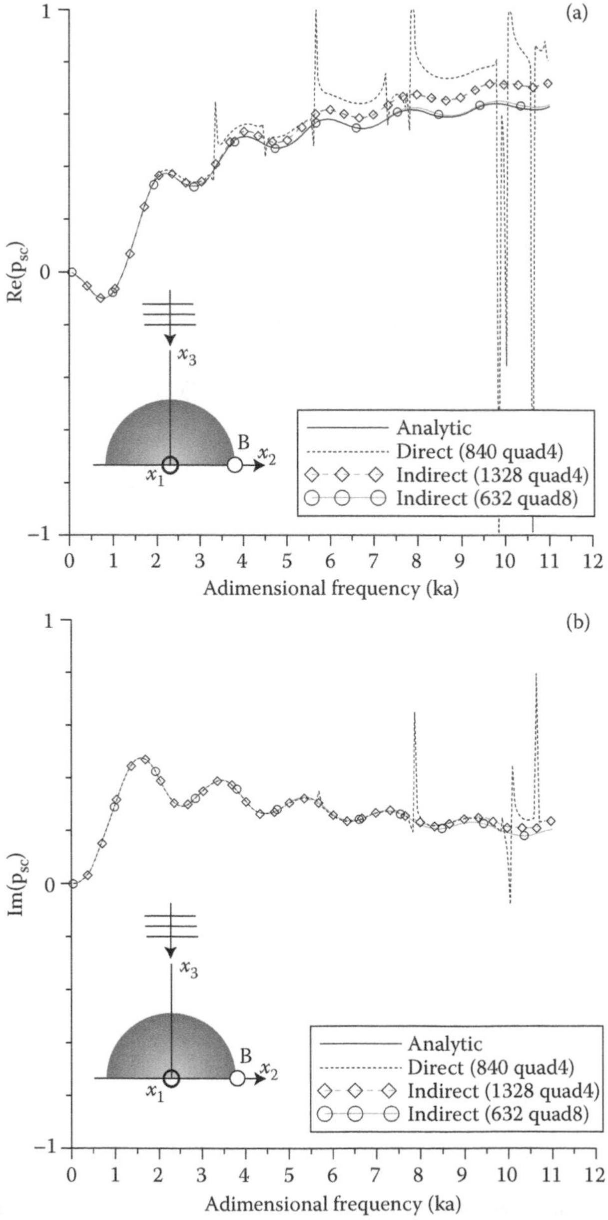

Figure 7.41 Pressure scattered by a half-rigid sphere as a function of adimensional frequency at point B. Comparisons between three approaches (analytical solution, direct VBEM, and indirect VBEM) and two topologies of boundary elements 4-noded (quad4) and 8-noded (quad8) quadrangles. (a) Real part and (b) imaginary part.

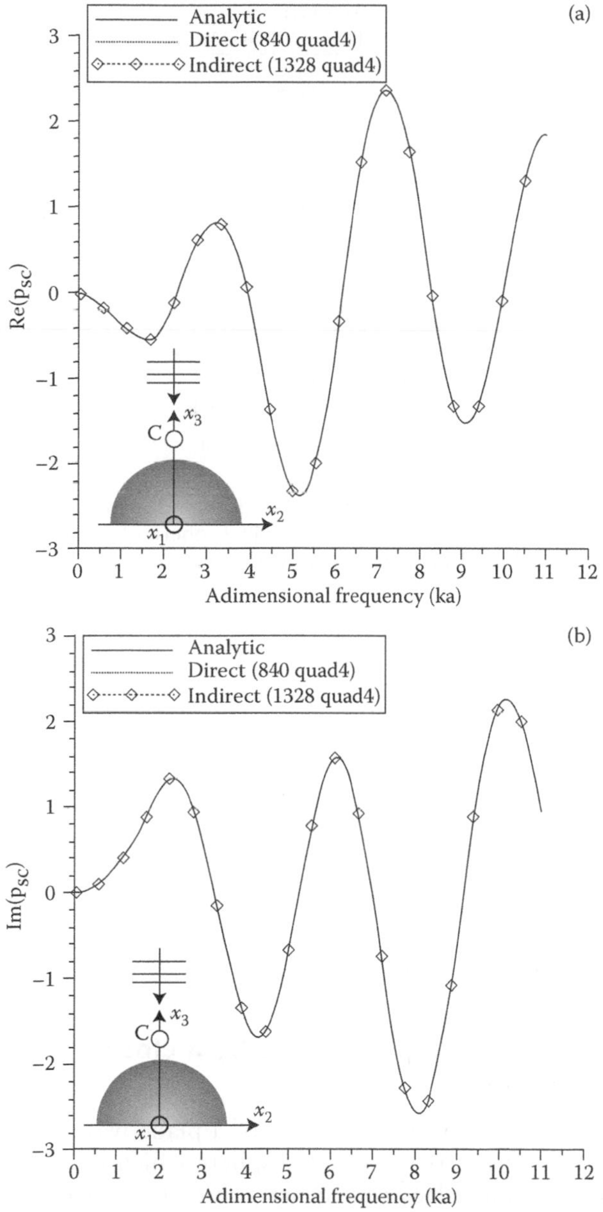

Figure 7.42 Pressure scattered by a half-rigid sphere as a function of adimensional frequency at point C. Comparisons between three approaches (analytical solution, direct VBEM, and indirect VBEM) and two topologies of boundary elements 4-noded (quad4) and 8-noded (quad8) quadrangles. (a) Real part and (b) imaginary part.

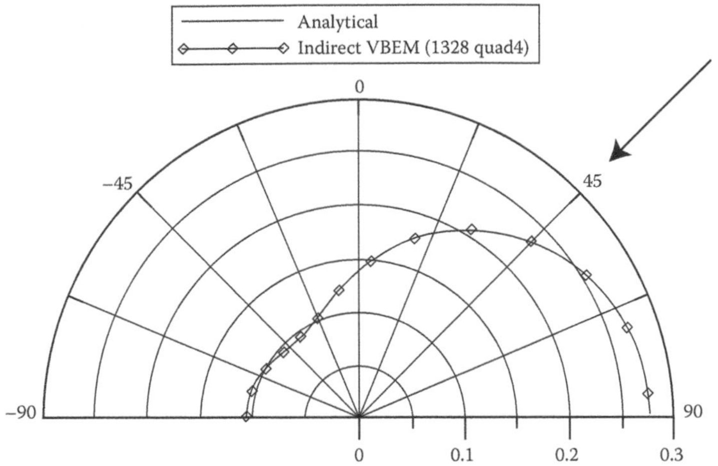

Figure 7.43 Comparison between numerical solutions and analytical results: calculation of the acoustic field scattered by a rigid sphere of radius $a = 0.3$ m at a distance $r = 3a$ and adimensional frequency $ka = 1$.

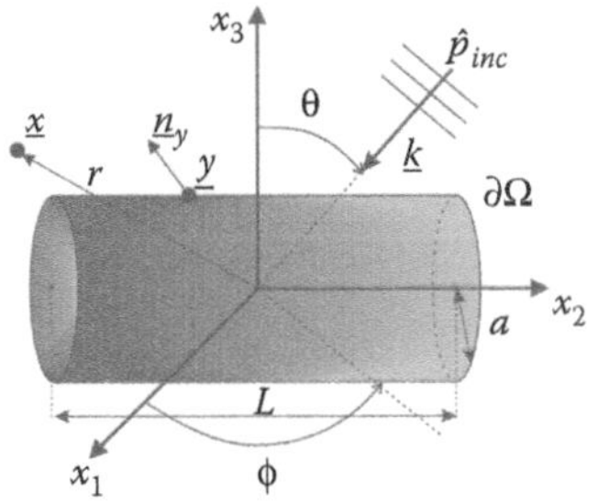

Figure 7.44 Rigid cylinder excited by a plane wave.

7.13.6 Scattering of a plane wave by a closed rigid cylinder

Consider now the case of the scattering of a plane wave by a closed rigid cylinder (Figure 7.44). Contrary to the sphere which is a regular surface,* this geometry exhibits edges and corner points. It is, therefore, interesting to investigate how the boundary element techniques perform in this case.

* Note that when the sphere is discretized into planar boundary elements, nodes are corner-like (the normal is not defined). However, the free coefficients which are required to evaluate the nodal parietal sound pressure are very close to 0.5 when the discretization scheme is small enough to approach the geometry of the sphere.

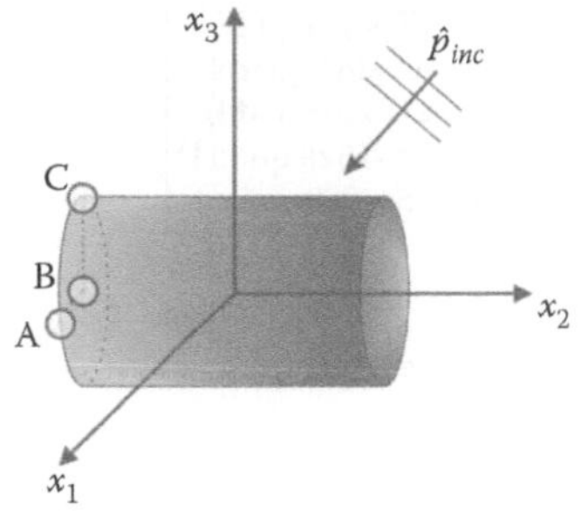

Figure 7.45 Studied configuration.

Let us consider the case of a rigid cylinder of radius a and length L excited by an incoming oblique incidence plane wave $\hat{p}_{inc}(\underline{x}) = \hat{A}\exp(-i\underline{k} \cdot x)$. Its surface is referred to as $\partial\Omega$. The equations that govern the acoustical behavior of the system are the same as in Section 7.13.4 and will not therefore be repeated.

The numerical results obtained with the direct VBEM and the indirect VBEM are compared for the scattered pressure at selected points on the cylinder boundary in order to see how the indirect VBEM allows for eliminating the irregular frequencies. We consider a cylinder of radius $a = 0.1$ m and length $L = 0.3$ m excited by a plane wave propagating along the direction defined by spherical angles $(\theta = \pi/4, \phi = \pi/4)$. We evaluate the scattered sound pressure at three points A, B, and C on the boundary $\partial\Omega$ (see Figure 7.45). Since no analytical solution is available for this case, the VBEM solutions are compared to a full FEM solution obtained with a commercial software (Virtual Lab 12, © LMS & Siemens), which uses an automatically matched layer (AML) approach to account for the radiation condition. The AML is an optimal implementation of the perfectly matched layer (PML) technique, which was mentioned in the introduction of this book. The PML is an alternative for emulating the Sommerfeld radiation condition in the numerical solution of exterior problems (wave radiation and scattering problems).

The results are shown in Figures 7.46 through 7.48. An excellent agreement is observed between the AML technique and the VBEM results. However, as for the sphere examples, the calculations using the direct VBEM exhibit spurious spikes around the resonance frequencies of the interior problem, whereas those based on the indirect VBEM allow for an elimination of these numerical artifacts. Additional calculations of pressure frequency spectra at points in the external medium (not shown here) reveal that, for this configuration, the irregular frequencies are not visible for these points. Finally, note that in the case of the direct VBEM, the quality of the prediction is sensitive to the mesh refinement.

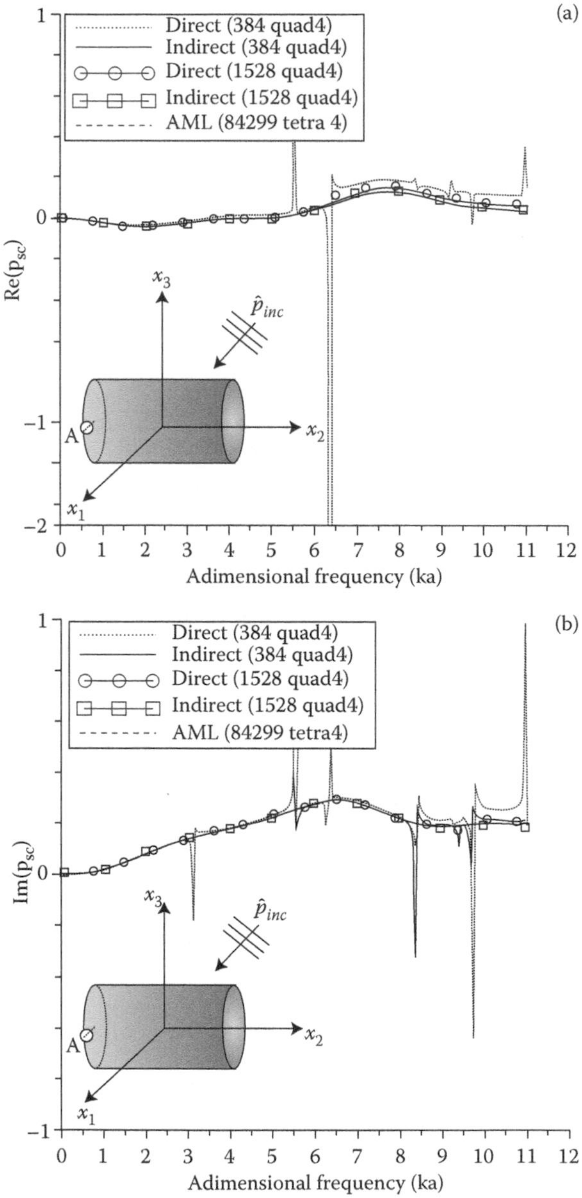

Figure 7.46 Pressure scattered by a cylinder as a function of adimensional frequency at point A. Comparisons between two approaches (direct and indirect VBEM) for two 4-noded boundary elements mesh refinements. (a) Real part and (b) Imaginary part.

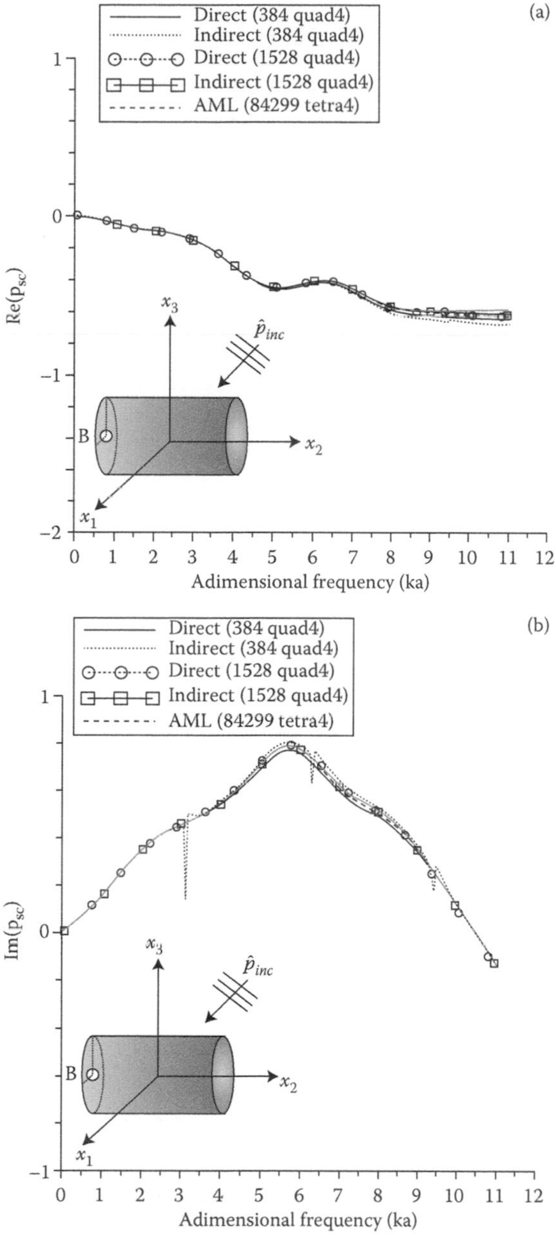

Figure 7.47 Pressure scattered by a cylinder as a function of adimensional frequency at point B. Comparisons between two approaches (direct and indirect VBEM) for two 4-noded boundary elements mesh refinements. (a) Real part and (b) imaginary part.

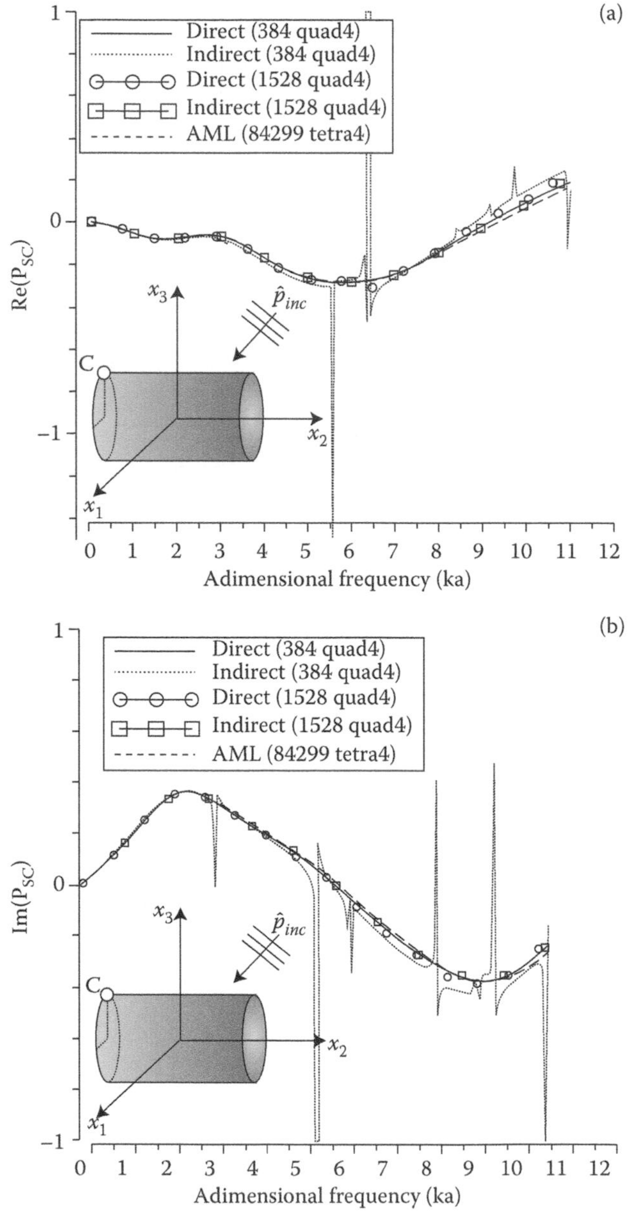

Figure 7.48 Pressure scattered by a cylinder as a function of adimensional frequency at point C. Comparisons between two approaches (direct and indirect VBEM) for two 4-noded boundary elements mesh refinements. (a) Real part and (b) imaginary part.

7.14 CONCLUSION

This chapter has provided a detailed description of the BEM in the context of acoustic applications. The various formulations (direct, indirect, and variational) have been presented with an emphasis on their numerical implementation and techniques to alleviate the difficulties associated with the calculation of singular integrals of various orders (weak to hypersingular). Several examples have been provided to demonstrate the efficiency of the method and issues related to the non-uniqueness problem. Chapter 8 will continue this analysis by tackling the fully-coupled fluid-structure interaction problem.

APPENDIX 7A: PROOF OF $C^{+}(\underline{x}) = 1 + \int_{\partial\Omega} \partial G_0(\underline{x}, \underline{y})/\partial n_y dS_y$

We start from the integral formulation of the exterior problem

$$\hat{p}(\underline{x}) = \int_{\partial\Omega} \left[\hat{p}(\underline{y}) \frac{\partial \hat{G}(\underline{x}, \underline{y})}{\partial n_y} - \hat{G}(\underline{x}, \underline{y}) \frac{\partial \hat{p}(\underline{y})}{\partial n_y} \right] dS_y \tag{7.249}$$

where $\underline{n}_y$ is the normal that points into the exterior domain. For a point $\underline{x}$ on $\partial\Omega$, we can write

$$\hat{p}(x) = \int_{\partial\Omega^*} \left[\hat{p}(\underline{y}) \frac{\partial \hat{G}(\underline{x}, \underline{y})}{\partial n_y} - \hat{G}(\underline{x}, \underline{y}) \frac{\partial \hat{p}}{\partial n_y} \right] dS_y + \hat{p}(\underline{x}) \lim_{\varepsilon \to 0} \int_{S_\varepsilon^-} \frac{\partial G_0(\underline{x}, \underline{y})}{\partial n_y} dS_y \tag{7.250}$$

so that

$$C^{+}(\underline{x}) = 1 - \lim_{\varepsilon \to 0} \int_{S_\varepsilon^-} \frac{\partial G_0(\underline{x}, \underline{y})}{\partial n_y} dS_y \tag{7.251}$$

In addition, we have shown that (cf. Equation 7.25)

$$C^{-}(\underline{x}) = 1 + \lim_{\varepsilon \to 0} \int_{S_\varepsilon^+} \frac{\partial G_0(\underline{x}, \underline{y})}{\partial n_y} dS_y \tag{7.252}$$

In Equation 7.251, the normal $\underline{n}_y$ is the outward normal to the interior domain, which is also the normal that points into the exterior domain.

Let $S_\varepsilon = S_\varepsilon^+ \cup S_\varepsilon^-$ be the surface of the sphere centered at point $\underline{x}$ and radius ε. Given that $\underline{x}$ is inside the sphere, we have

$$\int_{S_\varepsilon} \frac{\partial G_0}{\partial n_y} dS_y = -1 \tag{7.253}$$

we see that

$$C^+(\underline{x}) + C^-(\underline{x}) = 2 + \lim_{\varepsilon \to 0} \int_{S_\varepsilon^+} \frac{\partial G_0(\underline{x}, y)}{\partial n_y} dS_y - \lim_{\varepsilon \to 0} \int_{S_\varepsilon^-} \frac{\partial G_0(\underline{x}, y)}{\partial n_y} dS_y \tag{7.254}$$

Since the normal $\underline{n}_y$ of S_ε^- points inside Ω^-, it follows, by changing the sign of the normal that

$$C^+(\underline{x}) + C^-(\underline{x}) = 1 \tag{7.255}$$

Therefore, using the expression (7.31) for C^-, we get

$$C^-(\underline{x}) = -\int_{\partial\Omega} \frac{\partial G_0}{\partial n_y} dS_y \tag{7.256}$$

it follows that

$$C^+(\underline{x}) = 1 + \int_{\partial\Omega} \frac{\partial G_0}{\partial n_y} dS_y \tag{7.257}$$

In these expressions, $\underline{n}_y$ is the outward normal to the interior domain (inward normal for the exterior problem).

This completes the proof.

APPENDIX 7B: CONVERGENCE OF THE INTEGRAL INVOLVING THE NORMAL DERIVATIVE OF THE GREEN'S FUNCTION IN EQUATION 7.20

For a proof that in Equation 7.20, the integral does not need to be taken in the principal value sense, let ∂Ω be a smooth surface at point $\underline{x}$. Let $(\underline{e}_1, \underline{e}_2, \underline{e}_3)$ be the local axis system of origin $\underline{x}$. Since ∂Ω is smooth at point $\underline{x}$, it can be approximated locally with a portion of an ellipsoid bowl of center $\underline{x}_0$ going through $\underline{x}$ (see Figure 7.49). Then e_ε can be considered as the intersection of a bowl of radius ε and the ellipsoid. The equation of an ellipsoid of center $\underline{x}_0(0, 0, x_{03})$ in this axis system can be written as

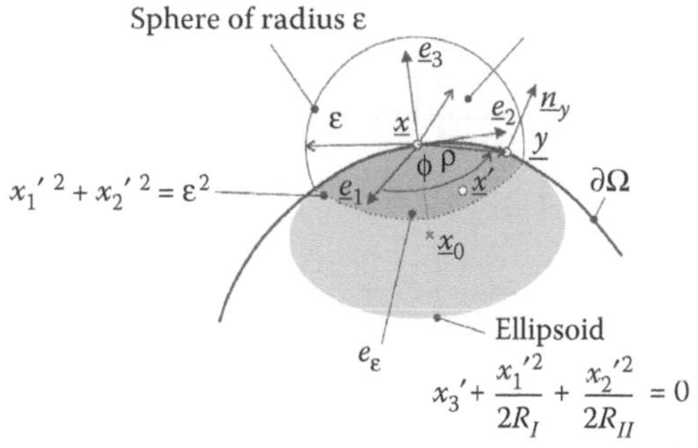

Figure 7.49 Convergence of the improper integrals for a singular point $\underline{x}$ on the boundary $\partial\Omega$.

$$\frac{x_1'^2}{a^2} + \frac{x_2'^2}{b^2} + \frac{(x_3' - x_{03})^2}{c^2} = 1 \tag{7.258}$$

where $x_{30} < 0$. Since the point $\underline{x}'$ of coordinates $(x_1' = 0, x_2' = 0, x_3' = 0)$ in the local axis system $(\underline{e}_1, \underline{e}_2, \underline{e}_3)$ belongs to the ellipsoid, we have

$$\frac{x_1'^2}{a^2} + \frac{x_2'^2}{b^2} + \frac{(x_3' + c)^2}{c^2} = 1 \tag{7.259}$$

We can assume that over δS, x_3' is very small so that

$$\frac{x_1'^2}{a^2} + \frac{x_2'^2}{b^2} + \frac{c^2}{c^2}\left(1 + \frac{x_3'}{c}\right)^2 - 1 \approx \frac{x_1'^2}{a^2} + \frac{x_2'^2}{b^2} + \frac{2x_3'}{c} \tag{7.260}$$

That is,

$$f(x_1', x_2', x_3') = x_3' + \frac{cx_1'^2}{2a^2} + \frac{cx_2'^2}{2b^2} = x_3' + \frac{x_1'^2}{2R_I} + \frac{x_2'^2}{2R_{II}} \approx 0 \tag{7.261}$$

since for an ellipsoïd, the principal radii of curvature at every point are given by $R_I = a^2/c$ and $R_{II} = b^2/c$.

The contour of e_ε is defined as the intersection of the sphere of center $\underline{x}$ and ellipsoid given by Equation 7.259. It is a curve of equation

$$x_1'^2 + x_2'^2 + \left(\frac{x_1'^2}{2R_I} + \frac{x_2'^2}{2R_{II}}\right)^2 = \varepsilon^2 \tag{7.262}$$

which is a circle of equation $x_1'^2 + x_2'^2 = \varepsilon^2$ when $\varepsilon \to 0$.

The normal at a point $\underline{y}$ of e_ε is given by the gradient of Equation 7.261

$$\begin{aligned}\underline{n}_y &\approx \frac{1}{\sqrt{1+\left(\dfrac{\partial f}{\partial x_1'}\right)^2+\left(\dfrac{\partial f}{\partial x_2'}\right)^2}}\left(\underline{e}_3+\frac{\partial f}{\partial x_1'}\underline{e}_1+\frac{\partial f}{\partial x_2'}\underline{e}_2\right)\\ &= \frac{1}{\sqrt{1+\left(\dfrac{x_1'}{R_I}\right)^2+\left(\dfrac{x_2'}{R_{II}}\right)^2}}\left(\underline{e}_3+\frac{x_1'}{R_I}\underline{e}_1+\frac{x_2'}{R_{II}}\underline{e}_2\right)\\ &\approx \left(\underline{e}_3+\frac{x_1'}{R_I}\underline{e}_1+\frac{x_2'}{R_{II}}\underline{e}_2\right)\end{aligned} \tag{7.263}$$

The normal derivative of the Green's function can be written as

$$\frac{\partial\hat{G}(\underline{x},\underline{y})}{\partial n_y} = -\frac{\underline{y}-\underline{x}}{r^3}\cdot\underline{n}_y(ikr+1)\exp(-ikr) \tag{7.264}$$

The vector $\underline{y}-\underline{x}$ in the axis system $(\underline{e}_1,\underline{e}_2,\underline{e}_3)$ is given by $x_1'\underline{e}_1+x_2'\underline{e}_2-(x_1'^2/2R_I+x_2'^2/2R_{II})\underline{e}_3$ and $r=|\underline{y}-\underline{x}|$.

Then

$$\begin{aligned}\frac{\underline{y}-\underline{x}}{r^3}\cdot\underline{n}_y &= \frac{1}{r^3}\left[x_1'\underline{e}_1+x_2'\underline{e}_2-\left(\frac{x_1'^2}{2R_I}+\frac{x_2'^2}{2R_{II}}\right)\underline{e}_3\right]\cdot\left(\underline{e}_3+\frac{x_1'}{R_I}\underline{e}_1+\frac{x_2'}{R_{II}}\underline{e}_2\right)\\ &= \frac{1}{r^3}\left(\frac{x_1'^2}{2R_I}+\frac{x_2'^2}{2R_{II}}\right)\end{aligned} \tag{7.265}$$

Using polar coordinates $x'=\rho\cos\phi$, $y'=\rho\sin\phi$

$$r = |\underline{y}-\underline{x}| = \sqrt{x'^2+y'^2+\left(\frac{x'^2}{2R_I}+\frac{y'^2}{2R_{II}}\right)^2} \approx \rho \tag{7.266}$$

and Equation 7.264 becomes

$$\frac{\partial\hat{G}(\underline{x},\underline{y})}{\partial n_y} \approx -\frac{1}{\rho}\left(\frac{\cos\phi^2}{2R_I}+\frac{\sin\phi^2}{2R_{II}}\right)(ik\rho+1)\exp(-ik\rho) \tag{7.267}$$

We see that $\partial\hat{G}(\underline{x}, y)/\partial n_y$ has a weak $(1/\rho)$ singularity. The region e_ε over which we must integrate is defined by $\phi \in [0,2\pi]$, $\rho \in [0,\varepsilon]$. Then,

$$\int_{e_\varepsilon} \hat{p}(\underline{y}) \frac{\partial \hat{G}(\underline{x}, \underline{y})}{\partial n_y} dS_y = \int_0^{2\pi}\int_0^{\varepsilon} -\frac{1}{\rho}\left(\frac{\cos\phi^2}{2R_I} + \frac{\sin\phi^2}{2R_{II}}\right)(ik\rho + 1)\exp(-ik\rho)\hat{p}(\rho,\phi)\rho d\rho d\phi \tag{7.268}$$

In addition,

$$\begin{aligned}
&\int_0^{2\pi}\int_0^{\varepsilon} -\frac{1}{\rho}\left(\frac{\cos\phi^2}{2R_I} + \frac{\sin\phi^2}{2R_{II}}\right)(ik\rho + 1)\exp(-ik\rho)\,\hat{p}(\rho,\phi)\rho d\rho d\phi \\
&= -\int_0^{2\pi}\int_0^{\varepsilon}\left(\frac{\cos\phi^2}{2R_I} + \frac{\sin\phi^2}{2R_{II}}\right)ik\rho\exp(-ik\rho)\,\hat{p}(\rho,\phi)d\rho d\phi \\
&\quad - \int_0^{2\pi}\int_0^{\varepsilon}\left(\frac{\cos\phi^2}{2R_I} + \frac{\sin\phi^2}{2R_{II}}\right)\exp(-ik\rho)\,\hat{p}(\rho,\phi)d\rho d\phi
\end{aligned} \tag{7.269}$$

Over e_ε, $|\hat{p}(\rho,\phi)|$ is bounded so that

$$\exists(C_1,C_2) > 0;\quad \begin{aligned}
&\left|\left(\frac{\cos\phi^2}{2R_I} + \frac{\sin\phi^2}{2R_{II}}\right)ik\rho\exp(-ik\rho)\,\hat{p}(\rho,\phi)\right| < C_1\left|\left(\frac{\cos\phi^2}{2R_I} + \frac{\sin\phi^2}{2R_{II}}\right)k\rho\right| \\
&\left|\left(\frac{\cos\phi^2}{2R_I} + \frac{\sin\phi^2}{2R_{II}}\right)\exp(-ik\rho)\,\hat{p}(\rho,\phi)\right| < C_2\left|\left(\frac{\cos\phi^2}{2R_I} + \frac{\sin\phi^2}{2R_{II}}\right)\right|
\end{aligned} \tag{7.270}$$

Besides,

$$\begin{aligned}
\int_0^{2\pi}\int_0^{\varepsilon}\left(\frac{\cos\phi^2}{2R_I} + \frac{\sin\phi^2}{2R_{II}}\right)k\rho d\rho d\phi &= \int_0^{2\pi} k\frac{\varepsilon^2}{2}\left(\frac{\cos\phi^2}{2R_I} + \frac{\sin\phi^2}{2R_{II}}\right)d\phi \\
&= k\frac{\varepsilon^2}{2}\pi\left(\frac{1}{2R_I} + \frac{1}{2R_{II}}\right)
\end{aligned} \tag{7.271}$$

and

$$\int_0^{2\pi}\int_0^{\varepsilon}\left(\frac{\cos\phi^2}{2R_I}+\frac{\sin\phi^2}{2R_{II}}\right)d\rho d\phi = \varepsilon\pi\left(\frac{1}{2R_I}+\frac{1}{2R_{II}}\right) \tag{7.272}$$

Both expressions in Equations 7.271 and 7.272 tend to zero when $\varepsilon \to 0$ then $\int_{\partial\Omega - e_\varepsilon} \hat{p}(\underline{y})\partial\hat{G}(\underline{x},\underline{y})/\partial n_y dS_y$ converges when $\varepsilon \to 0$ and there is no need to take the integral in the principal value sense. The same is true for the third equation in Equation 7.32.

APPENDIX 7C: EXPRESSIONS OF THE REDUCED SHAPE FUNCTIONS

We provide here the reduced shape functions $\tilde{N}_i$ for the geometry or the solution.

7C.1 Linear triangular element

For a linear 3-noded triangle the shape functions are given by $N_1(\underline{\eta}) = 1-\eta_1-\eta_2, N_2(\underline{\eta}) = \eta_1, N_3(\underline{\eta}) = \eta_2$. The reduced shape functions are given by

$$\begin{cases} \tilde{N}_1(\rho,\alpha,\xi_1,\xi_2) = -(\cos\alpha+\sin\alpha) \\ \tilde{N}_2(\rho,\alpha,\xi_1,\xi_2) = \cos\alpha \\ \tilde{N}_3(\rho,\alpha,\xi_1,\xi_2) = \sin\alpha \end{cases} \tag{7.273}$$

7C.2 Linear quadrangle element

For a linear quad-4 element with the following shape functions:

$$\begin{cases} N_1 = \frac{1}{4}(1-\eta_1)(1-\eta_2) \\ N_2 = \frac{1}{4}(1+\eta_1)(1-\eta_2) \\ N_3 = \frac{1}{4}(1+\eta_1)(1+\eta_2) \\ N_4 = \frac{1}{4}(1-\eta_1)(1+\eta_2) \end{cases} \tag{7.274}$$

We have

$$\begin{cases} \tilde{N}_1(\rho,\alpha,\xi_1,\xi_2) = \frac{1}{4}[(\xi_1 - 1)\sin\alpha + (\xi_2 - 1)\cos\alpha + \rho\cos\alpha\sin\alpha] \\ \tilde{N}_2(\rho,\alpha,\xi_1,\xi_2) = \frac{1}{4}[-(\xi_1 + 1)\sin\alpha + (1 - \xi_2)\cos\alpha - \rho\cos\alpha\sin\alpha] \\ \tilde{N}_3(\rho,\alpha,\xi_1,\xi_2) = \frac{1}{4}[(\xi_1 + 1)\sin\alpha + (\xi_2 + 1)\cos\alpha + \rho\cos\alpha\sin\alpha] \\ \tilde{N}_4(\rho,\alpha,\xi_1,\xi_2) = \frac{1}{4}[(1 - \xi_1)\sin\alpha - (\xi_2 + 1)\cos\alpha - \rho\cos\alpha\sin\alpha] \end{cases} \tag{7.275}$$

APPENDIX 7D: CALCULATION OF $\int_{\partial\Omega}\int_{\partial\Omega} \hat{\mu}(\underline{x})\partial^2 \hat{G}/\partial n_x \partial n_y \hat{\mu}(\underline{y}) dS_x dS_y = \langle\hat{\mu}\rangle[\hat{D}(\omega)]\{\hat{\mu}\}$

Consider two boundary elements $\partial\Omega^e$ and $\partial\Omega^{e'}$. We are interested in the discretization of $[\underline{n}_x \times \underline{\nabla}_x \hat{\mu}(\underline{x})] \cdot [\underline{n}_y \times \underline{\nabla}_y \hat{\mu}(\underline{y})]$ for $\underline{x} \in \partial\Omega^e$ and $\underline{y} \in \partial\Omega^{e'}$.

A point in space can be described in terms of parametric coordinates as $\underline{x} = x_i(\xi_1,\xi_2,\xi_3)\underline{e}_i$ where $(\underline{e}_1,\underline{e}_2,\underline{e}_3)$ is an orthogonal basis. The gradient of a field with respect to x is given by

$$\underline{\nabla}_{\underline{\xi}}(.) = [J]\underline{\nabla}_{\underline{x}}(.) \tag{7.276}$$

where $[J]$ is the Jacobian matrix

$$[J] = \begin{bmatrix} \frac{\partial x_1}{\partial \xi_1} & \frac{\partial x_2}{\partial \xi_1} & \frac{\partial x_3}{\partial \xi_1} \\ \frac{\partial x_1}{\partial \xi_2} & \frac{\partial x_2}{\partial \xi_2} & \frac{\partial x_3}{\partial \xi_2} \\ \frac{\partial x_1}{\partial \xi_3} & \frac{\partial x_2}{\partial \xi_3} & \frac{\partial x_3}{\partial \xi_3} \end{bmatrix} = \begin{bmatrix} \underline{a}_1^T \\ \underline{a}_2^T \\ \underline{a}_3^T \end{bmatrix} \tag{7.277}$$

and $\underline{a}_1 = \partial\underline{x}/\partial\xi_1$; $\underline{a}_2 = \partial\underline{x}/\partial\xi_2$ and $\underline{a}_3 = \partial\underline{x}/\partial\xi_3$. Conversely, the gradient in physical coordinates can be expressed in terms of parametric coordinates as

$$\underline{\nabla}_{\underline{x}}(.) = [J]^{-1}\underline{\nabla}_{\underline{\xi}}(.) \tag{7.278}$$

For a point $\underline{x} \in \partial\Omega^e$, ξ_3 can be set to 0. The transformation $\underline{x} = x_i^e(\xi_1,\xi_2)\underline{e}_i$ then maps the element $\partial\Omega^e$ into the reference element R^e. $\underline{a}_1^e = \partial\underline{x}/\partial\xi_1$; $\underline{a}_2^e = \partial\underline{x}/\partial\xi_2$ are the tangent vectors at point $\underline{x}$ and $\underline{a}_3^e = \underline{a}_1^e \times \underline{a}_2^e$

is orthogonal to the plane defined by the tangent vectors $\underline{a}_1^e$ and $\underline{a}_2^e$. The unit normal vector $\underline{n}_x$ at point $\underline{x}$ is given by

$$\underline{n}_x = \frac{\underline{a}_3^e}{|j^e|} = \frac{\underline{a}_1^e \times \underline{a}_2^e}{|j^e|} \tag{7.279}$$

where $|j^e| = |\underline{a}_1^e \times \underline{a}_2^e|$. In addition,

$$[J^e]^{-1} = \frac{1}{|J^e|}[\underline{a}_2^e \times \underline{a}_3^e \quad \underline{a}_3^e \times \underline{a}_1^e \quad \underline{a}_1^e \times \underline{a}_2^e] \tag{7.280}$$

with

$$|J^e| = (\underline{a}_1^e \times \underline{a}_2^e) \cdot \underline{a}_3^e = |j^e|^2 \tag{7.281}$$

The nodal approximation of the pressure jumps $\hat{\mu}$ for $\underline{x} \in \partial\Omega^e$ and $\underline{y} \in \partial\Omega^{e'}$ is given by

$$\begin{cases} \hat{\mu}(\underline{y}) = \langle N(\eta_1, \eta_2)\rangle \{\hat{\mu}^e\} \\ \hat{\mu}(\underline{x}) = \langle N(\xi_1, \xi_2)\rangle \{\hat{\mu}^{e'}\} \end{cases} \tag{7.282}$$

where (ξ_1,ξ_2) and (η_1, η_2) are the coordinates of points $\underline{x}$ and $\underline{y}$ in the reference space, respectively. Vectors $\{\hat{\mu}^e\}$ and $\{\hat{\mu}^{e'}\}$ represent the nodal values of $\hat{\mu}$ on elements $\partial\Omega^e$ and $\partial\Omega^{e\prime}$, respectively.

Consequently

$$\begin{aligned} \underline{n}_x \times \underline{\nabla}_x \hat{\mu}(\underline{x}) &= \frac{\underline{a}_3^e}{|j^e|} \times \frac{1}{|j^e|^2}[\underline{a}_2^e \times \underline{a}_3^e \quad \underline{a}_3^e \times \underline{a}_1^e \quad \underline{a}_3^e]\, \underline{\nabla}_\xi \hat{\mu}(\underline{\xi}) \\ &= \frac{1}{|j^e|^3}[\underline{a}_3^e \times (\underline{a}_2^e \times \underline{a}_3^e) \quad \underline{a}_3^e \times (\underline{a}_3^e \times \underline{a}_1^e) \quad 0]\, \underline{\nabla}_\xi \hat{\mu}(\underline{\xi}) \end{aligned} \tag{7.283}$$

Using the identity $\underline{a} \times (\underline{b} \times \underline{c}) = (\underline{a} \cdot \underline{c})\underline{b} - (\underline{a} \cdot \underline{b})\underline{c}$ and the fact $\underline{a}_1^e$, $\underline{a}_2^e$, and $\underline{a}_3^e$ are orthogonal, we get

$$\underline{n}_x \times \underline{\nabla}_x \hat{\mu}(\underline{x}) = \frac{1}{|j^e|^3}\left[|j^e|^2 \underline{a}_2^e \quad -|j^e|^2 \underline{a}_1^e \quad 0\right] \underline{\nabla}_\xi \hat{\mu}(\underline{\xi}) \tag{7.284}$$

That is,

$$\underline{n}_x \times \underline{\nabla}_x \hat{\mu}(\underline{x}) = \frac{\underline{a}_2^e}{|j^e|}\frac{\partial \hat{\mu}(\underline{\xi})}{\partial \xi_1} - \frac{\underline{a}_1^e}{|j|^e}\frac{\partial \hat{\mu}(\underline{\xi})}{\partial \xi_2} \tag{7.285}$$

Similarly, it can be proved that

$$\underline{n}_y \times \nabla_y \hat{\mu}(\underline{y}) = \frac{\underline{a}_2^{e'}}{|j^{e'}|}\frac{\partial \hat{\mu}(\underline{\eta})}{\partial \eta_1} - \frac{\underline{a}_1^{e'}}{|j^{e'}|}\frac{\partial \hat{\mu}(\underline{\eta})}{\partial \eta_2} \tag{7.286}$$

Then, the following discretized form is obtained:

$$\begin{aligned}
&(\underline{n}_x \times \underline{\nabla}_x \hat{\mu}(\underline{x})) \cdot (\underline{n}_y \times \underline{\nabla}_y \hat{\mu}(\underline{y})) \\
&= \Big[(\underline{a}_2^e \cdot \underline{a}_2^{e'})\langle \hat{\mu}^e \rangle \{N_{,\xi_1}\}\langle N_{,\eta_1}\rangle \{\hat{\mu}^{e'}\} + (\underline{a}_1^e \cdot \underline{a}_1^{e'})\langle \hat{\mu}^e \rangle \{N_{,\xi_2}\}\langle N_{,\eta_2}\rangle \{\hat{\mu}^{e'}\} \\
&\quad - (\underline{a}_1^e \cdot \underline{a}_2^{e'})\langle \hat{\mu}^e \rangle \{N_{,\xi_2}\}\langle N_{,\eta_1}\rangle \{\hat{\mu}^{e'}\} - (\underline{a}_2^e \cdot \underline{a}_1^{e'})\langle \hat{\mu}^e \rangle \{N_{,\xi_1}\}\langle N_{,\eta_2}\rangle \{\hat{\mu}^{e'}\}\Big]\frac{1}{|j^e||j^{e'}|} \\
&= \langle \hat{\mu}^e \rangle \Big[(\underline{a}_2^e \cdot \underline{a}_2^{e'})\{N_{,\xi_1}\}\langle N_{,\eta_1}\rangle + (\underline{a}_1^e \cdot \underline{a}_1^{e'})\{N_{,\xi_2}\}\langle N_{,\eta_2}\rangle - (\underline{a}_1^e \cdot \underline{a}_2^{e'})\{N_{,\xi_2}\}\langle N_{,\eta_1}\rangle \\
&\quad - (\underline{a}_2^e \cdot \underline{a}_1^{e'})\{N_{,\xi_1}\}\langle N_{,\eta_2}\rangle\Big]\{\hat{\mu}^{e'}\}\frac{1}{|j^e||j^{e'}|}
\end{aligned} \tag{7.287}$$

where $N_{,\xi}$ denotes $\partial N/\partial \xi$.

Likewise, it can be shown that

$$k^2(\underline{n}_x \cdot \underline{n}_y) = k^2 \frac{\underline{a}_3^e}{|j^e|} \cdot \frac{\underline{a}_3^{e'}}{|j^{e'}|} \tag{7.288}$$

and the final discretized form can be written as

$$\begin{aligned}
&\int_{\partial\Omega^e}\int_{\partial\Omega^{e'}} \hat{\mu}(\underline{x})\frac{\partial^2 \hat{G}(\underline{x},\underline{y})}{\partial n_x \partial n_y}\hat{\mu}(\underline{y})\, dS_x dS_y \\
&= \int_{\partial\Omega^e}\int_{\partial\Omega^{e'}} \Big[k^2(\underline{n}_x \cdot \underline{n}_y)\hat{\mu}(\underline{x})\hat{\mu}(\underline{y}) - (\underline{n}_x \times \underline{\nabla}_x \hat{\mu}(\underline{x})) \cdot (\underline{n}_y \times \underline{\nabla}_y \hat{\mu}(\underline{y}))\Big]\hat{G}(\underline{x},\underline{y}) dS_x dS_y \\
&= \langle \hat{\mu}^e \rangle \Bigg[\int_{\partial\Omega^r}\int_{\partial\Omega^r} \Big[k^2(\underline{a}_3^e \cdot \underline{a}_3^{e'})\{N(\underline{\xi})\}\langle N(\underline{\eta})\rangle + (\underline{a}_2^e \cdot \underline{a}_2^{e'})\{N_{\xi_1}(\underline{\xi})\}\langle N_{\eta_1}(\underline{\eta})\rangle \\
&\quad + (\underline{a}_1^e \cdot \underline{a}_1^{e'})\{N_{\xi_2}(\underline{\xi})\}\langle N_{\eta_2}(\underline{\eta})\rangle - (\underline{a}_1^e \cdot \underline{a}_2^{e'})\{N_{\xi_2}(\underline{\xi})\}\langle N_{\eta_1}(\underline{\eta})\rangle - (\underline{a}_2^e \cdot \underline{a}_1^{e'}) \\
&\quad \times \{N_{\xi_1}(\underline{\xi})\}\langle N_{\eta_2}(\underline{\eta})\rangle\Big] \times \hat{G}(\underline{x}(\underline{\xi}),\underline{y}(\underline{\eta}))d\xi_1 d\xi_2 d\eta_1 d\eta_2\Bigg]\{\hat{\mu}^{e'}\} = \langle \hat{\mu}^e \rangle [\mathcal{D}^{e,e'}(\omega)]\{\hat{\mu}^{e'}\}
\end{aligned} \tag{7.289}$$

for $\underline{x} \in \partial\Omega^e$ and $\underline{y} \in \partial\Omega^{e'}$.

After assembling, we get

$$\int_{\partial\Omega}\int_{\partial\Omega} \hat{\mu}(\underline{x}) \frac{\partial^2 \hat{G}(\underline{x},\underline{y})}{\partial n_x \partial n_y} \hat{\mu}(\underline{y}) dS_x dS_y = \langle\hat{\mu}\rangle[\hat{\mathcal{D}}(\omega)]\{\hat{\mu}\} \tag{7.290}$$

where

$$[\hat{\mathcal{D}}(\omega)] = \sum_{e,e'} \langle\hat{\mu}^e\rangle[\hat{\mathcal{D}}^{ee'}(\omega)]\{\hat{\mu}^{e'}\} \tag{7.291}$$

APPENDIX 7E: CALCULATION OF $C^+(\underline{x})$ FOR A STRUCTURE IN CONTACT WITH AN INFINITE RIGID BAFFLE

We start from the integral formulation of the exterior baffled problem

$$\hat{p}(\underline{x}) = \int_{\partial\Omega} \hat{p}(\underline{y}) \frac{\partial \hat{G}_b(\underline{x},\underline{y})}{\partial n_y} dS_y - \int_{\partial\Omega} \hat{G}_b(\underline{x},\underline{y}) \frac{\partial \hat{p}}{\partial n_y}(\underline{y}) dS_y \tag{7.292}$$

For a point $\underline{x}$ on $\partial\Omega$ and off the infinite rigid plane Σ, we have

$$C^+(\underline{x})\hat{p}(\underline{x}) = \int_{\partial\Omega} \hat{p}(\underline{y}) \frac{\partial \hat{G}_b(\underline{x},\underline{y})}{\partial n_y} dS_y - \int_{\partial\Omega} \hat{G}_b(\underline{x},\underline{y}) \frac{\partial \hat{p}}{\partial n_y}(\underline{y}) dS_y \tag{7.293}$$

where $C^+(\underline{x})$ can be calculated by Equation 7.145 using the integral of the normal derivative of the singular part of the baffled Green's function* over a closed surface $\partial\Omega \cup \partial\Omega_c$ (see Figure 7.15)

$$C^+(\underline{x}) = 1 + \int_{\partial\Omega+\partial\Omega_c} \frac{\partial G_0(\underline{x},\underline{y})}{\partial n_y} dS_y \tag{7.294}$$

If $\underline{x}$ is on $\partial\Omega$ and also in contact with the infinite rigid baffle Σ, then we can follow the procedure used in Section 7.2.2 (see Figure 7.50). We then have

* It is given by $\frac{\partial G_0(\underline{x},\underline{y})}{\partial n_y}$. The part related to the image source does not contribute to the integral since the distance between $\underline{x}$ and its image with respect to the baffle is finite.

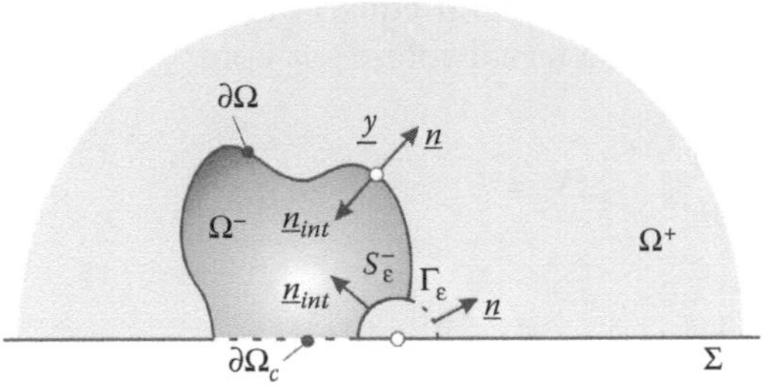

Figure 7.50 Body sitting on an infinite rigid baffle Σ.

$$C^{+}(\underline{x}) = -\lim_{\varepsilon\to 0}\int_{S_\varepsilon^-}\frac{\partial G_b(\underline{x},\underline{y})}{\partial n_y}dS_y = \lim_{\varepsilon\to 0}\int_{S_\varepsilon^-}\frac{\partial G_b(\underline{x},\underline{y})}{\partial n_{y,\text{int}}}dS_y \tag{7.295}$$

where $\underline{n}_{y,\text{int}}$is the normal pointing into Ω^-. For $\varepsilon \to 0$, r and r' become equal (see Section 7.8 for notations) and

$$G_b(\underline{x},\underline{y}) = \frac{2}{4\pi r} \tag{7.296}$$

and

$$C^{+}(\underline{x}) = 2\lim_{\varepsilon\to 0}\int_{S_\varepsilon^-}\frac{\partial}{\partial n_{y,\text{int}}}\left(\frac{1}{4\pi r}\right)dS_y \tag{7.297}$$

If we recall that the integral of $\partial/\partial n(1/r)$ over a hemisphere is equal to 2π namely

$$\int_{S_\varepsilon^-}\frac{\partial}{\partial n_{y,\text{int}}}\left(\frac{1}{r}\right)dS_y + \int_{\Gamma_\varepsilon}\frac{\partial}{\partial n_y}\left(\frac{1}{r}\right)dS_y = 2\pi \tag{7.298}$$

where $S_\varepsilon^- \cup \Gamma_\varepsilon$ constitutes the hemisphere (see Figure 7.50), we then have

$$\lim_{\varepsilon\to 0}\int_{S_\varepsilon^-}\frac{\partial}{\partial n_{y,\text{int}}}\left(\frac{1}{r}\right)dS_y = 2\pi - \lim_{\varepsilon\to 0}\int_{\Gamma_\varepsilon}\frac{\partial}{\partial n_y}\left(\frac{1}{r}\right)dS_y \tag{7.299}$$

Applying the boundary integral equation for Laplace's equation with a constant solution in the external volume at point $\underline{x}$, we have

$$\lim_{\varepsilon\to 0} \int_{\partial\Omega+\partial\Omega_c+\Gamma_\varepsilon} \frac{\partial}{\partial n_y}\left(\frac{1}{r}\right) dS_y = 0 \tag{7.300}$$

$$\lim_{\varepsilon\to 0} \int_{\Gamma_\varepsilon} \frac{\partial}{\partial n_y}\left(\frac{1}{r}\right) dS_y = - \int_{\partial\Omega+\partial\Omega_c} \frac{\partial}{\partial n_y}\left(\frac{1}{r}\right) dS_y \tag{7.301}$$

Thus,

$$\begin{aligned} C^+(\underline{x}) &= \frac{2}{4\pi}\left(2\pi + \int_{\partial\Omega+\partial\Omega_c} \frac{\partial}{\partial n_y}\left(\frac{1}{r}\right) dS_y\right) \\ &= 1 + 2 \int_{\partial\Omega+\partial\Omega_c} \frac{\partial G_0(\underline{x},\underline{y})}{\partial n_y} dS_y \end{aligned} \tag{7.302}$$

APPENDIX 7F: NUMERICAL QUADRATURE OF $I = \int_{-1}^{1}\int_{-1}^{1}\int_{-1}^{1}\int_{-1}^{1} F(\xi_1,\xi_2,\eta_1,\eta_2)/R \, d\xi_1 d\xi_2 d\eta_1 d\eta_2$

This appendix provides a summary of a numerical algorithm to calculate efficiently a double surface integral involving a $1/R$ singularity over unit quadrangles. The details are provided in the article by Wang and Atalla (1997).

Let $r = \sqrt{(\xi_1 - \eta_1)^2 + (\xi_2 - \eta_2)^2}$ and $f(\xi_1,\xi_2,\eta_1,\eta_2) = (F(\xi_1,\xi_2,\eta_1,\eta_2)r/R)$ we then have to calculate the following integral:

$$I = \int_{-1}^{1}\int_{-1}^{1}\int_{-1}^{1}\int_{-1}^{1} \frac{f(\xi_1,\xi_2,\eta_1,\eta_2)}{r} d\xi_1 d\xi_2 d\eta_1 d\eta_2 \tag{7.303}$$

Wang and Atalla provided an integration rule of the type

$$I = \sum_{i,j,k,l=1}^{N_g} W^g_{ijkl} f(\xi^g_{1,i}, \xi^g_{2,i}, \eta^g_{1,k}, \eta^g_{2,l}) \tag{7.304}$$

The following table summarizes the associated integration points together with the integration weights.

$N_g = 2^4$ quadrature rule (2 points in each dimension)
Let $a = 0.5773502692$, $b = 0.5$ and

w_1	3.37581706047058	w_2	0.957757115364075	w_3	0.655087828636169

id	ξ_1^g	ξ_2^g	η_1^g	η_2^g	W^g
1	a	a	b	b	w_1
2	a	a	b	$-b$	w_2
3	a	a	$-b$	b	w_2
4	a	a	$-b$	$-b$	w_3
5	a	$-a$	b	b	w_2
6	a	$-a$	b	$-b$	w_1
7	a	$-a$	$-b$	b	w_3
8	a	$-a$	$-b$	$-b$	w_2
9	$-a$	a	b	b	w_2
10	$-a$	a	b	$-b$	w_3
11	$-a$	a	$-b$	b	w_1
12	$-a$	a	$-b$	$-b$	w_2
13	$-a$	$-a$	b	b	w_3
14	$-a$	$-a$	b	$-b$	w_2
15	$-a$	$-a$	$-b$	b	w_2
16	$-a$	$-a$	$-b$	$-b$	w_1

$N_g = 3^4$ quadrature rule (3 points in each dimension)
Let $a = 0.7745966692$, $b = 0$, $c = 0.7071067812$ and

w_1	0.696945071220398	w_2	0.163331389427185	w_3	8.716521412134171 10^{-2}
w_4	0.142054870724678	w_5	8.092511445283890 10^{-2}	w_6	6.499387323856354 10^{-2}
w_7	0.352819710969925	w_8	1.13206529617310	w_9	0.219317153096199
w_{10}	0.285526424646378	w_{11}	0.127541959285736	w_{12}	0.143435299396515
w_{13}	0.359252929687500	w_{14}	0.597731828689575	w_{15}	1.84885013103485

	ξ_1^g	ξ_2^g	η_1^g	η_2^g	W_g		ξ_1^g	ξ_2^g	η_1^g	η_2^g	W_g
1	a	a	c	c	w_1	7	a	a	$-c$	c	w_3
2	a	a	c	b	w_2	8	a	a	$-c$	b	w_5
3	a	a	c	$-c$	w_3	9	a	a	$-c$	$-c$	w_6
4	a	a	b	c	w_2	10	a	b	c	c	w_7
5	a	a	b	b	w_4	11	a	b	c	b	w_8
6	a	a	b	$-c$	w_5	12	a	b	c	$-c$	w_7

	ξ_1^g	ξ_2^g	η_1^g	η_2^g	W_g		ξ_1^g	ξ_2^g	η_1^g	η_2^g	W_g
13	a	b	b	c	w_9	47	b	$-a$	c	b	w_9
14	a	b	b	b	w_{10}	48	b	$-a$	c	$-c$	w_7
15	a	b	b	$-c$	w_9	49	b	$-a$	b	c	w_{12}
16	a	b	$-c$	c	w_{11}	50	b	$-a$	b	b	w_{10}
17	a	b	$-c$	b	w_{12}	51	b	$-a$	b	$-c$	w_8
18	a	b	$-c$	$-c$	w_{11}	52	b	$-a$	$-c$	c	w_{11}
19	a	$-a$	c	c	w_3	53	b	$-a$	$-c$	b	w_9
20	a	$-a$	c	b	w_2	54	b	$-a$	$-c$	$-c$	w_7
21	a	$-a$	c	$-c$	w_1	55	$-a$	a	c	c	w_3
22	a	$-a$	b	c	w_5	56	$-a$	a	c	b	w_5
23	a	$-a$	b	b	w_4	57	$-a$	a	c	$-c$	w_6
24	a	$-a$	b	$-c$	w_2	58	$-a$	a	b	c	w_2
25	a	$-a$	$-c$	c	w_6	59	$-a$	a	b	b	w_4
26	a	$-a$	$-c$	b	w_5	60	$-a$	a	b	$-c$	w_5
27	a	$-a$	$-c$	$-c$	w_3	61	$-a$	a	$-c$	c	w_1
28	b	a	c	c	w_7	62	$-a$	a	$-c$	b	w_2
29	b	a	c	b	w_9	63	$-a$	a	$-c$	$-c$	w_3
30	b	a	c	$-c$	w_{11}	64	$-a$	b	c	c	w_{11}
31	b	a	b	c	w_8	65	$-a$	b	c	b	w_{12}
32	b	a	b	b	w_{10}	66	$-a$	b	c	$-c$	w_{11}
33	b	a	b	$-c$	w_{12}	67	$-a$	b	b	c	w_9
34	b	a	$-c$	c	w_7	68	$-a$	b	b	b	w_{10}
35	b	a	$-c$	b	w_9	69	$-a$	b	b	$-c$	w_9
36	b	a	$-c$	$-c$	w_{11}	70	$-a$	b	$-c$	c	w_7
37	b	b	c	c	w_{13}	71	$-a$	b	$-c$	b	w_8
38	b	b	c	b	w_{14}	72	$-a$	b	$-c$	$-c$	w_7
39	b	b	c	$-c$	w_{13}	73	$-a$	$-a$	c	c	w_6
40	b	b	b	c	w_{14}	74	$-a$	$-a$	c	b	w_5
41	b	b	b	b	w_{15}	75	$-a$	$-a$	c	$-c$	w_3
42	b	b	b	$-c$	w_{14}	76	$-a$	$-a$	b	c	w_5
43	b	b	$-c$	c	w_{13}	77	$-a$	$-a$	b	b	w_4
44	b	b	$-c$	b	w_{14}	78	$-a$	$-a$	b	$-c$	w_2
45	b	b	$-c$	$-c$	w_{13}	79	$-a$	$-a$	$-c$	c	w_3
46	b	$-a$	c	c	w_{11}	80	$-a$	$-a$	$-c$	b	w_2
						81	$-a$	$-a$	$-c$	$-c$	w_1

$N_g = 4^4$ quadrature rule (4 points in each dimension)

Let $a = 0.8611363116$, $b = 0.3399810436$, $c = 0.8090169944$, $d = 0.3090169944$ and

w_1	0.186801761388779	w_2	3.979524224996567 10^{-2}	w_3	3.183203190565109 10^{-2}
w_4	1.282163243740797 10^{-2}	w_5	5.017421394586563 10^{-2}	w_6	3.041967377066612 10^{-2}
w_7	1.630745269358158 10^{-2}	w_8	2.656092867255211 10^{-2}	w_9	1.438716053962708 10^{-2}
w_{10}	9.017957374453545 10^{-3}	w_{11}	0.110350176692009	w_{12}	0.382629752159119
w_{13}	0.106654748320580	w_{14}	2.600543759763241 10^{-2}	w_{15}	7.289469242095947 10^{-2}
w_{16}	0.106035530567169	w_{17}	9.223712235689163 10^{-2}	w_{18}	4.017712920904160 10^{-2}
w_{19}	4.632039740681648 10^{-2}	w_{20}	7.514457404613495 10^{-2}	w_{21}	5.995707586407661 10^{-2}
w_{22}	3.449163585901260 10^{-2}	w_{23}	2.324717864394188 10^{-2}	w_{24}	3.254054486751556 10^{-2}
w_{25}	3.079050593078136 10^{-2}	w_{26}	1.943890750408173 10^{-2}	w_{27}	0.123726531863213
w_{28}	0.255256652832031	w_{29}	0.145255103707314	w_{30}	5.570301041007042 10^{-2}
w_{31}	0.800638735294342	w_{32}	0.257773995399475	w_{33}	6.846934556961060 10^{-2}
w_{34}	0.177821427583694	w_{35}	7.563985139131546 10^{-2}	w_{36}	4.652449116110802 10^{-2}

	ξ_1^g	ξ_2^g	η_1^g	η_2^g	W_g		ξ_1^g	ξ_2^g	η_1^g	η_2^g	W_g
1	a	a	c	c	w_1	8	a	a	d	$-c$	w_7
2	a	a	c	d	w_2	9	a	a	$-d$	c	w_3
3	a	a	c	$-d$	w_3	10	a	a	$-d$	d	w_6
4	a	a	c	$-c$	w_4	11	a	a	$-d$	$-d$	w_8
5	a	a	d	c	w_2	12	a	a	$-d$	$-c$	w_9
6	a	a	d	d	w_5	13	a	a	$-c$	c	w_4
7	a	a	d	$-d$	w_6	14	a	a	$-c$	d	w_7
15	a	a	$-c$	$-d$	w_9	55	a	$-a$	d	$-d$	w_5
16	a	a	$-c$	$-c$	w_{10}	56	a	$-a$	d	$-c$	w_2
17	a	b	c	c	w_{11}	57	a	$-a$	$-d$	c	w_9
18	a	b	c	d	w_{12}	58	a	$-a$	$-d$	d	w_8
19	a	b	c	$-d$	w_{13}	59	a	$-a$	$-d$	$-d$	w_6
20	a	b	c	$-c$	w_{14}	60	a	$-a$	$-d$	$-c$	w_3
21	a	b	d	c	w_{15}	61	a	a	$-c$	c	w_{10}
22	a	b	d	d	w_{16}	62	a	$-a$	$-c$	d	w_9
23	a	b	d	$-d$	w_{17}	63	a	$-a$	$-c$	$-d$	w_7
24	a	b	d	$-c$	w_{18}	64	a	$-a$	$-c$	$-c$	w_4
25	a	b	$-d$	c	w_{19}	65	b	a	c	c	w_{11}
26	a	b	$-d$	d	w_{20}	66	b	a	c	d	w_{15}
27	a	b	$-d$	$-d$	w_{21}	67	b	a	c	$-d$	w_{19}
28	a	b	$-d$	$-c$	w_{22}	68	b	a	c	$-c$	w_{23}
29	a	b	$-c$	c	w_{23}	69	b	a	d	c	w_{12}
30	a	b	$-c$	d	w_{24}	70	b	a	d	d	w_{16}
31	a	b	$-c$	$-d$	w_{25}	71	b	a	d	$-d$	w_{20}
32	a	b	$-c$	$-c$	w_{26}	72	b	a	d	$-c$	w_{24}
33	a	$-b$	c	c	w_{14}	73	b	a	$-d$	c	w_{13}
34	a	$-b$	c	d	w_{13}	74	b	a	$-d$	d	w_{17}

	ξ_1^g	ξ_2^g	η_1^g	η_2^g	W_g		ξ_1^g	ξ_2^g	η_1^g	η_2^g	W_g
35	a	$-b$	c	$-d$	w_{12}	75	b	a	$-d$	$-d$	w_{21}
36	a	$-b$	c	$-c$	w_{11}	76	b	a	$-d$	$-c$	w_{25}
37	a	$-b$	d	c	w_{18}	77	b	a	$-c$	c	w_{14}
38	a	$-b$	d	d	w_{17}	78	b	a	$-c$	d	w_{18}
39	a	$-b$	d	$-d$	w_{16}	79	b	a	$-c$	$-d$	w_{22}
40	a	$-b$	d	$-c$	w_{15}	80	b	a	$-c$	$-c$	w_{26}
41	a	$-b$	$-d$	c	w_{22}	81	b	b	c	c	w_{27}
42	a	$-b$	$-d$	d	w_{21}	82	b	b	c	d	w_{28}
43	a	$-b$	$-d$	$-d$	w_{20}	83	b	b	c	$-d$	w_{29}
44	a	$-b$	$-d$	$-c$	w_{19}	84	b	b	c	$-c$	w_{30}
45	a	$-b$	$-c$	c	w_{26}	85	b	b	d	c	w_{28}
46	a	$-b$	$-c$	d	w_{25}	86	b	b	d	d	w_{31}
47	a	$-b$	$-c$	$-d$	w_{24}	87	b	b	d	$-d$	w_{32}
48	a	$-b$	$-c$	$-c$	w_{23}	88	b	b	d	$-c$	w_{33}
49	a	$-a$	c	c	w_4	89	b	b	$-d$	c	w_{29}
50	a	$-a$	c	d	w_3	90	b	b	$-d$	d	w_{32}
51	a	$-a$	c	$-d$	w_2	91	b	b	$-d$	$-d$	w_{34}
52	a	$-a$	c	$-c$	w_1	92	b	b	$-d$	$-c$	w_{35}
53	a	$-a$	d	c	w_7	93	b	b	$-c$	c	w_{30}
54	a	$-a$	d	d	w_6	94	b	b	$-c$	d	w_{33}
95	b	b	$-c$	$-d$	w_{35}	134	$-b$	a	d	d	w_{17}
96	b	b	$-c$	$-c$	w_{36}	135	$-b$	a	d	$-d$	w_{21}
97	b	$-b$	c	c	w_{30}	136	$-b$	a	d	$-c$	w_{25}
98	b	$-b$	c	d	w_{29}	137	$-b$	a	$-d$	c	w_{12}
99	b	$-b$	c	$-d$	w_{28}	138	$-b$	a	$-d$	d	w_{16}
100	b	$-b$	c	$-c$	w_{27}	139	$-b$	a	$-d$	$-d$	w_{20}
101	b	$-b$	d	c	w_{33}	140	$-b$	a	$-d$	$-c$	w_{24}
102	b	$-b$	d	d	w_{32}	141	$-b$	a	$-c$	c	w_{11}
103	b	$-b$	d	$-d$	w_{31}	142	$-b$	a	$-c$	d	w_{15}
104	b	$-b$	d	$-c$	w_{28}	143	$-b$	a	$-c$	$-d$	w_{19}
105	b	$-b$	$-d$	c	w_{35}	144	$-b$	a	$-c$	$-c$	w_{23}
106	b	$-b$	$-d$	d	w_{34}	145	$-b$	b	c	c	w_{30}
107	b	$-b$	$-d$	$-d$	w_{32}	146	$-b$	b	c	d	w_{33}
108	b	$-b$	$-d$	$-c$	w_{29}	147	$-b$	b	c	$-d$	w_{35}
109	b	$-b$	$-c$	c	w_{36}	148	$-b$	b	c	$-c$	w_{36}
110	b	$-b$	$-c$	d	w_{35}	149	$-b$	b	d	c	w_{29}
111	b	$-b$	$-c$	$-d$	w_{33}	150	$-b$	b	d	d	w_{32}
112	b	$-b$	$-c$	$-c$	w_{30}	151	$-b$	b	d	$-d$	w_{34}
113	b	$-a$	c	c	w_{23}	152	$-b$	b	d	$-c$	w_{35}
114	b	$-a$	c	d	w_{19}	153	$-b$	b	$-d$	c	w_{28}

	ξ_1^g	ξ_2^g	η_1^g	η_2^g	W_g		ξ_1^g	ξ_2^g	η_1^g	η_2^g	W_g
115	b	$-a$	c	$-d$	w_{15}	154	$-b$	b	$-d$	d	w_{31}
116	b	$-a$	c	$-c$	w_{11}	155	$-b$	b	$-d$	$-d$	w_{32}
117	b	$-a$	d	c	w_{24}	156	$-b$	b	$-d$	$-c$	w_{33}
118	b	$-a$	d	d	w_{20}	157	$-b$	b	$-c$	c	w_{27}
119	b	$-a$	d	$-d$	w_{16}	158	$-b$	b	$-c$	d	w_{28}
120	b	$-a$	d	$-c$	w_{12}	159	$-b$	b	$-c$	$-d$	w_{29}
121	b	$-a$	$-d$	c	w_{25}	160	$-b$	b	$-c$	$-c$	w_{30}
122	b	$-a$	$-d$	d	w_{21}	161	$-b$	$-b$	c	c	w_{36}
123	b	$-a$	$-d$	$-d$	w_{17}	162	$-b$	$-b$	c	d	w_{35}
124	b	$-a$	$-d$	$-c$	w_{13}	163	$-b$	$-b$	c	$-d$	w_{33}
125	b	$-a$	$-c$	c	w_{26}	164	$-b$	$-b$	c	$-c$	w_{30}
126	b	$-a$	$-c$	d	w_{22}	165	$-b$	$-b$	d	c	w_{35}
127	b	$-a$	$-c$	$-d$	w_{18}	166	$-b$	$-b$	d	d	w_{34}
128	b	$-a$	$-c$	$-c$	w_{14}	167	$-b$	$-b$	d	$-d$	w_{32}
129	$-b$	a	c	c	w_{14}	168	$-b$	$-b$	d	$-c$	w_{29}
130	$-b$	a	c	d	w_{18}	169	$-b$	$-b$	$-d$	c	w_{33}
131	$-b$	a	c	$-d$	w_{22}	170	$-b$	$-b$	$-d$	d	w_{32}
132	$-b$	a	c	$-c$	w_{26}	171	$-b$	$-b$	$-d$	$-d$	w_{31}
133	$-b$	a	d	c	w_{13}	172	$-b$	$-b$	$-d$	$-c$	w_{28}
173	$-b$	$-b$	$-c$	c	w_{30}	215	$-a$	b	d	$-d$	w_{21}
174	$-b$	$-b$	$-c$	d	w_{29}	216	$-a$	b	d	$-c$	w_{22}
175	$-b$	$-b$	$-c$	$-d$	w_{28}	217	$-a$	b	$-d$	c	w_{15}
176	$-b$	$-b$	$-c$	$-c$	w_{27}	218	$-a$	b	$-d$	d	w_{16}
177	$-b$	$-a$	c	c	w_{26}	219	$-a$	b	$-d$	$-d$	w_{17}
178	$-b$	$-a$	c	d	w_{22}	220	$-a$	b	$-d$	$-c$	w_{18}
179	$-b$	$-a$	c	$-d$	w_{18}	221	$-a$	b	$-c$	c	w_{11}
180	$-b$	$-a$	c	$-c$	w_{14}	222	$-a$	b	$-c$	d	w_{12}
181	$-b$	$-a$	d	c	w_{25}	223	$-a$	b	$-c$	$-d$	w_{13}
182	$-b$	$-a$	d	d	w_{21}	224	$-a$	b	$-c$	$-c$	w_{14}
183	$-b$	$-a$	d	$-d$	w_{17}	225	$-a$	$-b$	c	c	w_{26}
184	$-b$	$-a$	d	$-c$	w_{13}	226	$-a$	$-b$	c	d	w_{25}
185	$-b$	$-a$	$-d$	c	w_{24}	227	$-a$	$-b$	c	$-d$	w_{24}
186	$-b$	$-a$	$-d$	d	w_{20}	228	$-a$	$-b$	c	$-c$	w_{23}
187	$-b$	$-a$	$-d$	$-d$	w_{16}	229	$-a$	$-b$	d	c	w_{22}
188	$-b$	$-a$	$-d$	$-c$	w_{12}	230	$-a$	$-b$	d	d	w_{21}
189	$-b$	$-a$	$-c$	c	w_{23}	231	$-a$	$-b$	d	$-d$	w_{20}
190	$-b$	$-a$	$-c$	d	w_{19}	232	$-a$	$-b$	d	$-c$	w_{19}
191	$-b$	$-a$	$-c$	$-d$	w_{15}	233	$-a$	$-b$	$-d$	c	w_{18}
192	$-b$	$-a$	$-c$	$-c$	w_{11}	234	$-a$	$-b$	$-d$	d	w_{17}
193	$-a$	a	c	c	w_4	235	$-a$	$-b$	$-d$	$-d$	w_{16}
194	$-a$	a	c	d	w_7	236	$-a$	$-b$	$-d$	$-c$	w_{15}

	ξ_1^g	ξ_2^g	η_1^g	η_2^g	W_g		ξ_1^g	ξ_2^g	η_1^g	η_2^g	W_g
195	$-a$	a	c	$-d$	w_9	237	$-a$	$-b$	$-c$	c	w_{14}
196	$-a$	a	c	$-c$	w_{10}	238	$-a$	$-b$	$-c$	d	w_{13}
197	$-a$	a	d	c	w_3	239	$-a$	$-b$	$-c$	$-d$	w_{12}
198	$-a$	a	d	d	w_6	240	$-a$	$-b$	$-c$	$-c$	w_{11}
199	$-a$	a	d	$-d$	w_8	241	$-a$	$-a$	c	c	w_{10}
200	$-a$	a	d	$-c$	w_9	242	$-a$	$-a$	c	d	w_9
201	$-a$	a	$-d$	c	w_2	243	$-a$	$-a$	c	$-d$	w_7
202	$-a$	a	$-d$	d	w_5	244	$-a$	$-a$	c	$-c$	w_4
203	$-a$	a	$-d$	$-d$	w_6	245	$-a$	$-a$	d	c	w_9
204	$-a$	a	$-d$	$-c$	w_7	246	$-a$	$-a$	d	d	w_8
205	$-a$	a	$-c$	c	w_1	247	$-a$	$-a$	d	$-d$	w_6
206	$-a$	a	$-c$	d	w_2	248	$-a$	$-a$	d	$-c$	w_3
207	$-a$	a	$-c$	$-d$	w_3	249	$-a$	$-a$	$-d$	c	w_7
208	$-a$	a	$-c$	$-c$	w_4	250	$-a$	$-a$	$-d$	d	w_6
209	$-a$	b	c	c	w_{23}	251	$-a$	$-a$	$-d$	$-d$	w_5
210	$-a$	b	c	d	w_{24}	252	$-a$	$-a$	$-d$	$-c$	w_2
211	$-a$	b	c	$-d$	w_{25}	253	$-a$	$-a$	$-c$	c	w_4
212	$-a$	b	c	$-c$	w_{26}	254	$-a$	$-a$	$-c$	d	w_3
213	$-a$	b	d	c	w_{19}	255	$-a$	$-a$	$-c$	$-d$	w_2
214	$-a$	b	d	d	w_{20}	256	$-a$	$-a$	$-c$	$-c$	w_1

APPENDIX 7G: PROOF OF FORMULA

$$FP\int_{\partial\Omega}\left[\hat{\mu}(\underline{y})\frac{\partial^2\hat{G}(\underline{x},\underline{y})}{\partial n_x\partial n_y}\right]dS_y = \int_{\partial\Omega}\Big(k^2\underline{n}_x\cdot\underline{n}_y\hat{G}(\underline{x},\underline{y})\hat{\mu}(\underline{y}) + \left[\underline{n}_x\times\underline{\nabla}_x\hat{G}(\underline{x},\underline{y})\right]\cdot\left[\underline{n}_y\times\underline{\nabla}_y\hat{\mu}(\underline{y})\right]\Big)dS_y$$

Detailed proofs of this formula are given in Hamdi (1982) and Pierce (1993). Recall that the left-hand side is a hyper-singular integral when $\underline{x}\in\partial\Omega$. It only exists if interpreted in the sense of Hadamard finite part which is why the symbol *FP* is used. This appendix summarizes the main steps of the proof. Using nabla operator* notation (see Appendix 3B), we can write

$$\frac{\partial^2(\cdot)}{\partial n_x\partial n_y} = (\underline{n}_x\cdot\underline{\nabla}_x)(\underline{n}_y\cdot\underline{\nabla}_y) \tag{7.305}$$

* Nabla operator is considered as a vector in the mathematical operations but it actually operates on a quantity after it. $(\underline{n}_x\cdot\underline{\nabla}_x)$ is considered as a scalar in the mathematical operations but again it operates on a quantity after it.

Let $\underline{a}, \underline{b}, \underline{c}, \underline{d}$ be four vectors of $\mathbb{R}^3$. We then have the usual vector analysis formula

$$\underline{b} \times (\underline{c} \times \underline{d}) = (\underline{b} \cdot \underline{d})\underline{c} - (\underline{b} \cdot \underline{c})\underline{d} \tag{7.306}$$

Let us take the dot product of $\underline{a}$ with Equation 7.306. We get

$$\underline{a} \cdot (\underline{b} \times (\underline{c} \times \underline{d})) = (\underline{a} \cdot \underline{c})(\underline{b} \cdot \underline{d}) - (\underline{a} \cdot \underline{d})(\underline{b} \cdot \underline{c}) \tag{7.307}$$

We also have

$$\underline{a} \cdot (\underline{b} \times (\underline{c} \times \underline{d})) = (\underline{a} \times \underline{b}) \cdot (\underline{c} \times \underline{d}) \tag{7.308}$$

So that

$$(\underline{a} \cdot \underline{c})(\underline{b} \cdot \underline{d}) = (\underline{a} \times \underline{b}) \cdot (\underline{c} \times \underline{d}) + (\underline{a} \cdot \underline{d})(\underline{b} \cdot \underline{c}) \tag{7.309}$$

From Equation 7.307, we also have

$$\underline{b} \cdot (\underline{d} \times (\underline{a} \times \underline{c})) = (\underline{b} \cdot \underline{a})(\underline{d} \cdot \underline{c}) - (\underline{b} \cdot \underline{c})(\underline{d} \cdot \underline{a}) \tag{7.310}$$

Adding Equations 7.309 and 7.310, we get

$$(\underline{a} \cdot \underline{c})(\underline{b} \cdot \underline{d}) = (\underline{a} \cdot \underline{b})(\underline{c} \cdot \underline{d}) + (\underline{a} \times \underline{b}) \cdot (\underline{c} \times \underline{d}) - \underline{b} \cdot (\underline{d} \times (\underline{a} \times \underline{c})) \tag{7.311}$$

Identifying $\underline{a} = \underline{n}_x, \underline{c} = \underline{\nabla}_x, \underline{b} = \underline{n}_y, \underline{d} = \underline{\nabla}_y$, Equation 7.305 becomes

$$\begin{aligned}(\underline{n}_x \cdot \underline{\nabla}_x)(\underline{n}_y \cdot \underline{\nabla}_y) = (\underline{n}_x \cdot \underline{n}_y)(\underline{\nabla}_x \cdot \underline{\nabla}_y) + (\underline{n}_x \times \underline{n}_y) \cdot (\underline{\nabla}_x \times \underline{\nabla}_y) \\ - \underline{n}_y \cdot (\underline{\nabla}_y \times (\underline{n}_x \times \underline{\nabla}_x))\end{aligned} \tag{7.312}$$

Remembering that $\underline{\nabla}_x \hat{G}(\underline{x}, \underline{y}) = -\underline{\nabla}_y \hat{G}(\underline{x}, \underline{y})$, the application of Equation 7.312 leads to

$$\frac{\partial^2 \hat{G}(\underline{x}, \underline{y})}{\partial n_x \partial n_y} = -(\underline{n}_x \cdot \underline{n}_y)\nabla_x^2 \hat{G}(\underline{x}, \underline{y}) - \underline{n}_y \cdot \left(\underline{\nabla}_y \times (\underline{n}_x \times \underline{\nabla}_x \hat{G}(\underline{x}, \underline{y}))\right) \tag{7.313}$$

or since $\hat{G}(\underline{x}, \underline{y})$ is solution of Helmholtz equation for $\underline{x} \neq \underline{y}$

$$\frac{\partial^2 \hat{G}(\underline{x}, \underline{y})}{\partial n_x \partial n_y} = (\underline{n}_x \cdot \underline{n}_y) k^2 \hat{G}(\underline{x}, \underline{y}) - \underline{n}_y \cdot \left(\underline{\nabla}_y \times (\underline{n}_x \times \underline{\nabla}_x \hat{G}(\underline{x}, \underline{y}))\right) \tag{7.314}$$

Using Equation 7.314 in the integral $FP\int_{\partial\Omega}\left[\hat{\mu}(\underline{y})\partial^2\hat{G}(\underline{x},\underline{y})/\partial n_x \partial n_y\right]dS_y$, we end up with

$$FP\int_{\partial\Omega}\left[\hat{\mu}(\underline{y})\frac{\partial^2\hat{G}(\underline{x},\underline{y})}{\partial n_x \partial n_y}\right]dS_y = \int_{\partial\Omega}\Bigg(k^2\underline{n}_x\cdot\underline{n}_y\,\hat{G}(\underline{x},\underline{y})\hat{\mu}(\underline{y})$$
$$-\underline{n}_y\cdot\left(\underline{\nabla}_y\times(\underline{n}_x\times\underline{\nabla}_x\hat{G}(\underline{x},\underline{y}))\right)\hat{\mu}(\underline{y})\Bigg)dS_y$$
$$= I_1 - I_2 \tag{7.315}$$

Let us now consider the vector form $\underline{\Theta} = \underline{n}_x \times \underline{\nabla}_x\hat{G}(\underline{x},\underline{y})$. The term I_2 in Equation 7.315 can be written as

$$I_2 = \int_{\partial\Omega}\hat{\mu}(\underline{y})\underline{n}_y\cdot\left(\underline{\nabla}_y\times\underline{\Theta}(\underline{x},\underline{y})\right)dS_y \tag{7.316}$$

For $\partial\Omega$ sufficiently regular and $\hat{\mu}(\underline{y}) \in C^1(\partial\Omega)$, we have (using Equation 3.118)

$$\hat{\mu}(\underline{y})\underline{\nabla}_y\times\underline{\Theta}(\underline{x},\underline{y}) = \underline{\nabla}_y\times\left(\hat{\mu}(\underline{y})\underline{\Theta}(\underline{x},\underline{y})\right) - \underline{\nabla}_y\hat{\mu}(\underline{y})\times\underline{\Theta}(\underline{x},\underline{y}) \tag{7.317}$$

Thus, Equation 7.316 becomes

$$I_2 = \int_{\partial\Omega}\underline{n}_y\cdot\underline{\nabla}_y\times\left(\hat{\mu}(\underline{y})\underline{\Theta}(\underline{x},\underline{y})\right)dS_y - \int_{\partial\Omega}\underline{n}_y\cdot\underline{\nabla}_y\hat{\mu}(\underline{y})\times\underline{\Theta}(\underline{x},\underline{y})dS_y \tag{7.318}$$

Using Stoke's theorem (Equation 3.120), the first integral of Equation 7.317 becomes a contour integral, that is,

$$\int_{\partial\Omega}\underline{n}_y\cdot\underline{\nabla}_y\times(\hat{\mu}(\underline{y})\underline{\Theta}(\underline{x},\underline{y}))dS_y = \int_{\partial^2\Omega}\hat{\mu}(\underline{y})\underline{\Theta}(\underline{x},\underline{y})ds_y \tag{7.319}$$

Using the properties of the triple product,* the second integral of Equation 7.318 can be rewritten as

$$\int_{\partial\Omega}\underline{n}_y\cdot\left(\underline{\nabla}_y\hat{\mu}(\underline{y})\times\underline{\Theta}(\underline{x},\underline{y})\right)dS_y = \int_{\partial\Omega}\underline{\Theta}(\underline{x},\underline{y})\cdot\left(\underline{n}_y\times\underline{\nabla}_y\hat{\mu}(\underline{y})\right)dS_y \tag{7.320}$$

* $\underline{a}\cdot(\underline{b}\times\underline{c}) = \underline{b}\cdot(\underline{c}\times\underline{a}) = \underline{c}\cdot(\underline{a}\times\underline{b})$

Finally, I_2 can be rewritten as

$$I_2 = \int_{\partial^2\Omega} \hat{\mu}(y)\underline{n}_x \times \underline{\nabla}_x\hat{G}(\underline{x},\underline{y})ds_y - \int_{\partial\Omega}\left[\underline{n}_x \times \underline{\nabla}_x\hat{G}(\underline{x},\underline{y})\right]\cdot\left[\underline{n}_y \times \underline{\nabla}_y\hat{\mu}(\underline{y})\right]dS_y \tag{7.321}$$

For a closed surface $\partial\Omega$, the contour $\partial^2\Omega$ shrinks to a point and the first term of Equation 7.321 vanishes and therefore

$$FP\int_{\partial\Omega}\left[\hat{\mu}(\underline{y})\frac{\partial^2\hat{G}(\underline{x},\underline{y})}{\partial n_x\partial n_y}\right]dS_y = \int_{\partial\Omega}\Big(k^2\underline{n}_x\cdot\underline{n}_y\hat{G}(\underline{x},\underline{y})\hat{\mu}(\underline{y}) + \left[\underline{n}_x \times \underline{\nabla}_x\hat{G}(\underline{x},\underline{y})\right]\cdot\left[\underline{n}_y \times \underline{\nabla}_y\hat{\mu}(\underline{y})\right]\Big)dS_y \tag{7.322}$$

which completes the proof.

APPENDIX 7H: SIMPLE CALCULATION OF THE RADIATED ACOUSTIC POWER BY A BAFFLED PANEL BASED ON RAYLEIGH'S INTEGRAL

This appendix presents a simple approach for calculating the sound power radiated by a baffled panel. The method is based on a calculation of Rayleigh's integral using a discretization of the radiating surface into patches (finite elements with constant interpolation functions).

7H.1 Calculation of the radiated acoustic power

The acoustic power radiated by a flat vibrating surface $\partial\Omega_w$ with normal velocity field distribution $\hat{v}_n$, embedded in a rigid baffle can be calculated using

$$\Pi_{rad} = \frac{1}{2}\Re\left(\int_{\partial\Omega_w}\hat{p}\hat{v}_n^*dS\right) \tag{7.323}$$

with

$$\hat{p}(\underline{x}) = i\omega\rho_0\int_{\partial\Omega_w}\hat{v}_n(\underline{y})\hat{G}_b(\underline{x},\underline{y})dS_y \tag{7.324}$$

where $\hat{G}_b(\underline{x},\underline{y}) = \exp(-ikR)/2\pi R$ is the baffled Green's function.

Assuming that $\partial\Omega_w$ is discretized into patches such that $\partial\Omega_w = \bigcup_{i=1}^{N} \partial\Omega_{w,i}$ (Figure 7.51), Equation 7.324 can be written at a point $\underline{x}_i$ of $\partial\Omega_{w,i}$

$$\hat{p}(\underline{x}_i) = i\omega\rho_0 \sum_{\substack{j=1 \\ j\neq i}}^{N} \int_{\partial\Omega_{w,j}} \hat{v}_n(\underline{y})\hat{G}_b(\underline{x}_i, \underline{y})dS_y + i\omega\rho_0 \int_{\partial\Omega_{w,i}} \hat{v}_n(\underline{y})\hat{G}_b(\underline{x}_i, \underline{y})dS_y \tag{7.325}$$

For small patches, the normal velocity is assumed to be constant over each patch and

$$\begin{aligned} \hat{p}_i = \hat{p}(\underline{x}_i) &= i\omega\rho_0 \sum_{\substack{j=1 \\ j\neq i}}^{N} \hat{v}_{n,j}\partial\Omega_{w,j} \frac{\exp(-ikR_{ij})}{2\pi R_{ij}} + i\omega\rho_0\hat{v}_{n,i} \int_{\partial\Omega_{w,i}} \frac{\exp(-ikR)}{2\pi R} dS_y \\ &= \sum_{\substack{j=1 \\ j\neq i}}^{N} \hat{v}_{n,j}\hat{Z}_{rad,ij} + \hat{v}_{n,i}\hat{Z}_{rad,ii} \end{aligned} \tag{7.326}$$

where R_{ij} is the distance between the center of two patches $\partial\Omega_{w,i}$ and $\partial\Omega_{w,j}$.

$$\hat{Z}_{rad,ij} = i\omega\rho_0 c_0 k \frac{\exp(-ikR_{ij})}{2\pi R_{ij}} \tag{7.327}$$

$$\hat{Z}_{rad,ii} = i\omega\rho_0 \int_{\partial\Omega_{w,i}} \frac{\exp(-ikR)}{2\pi R} dS_y \approx \rho_0 c_0(1 - \exp(-ika)) \tag{7.328}$$

with $a = \sqrt{\partial\Omega_{w,i}/\pi}$.

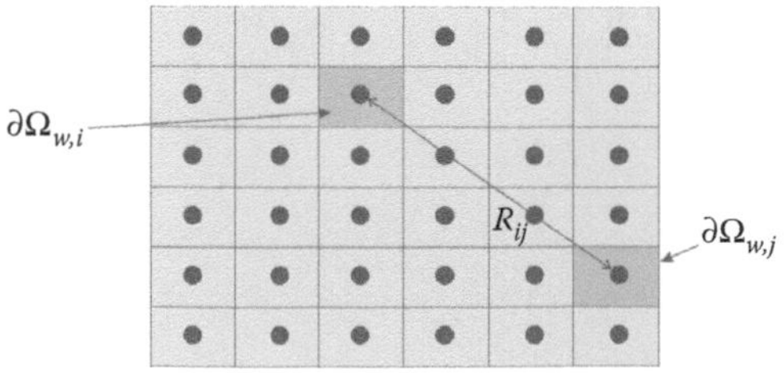

Figure 7.51 Use of patches to discretize the panel surface.

The radiated power is finally given by

$$\begin{aligned}\Pi_{rad} &= \frac{1}{2}\Re\left(\int_{\partial\Omega_w} \hat{p}\hat{v}_n^* dS\right)\\ &= \frac{1}{2}\sum_{i=1}^{N}\sum_{j=1,j\neq i}^{N} \hat{v}_{n,j}\Re(\hat{Z}_{rad,ij})\hat{v}_{n,i}^*\partial\Omega_{w,j} + \frac{1}{2}\left|\hat{v}_{n,i}\right|^2\partial\Omega_{w,i}\sum_{i=1}^{N}\Re(\hat{Z}_{rad,ii})\end{aligned} \tag{7.329}$$

with

$$\Re(\hat{Z}_{rad,ij}) = \rho_0 c_0 k\frac{\sin(kR_{ij})}{2\pi R_{ij}}\partial\Omega_{w,i} \tag{7.330}$$

$$\Re(\hat{Z}_{rad,ii}) = \rho_0 c_0(1 - \cos(ka)) \tag{7.331}$$

Remarks: At low frequencies

$$\Re(\hat{Z}_{rad,ij}) \approx \frac{\rho_0 c_0 k^2}{2\pi}\partial\Omega_{w,i} \tag{7.332}$$

In consequence, we can use the following approximation for the radiated power:

$$\begin{aligned}\Pi_{rad} &\approx \frac{1}{2}\sum_{i=1}^{N}\sum_{j=1}^{N}\hat{v}_{n,j}\Re(\hat{Z}_{rad,ij})\hat{v}_{n,i}^*\partial\Omega_{w,i}\\ &\approx \frac{\rho_0\omega^2}{4\pi c_0}\sum_{i=1}^{N}\sum_{j=1}^{N}\Re\left(\hat{v}_{n,i}\hat{v}_{n,j}^*\right)\partial\Omega_{w,i}\partial\Omega_{w,j}\frac{\sin(kR_{ij})}{kR_{ij}}\end{aligned} \tag{7.333}$$

with $\sin(kR_{ij})/kR_{ij} = 1$ for $i = j$.

7H.2 Calculation of the panel mean square velocity

Using the same discretization of the panel surface into patches, the panel mean square velocity can be written as

$$\left\langle V^2\right\rangle = \frac{1}{2\partial\Omega_w}\left(\int_{\partial\Omega_w}\left|\hat{v}_n^2\right|dS\right) = \frac{1}{2\partial\Omega_w}\sum_{i=1}^{N}\left|\hat{v}_{n,i}\right|^2\partial\Omega_{w,i} \tag{7.334}$$

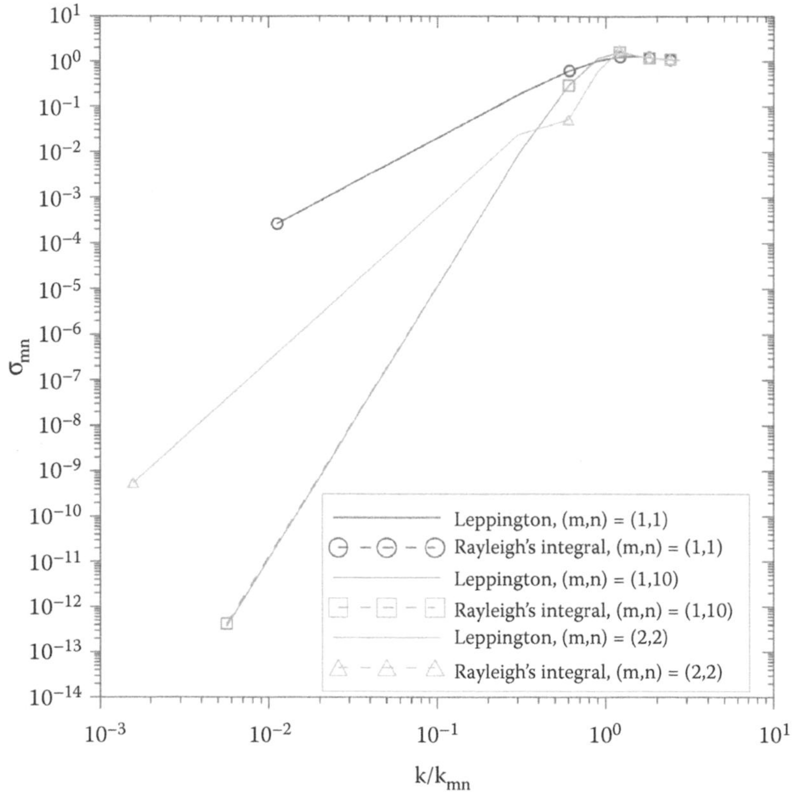

Figure 7.52 Radiation efficiency σ_{mn} of a square simply supported panel for various mode orders. Comparisons between numerical estimate of Rayleigh's integral and Leppington formula.

7H.3 Validation example

We consider the estimation of the modal radiation efficiency of a simply supported square panel. Figure 7.52 compares the Rayleigh's integral-based solution with the asymptotic formulas of Leppington et al. (1982) for various modal orders. A mesh criterion of 6 elements per wavelength is used for the numerical solution. Excellent comparison is observed between the two methods.

Second, we estimate the radiation impedance of a circular piston normalized by $\rho_0 c_0$. Figure 7.53 shows the comparison, for the real and imaginary parts, of the numerical estimates using Rayleigh's integral and the analytical solution (Appendix 7I). Again good comparison is observed.

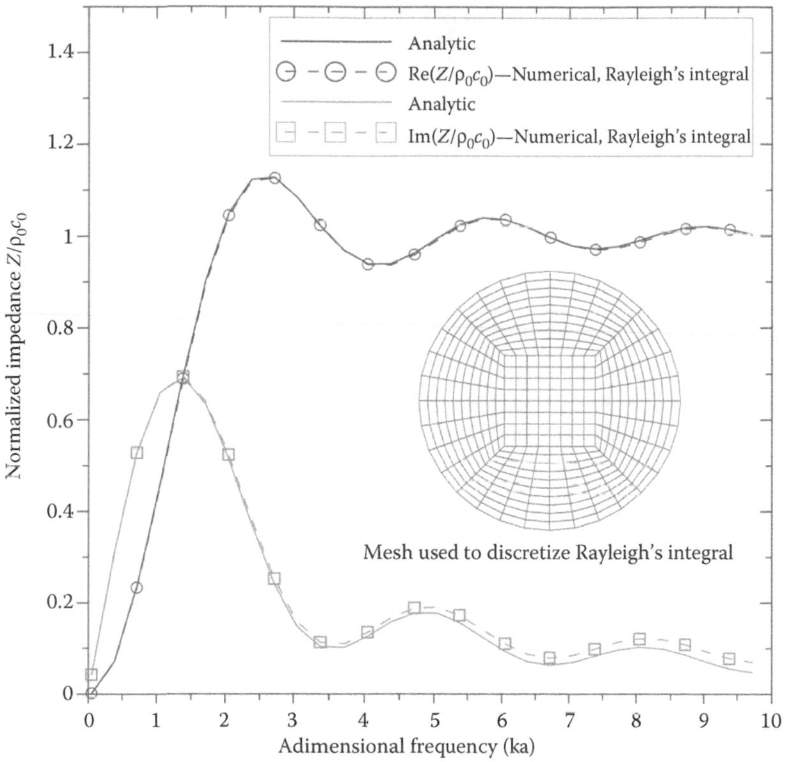

Figure 7.53 Normalized impedance of a circular piston of radius a = 1 m. Comparisons between numerical estimate of Rayleigh's integral and Leppington formula.

APPENDIX 7I: RADIATION OF A BAFFLED CIRCULAR PISTON

Consider a baffled flat circular piston of radius a vibrating with velocity $\hat{v}_n$ (Figure 7.54).

7I.1 Calculation of the sound pressure along the z-axis

The sound pressure radiated at a distance r along the x_3 axis is given by Rayleigh's integral (Kinsler and Frey 2000)

$$\hat{p}_{ax}(r) = \frac{i\rho_0 c_0 k \hat{v}_n}{2\pi} \int_0^a \frac{\exp\left(-ik\sqrt{r^2+\rho^2}\right)}{\sqrt{r^2+\rho^2}} 2\pi\rho d\rho \tag{7.335}$$

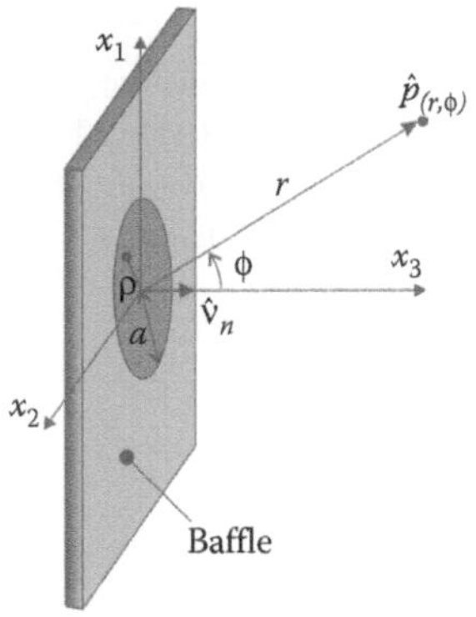

Figure 7.54 Circular piston of radius a embedded in a rigid baffle vibrating with uniform normal velocity $\hat{v}_n$ and radiating in a semi-infinite fluid.

which amounts to

$$\hat{p}_{ax}(r) = \rho_0 c_0 \hat{v}_n \left[1 - \exp\left(-ik\left(\sqrt{r^2 + a^2} - r\right)\right)\right]\exp(-ikr) \tag{7.336}$$

The amplitude is given by

$$\left|\hat{p}_{ax}(r)\right| = 2\rho_0 c_0 \left|\hat{v}_n \sin\left(\frac{kr}{2}\left(\sqrt{1 + \left(\frac{a}{r}\right)^2} - 1\right)\right)\right| \tag{7.337}$$

For $a \ll r$, Equation 7.337 becomes

$$\left|\hat{p}_{ax}(r)\right| = 2\rho_0 c_0 \left|\hat{v}_n \sin\left(\frac{ka}{4}\frac{a}{r}\right)\right| \tag{7.338}$$

71.2 Calculation of the pressure directivity function

In the far field, $ka \ll (r/a)$ and Equation 7.338 can be rewritten as

$$\left|\hat{p}_{ax}(r)\right| = \frac{\rho_0 c_0 k \hat{v}_n}{2\pi r}\pi a^2 \tag{7.339}$$

In the far field, it can be shown that (Kinsler and Frey 2000)

$$\hat{p}(r,\phi) = i\rho_0 \omega \pi a^2 \hat{v}_n \frac{\exp(-ikr)}{2\pi r}\mathfrak{D}(\phi, ka) \tag{7.340}$$

where

$$\mathfrak{D}(\phi, ka) = \frac{2J_1(ka\sin\phi)}{ka\sin\phi} \tag{7.341}$$

is the pressure directivity function. $J_1(x)$ is the Bessel function of order 1.

We note that

$$\left|\hat{p}(r,\phi)\right| = \left|\hat{p}_{ax}(r)\right|\mathfrak{D}(\phi, ka) \tag{7.342}$$

71.3 Specific radiation impedance

The specific radiation impedance of a baffled circular piston is given by (Kinsler and Frey 2000)

$$\mathcal{Z}_{rad} = \rho_0 c_0\left(1 - \frac{2J_1(2ka)}{2ka} + i\frac{2S_1(2ka)}{2ka}\right) \tag{7.343}$$

where $S_1(x)$ is the first-order Struve function (Abramowitz and Stegun 1965).

71.4 MATLAB® scripts

The following MATLAB scripts compare the analytical results with a finite element discretization of Rayleigh's integral (patches with constant shape functions) regarding the directivity function (`DirectivityCircularPiston.m`) and the radiation impedance (`SigmaCircularPiston.m`).

```
function DirectivityCircularPiston
%==============================================================================
% Radiation of a baffled circular Piston
% Use of the Rayleigh's integral
% Calculation of the pressure directivity function
%
% Zero order elements
%
%
%==============================================================================
clear all; close all; clc;
warning off
fprintf('\n=============================================================')
fprintf('\n\t Radiation of a baffled circular Piston');
fprintf('\n Use of the Rayleigh' 'integral using zero order quad elements');
fprintf('\n Comparison with the analytical solution');
fprintf('\n============================================================\n\n')
clc;
close all;


rho0 = 1.213;
c0 = 342;
Z0 = rho0*c0;
```

```
v0=1;   % Arbitrary
a=1;        % piston radius
Surf=pi*a^2;

ka=input(' input ka for directivity calculation:');
k=ka/a;
w=k*c0;

%% mesh the plate and plot the mesh
% data to define the mesh
ratio=0.7;      % 1 : size of the square with respect to the radius
np=16 *4;       % number of nodes on the circumference (should be divisible by
4)
nr=5*4;         % number of nodes between the internal square and the external
circle
[nele, x,y,Se]=meshDisk(a,ratio,np,nr);

fprintf('\n\n meshed surface/exact surface = %f \n', sum(Se)/Surf)

%%
% Compute the directivity function

kR=30;
teta=[-0.999:0.001:0.999]*pi/2;  % receiver location; assume in the far field
R= kR/k;
phi=pi/5;
x0=R*cos(teta); y0=R*sin(teta)*0; z0= R*sin(teta);  % location of the points
monopole= Surf * (1i*w*rho0*v0)*exp(-1i*k*R)/(2*pi*R);  % Pressure radiated by
a monopole in the far field
for it=1:length(teta)
    theta=teta(it);
    r=sqrt((x-x0(it)).^2+(y-y0(it)).^2 + z0(it)^2); % distance between center
element-receiver
    G=exp(-1i*k*r)./(2*pi*r);                    % Rayleigh's integral
    p=sum(G.*Se)*(1i*w*rho0*v0);


    D = abs(p / monopole);
    Dx(it)=D.*cos(theta);
    Dy(it)=D.*sin(theta);
end


figure(2)
plot(Dy,Dx', 'b--')
title([' Directivity for ka=', num2str(ka)]);


%%
% Analytical solution
%
figure(2)
hold on


x= ka.*cos(teta);
D=Directivity(x);       % directivity
D=abs(D);
x=D.*cos(teta);
y=D.*sin(teta);
plot(y,x', 'r')
legend('Rayleigh', 'Analytical')
```

```
function j0=Directivity(x)

if (x<eps)
    j0=ones(1,length(x));
else
    j0=2*bessel(1,x)./x;
end

function SigmaCircularPiston
%==============================================================================
% Radiation of a baffled circular Piston
% Use of the Rayleigh's integral
%
% Zero order elements
%
%
%==============================================================================
clear all; close all; clc;
warning off
fprintf('\n=================================================================')
fprintf('\n\t Resistance and reactance of a baffled circular Piston');
fprintf('\n Use of the Rayleigh' 'integral using zero order quad 4 elements');
fprintf('\n Comparison with the analytical solution');
fprintf('\n===========================================================\n\n')
clc;
close all;

kamax= 10;     % max normalized wavenumber (limits frequency)
ro0 = 1.213;
c0 = 342;
Z0 = ro0*c0;

a=1;        % piston radius (arbitrary)
S=pi*a^2;   % surface of piston

np=16 *4;
nr=5*4;

%% mesh the plate and plot the mesh
%% data to define the mesh
n=2;                        % increase n to refine the mesh
ratio=0.7;      % 1 : size of the square with respect to the radius
np=16 * n;   % number of nodes on the circumference (should be divisible by 4)
nr=5*n;     % number of nodes between the internal square and the external
circle
[nele, x,y,Se]=meshDisk(a,ratio,np,nr);

fprintf('\n\n meshed surface/exact surface = %f \n', sum(Se)/S)

%%
% Compute the radiation impedance

k=[0.05:kamax/30:kamax] / a;
w=k*c0;
nfreq = length(w);  % number of frequency

% get vquad and Wrad
```

```
vquad = zeros(1,nfreq);

R = zeros(1,nfreq);
X = zeros(1,nfreq);  % reactance

Np=nele;

for ipt1 = 1:Np
        x1 = x(ipt1);
        y1 = y(ipt1);
        Si=Se(ipt1);
        for ipt2 = 1:Np
            x2 = x(ipt2);
            y2 = y(ipt2);
            Sj=Se(ipt2);
            rij = sqrt((x1-x2)^2+(y1-y2)^2);
            a0=sqrt(Se(ipt2)/pi);
            if ipt2 == ipt1
                Zij= ro0*c0*(1-exp(-1i.*k*a0));
            else
                Zij=1i*ro0*c0.*k*Sj.*exp(-1i.*k*rij)/(2*pi*rij);
            end
            R = R + real(Zij)*Si;
            X = X+ imag(Zij)*Si;
        end
end
R=R./(ro0*c0*S);   % resistance
X=X./(ro0*c0*S);   % reactance
figure(2)
plot(k*a,R, 'b','linewidth',3); hold on;
xlabel('Ka');
ylabel('R (normalized resistance)')


figure(3)
plot(k*a,X, 'b','linewidth',3); hold on;
xlabel('Ka');
ylabel('X(Normalized Reactance)')


%%
% Analytical solution
%
figure(2)
hold on


ka=k*a;
x=2*ka;
j1_ka=bessel_func(x);        % actually j1(ka)/ka
R_theo= 1- 2*j1_ka;      % See for example Kinsler & Frei


h1_ka=struve(1,x);
X_theo= h1_ka ./ka;        % See for example Kinsler & Frei


plot(k*a,R_theo, 'r:','linewidth',3);
legend('Rayleigh', 'Analytical')

figure(3)
plot(k*a,X_theo, 'r:','linewidth',3);
```

```
legend('Rayleigh', 'Analytical')

function j1=bessel_func(x)


if (x<eps)
    j1=ones(1,length(x));
else
    j1=bessel(1,x)./x;
end


%
% Here an automatic version (from Matlab exchange files)
function f=struve(v,x,n)
% Calculates the Struve Function
%
% struve(v,x)
% struve(v,x,n)
%
% H_v(x) is the struve function and n is the length of
% the series calculation (n=100 if unspecified)
%
% from: Abramowitz and Stegun: Handbook of Mathematical Functions
%       http://www.math.sfu.ca/~cbm/aands/page_496.htm

if nargin<3
n=100;
end
k=0:n;

x=x(:)';
k=k(:);


xx=repmat(x,length(k),1);
kk=repmat(k,1,length(x));


TOP=(-1).^kk;
BOT=gamma(kk+1.5).*gamma(kk+v+1.5);
RIGHT=(xx./2).^(2.*kk+v+1);
FULL=TOP./BOT.*RIGHT;

f=sum(FULL);


function [nele,x,y,Surf]=meshDisk(rayon,ratio,np,nr);
% mesh a disk defined in the plane  x,y
% data to define the mesh
% ratio=0.7;              % 1 : size of the square with respect to the square
% np=40; % number of nodes on the circumference (should be divisible by 4)
% nr=5; number of nodes between the internal square and the external circle
%%%%%%%%%%%%%%%%%%%%%%%%%%%%%%%%%%%%%%%%%%%%%%%%%%%%%%%%%%%
nx=np/4+1;
ny=nx;


dx=ratio*rayon/(nx-1);
dy=ratio*rayon/(ny-1);
dtheta=2*pi/np;


%%
%Definition of positions
%///////////////////////////
```

```
nnpt=nx*ny+np*nr;
nept=(nx-1)*(ny-1)+np*nr;

in = 0;
k=0;
%  Construction of the internal square
for i=0:nx-1
      for j=0:ny-1
            in = in+1;
            nodes(in).id=k*nnpt+i*ny+j+1;
            nodes(in).xyz=[-ratio*rayon*0.5+i*dx;-ratio*rayon/2+j*dy;0];
            nodes(in).group = 0;
            nodes(in).ref = 0;
      end
end
for i=0 :nr-1
%
% Calculation of position for the left part
      for j=0:np/4 -1
            xc=-ratio*rayon/2;
            yc=-ratio*rayon/2+j*dy;
            xp=rayon*cos(5./4.*pi-j*dtheta);
            yp=rayon*sin(5./4.*pi-j*dtheta);
            dxt=(xp-xc)/nr;
            dyt=(yp-yc)/nr;
            in=in+1;
            nodes(in).id=k*nnpt+nx*ny+i*np+j+1;
            nodes(in).xyz=[(i+1)*dxt+xc;(i+1)*dyt+yc;0];
      end
      % Calculation of position for the upper part
      for j=0 : np/4-1
            xc=-ratio*rayon/2+j*dx;
            yc=ratio*rayon/2;
            xp=rayon*cos(3./4.*pi-j*dtheta);
            yp=rayon*sin(3./4.*pi-j*dtheta);
            dxt=(xp-xc)/nr;
            dyt=(yp-yc)/nr;
            in = in + 1;
            nodes(in).id=k*nnpt+nx*ny+i*np+np/4+j+1;
            nodes(in).xyz=[(i+1)*dxt+xc,(i+1)*dyt+yc,0];
            nodes(in).group = 0;
            nodes(in).ref = 0;
      end
      % Calcul of positions for the right part
      for j=0:np/4-1
            xc=ratio*rayon/2;
            yc=ratio*rayon/2-j*dy;
            xp=rayon*cos(1./4.*pi-j*dtheta);
            yp=rayon*sin(1./4.*pi-j*dtheta);
            dxt=(xp-xc)/nr;
            dyt=(yp-yc)/nr;
            in=in+1;
            nodes(in).id=k*nnpt+nx*ny+i*np+2*np/4+j+1;
            nodes(in).xyz=[(i+1)*dxt+xc;(i+1)*dyt+yc;0];
            nodes(in).group = 0;
            nodes(in).ref = 0;
      end
      % Calculation of positions for the lower part
      for j=0:np/4-1;
            xc=ratio*rayon/2-j*dx;
            yc=-ratio*rayon/2;
            xp=rayon*cos(-1./4.*pi-j*dtheta);
```

```
            yp=rayon*sin(-1./4.*pi-j*dtheta);
            dxt=(xp-xc)/nr;
            dyt=(yp-yc)/nr;
            in=in+1;
            nodes(in).id=k*nnpt+nx*ny+i*np+3*np/4+j+1;
            nodes(in).xyz=[(i+1)*dxt+xc;(i+1)*dyt+yc;0];
            nodes(in).group = 0;
            nodes(in).ref = 0;
      end


end
nnode = in;


%%
% definition of elements
%


ie=0;
elem_type=16;
% definition of internal square
for i=0:nx-2
    for j=0 : ny-2
     ie = ie + 1;
     elemnets(ie).id=nept*k+i*(ny-1)+j+1;
     elements(ie).prop = 1;
     elements(ie).type = elem_type;
     elements(ie).topo = 4;
     elements(ie).ien(1)= nnpt*k+i*ny+j+1;
     elements(ie).ien(2)= nnpt*k+(i+1)*ny+j+1;
     elements(ie).ien(3)= nnpt*k+(i+1)*ny+j+2;
     elements(ie).ien(4)= nnpt*k+i*ny+j+2;
     end
end

% definition of left row
for i=0 : np/4 -1
     ie = ie + 1;
     elements(ie).id=nept*k+(nx-1)*(ny-1)+i+1;
     elements(ie).prop = 1;
     elements(ie).type = elem_type;
     elements(ie).topo = 4;
     elements(ie).ien(1)= nnpt*k+nx*ny+i+1;
     elements(ie).ien(2)= nnpt*k+i+1;
     elements(ie).ien(3)= nnpt*k+i+2;
     elements(ie).ien(4)= nnpt*k+nx*ny+i+2;
end

% definition of upper row
for i=0 : np/4 -1
     ie = ie + 1;
     elements(ie).id=nept*k+(nx-1)*(ny-1)+np/4+i+1;
     elements(ie).prop = 1;
     elements(ie).type = elem_type;
     elements(ie).topo = 4;
     elements(ie).ien(1)= nnpt*k+nx*ny+i+1+np/4;
     elements(ie).ien(2)= nnpt*k+(i+1)*ny;
     elements(ie).ien(3)= nnpt*k+(i+2)*ny;
     elements(ie).ien(4)= nnpt*k+nx*ny+i+2+np/4;
end
```

```
% definition of right row
for i=0 : np/4-1
     ie = ie + 1;
     elements(ie).id=nept*k+(nx-1)*(ny-1)+2*np/4+i+1;
     elements(ie).prop = 1;
     elements(ie).type = elem_type;
     elements(ie).topo = 4;
     elements(ie).ien(1)= nnpt*k+nx*ny+i+1+2*np/4;
     elements(ie).ien(2)= nnpt*k+nx*ny-i;
     elements(ie).ien(3)= nnpt*k+nx*ny-i-1;
     elements(ie).ien(4)= nnpt*k+nx*ny+i+2+2*np/4;
end

% definition of lower row


for i=0 : np/4-2
     ie = ie + 1;
     elements(ie).id=nept*k+(nx-1)*(ny-1)+3*np/4+i+1;
     elements(ie).prop = 1;
     elements(ie).type = elem_type;
     elements(ie).topo = 4;
     elements(ie).ien(1)= nnpt*k+nx*ny+i+1+3*np/4;
     elements(ie).ien(2)= nnpt*k+ny*(nx-i-1)+1;
     elements(ie).ien(3)= nnpt*k+ny*(nx-i-2)+1;
     elements(ie).ien(4)= nnpt*k+nx*ny+i+2+3*np/4;
 end
ie = ie + 1;
elements(ie).id=nept*k+(nx-1)*(ny-1)+np;
elements(ie).prop = 1;
elements(ie).type = elem_type;
elements(ie).topo = 4;
elements(ie).ien(1)= nnpt*k+nx*ny+np;
elements(ie).ien(2)= nnpt*k+ny+1;
elements(ie).ien(3)= nnpt*k+1;
elements(ie).ien(4)= nnpt*k+nx*ny+1;


%Construction of the ring
for j=0 : nr-2;
      for i=0 :np-2
            ie = ie + 1;
            elements(ie).id=nept*k+(nx-1)*(ny-1)+np*(j+1)+i+1;
            elements(ie).prop = 1;
            elements(ie).type = elem_type;
            elements(ie).topo = 4;
            elements(ie).ien(1)=  nnpt*k+nx*ny+np*(j+1)+i+1;
            elements(ie).ien(2)= nnpt*k+nx*ny+np*j+i+1;
            elements(ie).ien(3)= nnpt*k+nx*ny+np*j+i+2;
            elements(ie).ien(4)= nnpt*k+nx*ny+np*(j+1)+i+2;
      end
      ie = ie + 1;
      elements(ie).id=nept*k+(nx-1)*(ny-1)+np*(j+2);
      elements(ie).prop = 1;
      elements(ie).type = elem_type;
      elements(ie).topo = 4;
      elements(ie).ien(1)= nnpt*k+nx*ny+np*(j+2);
      elements(ie).ien(2)= nnpt*k+nx*ny+np*(j+1);
      elements(ie).ien(3)= nnpt*k+nx*ny+np*j+1;
      elements(ie).ien(4)= nnpt*k+nx*ny+np*(j+1)+1;
end

nele=ie;
```

```
%%
% Graph the mesh
%
%
node=nodes;
close all;
figure(1);
axis off;
hold on;
for ie= 1: nele  %plotting the sub-systems
    ien = elements(ie).ien;
    node1 = nodes(ien(1)).xyz;
    node2 = nodes(ien(2)).xyz;
    node3 = nodes(ien(3)).xyz;
    node4 = nodes(ien(4)).xyz;

    x=[node1(1),node2(1),node3(1),node4(1),node1(1)];  %nodes id in global
notation + the first node
    y=[node1(2),node2(2),node3(2),node4(2),node1(2)];
    z=[node1(3),node2(3),node3(3),node4(3),node1(3)];
    plot3(x,y,z);

end

%
% Compute the location of the center nodes and the surface of the elements
%


for ie= 1: nele  %plotting the sub-systems
    ien = elements(ie).ien;
    xy(1,:) = nodes(ien(1)).xyz;
    xy(2,:) = nodes(ien(2)).xyz;
    xy(3,:) = nodes(ien(3)).xyz;
    xy(4,:) = nodes(ien(4)).xyz;

    % center node
    x(ie)=(xy(1,1)+xy(2,1)+xy(3,1)+xy(4,1))/4;
    y(ie)=(xy(1,2)+xy(2,2)+xy(3,2)+xy(4,2))/4;
    % surface of the element
    x31=xy(3,1)-xy(1,1); y31=xy(3,2)-xy(1,2);
    x42=xy(4,1)-xy(2,1); y42=xy(4,2)-xy(2,2);
    Surf(ie)=(y42*x31-y31*x42)/2;
end
```

REFERENCES

Abramowitz, M. and I. A. Stegun 1965. *Handbook of Mathematical Functions with Formulas, Graphs, and Mathematical Tables*. New York, USA: Dover Publications.

Amini, S. and P. J. Harris 1990. A Comparison between various boundary integral formulations of the exterior acoustic problem. *Computer Methods in Applied Mechanics and Engineering* 84 (1): 59–75.

Amini, S. and D. T. Wilton 1986. An investigation of boundary element methods for the exterior acoustic problem. *Computer Methods in Applied Mechanics and Engineering* 54 (1): 49–65.

Amsallem, D. and C. Farhat 2012. Stabilization of projection-based reduced-order models. *International Journal for Numerical Methods in Engineering* 91: 343–456.

Atalla, N. and R. J. Bernhard 1994. Review of numerical solutions for low-frequency structural-acoustic problems. *Applied Acoustics* 43 (3): 271–94.

Bonnet, M. 1995. Équations intégrales et éléments de frontière: Applications en mécanique des solides et des fluides. CNRS éd.

Bruneau, M. 2006. *Fundamentals of Acoustics*. London; Newport Beach, CA: ISTE Publishing Company.

Burgschweiger, R., I. Schäfer, A. Mohsen, R. Piscoya, M. Ochmann, and B. Nolte. 2013. Results of an implementation of the dual surface method to treat the nonuniqueness in solving acoustic exterior problems using the boundary element method. In: *Proceedings of Meetings on Acoustics, 19:. 065060*, Montreal, QC, Canada: Acoustical Society of America, pp. 1–8.

Chen, K., J. Cheng, and P. J. Harris 2009. A new study of the Burton and Miller method for the solution of a 3D Helmholtz problem. *IMA Journal of Applied Mathematics* 74 (2): 163–77.

Ciskowski, R. D. and C. A. Brebbia. 1991. *Boundary Element Methods in Acoustics*. Southampton, UK: Computational Mechanics Publications.

Coyette, J. P., C. Lecomte, and J. L. Migeot. 1999. Calculation of vibro-acoustic frequency response functions using a single frequency boundary element solution and a pade expansion. *Acustica—Acta Acustica* 85: 371–77.

D'Amico, R., A. Pratellesi, M. Pierini, and M. Tournour. 2010. Efficient method to avoid fictitious eigenvalues for indirect BEM. In *Proceedings of ISMA 2010*, Leuven, Belgium.

Dunavant, D. 1985. High degree efficient symmetrical Gaussian quadrature rules for the triangle. *International Journal for Numerical Methods in Engineering* 21: 1129–48.

Guiggiani, M., G. Krishnasamy, T. J. Rudolphi, and F. J. Rizzo. 1992. A general algorithm for the numerical solution of hypersingular boundary integral equations. *Journal of Applied Mechanics-Transactions of the ASME* 59: 604–14.

Hamdi, M. A. 1982. *Formulation variationnelle par équations intégrales pour le calcul de champs acoustiques linéaires proches et lointains*. PhD Thesis, Université de Technologie de Compiègne.

Hamdi, M. A. 1988. Méthodes de Discrétisation Par Éléments Finis et Éléments Finis de Frontière. In Rayonnement Acoustique Des Structures, Eyrolle. Paris: Lesueur.

Hammer, P. C., O. P. Marlowe, and A. H. Stroud. 1956. Numerical integration over simplexes and cones. *Mathematical Tables and Other Aids to Computation* 10: 130–37.

Harris, P. J. 1992. A boundary element method for the Helmholtz equation using finite part integration. *Computer Methods in Applied Mechanics and Engineering* 95 (3): 331–42.

Harris, P. J., K. Chen, and J. Cheng. 2006. A weakly singular boundary integral formulation of the external Helmholtz problem valid for all wavenumbers. In: C. Constanda, Z. Nashed, and D. Rollins (Eds.), *Integral Methods in Science and Engineering*, Birkhäuser: Boston, pp. 79–87. http://link.springer.com/chapter/10.1007/0-8176-4450-4_8.

Hetmaniuk, U., R. Tezaur, and C. Farhat. 2012. Review and assessment of interpolatory model order reduction methods for frequency response structural dynamics and acoustics problems. *International Journal for Numerical Methods in Engineering* 90: 1636–62.

Hetmaniuk, U., R. Tezaur, and C. Farhat. 2013. An adaptive scheme for a class of interpolatory model reduction methods for frequency response problems. *International Journal for Numerical Methods in Engineering* 93: 1109–24.

Ingber, M. S. and C. E. Hickox. 1992. A modified Burton-Miller algorithm for treating the uniqueness of representation problem for exterior acoustic radiation and scattering problems. *Engineering Analysis with Boundary Elements* 9 (4): 323–29.

Junger, M. C. and D. Feit. 1972. *Sound Structures, and Their Interaction*. MIT Press: Cambridge, MA.

Kinsler, L.E. and A. R. Frey. 2000. *Fundamentals of Acoustics*. 4th ed. Wiley & Sons: Chichester, UK.

Leppington, F. G., E. G. Broadbent, and K. H. Heron. 1982. The acoustic radiation efficiency of rectangular panels. *Proceedings of the Royal Society of London. Series A, Mathematical and Physical Sciences* 382 (1783): 245–71.

Li, S. 2005. An efficient technique for multi-frequency acoustic analysis by boundary element method. *Journal of Sound and Vibration* 283 (3–5): 971–80.

Lyness, J. and D. Jespersen. 1975. Moderate degree symmetric quadrature rules for the triangle. *Journal of the Institute of Mathematics and Its Applications* 15 (1): 19–32.

Marburg, S. 2008. Discretization requirements: How many elements per wavelength are necessary? In: S. Marburg and B. Nolte (Eds.), *Computational Acoustics of Noise Propagation in Fluids—Finite and Boundary Element Methods*. Berlin Heidelberg: Springer, pp. 309–32. http://link.springer.com.ezproxy.usherbrooke.ca/chapter/10.1007/978-3-540-77448-8_12.

Marburg, S. and S. Amini. 2005. Cat's eye radiation with boundary elements: Comparative study on treatment of irregular frequencies. *Journal of Computational Acoustics* 13 (01): 21–45.

Marburg, S. and T.-W. Wu. 2008. Treating the phenomenon of irregular frequencies. In: S. Marburg and B. Nolte (Eds.), *Computational Acoustics of Noise Propagation in Fluids—Finite and Boundary Element Methods*. Berlin, Heidelberg: Springer, pp. 411–34. http://link.springer.com.ezproxy.usherbrooke.ca/chapter/10.1007/978-3-540-77448-8_16.

Maue, A.-W. 1949. Zur Formulierung eines allgemeinen Beugungs-problems durch eine Integralgleichung. *Zeitschrift für Physik* 126 (7–9): 601–18.

Mohsen, A., R. Piscoya, and M. Ochmann. 2011. The application of the dual surface method to treat the nonuniqueness in solving acoustic exterior problems. *Acta Acustica United with Acustica* 97 (4): 699–707.

Panich, I. O. 1965. On the question of the solvability of the exterior boundary values problem for the wave equation and Maxwell's equation (in Russian). *Uspekhi Mat. Nauk. (Advanced Mathematical Sciences)* 20 (1): 221–26.

Pierce, A. D. 1993. *Variational Formulations in Acoustics Radiation and Scattering*. Academic Press: San Diego, CA. Physical Acoustics, XXII.

Qin, X., J. Zhang, G. Xie, F. Zhou, and G. Li. 2011. A general algorithm for the numerical evaluation of nearly singular integrals on 3D boundary element. *Journal of Computational and Applied Mathematics* 235 (14): 4174–86.

Quevat, J. P., M. A. Hamdi, and V. Martin. 1988. Fréquences Parasites Pour Les Vibrations Des Systèmes Couplés. In: *Calcul Des Structures et Intelligence Artificielle*. Méthodes Numériques Dans Les Sciences de L'ingénieur. Pluralis, pp. 183–95.

Raveendra, S. T. 1999. An efficient indirect boundary element technique for multi-frequency acoustic analysis. *International Journal for Numerical Methods in Engineering* 44 (1): 59–76.

Rumpler, R. and P. Göransson. 2013. A finite element solution strategy based on padé approximants for fast multiple frequency sweeps of multivariate problems. In: *Proceedings of Meetings on Acoustics, 19:. 065003*, Montreal, QC, Canada: Acoustical Society of America, pp. 1–9.

Schenck, H. A. 1968. Improved integral formulation for acoustic radiation problems. *The Journal of the Acoustical Society of America* 44 (1): 41–58.

Seybert, A. F. and T. W. Wu. 1989. Modified Helmholtz integral equation for bodies sitting on an infinite plane. *The Journal of the Acoustical Society of America* 85 (1): 19–23.

Sgard, F. 1995. Numerical study of the vibro-acoustic behavior of a plate-cavity system radiating in mean flow subjected to different kinds of excitations. Etude Numérique Du Comportement Vibro-Acoustique D'un Système Plaque-Cavité Rayonnant Dans Un Écoulement Uniforme, Pour Différents Types D'excitation(in French). PhD, University of Sherbrooke, Sherbrooke, QC, Canada.

Shannon, C. 1949. Communication in the presence of noise. *Proceedings of the Institute of Radio Engineers* 37 (1): 10–21.

Silva, J. J. do Rêgo. 1994. *Acoustic and Elastic Wave Scattering Using Boundary Elements*. Vol. 18. Topics in Engineering. Southampton, UK: Computational Mechanics Publications.

Stallybrass, M. P. 1967. On a pointwise variational principle for the approximate solution of linear boundary value problems. *Journal of Mathematics and Mechanics* 16 (11): 1247–86.

Tournour, M. A. and N. Atalla. 1998. State-of-the-Art of FEM/BEM for the acoustic and vibration response of an elastic box. *Noise Control Engineering Journal* 46: 83–90.

Vanhille, C. and A. Lavie. 1998. An efficient tool for multi-frequency analysis in acoustic scattering or radiation by boundary element method. *Acustica—Acta Acustica* 84: 884–93.

Von Estorff, O. 2003. Efforts to reduce computation time in numerical acoustics an overview. *Acta Acustica United with Acustica* 89 (1): 1–13.

Von Estorff, O. and O. Zaleski. 2003. Efficient acoustic calculations by the BEM and frequency interpolated transfer functions. *Engineering Analysis with Boundary Elements* 27 (7): 683–94.

Wang, W. and N. Atalla. 1997. A numerical algorithm for double surface integrals over quadrilaterals with a 1/R singularity. *Communications in Numerical Methods in Engineering* 13 (11): 885–90.

Wright, M. C. M. 2005. "Chapter 3." *In Lecture Notes on the Mathematics of Acoustics*. 1st ed. London, UK: Imperial College Press, p. 308.

Wu, T. W., W. L. Li, and A. F. Seybert. 1993. An efficient boundary element algorithm for multi-frequency acoustical analysis. *Journal of the Acoustical Society of America* 94 (1): 447–52.

Zhang, Z., N. Vlahopoulos, T. Allen, and K. Y. Zhang. 2001. A source reconstruction process based on an indirect variational boundary element formulation. *Engineering Analysis with Boundary Elements* 25 (2): 93–114.

Chapter 8

Problem of exterior coupling

8.1 INTRODUCTION

This chapter deals with numerical methods used for solving exterior coupling problems. By exterior coupling we mean the coupling of a vibrating structure with an exterior unbounded fluid domain. The formulations used are generally based on the coupling of the FEM for the structure and the BEM for the exterior fluid. We present in this chapter two classes of formulations: FEM-Standard Boundary Element Method (SBEM)* and FEM-VBEM†. We will focus on formulations leading to symmetric coupled systems.

The methods presented in this chapter can be combined with the FEM presented in Chapter 6 to solve problems of exterior/interior coupling. In this category of problems, the structure couples two fluid domains. The interior domain is bounded, whereas the exterior domain is unbounded. This is, for example, the case of a structure coupled to an internal cavity and radiating in an unbounded heavy fluid.

8.2 EQUATIONS OF THE PROBLEM OF FLUID-STRUCTURE EXTERIOR COUPLING

The configuration of interest is depicted in Figure 8.1. The structure Ω_s is supposed to be linear elastic and the fluid Ω_f is supposed to be homogeneous, nonviscous, and at rest.

We are seeking the harmonic response ($\exp(i\omega t)$) of the structure in the presence of the exterior fluid. The classic solution methodology uses the FEM for the structure and the boundary integral formulation for the exterior problem (see Chapter 7).

* *Standard BEM*: This is the boundary element method associated with the SBIE presented in Chapter 7.

† *Variational Boundary Element Method*: This is the boundary element method associated with the VBIE presented in Chapter 7.

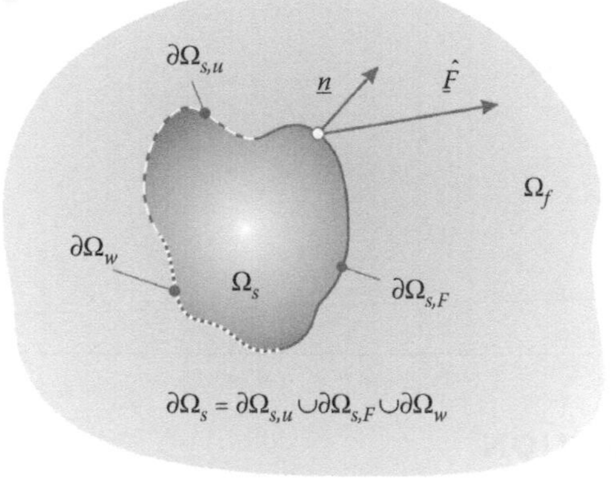

Figure 8.1 Exterior coupled fluid-structure problem.

Using the notations of Chapters 6 and 7, the equations for the structural displacement and the acoustic pressure in the fluid can be written as

$$\underline{\nabla} \cdot \underline{\underline{\hat{\sigma}}} + \rho_s \omega^2 \underline{\hat{u}} = 0 \quad \text{in } \Omega_s \tag{8.1a}$$

$$\underline{\underline{\hat{\sigma}}} \cdot \underline{n} = \hat{\underline{F}} \quad \text{over } \partial\Omega_{s,F} \tag{8.1b}$$

$$\underline{\underline{\hat{\sigma}}} \cdot \underline{n} = -\hat{p}\underline{n} \quad \text{over } \partial\Omega_w \tag{8.1c}$$

$$\underline{\hat{u}} = \underline{\bar{u}} \quad \text{over } \partial\Omega_{s,u} \tag{8.1d}$$

$$\nabla^2 \hat{p} + k^2 \hat{p} = 0 \quad \text{in } \Omega_f \tag{8.1e}$$

$$\frac{\partial \hat{p}}{\partial n} = \rho_0 \omega^2 \underline{\hat{u}} \cdot \underline{n} \quad \text{over } \partial\Omega_w \tag{8.1f}$$

$$\lim_{r \to \infty} \left(\frac{\partial \hat{p}}{\partial r} + ik\hat{p} \right) r = 0 \tag{8.1g}$$

where $\underline{\hat{u}}$ is the structural displacement field, $\underline{\underline{\hat{\sigma}}}$ is the structural stress tensor, ρ_s is the structure density, $\hat{p}$ is the acoustic pressure in the fluid, ρ_0 is the fluid density, and $k = \omega/c_0$ is the wavenumber in Ω_f.

Equation 8.1a represents the linear elastodynamic equation.

Equation 8.1b corresponds to the external force per unit area applied on the structure.

Equation 8.1c corresponds to the continuity of the normal stresses on the wetted surface $\partial\Omega_w$. The right-hand side represents the force per unit area acting on the structure due to its sound radiation into Ω_f.
Equation 8.1d corresponds to the imposed displacement vector.
Equation 8.1e is Helmholtz equation.
Equation 8.1f describes the continuity of fluid and structural normal displacements on the surface of the structure $\partial\Omega_s$. Together with Equation 8.1c, they describe the coupling between the structure and the exterior fluid.
Equation 8.1g is the Sommerfeld radiation condition.

The integral formulation of this problem is established in two steps:

- A Galerkin's weak formulation of the equations of the vibrating structure induced by the acoustic pressure field is derived.
- An integral representation of the acoustic field induced by the displacement of the vibrating structure is written. Two approaches are considered: the standard formulation (SBEM) and the variational formulation (VBEM).

8.3 VARIATIONAL FORMULATION OF THE STRUCTURE EQUATIONS

This step is identical to problems of interior coupling. It is detailed in Chapter 6. Let us recall the main results. Let $\delta\underline{u}$ be a regular function defined over Ω_s. For all admissible $(\underline{u}, \delta\underline{u})$, the variational formulation of the harmonic response of the structure subjected to the actions of the acoustic pressure in Ω_f can be written as

$$\int_{\Omega_s} \underline{\underline{\hat{\sigma}}}(\hat{u}) : \underline{\underline{\hat{\varepsilon}}}(\delta\underline{\hat{u}})\,dV - \int_{\Omega_s} \rho_s\omega^2\,\underline{\hat{u}}\cdot\delta\underline{\hat{u}}\,dV + \int_{\partial\Omega_w} \hat{p}\underline{n}\cdot\delta\underline{\hat{u}}\,dS = \int_{\partial\Omega_{s,F}} \delta\underline{\hat{u}}\cdot\underline{\hat{F}}\,dS \quad \forall\delta\underline{\hat{u}} \tag{8.2}$$

where $\underline{\underline{\hat{\varepsilon}}}$ is the structural strain tensor. The finite element discretization of this equation involves the following matrices and vectors (see Chapter 6):

$$\int_{\Omega_s} \underline{\underline{\hat{\sigma}}}(\underline{\hat{u}}) : \underline{\underline{\hat{\varepsilon}}}(\delta\underline{\hat{u}})\,dV = \langle\delta\hat{u}\rangle[K]\{\hat{u}\} \tag{8.3}$$

$$\int_{\Omega_S} \rho_s\underline{\hat{u}}\cdot\delta\underline{\hat{u}}\,dV = \langle\delta\hat{u}\rangle[M]\{\hat{u}\} \tag{8.4}$$

$$\int_{\partial\Omega_w} \hat{p}\underline{n} \cdot \delta\underline{\hat{u}}\, dS = \langle \delta\hat{u} \rangle [C_{up}] \{\hat{p}\} \tag{8.5}$$

$$\int_{\partial\Omega_{s,F}} \delta\underline{\hat{u}} \cdot \underline{\hat{F}}\, dS = \langle \delta\hat{u} \rangle \{\hat{F}\} \tag{8.6}$$

8.4 FEM–BEM COUPLING

8.4.1 Direct formulation

Using the acoustic pressure as the main variable, the regularized direct integral formulation of the exterior problem is given in Section 7.2.2 of Chapter 7:

$$\hat{p}(\underline{x}) = \int_{\partial\Omega_w} \left[\hat{p}(\underline{y}) \frac{\partial \hat{G}(\underline{x},\underline{y})}{\partial n_y} - \frac{\partial G_0(\underline{x},\underline{y})}{\partial n_y} \hat{p}(\underline{x}) \right] dS_y - \int_{\partial\Omega_w} \hat{G}(\underline{x},\underline{y}) \frac{\partial \hat{p}(\underline{y})}{\partial n_y} dS_y \quad \forall \underline{x} \in \partial\Omega_w \tag{8.7}$$

where $\hat{G}(\underline{x},\underline{y})$ is the free-space fundamental solution* or free-space Green's function and $G_0(\underline{x},\underline{y})$ is the corresponding static solution. $\hat{G}(\underline{x},\underline{y})$ and $G_0(\underline{x},\underline{y})$ are given by Equations 7.3 and 7.7, respectively.

Recall (see Chapter 7) that the associated normal derivatives are given by

$$\frac{\partial \hat{G}(\underline{x},\underline{y})}{\partial n_y} = -\left(\frac{1 + ikr}{4\pi r^2} \right) \exp(-ikr) \frac{\partial r}{\partial n_y} \tag{8.8}$$

and

$$\frac{\partial G_0(\underline{x},\underline{y})}{\partial n_y} = -\frac{1}{4\pi r^2} \frac{\partial r}{\partial n_y} \tag{8.9}$$

where $r = |\underline{y} - \underline{x}|$ is the Euclidian distance between the location $\underline{x}$ of the source and the current point $\underline{y}$ and $\partial r / \partial n_y$ is the normal derivative of r at $\underline{y}$:

* Referred to as $\hat{G}_\infty(\underline{x},\underline{y})$ in Chapter 7.

$$\frac{\partial r}{\partial n_y} = \frac{\underline{y} - \underline{x}}{r} \cdot \underline{n}_y \tag{8.10}$$

Substituting Neumann boundary condition (Equation 8.1e) into Equation 8.7 leads to:

$$\hat{p}(\underline{x}) = \int_{\partial\Omega_w} \left[\hat{p}(\underline{y}) \frac{\partial \hat{G}(\underline{x}, \underline{y})}{\partial n_y} - \frac{\partial G_0(\underline{x}, \underline{y})}{\partial n_y} \hat{p}(\underline{x}) \right] dS_y - \int_{\partial\Omega_w} \rho_0 \omega^2 \hat{G}(\underline{x}, \underline{y}) \underline{\hat{u}}(\underline{y}) \cdot \underline{n}_y \, dS_y \tag{8.11}$$

The variational formulation of the coupled problem consists in finding $\underline{\hat{u}}$ and $\hat{p}$ regular admissible functions satisfying both Equations 8.2 and 8.11 for all kinematically admissible variations $\delta\underline{\hat{u}}$ of the structure displacement field.

The boundary element discretization of Equation 8.11, based on the collocation method involves the following matrices (see Chapter 7):

$$-\hat{p}(\underline{x}) + \int_{\partial\Omega_w} \left[\hat{p}(\underline{y}) \frac{\partial \hat{G}(\underline{x}, \underline{y})}{\partial n_y} - \frac{\partial G_0(\underline{x}, \underline{y})}{\partial n_y} \hat{p}(\underline{x}) \right] dS_y = [\hat{\mathcal{A}}_c(\omega)]\{\hat{p}\} \tag{8.12}$$

$$\int_{\partial\Omega_w} \hat{G}(\underline{x}, \underline{y}) \underline{\hat{u}}(\underline{y}) \cdot \underline{n}_y \, dS_y = [\hat{\mathcal{B}}_c(\omega)]\{\hat{u}\} \tag{8.13}$$

The matrices $[\hat{\mathcal{A}}_c(\omega)]$ and $[\hat{\mathcal{B}}_c(\omega)]$ are complex-valued, nonsymmetric, full, and frequency-dependent (= function of circular frequency ω).

Invoking the stationarity of Equation 8.2 and using Equation 8.11 leads to the following coupled linear system:

$$\begin{pmatrix} [K] - \omega^2[M] & [C_{up}] \\ -\rho_0\omega^2[\hat{\mathcal{B}}_c(\omega)] & [\hat{\mathcal{A}}_c(\omega)] \end{pmatrix} \begin{Bmatrix} \{\hat{u}\} \\ \{\hat{p}\} \end{Bmatrix} = \begin{Bmatrix} \{\hat{F}\} \\ \{0\} \end{Bmatrix} \tag{8.14}$$

Matrices $[M]$, $[K]$, $[C_{up}]$, and $\{\hat{F}\}$ are given by Equations 8.3 through 8.6.

Contrary to the interior problem, this system is nonsymmetric and contains frequency-dependent matrices. Consequently, matrices $[\hat{\mathcal{A}}_c(\omega)]$ and $[\hat{\mathcal{B}}_c(\omega)]$ must be evaluated at each calculation frequency. The numerical implementation of such a formulation can prove to be costly (see Chapter 7 for a discussion on possible acceleration of the direct solution using frequency interpolation methods). Two other methods can be used in order to reduce the size of system (8.14):

- Elimination of the structure variable
- Elimination of the fluid variable

8.4.2 Condensation of the structure variable

Let us recall (see Chapters 5 and 6) the definition of the in vacuo structural normalized modal basis $\left\{[\Omega_s^2],[\Phi_s]\right\}$ truncated at the order $n_{m,s}$ ($n_{m,s} \ll N_s$, N_s being the order of matrices $[M]$ and $[K]$):

$$\{\hat{u}\} = [\Phi_s]\{\hat{u}_m\};\ \ [\Phi_s]^T[K][\Phi_s] = [\Omega_s^2];\ \ [\Phi_s]^T[M][\Phi_s] = [I_{n_{m,s}}]$$
$$\{\hat{F}_m\} = [\Phi_s]^T\{\hat{F}\} \tag{8.15}$$

where $\{\hat{u}_m\}$ and $\{\hat{F}_m\}$ are the generalized displacement vector and generalized force vector (modal coordinates vectors). $[\Phi_s]$ is a rectangular matrix of dimensions $N_s \times n_{m,s}$ containing the $n_{m,s}$ first structural eigenvectors:

$$[\Phi_s] = \left\langle\{\Phi_1\},\{\Phi_2\},\ldots,\{\Phi_{n_{m,s}}\}\right\rangle \tag{8.16}$$

and $[\Omega_s^2]$ is a diagonal matrix of dimension $n_{m,s} \times n_{m,s}$ containing the $n_{m,s}$ first structural eigenvalues:

$$[\Omega_s^2] = \begin{pmatrix} \omega_{s,1}^2 & 0 & \cdots & 0 \\ 0 & \omega_{s,2}^2 & \ddots & \vdots \\ \vdots & \ddots & \ddots & 0 \\ 0 & \cdots & 0 & \omega_{s,n_{m,s}}^2 \end{pmatrix} \tag{8.17}$$

By projecting the first line of Equation 8.14 over the truncated structural modal basis $\left\{[\Omega_s^2],[\Phi_s]\right\}$, we get

$$\begin{pmatrix} [\hat{\Omega}_s^2] - \omega^2[I_{n_{m,s}}] & [\Phi_s]^T[C_{up}] \\ -\rho_0\omega^2[\hat{\mathcal{B}}_c(\omega)] & [\hat{\mathcal{A}}_c(\omega)] \end{pmatrix} \begin{Bmatrix} \{\hat{u}_m\} \\ \{\hat{p}\} \end{Bmatrix} = \begin{Bmatrix} \{\hat{F}_m\} \\ \{0\} \end{Bmatrix} \tag{8.18}$$

Here, the notation $[\hat{\Omega}_s^2]$ indicates that matrix $[\Omega_s^2]$ has been corrected to account for the modal damping coefficients (see Chapter 6).

Let

$$[\hat{S}_{s,m}(\omega)] = i\omega[\hat{Z}_{s,m}(\omega)] = i\omega\left[\frac{[\hat{\Omega}_s^2]}{i\omega} + i\omega[I_{n_{m,s}}]\right] \tag{8.19}$$

$[\hat{Z}_{s,m}(\omega)]$ represents the modal structural mechanical impedance matrix.

From the first line of Equation 8.18, we write

$$\{\hat{u}_m\} = [\hat{S}_{s,m}(\omega)]^{-1}\left[\{\hat{F}_m\} - [\Phi_s]^T[C_{up}]\{\hat{p}\}\right] \tag{8.20}$$

It should be noted that $[\hat{S}_{s,m}(\omega)]$ is a diagonal matrix and thus the calculation of its inverse is straightforward.

Using Equation 8.20, the second line of Equation 8.18 can be written as

$$\left[[\hat{\mathcal{A}}_c(\omega)] + \omega^2\rho_0[\hat{\mathcal{B}}_c(\omega)][\Phi_s][\hat{S}_{s,m}(\omega)]^{-1}[\Phi_s]^T[C_{up}]\right]\{\hat{p}\}$$
$$= \omega^2\rho_0[\hat{\mathcal{B}}_c(\omega)][\Phi_s][\hat{S}_{s,m}(\omega)]^{-1}\{\hat{F}_m\} \tag{8.21}$$

We end up with a complex-valued nonsymmetric system but with a dimension smaller than the initial system (= of the order of the number of nodes of the BEM mesh of boundary $\partial\Omega_w$).

Once the pressure nodal vector spectrum has been computed, the displacement nodal vector can be calculated from Equation 8.20. The mean kinetic energy of the structure can then be evaluated using:

$$\langle E_c\rangle = \frac{1}{2}\omega^2\int_{\Omega_s}\rho_s(\underline{\hat{u}}\cdot\underline{\hat{u}}^*)\,dV = \frac{1}{2}\omega^2\langle\hat{u}\rangle M\{\hat{u}^*\} = \frac{1}{2}\omega^2\langle\hat{u}_m\rangle\{\hat{u}_m^*\} \tag{8.22}$$

The acoustic power radiated by the structure in the exterior fluid is given by

$$\Pi_{rad} = \frac{1}{2}\Re\left(\int_{\partial\Omega_w}\hat{p}(i\omega\underline{\hat{u}}\cdot\underline{n})^*\,dS\right) = \frac{\omega}{2}\Im\left(\langle\hat{u}^*\rangle[C_{up}]\{\hat{p}\}\right)$$
$$= \frac{\omega}{2}\Im\left(\langle\hat{u}_m^*\rangle[\Phi_s]^T[C_{up}]\{\hat{p}\}\right) \tag{8.23}$$

Finally, the pressure at a point $\underline{x}$ in the exterior fluid is given by (see Chapter 7):

$$\hat{p}(\underline{x}) = \int_{\partial\Omega_w} \hat{p}(\underline{y}) \frac{\partial \hat{G}(\underline{x},\underline{y})}{\partial n_y} dS_y - \int_{\partial\Omega_w} \hat{G}(\underline{x},\underline{y}) \frac{\partial \hat{p}}{\partial n_y} dS_y$$
$$= [\hat{\mathcal{A}}_c'(\omega)]\{\hat{p}\} - \rho_0\omega^2[\hat{\mathcal{B}}_c(\omega)][\Phi_s]\{\hat{u}_m\} \quad (8.24)$$

Here, matrix $[\hat{\mathcal{A}}'(\omega)]$ comes out from the discretization of $\int_{\partial\Omega_w} \hat{p}(\underline{y})\partial\hat{G}(\underline{x},\underline{y})/\partial n_y \, dS_y$ (note the difference with matrix $[\hat{\mathcal{A}}(\omega)]$ given in Equation 8.12).

8.4.3 Condensation of the fluid variable: Calculation of the radiation impedance matrix

The previous approach has the drawback to lead to a large size nonsymmetric complex-valued system. A more efficient alternative is based on the condensation of the fluid variable in Equation 8.14. We start from the second line of Equation 8.14 to write

$$\{\hat{p}\} = \rho_0\omega^2[\hat{\mathcal{A}}_c(\omega)]^{-1}[\hat{\mathcal{B}}_c(\omega)]\{\hat{u}\} \quad (8.25)$$

In Equation 8.25, matrix $[\hat{\mathcal{A}}_c(\omega)]$ must be inverted at each calculation frequency.

Substituting Equation 8.25 in the first line of Equation 8.14 leads to

$$\left[-\omega^2[M] + [K] + \rho_0\omega^2[C_{up}][\hat{\mathcal{A}}_c(\omega)]^{-1} [\hat{\mathcal{B}}_c(\omega)] \right]\{\hat{u}\} = \{\hat{F}\} \quad (8.26)$$

Let

$$[\hat{\mathcal{Z}}_{rad}(\omega)] = -i\omega \, \rho_0[C_{up}][\hat{\mathcal{A}}_c(\omega)]^{-1}[\hat{\mathcal{B}}_c(\omega)] \quad (8.27)$$

Equation 8.14 takes the classic form

$$[-\omega^2[M] + i\omega[\hat{\mathcal{Z}}_{rad}(\omega)] + [K]]\{\hat{u}\} = \{\hat{F}\} \quad (8.28)$$

Matrix $[\hat{\mathcal{Z}}_{rad}(\omega)]$ represents the acoustic radiation impedance matrix of the structure. It embodies the effect of the fluid loading on the dynamic response of the structure. In order to gain insight into its physical meaning, let us decompose $[\hat{\mathcal{Z}}_{rad}(\omega)]$ as

$$[\hat{\mathcal{Z}}_{rad}(\omega)] = \Re([\hat{\mathcal{Z}}_{rad}(\omega)]) + i\Im([\hat{\mathcal{Z}}_{rad}(\omega)]) \tag{8.29}$$

and let

$$\Im[\hat{\mathcal{Z}}_{rad}(\omega)] = \omega[\hat{\mathcal{M}}_a(\omega)] \tag{8.30}$$

Equation 8.28 becomes

$$\left[-\omega^2[M + \hat{\mathcal{M}}_a(\omega)] + i\omega\Re([\hat{\mathcal{Z}}_{rad}(\omega)]) + [K]\right]\{\hat{u}\} = \{\hat{F}\} \tag{8.31}$$

We clearly see that, at low frequency, matrix $[\hat{\mathcal{M}}_a(\omega)]$ is related to an added mass matrix $[\hat{\mathcal{M}}_a(0)]$ of the fluid on the structure. In addition, the real part of the acoustic radiation impedance matrix contributes to the damping of the structure: this is radiation damping. To see it more clearly, substitute Equation 8.25 in the expression of the acoustic power radiated by the structure in the exterior fluid:

$$\begin{aligned}\Pi_{rad} &= \frac{1}{2}\Re\left(\int_{\partial\Omega_w} \hat{p}(i\omega\hat{\underline{u}}\cdot\underline{n})^* \, dS\right) = -\frac{1}{2}\Re(i\omega\langle\hat{u}^*\rangle[C_{up}]\{\hat{p}\}) \\ &= \frac{\omega^2}{2}\Re\left(\langle\hat{u}^*\rangle\left[-i\omega\rho_0[C_{up}][\hat{\mathcal{A}}_c(\omega)]^{-1}[\hat{\mathcal{B}}_c(\omega)]\right]\{\hat{u}\}\right)\end{aligned} \tag{8.32}$$

and using the definition of the acoustic radiation impedance matrix, Equation 8.27, we end up with

$$\Pi_{rad} = \frac{\omega^2}{2}\Re\left[\langle\hat{u}^*\rangle[\hat{\mathcal{Z}}_{rad}(\omega)]\{\hat{u}\}\right] \tag{8.33}$$

Remarks:

1. The dimension of the system given by Equation 8.31 can be further reduced by using the in-vacuo modal expansion of the structural displacement. We then get

$$\left[-\omega^2[I_{n_{m,s}}] + i\omega[\hat{\mathcal{Z}}_{rad,m}(\omega)] + [\hat{\Omega}_s^2]\right]\{\hat{u}_m\} = \{\hat{F}_m\} \tag{8.34}$$

where the projected (modal) acoustic radiation impedance matrix $[\hat{\mathcal{Z}}_{rad,m}(\omega)]$ is given by

$$[\hat{\mathcal{Z}}_{rad,m}(\omega)] = [\Phi_s]^T[\hat{\mathcal{Z}}_{rad}(\omega)][\Phi_s] \tag{8.35}$$

This leads to a much smaller system of size $n_{m,s} \times n_{m,s}$.

The acoustic power radiated by the structure is calculated using the modal coordinates as

$$\Pi_{rad} = \frac{\omega^2}{2}\Re\left(\left\langle \hat{u}_m^* \right\rangle [\hat{\mathcal{Z}}_{rad,m}(\omega)]\{\hat{u}_m\}\right) \tag{8.36}$$

2. In the case of an incompressible fluid ($k = 0$), the effect of the fluid reduces to an added mass effect on the structure

$$\left[-\omega^2[M + \hat{\mathcal{M}}_a(0)] + [K]\right]\{\hat{u}\} = \{\hat{F}\} \tag{8.37}$$

with

$$[\hat{\mathcal{M}}_a(0)] = -\rho_0[C_{up}][\hat{\mathcal{A}}_c(0)]^{-1}[\hat{\mathcal{B}}_c(0)] \tag{8.38}$$

3. For cases of strong coupling, we can use the eigenvalue problem associated with Equation 8.37 to accelerate the convergence of the modal expansion. That is, we use the modes of the structure accounting for the fluid-added mass effect. We then project the coupled system onto these modes and solve in terms of modal coordinates.
4. As mentioned earlier, the evaluation of the radiation impedance matrix necessitates the inversion of matrix $[\hat{\mathcal{A}}_c(\omega)]$ at each frequency step (see Equation 8.27). However, the changes of matrix $[\hat{\mathcal{Z}}_{rad}(\omega)]$ with frequency are usually smooth. In consequence, this matrix is only calculated at a few master frequencies and its values interpolated at intermediate frequencies. To circumvent storage requirements issues, the modal version of the matrix (Equation 8.35) is usually used.

8.5 FEM–VBEM COUPLING

8.5.1 Indirect formulation: Case of thin structures

In order to achieve a symmetric coupled formulation, we are going to use the variational indirect integral representation for the exterior problem (see

Chapter 7). For all kinematically admissible variation of the double-layer potential $\delta\hat{\mu}$, we have shown that (see Section 7.7.1)

$$\rho_0\omega^2 \int_{\partial\Omega_w} \delta\mu(\underline{x})(\underline{\hat{u}} \cdot \underline{n}_x)\, dS_x = \int_{\partial\Omega_w}\int_{\partial\Omega_w} k^2(\underline{n}_x \cdot \underline{n}_y)\hat{G}(\underline{x},\underline{y})\delta\hat{\mu}(\underline{x})\hat{\mu}(\underline{y})\, dS_x\, dS_y$$
$$- \int_{\partial\Omega_w}\int_{\partial\Omega_w} \left[\left(\underline{n}_x \times \underline{\nabla}_x\delta\hat{\mu}(\underline{x})\right) \cdot \left(\underline{n}_y \times \underline{\nabla}_y\hat{\mu}(\underline{y})\right)\right]\hat{G}(\underline{x},\underline{y})\, dS_x\, dS_y \tag{8.39}$$

Using the regularized boundary integral equation, the acoustic pressure in the exterior domain and on the wetted part of the structure are respectively given by

$$\hat{p}(\underline{x}) = \int_{\partial\Omega_w} \hat{\mu}(\underline{y})\frac{\partial\hat{G}(\underline{x},\underline{y})}{\partial n_y}\, dS_y \quad \text{for } \underline{x} \notin \partial\Omega_w \tag{8.40}$$

and

$$\hat{p}(\underline{x}) = \int_{\partial\Omega_w} \left[\hat{\mu}(\underline{y})\frac{\partial\hat{G}(\underline{x},\underline{y})}{\partial n_y} - \hat{\mu}(\underline{x})\frac{\partial G_0(\underline{x},\underline{y})}{\partial n_y}\right] dS_y \quad \text{for } \underline{x} \in \partial\Omega_w \tag{8.41}$$

The discretization of Equation 8.39 with boundary elements involves the following matrices (see Chapter 7):

$$\int_{\partial\Omega_w}\int_{\partial\Omega_w} k^2(\underline{n}_x \cdot \underline{n}_y)\hat{G}(\underline{x},\underline{y})\delta\hat{\mu}(\underline{x})\hat{\mu}(\underline{y})\, dS_x\, dS_y$$
$$- \int_{\partial\Omega_w}\int_{\partial\Omega_w} \left[\left(\underline{n}_x \times \underline{\nabla}_x\delta\hat{\mu}(\underline{x})\right) \cdot \left(\underline{n}_y \times \underline{\nabla}_y\hat{\mu}(\underline{y})\right)\right]\hat{G}(\underline{x},\underline{y})\, dS_x\, dS_y = \langle\delta\hat{\mu}\rangle[\hat{\mathcal{D}}(\omega)]\{\hat{\mu}\} \tag{8.42}$$

and

$$\int_{\partial\Omega_w} \delta\mu(\underline{x})(\underline{\hat{u}} \cdot \underline{n}_x)\, dS_x = \langle\, \delta\hat{\mu}\,\rangle[C_{u\mu}]^T\{\hat{u}\} \tag{8.43}$$

Note that matrix $[\hat{\mathcal{D}}(\omega)]$ is complex-valued, full but symmetric. Furthermore, note the equivalence between matrix $[C_{u\mu}]$ in Equation 8.43 and that in Equation 8.5.

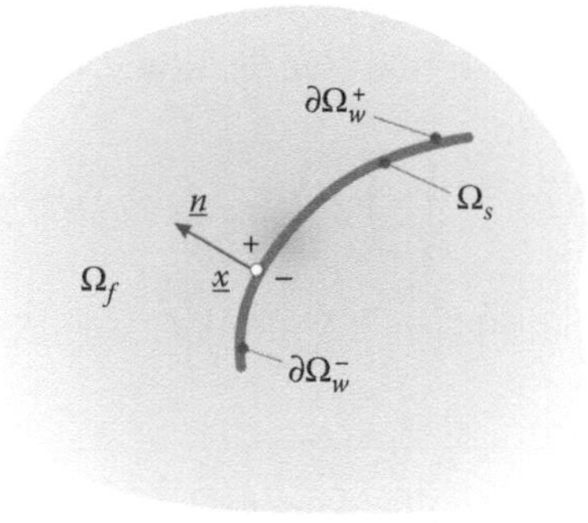

Figure 8.2 Notations used for a thin structure.

In order to calculate the fluid-loading term seen by the structure, $\int_{\partial\Omega_w} \hat{p}\underline{n} \cdot \delta\underline{\hat{u}}\, dS$, we have to determine the relationship between the sound pressure $\hat{p}$ and the variable $\hat{\mu}$. We are going to obtain this relationship by assuming that the structure is thin and that it is in contact with the fluid on both faces (see Figure 8.2). In this case, $\hat{\mu}$ can be interpreted as the pressure jump across $\partial\Omega_w$.

Let us denote the two faces of the boundary $\partial\Omega_w$ by $\partial\Omega_w^+$ and $\partial\Omega_w^-$ and let $\hat{p}^+$ and $\hat{p}^-$ be the pressure on these two faces. We can then write

$$\int_{\partial\Omega_w} \hat{p}(\underline{x})(\delta\underline{\hat{u}} \cdot \underline{n}_x)\, dS_x = \int_{\partial\Omega_w^+} \hat{p}^+(\underline{x})(\delta\underline{\hat{u}} \cdot \underline{n}_x)\, dS_x - \int_{\partial\Omega_w^-} \hat{p}^-(\underline{x})(\delta\underline{\hat{u}} \cdot \underline{n}_x)\, dS_x$$
$$= \int_{\partial\Omega_w^+} \hat{\mu}(\underline{x})(\delta\underline{u} \cdot \underline{n}_x)\, dS_x = \langle\delta\hat{u}\rangle[C_{u\mu}]\{\hat{\mu}\} \tag{8.44}$$

The stationarity of Equation 8.2 and Equation 8.39 leads to the following coupled system:

$$\begin{pmatrix} [K] - \omega^2[M] & [C_{u\mu}] \\ [C_{u\mu}]^T & -\dfrac{1}{\rho_0\omega^2}[\hat{\mathcal{D}}(\omega)] \end{pmatrix} \begin{Bmatrix} \{\hat{u}\} \\ \{\hat{\mu}\} \end{Bmatrix} = \begin{Bmatrix} \{\hat{F}\} \\ \{0\} \end{Bmatrix} \tag{8.45}$$

where matrices $[M]$, $[K]$, $[C]$, and $\{\hat{F}\}$ are given by Equations 8.3 through 8.6.

Contrary to the case where the direct integral formulation is used for the fluid, this system is symmetric. In addition, the condensation of variable $\hat{\mu}$ in Equation 8.45 leads to

$$\{\hat{\mu}\} = \rho_0\omega^2[\hat{\mathcal{D}}(\omega)]^{-1}[C_{u\mu}]^T\{\hat{u}\} \tag{8.46}$$

where matrix $[\hat{\mathcal{D}}(\omega)]$ must be inverted at each calculation frequency. Consequently, the solution of the coupled problem is equivalent to the resolution of Equation 8.28 where this time the acoustic radiation impedance matrix reads

$$[\hat{\mathcal{Z}}_{rad}(\omega)] = -i\rho_0\omega[C_{u\mu}][\hat{\mathcal{D}}(\omega)]^{-1}[C_{u\mu}]^T \tag{8.47}$$

Thus, with this formulation the acoustic radiation impedance matrix is symmetric. In addition, writing this matrix in terms of its real and imaginary part leads to

$$\left[-\omega^2[M + \hat{\mathcal{M}}_a(\omega)] + i\omega\Re\left([\hat{\mathcal{Z}}_{rad}(\omega)]\right) + [K]\right]\{\hat{u}\} = \{\hat{F}\} \tag{8.48}$$

where this time the added mass matrix, given by Equation 8.30 is symmetric.

Once Equation 8.48 has been solved, the pressure jump nodal vector is recovered from Equation 8.46. Then, the pressure at all points of the exterior domain and at the surface of the structure is recalculated from the discretized forms of Equations 8.40 and 8.41. The acoustic power radiated by the structure is then given by

$$\Pi_{rad} = \frac{1}{2}\Re\left(\int_{\partial\Omega_w^+} \hat{\mu}(i\omega\underline{\hat{u}} \cdot \underline{n})^* \, dS\right) = \frac{-1}{2}\Re\left(i\omega\langle\hat{u}\rangle^*[C_{u\mu}]\{\hat{\mu}\}\right) \tag{8.49}$$

Note that the acoustic power radiated by the structure can be directly calculated from Equation 8.33. However, since $[\hat{\mathcal{Z}}_{rad}(\omega)]$ is symmetric, Equation 8.33 reduces to

$$\Pi_{rad} = \frac{\omega^2}{2}\langle\hat{u}^*\rangle\Re([\hat{\mathcal{Z}}_{rad}(\omega)])\{\hat{u}\} \tag{8.50}$$

Thus, the real part of $[\hat{\mathcal{Z}}_{rad}(\omega)]$ governs the acoustic power radiated by the structure.

Remarks:

1. The dimensions of the system given by Equation 8.48 can be further reduced by using a modal expansion of the structural displacement in

terms of the associated in vacuo modal basis (see remark 1 of previous section). This leads to Equation 8.34 where the projected acoustic radiation impedance matrix is given by Equation 8.35. The radiated acoustic power is directly calculated from expression Equation 8.36. Again the multifrequency solution can be accelerated by interpolating the modal impedance matrix from its evaluation at a few master frequencies.

2. In the case of an incompressible fluid, ($k = 0$), the effect of the fluid reduces to an added mass effect on the structure where

$$[\hat{\mathcal{M}}_a(0)] = -\rho_0[C_{u\mu}][\hat{\mathcal{D}}(0)]^{-1}[C_{u\mu}]^T \tag{8.51}$$

In terms of modal coordinates, the modal added mass matrix is given by

$$[\hat{\mathcal{M}}_{a,m}] = [\Phi_s]^T[\hat{\mathcal{M}}_a(0)][\Phi_s] \tag{8.52}$$

Note that $[\hat{\mathcal{M}}_{a,m}]$ is symmetric and positive definite.

3. The numerical implementation of the FEM/VBEM approach can be carried out in multiple ways. One way to do it is to build one structural mesh and one BEM mesh whose nodes are coincident physically but which have different indexes. Each mesh uses its own node numbering system from 1 to N_s for the structural mesh and from 1 to N_f for the BEM mesh. Correspondence tables between the structural node index and the BEM fluid node index together with the structural element index and the BEM fluid element index are built. Each node can have several degrees of freedom depending on the nature of the element. For a structural thin shell element, the number of degrees of freedom per node is 6 (3 displacements, 3 rotations). For an acoustic BEM element, the number of degrees of freedom per node is 1 (pressure [direct formulation], pressure jump or normal pressure gradient jump [indirect formulation]) or 2 (pressure jump and normal pressure gradient jump [indirect formulation]). The fluid matrix $[\hat{\mathcal{D}}(\omega)]$ is calculated in the fluid nodes numbering system. To calculate the fluid-structure coupling matrix, the BEM mesh is swept and for each BEM element, the facing structural element is found in order to assemble the corresponding matrix coefficients at the right place. Equation 8.45 can be solved directly in terms of variables $\{\hat{u}\}$ and $\{\hat{\mu}\}$ or using the reduced form given by Equations 8.46 and 8.48.

8.5.2 Indirect formulation: General case

The previous formulation is only valid for thin structures. In the general case, the calculation of the fluid loading seen by the structure requires the use of the general relationship Equation 8.41 between the pressure $\hat{p}$ and the double-layer potential $\hat{\mu}$.

Using Equation 8.41, we can write the fluid-loading term as

$$\int_{\partial\Omega_w} \hat{p}(\underline{x})(\delta\underline{\hat{u}} \cdot \underline{n}_x) dS_x$$

$$= \int_{\partial\Omega_w} \int_{\partial\Omega_w} \left[\hat{\mu}(\underline{y}) \frac{\partial \hat{G}(\underline{x}, \underline{y})}{\partial n_y} - \hat{\mu}(\underline{x}) \frac{\partial G_0(\underline{x}, \underline{y})}{\partial n_y} \right] (\underline{n}_x \cdot \delta\underline{\hat{u}}(\underline{x})) dS_y \, dS_x \qquad (8.53)$$

The discretization of the right hand side leads to

$$\int_{\partial\Omega_w} \int_{\partial\Omega_w} \left[\hat{\mu}(\underline{y}) \frac{\partial \hat{G}(\underline{x}, \underline{y})}{\partial n_y} - \hat{\mu}(\underline{x}) \frac{\partial G_0(\underline{x}, \underline{y})}{\partial n_y} \right] (\underline{n}_x \cdot \delta\underline{\hat{u}}(\underline{x})) dS_y \, dS_x = \langle \delta\hat{u} \rangle \left[\hat{\mathcal{C}}_{u\mu}(\omega) \right] \{\hat{\mu}\} \qquad (8.54)$$

Combining the stationarity of Equation 8.2 and Equation 8.39 leads to the following coupled system:

$$\begin{pmatrix} [K] - \omega^2[M] & [\hat{C}_{u\mu}(\omega)] \\ [C_{u\mu}]^T & -\frac{1}{\rho_0 \omega^2}[\hat{\mathcal{D}}(\omega)] \end{pmatrix} \begin{Bmatrix} \{\hat{u}\} \\ \{\hat{\mu}\} \end{Bmatrix} = \begin{Bmatrix} \{\hat{F}\} \\ \{0\} \end{Bmatrix} \qquad (8.55)$$

We end up with a nonsymmetric coupled formulation. Moreover, the condensation of variable $\hat{\mu}$ in Equation 8.55 yields Equation 8.28 where the acoustic radiation impedance matrix is given this time by

$$[\hat{\mathcal{Z}}_{rad}(\omega)] = -i\rho_0\omega[\hat{\mathcal{C}}_{u\mu}(\omega)][\hat{\mathcal{D}}(\omega)]^{-1}[C_{u\mu}]^T \qquad (8.56)$$

Thus, in this formulation, the acoustic radiation impedance matrix is nonsymmetric. This is a drawback for the numerical resolution. The computation time and the memory required to solve the system are larger than for symmetric systems. The symmetric variational formulation presented in the next section is preferred.

8.5.3 FEM–VBEM symmetric direct formulation: General case

In order to make the coupled problem symmetric in the case of any structure, we use the direct integral formulation based on the acoustic pressure and its normal derivative.

Let y be a point on the boundary $\partial\Omega_w$, and $\underline{x}$, a point belonging to the fluid domain Ω_f. The classic direct integral representation of the acoustic pressure associated with the exterior problem allows us to write (see Chapter 7):

$$C^+(\underline{x})\hat{p}(\underline{x}) = \int_{\partial\Omega_w} \hat{p}(\underline{y})\frac{\partial\hat{G}(\underline{x},\underline{y})}{\partial n_y}dS_y - \rho_0\omega^2 \int_{\partial\Omega_w} \underline{\hat{u}}(\underline{y})\cdot \underline{n}_y\hat{G}(\underline{x},\underline{y})dS_y \tag{8.57}$$

with

$$\begin{cases} C^+(\underline{x}) = 1 & \forall \underline{x} \in \Omega_f \text{ and } \underline{x} \notin \partial\Omega_w \\ C^+(\underline{x}) = 1 + \displaystyle\int_{\partial\Omega_w} \frac{\partial G_0(\underline{x},\underline{y})}{\partial n_y}dS_y & \forall \underline{x} \in \partial\Omega_w \end{cases} \tag{8.58}$$

In the case of a structure with a smooth boundary, we have shown in Chapter 7 that for $\underline{x} \in \partial\Omega_w, C^+(\underline{x}) = 1/2$. In order to link the acoustic pressure and the normal particle velocity at the surface of the structure, assumed to be smooth, we have to calculate the normal gradient of the acoustic pressure on the surface of the structure (see Chapter 7):

$$\frac{1}{2}\frac{\partial\hat{p}(\underline{x})}{\partial n_x} = \text{FP}\int_{\partial\Omega_w} \hat{p}(\underline{y})\frac{\partial^2\hat{G}(\underline{x},\underline{y})}{\partial n_x \partial n_y}dS_y - \rho_0\omega^2 \int_{\partial\Omega_w} \underline{\hat{u}}(\underline{y})\cdot \underline{n}_y \frac{\partial\hat{G}(\underline{x},\underline{y})}{\partial n_x}dS_y \quad \forall \underline{x} \in \partial\Omega_w \tag{8.59}$$

where FP denotes the Hadamard finite part of the integral.

Substituting Neumann condition (Equation 8.1f) into Equation 8.59 and dividing by $\rho_0\omega^2$, the previous equation becomes

$$\frac{1}{2}\underline{\hat{u}}(\underline{x})\cdot \underline{n}_x = \frac{1}{\rho_0\omega^2}\text{FP}\int_{\partial\Omega_w} \hat{p}(\underline{y})\frac{\partial^2\hat{G}(\underline{x},\underline{y})}{\partial n_x \partial n_y}dS_y - \int_{\partial\Omega_w} \underline{\hat{u}}(\underline{y})\cdot \underline{n}_y \frac{\partial\hat{G}(\underline{x},\underline{y})}{\partial n_x}dS_y \quad \forall \underline{x} \in \partial\Omega_w \tag{8.60}$$

To avoid the calculation of the Hadamard finite part, we associate a variational formulation to Equation 8.60. Using the usual technique (see Chapter 7), we get

$$\frac{1}{2}\int_{\partial\Omega_w}\left(\hat{\underline{u}}(\underline{x})\cdot\underline{n}_x\right)\delta\hat{p}(\underline{x})\,dS_x = \frac{1}{\rho_0\omega^2}\int_{\partial\Omega_w}\int_{\partial\Omega_w}\hat{p}(\underline{y})\frac{\partial^2\hat{G}(\underline{x},\underline{y})}{\partial n_x\partial n_y}\delta\hat{p}(\underline{x})\,dS_y\,dS_x - \int_{\partial\Omega_w}\int_{\partial\Omega_w}\hat{\underline{u}}(\underline{y})\cdot\underline{n}_y\frac{\partial\hat{G}(\underline{x},\underline{y})}{\partial n_x}\delta\hat{p}(\underline{x})\,dS_y\,dS_x \tag{8.61}$$

for any admissible variation $\delta\hat{p}$ of $\hat{p}$. Therefore, the equations of the coupled system are given by Equations 8.2 and 8.61:

$$\int_{\Omega_s}\underline{\underline{\hat{\sigma}}}(\hat{\underline{u}}):\underline{\underline{\hat{\varepsilon}}}(\delta\hat{\underline{u}})\,dV - \int_{\Omega_s}\rho_s\omega^2\hat{\underline{u}}\cdot\delta\hat{\underline{u}}\,dV + \int_{\partial\Omega_w}\hat{p}\underline{n}\cdot\delta\hat{\underline{u}}\,dS = \int_{\partial\Omega_s/\partial\Omega_w}\delta\hat{\underline{u}}\cdot\hat{\underline{F}}\,dS$$

$$\frac{1}{2}\int_{\partial\Omega_w}\left(\hat{\underline{u}}(\underline{x})\cdot\underline{n}_x\right)\delta\hat{p}(\underline{x})\,dS_x = \frac{1}{\rho_0\omega^2}\int_{\partial\Omega_w}\int_{\partial\Omega_w}\hat{p}(\underline{y})\frac{\partial^2\hat{G}(\underline{x},\underline{y})}{\partial n_x\partial n_y}\delta\hat{p}(\underline{x})\,dS_y\,dS_x - \int_{\partial\Omega_w}\int_{\partial\Omega_w}\hat{\underline{u}}(\underline{y})\cdot\underline{n}_y\frac{\partial\hat{G}(\underline{x},\underline{y})}{\partial n_x}\delta\hat{p}(\underline{x})\,dS_y\,dS_x \tag{8.62}$$

for any admissible couple $\left(\delta\hat{\underline{u}},\delta\hat{p}\right)$.

In order to make this coupled system symmetric, let us write the fluid loading term seen by the structure as

$$\int_{\partial\Omega_w}\hat{p}(\underline{x})\left(\underline{n}_x\cdot\delta\hat{\underline{u}}(\underline{x})\right)dS_x = \frac{1}{2}\int_{\partial\Omega_w}\hat{p}(\underline{x})\left(\underline{n}_x\cdot\delta\hat{\underline{u}}(\underline{x})\right)dS_x + \frac{1}{2}\int_{\partial\Omega_w}\hat{p}(\underline{x})\left(\underline{n}_x\cdot\delta\hat{\underline{u}}(\underline{x})\right)dS_x \tag{8.63}$$

Using Equation 8.57 with $C^+(\underline{x}) = 1/2$ since the boundary is supposed to be smooth, we can write

$$\int_{\partial\Omega_w}\hat{p}(\underline{x})\left(\delta\hat{\underline{u}}(\underline{x})\cdot\underline{n}_x\right)dS_x = \frac{1}{2}\int_{\partial\Omega_w}\hat{p}(\underline{x})\left(\delta\hat{\underline{u}}(\underline{x})\cdot\underline{n}_x\right)dS_x + \int_{\partial\Omega_w}\int_{\partial\Omega_w}\hat{p}(\underline{y})\frac{\partial\hat{G}(\underline{x},\underline{y})}{\partial n_y}\left(\delta\hat{\underline{u}}(\underline{x})\cdot\underline{n}_x\right)dS_y\,dS_x - \rho_0\omega^2\int_{\partial\Omega_w}\int_{\partial\Omega_w}\left(\hat{\underline{u}}(\underline{y})\cdot\underline{n}_y\right)\hat{G}(\underline{x},\underline{y})\left(\delta\hat{\underline{u}}(\underline{x})\cdot\underline{n}_x\right)dS_y\,dS_x \tag{8.64}$$

The coupled system becomes

$$\int_{\Omega_s} \underline{\underline{\hat{\sigma}}}(\underline{\hat{u}}) : \underline{\underline{\hat{\varepsilon}}}(\delta\underline{\hat{u}})\,dV - \int_{\Omega_s} \rho_s\omega^2 \underline{\hat{u}} \cdot \delta\underline{\hat{u}}\,dV + \frac{1}{2}\int_{\partial\Omega_w} \hat{p}(\underline{x})\left(\underline{n}_x \cdot \delta\underline{\hat{u}}(\underline{x})\right)dS_x$$

$$+\int_{\partial\Omega_w}\int_{\partial\Omega_w} \hat{p}(\underline{y})\frac{\partial\hat{G}(\underline{x},\underline{y})}{\partial n_y}\left(\delta\underline{\hat{u}}(\underline{x})\cdot\underline{n}_x\right)dS_y\,dS_x$$

$$-\rho_0\omega^2\int_{\partial\Omega_w}\int_{\partial\Omega_w}\left(\underline{\hat{u}}(\underline{y})\cdot\underline{n}_y\right)\hat{G}(\underline{x},\underline{y})\left(\delta\underline{\hat{u}}(\underline{x})\cdot\underline{n}_x\right)dS_y\,dS_x$$

$$= \int_{\partial\Omega_s/\partial\Omega_w} \delta\underline{\hat{u}}(\underline{x})\cdot\underline{\hat{F}}(\underline{x})\,dS_x$$

$$\frac{1}{2}\int_{\partial\Omega_w}\left(\underline{\hat{u}}(\underline{x})\cdot\underline{n}_x\right)\delta\hat{p}(\underline{x})\,dS_x = \frac{1}{\rho_0\omega^2}\int_{\partial\Omega_w}\int_{\partial\Omega_w}\hat{p}(\underline{y})\frac{\partial^2\hat{G}(\underline{x},\underline{y})}{\partial n_x\partial n_y}\delta\hat{p}(\underline{x})\,dS_y\,dS_x$$

$$-\int_{\partial\Omega_w}\int_{\partial\Omega_w}\underline{\hat{u}}(\underline{y})\cdot\underline{n}_y\frac{\partial\hat{G}(\underline{x},\underline{y})}{\partial n_x}\delta\hat{p}(\underline{x})\,dS_y\,dS_x \tag{8.65}$$

for any admissible couple $\left(\delta\underline{\hat{u}},\delta\hat{p}\right)$.

We end up with a symmetric coupled formulation.

The boundary element discretization of Equation 8.65 involves three new matrices:

$$\frac{1}{2}\int_{\partial\Omega_w}\hat{p}(\underline{x})\left(\underline{n}_x\cdot\delta\underline{\hat{u}}(\underline{x})\right)dS_x = \langle\delta\hat{u}\rangle\left[C_{up}^{(1)}\right]\{\hat{p}\} \tag{8.66}$$

with $[C_{up}^{(1)}] = 1/2[C_{up}]$,$[C_{up}]$ being given by Equation 8.5.

$$\int_{\partial\Omega_w}\int_{\partial\Omega_w}\hat{p}(\underline{y})\frac{\partial\hat{G}(\underline{x},\underline{y})}{\partial n_y}\left(\delta\underline{\hat{u}}(\underline{x})\cdot\underline{n}_x\right)dS_y\,dS_x = \langle\delta\hat{u}\rangle[\hat{C}_{up}(\omega)]\{\hat{p}\} \tag{8.67}$$

$$\rho_0\int_{\partial\Omega_w}\int_{\partial\Omega_w}\left(\underline{\hat{u}}(\underline{y})\cdot\underline{n}_y\right)\hat{G}(\underline{x},\underline{y})\left(\delta\underline{\hat{u}}(\underline{x})\cdot\underline{n}_x\right)dS_y\,dS_x = \langle\delta\hat{u}\rangle[\hat{\mathcal{M}}_1(\omega)]\{\hat{u}\} \tag{8.68}$$

Note that matrix $[\hat{\mathcal{M}}_1(\omega)]$ is symmetric and frequency-dependent. The discretization of the other terms of the coupled system has been provided in the previous sections.

Writing the stationarity of Equation 8.65 leads to the coupled system

$$\begin{pmatrix} [K]-\omega^2[M+\hat{\mathcal{M}}_1(\omega)] & \left[C_{up}^{(1)}+\hat{\mathcal{C}}_{up}(\omega)\right] \\ \left[C_{up}^{(1)}+\hat{\mathcal{C}}_{up}(\omega)\right]^T & -\dfrac{1}{\rho_0\omega^2}\left[\hat{\mathcal{D}}(\omega)\right] \end{pmatrix} \begin{Bmatrix} \{\hat{u}\} \\ \{\hat{p}\} \end{Bmatrix} = \begin{Bmatrix} \{\hat{F}\} \\ \{0\} \end{Bmatrix} \quad (8.69)$$

Furthermore, the condensation of variable $\{\hat{p}\}$ in Equation 8.69 allows us to write this system under the classic form of Equation 8.28 where the acoustic radiation impedance matrix is given by

$$[\hat{\mathcal{Z}}_{rad}(\omega)] = -i\rho_0\omega[C_{up}^{(1)}+\hat{\mathcal{C}}_{up}(\omega)][\hat{\mathcal{D}}(\omega)]^{-1}[C_{up}^{(1)}+\hat{\mathcal{C}}_{up}(\omega)]^T + i\omega[\hat{\mathcal{M}}_1(\omega)] \quad (8.70)$$

Thus, in this formulation, the acoustic radiation impedance matrix is symmetric. This is advantageous regarding the numerical efficiency.

Once the system given by Equation 8.28 (with the acoustic radiation impedance matrix given by Equation 8.70) is solved, the pressure nodal vector can be calculated as

$$\{\hat{p}\} = \rho_0\omega^2[\hat{\mathcal{D}}(\omega)]^{-1}[C_{up}^{(1)}+\hat{\mathcal{C}}_{up}(\omega)]^T\{\hat{u}\} \quad (8.71)$$

The acoustic power radiated by the structure is given by

$$\Pi_{rad} = \frac{1}{2}\Re\left(\int_{\partial\Omega_w} \hat{p}(i\omega\underline{\hat{u}}\cdot\underline{n})^* dS\right) = \frac{\omega}{2}\Im\left(\langle\hat{u}^*\rangle[C_{up}]\{\hat{p}\}\right) \quad (8.72)$$

The pressure at all points in the exterior domain is calculated using the discretized form of Equation 8.57.

Remarks:

1. Using structural modal coordinates, the system given by Equation 8.28 transforms into Equation 8.34 where the projected acoustic radiation impedance matrix is given by Equation 8.35.
   The acoustic pressure is then given by

   $$\{\hat{p}\} = \rho_0\omega^2[\hat{\mathcal{D}}(\omega)]^{-1}[C_{up}^{(1)}+\hat{\mathcal{C}}_{up}(\omega)]^T[\Phi_s]\{\hat{u}_m\} \quad (8.73)$$

   and the acoustic power radiated by the structure reads as

   $$\Pi_{rad} = \frac{\omega}{2}\Im\left(\langle\hat{u}_m^*\rangle[\Phi_s]^T[C_{up}]\{\hat{p}\}\right) \quad (8.74)$$

2. In the case of an incompressible fluid, ($k = 0$), the effect of the fluid reduces to an added mass effect on the structure where

$$[\hat{\mathcal{M}}_a(0)] = -\rho_0[C_{up}^{(1)} + \hat{\mathcal{C}}_{up}^{(2)}(0)][\mathcal{D}(0)]^{-1}[C_{up}^{(1)} + \hat{\mathcal{C}}_{up}(0)]^T + [\hat{\mathcal{M}}_1(0)] \tag{8.75}$$

If modal coordinates are used, the modal added mass matrix $[\hat{\mathcal{M}}_{a,m}]$ is given by the Equation 8.52. It is symmetric and positive definite.

3. In the case of a thin structure (such as a plate), it can be verified that $[\hat{\mathcal{M}}_1(\omega)] = 0$ and $[\hat{\mathcal{C}}_{up}(\omega)] = 0$. Therefore, Equation 8.69 becomes Equation 8.45.

8.6 FEM–VBEM APPROACH FOR FLUID–POROELASTIC OR FLUID–FLUID EXTERIOR COUPLING

In this section, we are interested in configurations where an interior domain filled with a poroelastic material or a fluid is coupled to semi-infinite fluids (exterior domain). The boundaries of this interior domain in contact with the exterior fluids are assumed to be flat and embedded in rigid baffles. Typical applications are the sound absorption of a porous material lying on a ground, the sound transmission through a porous material sample or through a niche. In the following, we associate FEM–VBEM to the problem of interest where FEM is used for the interior domain (poroelastic material or fluid cavity) and VBEM is employed for the exterior domains. The details are available in Atalla et al. (2006). We present only the main steps to arrive at the final coupled system to be solved.

To illustrate the approach consider the case of a poroelastic material embedded in a rigid baffle located in the plane $x_3 = 0$ submitted to an acoustic excitation. The porous material is coupled to a semi-infinite fluid on one of its face (excitation side) and has specific boundary conditions on the other faces.

The weak integral form associated with the porous material is obtained (Allard and Atalla 2009) from Equations 2.22 and 2.23 in Section 2.4. It reads

$$\begin{aligned}
&\int_{\Omega_p}\left[\underline{\underline{\tilde{\sigma}}}^s(\underline{\hat{u}}^s):\underline{\underline{\hat{\varepsilon}}}^s(\delta\underline{\hat{u}}^s) - \omega^2\tilde{\rho}\underline{\hat{u}}^s\cdot\delta\underline{\hat{u}}^s\right]dV + \int_{\Omega_p}\left[\frac{\phi_p^2}{\omega^2\tilde{\rho}_{22}}\underline{\nabla}\hat{p}^f\cdot\underline{\nabla}\delta\hat{p}^f - \frac{\phi_p^2}{\tilde{R}}\hat{p}^f\delta\hat{p}^f\right]dV \\
&\quad - \int_{\Omega_p}\frac{\phi_p^2\rho_0}{\tilde{\rho}_{22}}\delta(\underline{\nabla}\hat{p}^f\cdot\underline{\hat{u}}^s)dV - \int_{\Omega_p}\phi_p\left(1+\frac{\tilde{Q}}{\tilde{R}}\right)\delta(\hat{p}^f\underline{\nabla}\cdot\underline{\hat{u}}^s)dV \\
&\quad - \int_{\partial\Omega_p}\phi_p[\underline{\hat{U}}^f\cdot\underline{n} - \underline{\hat{u}}^s\cdot\underline{n}]\delta\hat{p}^f\,dS - \int_{\partial\Omega_p}[\underline{\underline{\hat{\sigma}}}^t\cdot\underline{n}]\cdot\delta\underline{\hat{u}}^s\,dS = 0 \quad \forall(\delta\underline{\hat{u}}^s,\delta\hat{p}^f)
\end{aligned} \tag{8.76}$$

where Ω_p and $\partial\Omega_p$ refer to the poroelastic domain and its bounding surface. $\hat{\underline{u}}^s$ and $\hat{p}^f$ are the solid phase displacement vector and the interstitial pressure in the poroelastic medium, respectively. $\delta\hat{\underline{u}}^s$ and $\delta\hat{p}^f$ refer to their admissible variation, respectively. $\hat{\underline{U}}^f$ is the fluid macroscopic displacement vector. $\underline{n}$ denotes the unit normal vector external to the bounding surface $\partial\Omega_p$. $\tilde{\underline{\underline{\sigma}}}^s$ and $\hat{\underline{\underline{\varepsilon}}}^s$ are the in vacuo stress and strain tensors of the porous material. $\hat{\underline{\underline{\sigma}}}^t$ is the total stress tensor of the material given by $\tilde{\underline{\underline{\sigma}}}^s = \hat{\underline{\underline{\sigma}}}^t + \phi[1 + (\tilde{Q}/\tilde{R})]\hat{p}^f \underline{\underline{1}}$. Note that $\tilde{\underline{\underline{\sigma}}}^s$ accounts for structural damping in the skeleton through a complex Young's modulus $E_p(1 + i\eta_p)$.ϕ_p stands for the porosity, $\tilde{\rho}_{22}$ is the modified Biot's density of the fluid phase accounting for viscous dissipation, $\tilde{\rho}$ is an effective density given by $\tilde{\rho} = \tilde{\rho}_{11} - (\tilde{\rho}_{12}/\tilde{\rho}_{22})$ where $\tilde{\rho}_{11}$ is the modified Biot's density of the solid phase accounting for viscous dissipation. $\tilde{\rho}_{12}$ is the modified Biot's density which accounts for the interaction between the inertia forces of the solid and fluid phase together with viscous dissipation. $\tilde{Q}$ is an elastic coupling coefficient between the two phases, $\tilde{R}$ may be interpreted as the bulk modulus of the air occupying a fraction of the unit volume aggregate. For more information about the definitions of these coefficients, the reader is referred to Section 2.4.

In the presented formulation, the porous media couples to the semi-infinite fluid medium through the following boundary terms:

$$I_{\partial\Omega_p} = -\int_{\partial\Omega_p} [\hat{\underline{\underline{\sigma}}}^t \cdot \underline{n}]\delta\underline{u}^s \, dS - \int_{\partial\Omega_p} \phi_p(\hat{U}_n^f - \hat{u}_n^s)\delta\hat{p}^f \, dS \tag{8.77}$$

where $\hat{U}_n^f$ (respectively, $\hat{u}_n^s$) denotes the normal fluid phase displacement (respectively, normal solid phase displacement) of the poroelastic material. Since at the interface $\partial\Omega_p, \hat{\underline{\underline{\sigma}}}^t \cdot \underline{n} = -\hat{p}\underline{n} = -\hat{p}^f n$ (see Equation 2.44), Equation 8.77 becomes

$$I_{\partial\Omega_p} = \int_{\partial\Omega_p} \delta(\hat{p}^f \hat{u}_n^s) \, dS - \int_{\partial\Omega_p} \left[\phi_p(\hat{U}_n^f - \hat{u}_n^s) + \hat{u}_n^s\right]\delta\hat{p}^f \, dS \tag{8.78}$$

In the semi-infinite domain, the acoustic pressure $\hat{p}$ is the sum of the blocked pressure $\hat{p}_b$ and the radiated pressure $\hat{p}_r$. Applying the continuity of the normal displacement at the surface, Equation 8.78 becomes

$$I_{\partial\Omega_p} = \int_{\partial\Omega_p} \delta(\hat{p}^f \hat{u}_n^s) \, dS - \frac{1}{\rho_0\omega^2} \int_{\partial\Omega_p} \frac{\partial\hat{p}}{\partial n} \delta\hat{p} \, dS \tag{8.79}$$

Since $\partial\hat{p}/\partial n = \partial\hat{p}_r/\partial n$, the discretized form associated with the second term of Equation 8.79 reads

$$-\frac{1}{\rho_0\omega^2}\int_{\partial\Omega_p}\frac{\partial\hat{p}}{\partial n}\delta\hat{p}\,dS = -\frac{1}{\rho_0\omega^2}\langle\delta\hat{p}\rangle[C_{pp}^{(2)}]\left\{\frac{\partial\hat{p}_r}{\partial n}\right\} \tag{8.80}$$

where $[C_{pp}^{(2)}]$ is a surface-coupling matrix given by

$$[C_{pp}^{(2)}] = \int_{\partial\Omega_p}\{N^f\}\langle N^f\rangle dS \tag{8.81}$$

with $\{N^f\}$ the vector of the surface element shape functions.

The porous material being inserted into a rigid baffle, the radiated acoustic pressure is related to the normal velocity via Rayleigh's integral:

$$\hat{p}_r(x_1,x_2,x_3) = -\int_{\partial\Omega_p}\frac{\partial\hat{p}_r(y_1,y_2,0)}{\partial n}\hat{G}(x_1,x_2,x_3;y_1,y_2,0)\,dS_y \tag{8.82}$$

where $\hat{G}(x_1,x_2,x_3;y_1,y_2,0) = (\exp(-ikR)/2\pi R)$ is the baffled Green's function, $k = \omega/c_0$ is the acoustic wavenumber in the fluid, c_0 is the associated speed of sound and R is the distance between point (x_1, x_2, x_3) and $(y_1, y_2, 0)$: $R = \sqrt{(x_1 - y_1)^2 + (x_2 - y_2)^2 + x_3^2}$.

An associated integral form to Equation 8.82 is given by

$$\int_{\partial\Omega_p}\hat{p}_r(x_1,x_2,0)\delta\hat{p}(x_1,x_2,0)\,dS_x =$$

$$-\int_{\partial\Omega_p}\int_{\partial\Omega_p}\frac{\partial\hat{p}_r(y_1,y_2,0)}{\partial n}\hat{G}(x_1,x_2,0;y_1,y_2,0)\delta\hat{p}(x_1,x_2,0)\,dS_y\,dS_x \tag{8.83}$$

The associated discretized form is

$$\langle\delta\hat{p}\rangle[C_{pp}^{(2)}]\{\hat{p}_r\} = -\langle\delta\hat{p}\rangle[\hat{Z}_{rad}^{pp}]\left\{\frac{\partial\hat{p}_r}{\partial n}\right\} \tag{8.84}$$

where

$$[\hat{Z}_{rad}^{pp}] = \int\limits_{\partial\Omega_p}\int\limits_{\partial\Omega_p} \{N(\underline{x})\}\hat{G}(\underline{x},\underline{y})\langle N(\underline{y})\rangle dS_y\, dS_x \tag{8.85}$$

Since $\langle \delta\hat{p}\rangle$ is arbitrary, one gets

$$\left\{\frac{\partial \hat{p}_r}{\partial n}\right\} = -[\hat{Z}_{rad}^{pp}]^{-1}[C_{pp}^{(2)}]\{\hat{p}_r\} \tag{8.86}$$

Substituting Equation 8.86 into 8.80 and recalling that on the interface $\hat{p}^f = \hat{p} = \hat{p}_b + \hat{p}_r$, the discretized form of Equation 8.79 reads finally

$$I_{\partial\Omega_p} = \langle \delta\hat{u}_n^s\rangle[C_{up}]\{\hat{p}^f\} + \langle\delta\hat{p}^f\rangle[C_{up}]^T\{\hat{u}_n^s\} - \frac{1}{i\omega}\langle\delta\hat{p}^f\rangle[\hat{A}]\{\hat{p}^f - \hat{p}_b\} \tag{8.87}$$

Here,

$$[\hat{A}] = \frac{1}{i\omega\rho_0}[C_{pp}^{(2)}][\hat{Z}_{rad}^{pp}]^{-1}[C_{pp}^{(2)}] \tag{8.88}$$

is an admittance matrix. In consequence, the radiation of the porous medium into the semi-infinite fluid amounts to an admittance term added to the interface interstitial pressure degrees of freedom and to additional interface coupling terms between the solid phase and the interstitial pressure (first two terms in Equation 8.87). Note that the last term involving $\hat{p}_b$ is the excitation term and disappears in the case of free radiation.

Finally, the discretized form of Equation 8.76 combined with Equation 8.87 leads to the following linear system:

$$\begin{bmatrix} -\omega^2[\tilde{M}] + [\hat{K}] & -[C_{u^s p^f}^{(1)}] + [C_{up}] \\ -[C_{u^s p^f}^{(1)}]^T + [C_{up}]^T & \dfrac{[\tilde{H}]}{\omega^2} - \dfrac{[\hat{A}]}{i\omega} - [\tilde{Q}] \end{bmatrix} \begin{Bmatrix} \{\hat{u}^s\} \\ \{\hat{p}^f\} \end{Bmatrix} = \begin{Bmatrix} \{0\} \\ \{\hat{f}_f\} \end{Bmatrix} \tag{8.89}$$

with

$$\{\hat{f}_f\} = \frac{1}{i\omega}[\hat{A}]\{\hat{p}_b\} \tag{8.90}$$

where $\{\hat{u}^s\}$ and $\{\hat{p}^f\}$ represent the solid phase and the fluid phase global nodal vectors, respectively. $[\tilde{M}]$ and $[\hat{K}]$ represent equivalent mass and stiffness matrices for the solid phase, $[\tilde{H}]$ and $[\tilde{Q}]$ represent equivalent kinetic and compression energy matrices for the fluid phase, and finally $[C^{(1)}_{u^s p^f}]$ is a volume coupling matrix between the solid phase displacement and the fluid phase pressure. The detailed expressions for these matrices can be found in Allard and Atalla (2009).

The formulation is general and can be used to treat the case of an acoustic cavity coupled to a semi-infinite medium. Indeed, we only have to degenerate Equation 8.89 into an equivalent fluid by assuming the porous material as rigid framed and replacing the corresponding effective density and bulk modulus by those of the fluid contained in the cavity. The formulation has been presented in the case of a single porous material/external fluid interface. It can be reformulated for several interfaces provided that they are planar and embedded in a rigid infinite baffle (e.g., sound transmission through a porous material).

8.7 PRACTICAL CONSIDERATIONS FOR THE NUMERICAL IMPLEMENTATION

The wavelengths encountered in nondispersive acoustic media are much larger than the wavelengths in structures for subcritical frequencies. The structural response is sensitive to small geometrical details. The geometry needs thus to be finely meshed in order to capture correctly the structural response and stresses. The acoustical response is much less sensitive to these geometrical details provided that the structural features are much smaller than the acoustic wavelength; they can then be ignored for the BEM model. That means that the acoustic mesh can be chosen to be coarser than the structural mesh. In other words, the size of the acoustic finite elements can be chosen to be larger than the size of the structural finite elements. This, in turn, involves that incompatible meshes should be coupled on the interface. In this book, only compatible meshes are considered but techniques such as the Lagrange's multipliers presented in Section 5.2.5.1 can be used to account for incompatible meshes. Other approaches such as mapping techniques can also be used to map the displacement field onto the acoustic mesh. The reader can refer, for example, to Guerich and Hamdi (1999).

If we are interested in the sound radiation of a complex structure, we should keep in mind that only the external surface of the FE structural mesh radiates noise. To generate the acoustic boundary element mesh (structure skin), inner elements, one-dimensional elements, and structural details not important for the acoustical response (e.g., stiff ribs much smaller than the acoustic wavelength), should be removed using mesh cleaning tools. Holes and gaps between components whose size if larger than the element size should be filled to build the acoustic mesh. A coarsened acoustic mesh allows for better simulation quality and faster results since less details are involved.

8.8 EXAMPLES OF APPLICATIONS

We now present some application examples of fluid-structure coupled problems, which use the results of this chapter. We present calculations of free and forced responses of several structural shapes coupled with light or heavy fluids.

8.8.1 Eigenfrequencies of a free-free plate in water

In this example we illustrate the influence of fluid-structure coupling on the eigenfrequencies of a plate immersed in water (see Figure 8.3). An indirect VBEM is applied. Structural damping is neglected and the eigenvalue problem associated with Equation 8.48 where $k = 0$ is solved. The dimensions and the properties of the plate are indicated in Figure 8.3. Table 8.1 compares the present approach with published experimental and computed results (Sundqvist 1983; Olson and Bathe 1985). In the latter two references, the fluid domain is assumed cubical of lateral dimension 0.57 m and FEM is used. The used FEM and BEM meshes are given in Table 8.1. Overall, a good agreement is observed.

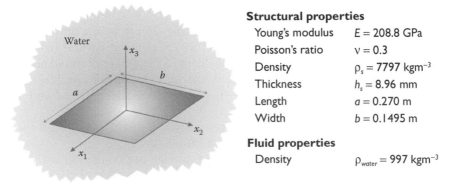

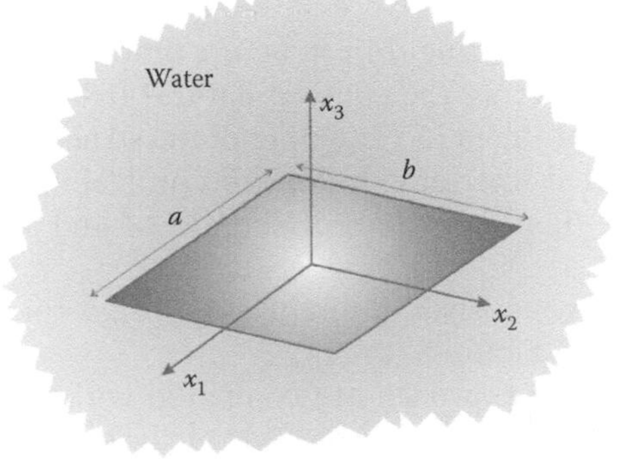

Structural properties

Young's modulus	$E = 208.8$ GPa
Poisson's ratio	$\nu = 0.3$
Density	$\rho_s = 7797$ kgm^{-3}
Thickness	$h_s = 8.96$ mm
Length	$a = 0.270$ m
Width	$b = 0.1495$ m

Fluid properties

Density	$\rho_{water} = 997$ kgm^{-3}

Figure 8.3 Unbaffled free-free plate immersed in water.

Table 8.1 Free-free plate: Natural frequencies

Source	*Mode #*	*Vacuum (Hz)*	Δf *(%)*	*Water (Hz)*	Δf *(%)*
Sundqvist-Experimental	1	641.000	(r)	497.000	(r)
	2	712.000	(r)	575.000	(r)
	3	1577.000	(r)	1293.000	(r)
	4	1766.000	(r)	1408.000	(r)
	5	2139.000	(r)	1758.000	(r)
Sundqvist-Adina (6 × 10 quad plate & 26 × 18 × 8 hexa fluid)	1	645.000	0.62	489.000	−1.61
	2	716.000	0.56	561.000	−2.43
	3	1585.000	0.51	1277.000	−1.24
	4	1766.000	0.00	1411.000	0.21
	5	2115.000	−1.12	1740.000	−1.02
Olson and Bathe (6 × 10 × 1 solid & 10 × 14 × 7 hexa 20 fluid)	1	656.000	2.34	516.000	3.82
	2	727.000	2.11	(na)	(na)
	3	1618.000	2.60	(na)	(na)
	4	1815.000	2.77	(na)	(na)
	5	2192.000	2.48	1844.000	4.89
Present method (6 × 10 × 1 solid & 12 × 20 quad 4 fluid (mesh #1)	1	644.730	0.58	503.349	1.28
	2	710.414	−0.22	578.392	0.59
	3	1568.565	−0.53	1306.722	1.06
	4	1747.856	−1.03	1421.624	0.97
	5	2084.977	−2.53	1755.429	−0.15

(r) reference, (na) not available.

8.8.2 Eigenfrequencies of a simply supported cylinder in water

We now consider the case of a curved vibrating thin structure vibrating in water. Again we are interested in predicting the coupled natural frequencies of this system. Both the exterior problem (no coupling of the structure with the interior fluid) and the coupled interior/exterior fluid problem are considered. These two problems are handled with a direct VBEM and an indirect VBEM, respectively.

More specifically, the configuration consists of simply supported thin hollow cylinder immersed in water (see Figure 8.4).

The results obtained with the VBEM approach are compared with the results obtained by Selmane and Lakis (1997), who used a hybrid finite element–shell theory method to compute the coupled natural frequencies of the system. The BEM mesh consists of 24 quadratic plate elements along

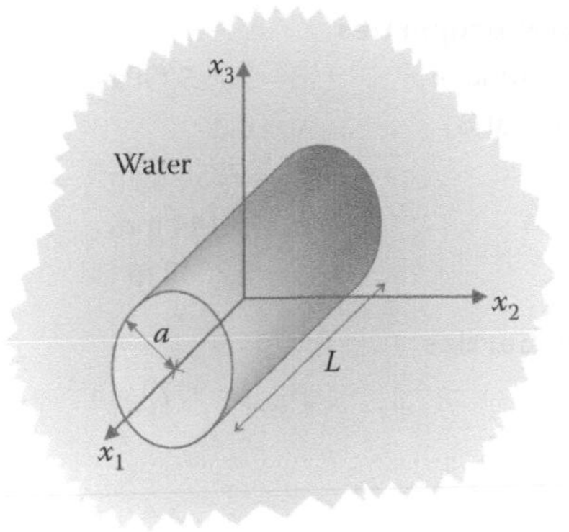

Structural properties

Young's modulus	$E = 219.8$ GPa
Poisson's ratio	$\nu = 0.3$
Density	$\rho_s = 7850$ kgm^{-3}
Thickness	$h_s = 2.35$ mm
Length	$L = 0.940$ m
Radius	$a = 0.235$ m

Fluid properties

Density	$\rho_{water} = 1000$ kgm^{-3}

Figure 8.4 Unbaffled simply supported open thin-walled cylinder immersed in water.

Table 8.2 Simply supported cylinder: Natural frequencies

Source	*Mode*	*In-vacuo (Hz)*	*Δf (%)*	*Water (exterior) (Hz)*	*Δf (%)*	*Water (exterior/ interior) (Hz)*	*Δf (%)*
Selmane and Lakis	4-1	659.000	(r)	331.400	(r)	251.400	(r)
	8-1	2187.000	(r)	1361.000	(r)	1064.000	(r)
Present method	4-1	657.049	–0.30	329.656	–0.53	247.835	–1.42
	8-1	2158.398	–1.31	1355.557	–0.40	1069.050	0.47

the circumference and 20 along the axis. Table 8.2 shows that the computed results agree well with those of Selmane and Lakis for the three considered configurations (in-vacuo, exterior coupling and exterior/interior coupling).

8.8.3 Eigenfrequencies of a free sphere in water

The third case is a free empty thin steel sphere immersed in water (Figure 8.5). Again, we are interested in the in vacuo and coupled eigenvalues of this system. This case was presented by Ding and Chen (1996), who made the computation using an analytical method. Here, a direct VBEM is used. The mesh consists of 16 × 16 quadratic shell elements. Table 8.3 shows that the computed results agree well with those of Ding and Chen.

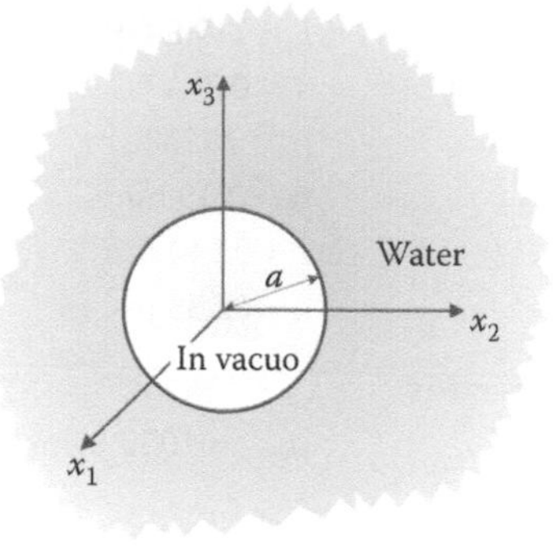

Structural properties

Young's modulus	$E = 206.8$ GPa
Poisson's ratio	$\nu = 0.3$
Density	$\rho_s = 7850$ kgm^{-3}
Thickness	$h_s = 9.144$ mm
Length	$a = 0.3048$ m

Fluid properties

Density	$\rho_{water} = 997$ kgm^{-3}

Figure 8.5 Free sphere immersed in water.

Table 8.3 Free sphere: Natural frequencies

Source	*Mode #*	*In vacuo (Hz)*	Δf *(%)*	*Water (Hz)*	Δf *(%)*
Ding and Chen 3D elastic theory	2	1958.298	(r)	1372.775	(r)
	3	2364.195	(r)	1701.566	(r)
	4	2509.009	(r)	1870.400	(r)
	5	2625.278	(r)	2022.002	(r)
	6	2748.335	(r)	2173.604	(r)
	7	2901.330	(r)	2346.963	(r)
	8	3101.842	(r)	2557.048	(r)
Ding and Chen plate theory	2	1971.004	0.65	1395.924	1.69
	3	2342.960	−0.90	1699.477	−0.12
	4	2510.053	0.04	1885.716	0.82
	5	2629.977	0.18	2038.711	0.83
	6	2755.819	0.27	2192.750	0.88
	7	2914.210	0.44	2369.416	0.96
	8	3121.162	0.62	2584.027	1.06
Present method	2	1967.584	0.47	1393.828	1.53
	3	2334.082	−1.27	1691.748	−0.58
	4	2496.673	−0.49	1870.740	0.02
	5	2612.725	−0.48	2020.109	−0.09
	6	2730.532	−0.65	2167.451	−0.28
	7	2878.539	−0.79	2338.399	−0.36
	8	3068.715	−1.07	2542.231	−0.58

8.8.4 Eigenfrequencies of a turbine in water

The case chosen here is a five-blades turbine runner, submerged in water (Figure 8.6). It has a diameter of 8 m and a height of 5.7 m. The in vacuo finite element model of the runner, with the top of the hub attached to a shaft, was developed and solved using the commercial program MSC/Nastran (© MSC Software). The modes were then computed using IDEAS (© Siemens) and Nastran, with the top of the hub attached to a shaft. The rotation of the structure was neglected in all the computation. The hub and the blades are modeled using quadratic tetrahedral solid elements, while the cone of the runner is represented by linear shell elements. The direct VBEM was then used for the computation of the added mass (Equation 8.75), considering only the areas that are in

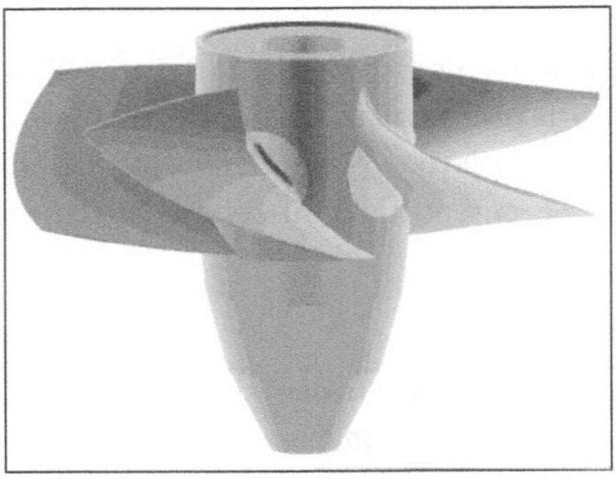

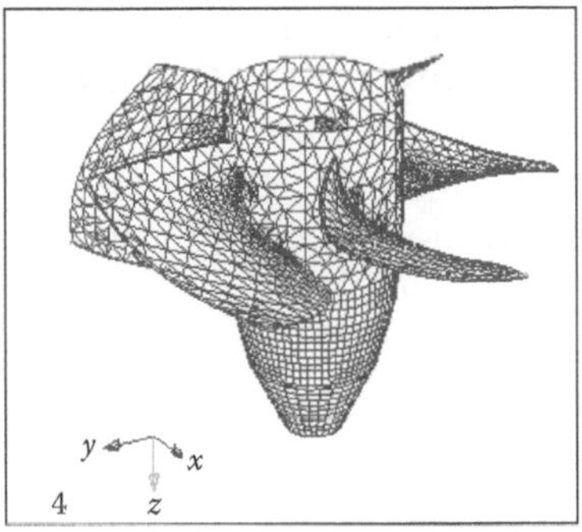

Figure 8.6 Hydraulic turbine runner in water.

Table 8.4 Turbine runner: Natural frequencies

Mode #	*In vacuo (Hz)*	*Water (Hz)*	*Δf (%)*
1	2.785	2.703	2.95
2	2.785	2.703	2.95
3	7.249	6.161	15.01
4	7.249	6.161	15.02
5	9.070	8.007	11.72
6	9.070	8.007	11.72
7	21.040	16.352	22.28
8	27.497	17.962	34.68
9	27.497	17.961	34.68
10	32.996	17.711	46.32
11	32.996	17.711	46.32
12	33.776	23.319	30.96
13	37.175	30.426	18.15
14	37.175	30.433	18.14

contact with water. The wet surface mesh was composed of 8090 elements and 4840 nodes. The results are shown in Table 8.4. They illustrate the importance of accounting for the added mass in the case of a structure submerged in a heavy fluid, even for such a large and heavy structure.

8.8.5 Sound radiation of an unbaffled plate

Consider the case of an unbaffled simply supported thin plate immersed in an infinite fluid medium. The plate is excited mechanically by a point force. We are interested in computing the forced response of this plate in a light fluid (air) and in a heavy fluid (water). The dimensions and the properties of the plate are indicated in Figure 8.7. The structural mesh consisted of 20×20 (respectively, 40×40) thin shell quad-4 elements in the case of air (respectively, in the case of water). The boundary element mesh was compatible with the structural mesh.

An indirect VBEM is used to solve the problem (Equation 8.55). Since the structure is thin the sound pressure in the infinite medium can be expressed with a double-layer potential. Note that the pressure jump $\hat{\mu}$ along the plate edges must be set to zero since the pressure is

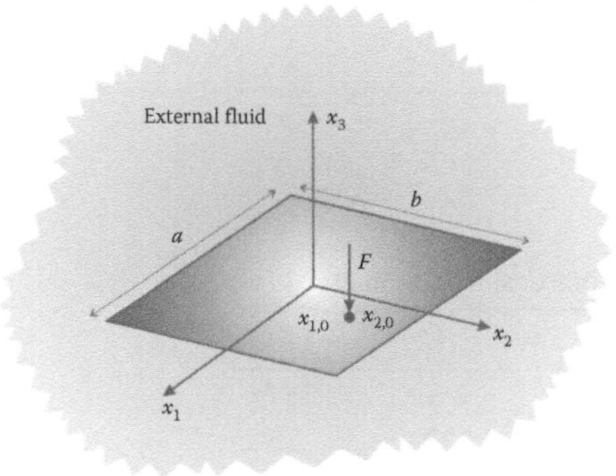

Structural properties

Young's modulus	$E = 200$ GPa
Poisson's ratio	$\nu = 0.28$
Density	$\rho_s = 7850$ kgm^{-3}
Loss factor	$\eta_s = 0.05$
Thickness	$h_s = 9$ mm
Length	$a = 0.6$ m
Width	$b = 0.6$ m

Fluid properties

Density	$\rho_{air} = 1.213$ kgm^{-3}
Sound speed	$c_{air} = 342.2$ ms^{-1}
Density	$\rho_{water} = 1000$ kgm^{-3}
Sound speed	$c_{water} = 1460$ ms^{-1}

Excitation

Position	$x_{1,0} = 0.06$ m, $x_{2,0} = 0.06$ m
Amplitude	$F = 1$N

Figure 8.7 Unbaffled simply supported plate excited mechanically and radiating in an infinite fluid.

continuous along the edges. In the numerical implementation, this condition is imposed using the Lagrange's multiplier technique. The problem is then condensed on the structural degrees of freedom as described in Equation 8.56. The results are compared to those obtained with a semi-analytical approach (Nelisse et al. 1996). Figures 8.8 and 8.9 present, respectively, the mean square velocity of the plate and its radiated power in air and water.

Note that the total power radiated by the plate is simply given by

$$\Pi_{rad} = \frac{1}{2}\Re\left(\int_{\partial\Omega_w} \hat{\mu}(i\omega\underline{\hat{u}} \cdot \underline{n})^* \, dS\right) = \frac{\omega}{2}\Im\left(\langle\hat{u}^*\rangle\left[C_{u\mu}\right]\{\hat{\mu}\}\right) \tag{8.91}$$

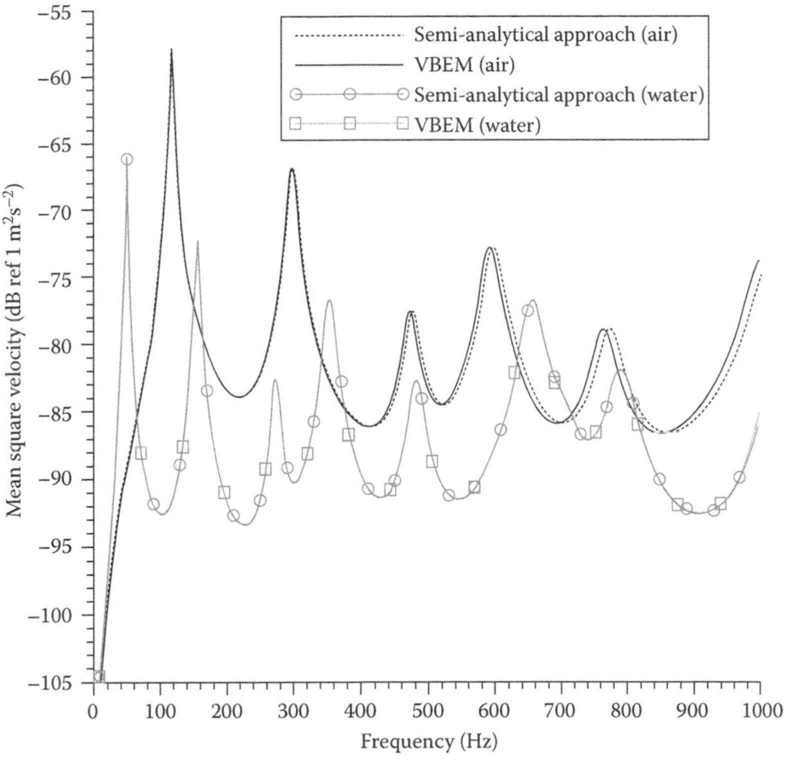

Figure 8.8 Mean square velocity of a simply supported plate excited mechanically and radiating in air or water as a function of frequency—comparisons between a semi-analytical approach and an indirect VBEM.

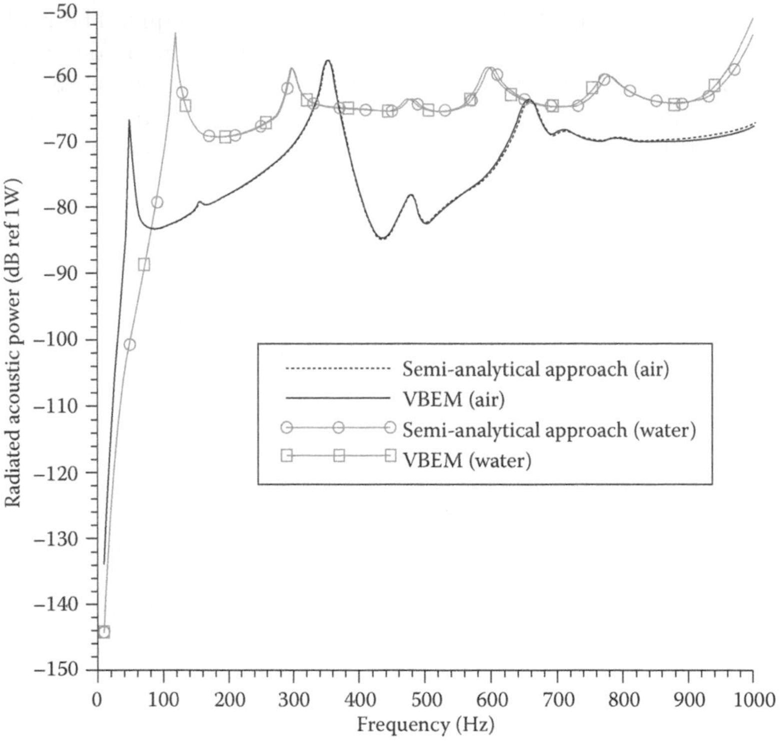

Figure 8.9 Sound power radiated by a simply supported plate excited mechanically and radiating in air or water as a function of frequency—comparisons between a semi-analytical approach and an indirect VBEM.

We see that there is a very good agreement between the VBEM/FEM approach and the semi-analytical approach. Moreover, one clearly sees the effect of the strong fluid loading for the plate immersed in water.

8.8.6 Scattering of a plane wave by a fluid-filled elastic sphere in an unbounded fluid medium

In Section 7.13.4, the case of the scattering of a plane wave by a rigid sphere in air has been considered. We now examine a similar configuration where the sphere is made up of an elastic material, filled with a fluid (fluid 1) and immersed in an unbounded fluid (fluid 2). The configuration together with the physical and geometrical parameters are described in Figure 8.10. $\Omega_{f,e}$, $\partial\Omega_{w,e}$, $\Omega_{f,i}$, $\partial\Omega_{w,i}$ denote the exterior domain, exterior domain boundary, interior domain, and interior domain boundary, respectively. $\underline{n}$ is the normal to the sphere pointing into domain $\Omega_{f,e}$. The equations of the coupled system are given by

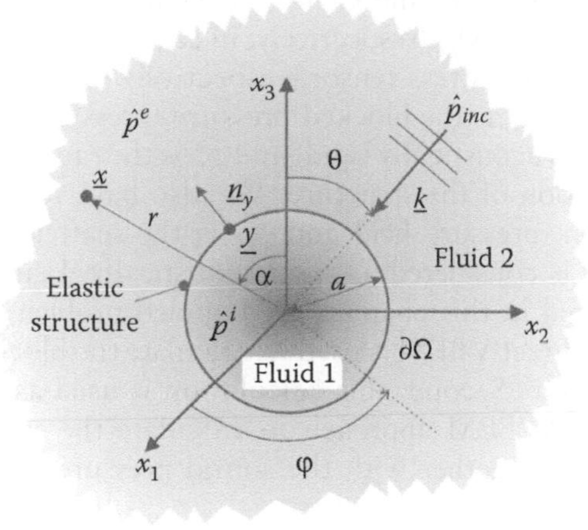

Figure 8.10 Free fluid-filled elastic sphere excited by a plane wave in an unbounded fluid medium.

$$\underline{\nabla} \cdot \underline{\underline{\hat{\sigma}}} + \rho_s \omega^2 \underline{\hat{u}} = 0 \quad \text{in } \Omega_s \tag{8.92a}$$

$$\underline{\underline{\hat{\sigma}}} \cdot \underline{n} = -\hat{p}^e \underline{n} \quad \text{over } \partial\Omega_{w,e} \tag{8.92b}$$

$$\underline{\underline{\hat{\sigma}}} \cdot \underline{n} = \hat{p}^i \underline{n} \quad \text{over } \partial\Omega_{w,i} \tag{8.92c}$$

$$\nabla^2 \hat{p}^e + k_e^2 \hat{p}^e = 0 \quad \text{in } \Omega_{f,e} \tag{8.92d}$$

$$\frac{\partial \hat{p}^e}{\partial n} = \rho_0 \omega^2 \underline{\hat{u}} \cdot \underline{n} \quad \text{over } \partial\Omega_{w,e} \tag{8.92e}$$

$$\nabla^2 \hat{p}^i + k_i^2 \hat{p}^i = 0 \quad \text{in } \Omega_{f,i} \tag{8.92f}$$

$$\frac{\partial \hat{p}^i}{\partial n} = \rho_0 \omega^2 \underline{\hat{u}} \cdot \underline{n} \quad \text{over } \partial\Omega_{w,i} \tag{8.92g}$$

$$\lim_{r\to\infty} \left(\frac{\partial \hat{p}_r^e}{\partial r} + ik\hat{p}_r^e \right) r = 0 \tag{8.92h}$$

$\hat{p}^e$ and $\hat{p}^i$ (respectively, k_e and k_i) are the total acoustic pressures (respectively, wavenumbers) in $\Omega_{f,e}$ (respectively, in $\Omega_{f,i}$). $\underline{\hat{u}}$ and $\underline{\underline{\hat{\sigma}}}$ are the structural displacement field and stress tensor, respectively. $\hat{p}^e$ can be expressed as $\hat{p}^e = \hat{p}^e_b + \hat{p}^e_r$ where $\hat{p}^e_b$ is the blocked pressure (pressure which would exist if the structure was acoustically rigid) and $\hat{p}^e_r$ is the radiated pressure generated by the vibration of the structure. We also have $\hat{p}^e_b = \hat{p}_{inc} + \hat{p}_{sc}$ where $\hat{p}_{inc}$ is the incident pressure field and $\hat{p}_{sc}$ is the scattered pressure in $\Omega_{f,e}$ when the sphere is considered as rigid acoustically. Using the variational approach presented in this chapter, the coupled problem is solved in two steps. First, an indirect VBEM is used to calculate the blocked pressure field acting on the sphere. Second, this calculation is used as input in a direct and indirect VBEM/FEM approach to calculate the structural displacement of the sphere together with the sound pressure radiated inside and outside the sphere.

The response of a thin elastic sphere can be calculated analytically (see Appendix 8A).

The first configuration of interest consists of a $h_s = 1$ mm thick steel sphere of radius $a = 0.3$ m filled with air and radiating in air (Figure 8.11). The properties of the steel sphere are $\rho_s = 7800$ kgm^{-3}, $E = 210$ GPa, $\nu = 0.31$, and $\eta_s = 0.001$. The air density and sound speed are $\rho_0 = 1.213$ kgm^{-3} and $c_0 = 342.2$ ms^{-1}, respectively. The sphere is excited by an incoming plane wave propagating along the x_3 axis. The structural mesh consists of 1680 thin shell quad-4 elements and the acoustical boundary element mesh consists of 1680 quad-4 elements (on both sides of the structure in the case of the direct VBEM and on the exterior side in the case of the indirect VBEM). Note that compatible meshes are used here. Figure 8.12 compares the mean square velocity of the sphere together with its acoustic

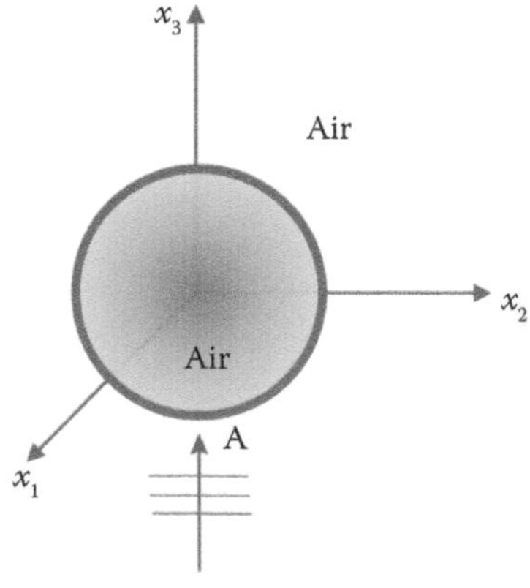

Figure 8.11 Free air-filled elastic sphere excited by a plane wave and radiating in an unbounded air medium.

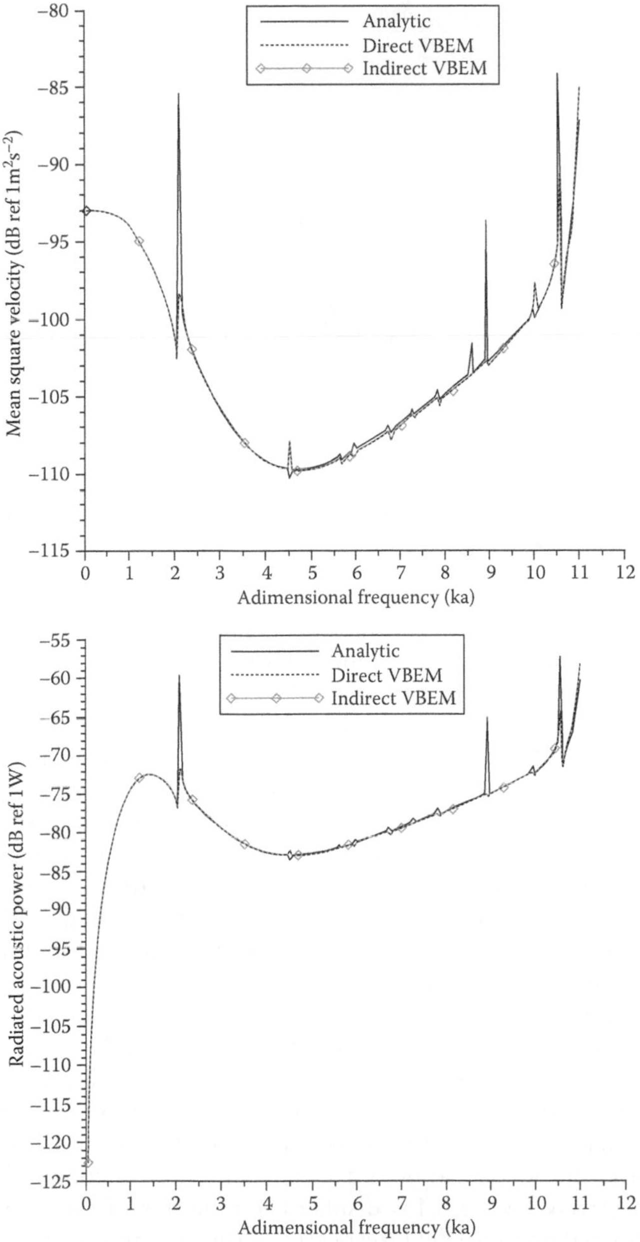

Figure 8.12 Response of a free air-filled elastic sphere excited by a plane in an unbounded air medium as a function of adimensional frequency. Comparisons between three approaches (analytical, direct VBEM, and indirect VBEM). Top—Mean square velocity; Bottom—Acoustic power radiated outside the sphere.

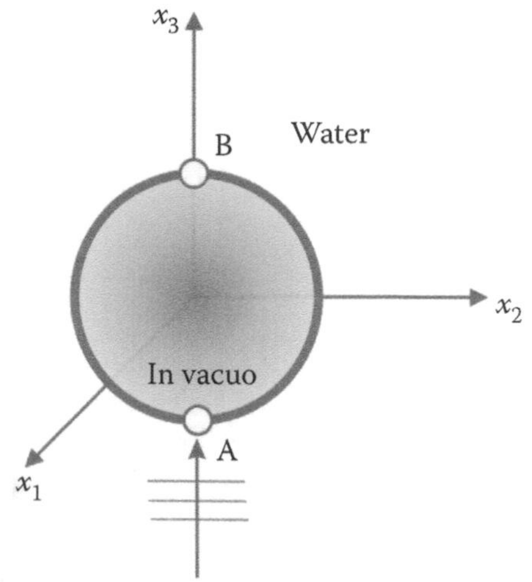

Figure 8.13 Free elastic sphere excited by a plane wave and radiating in an unbounded external water medium.

power radiated in the external fluid for the three approaches: analytical, direct VBEM, and indirect VBEM for the coupled problem. In the analytical approach a structural damping $\eta_a = 0.001$ for the internal fluid has been considered. Note that a finer frequency step has been utilized in the analytical calculations. All the results are in very good agreement.

The second configuration of interest is taken from Gaul et al. (2009). It consists of an elastic sphere of radius 5 m excited along the x_3 axis by a plane wave and radiating in water (Figure 8.13). The interior domain is supposed to be in vacuo. The properties of the steel sphere are $\rho_s = 7669$ kgm^{-3}, $E = 207$ GPa, $\nu = 0.3$, and $\eta_s = 0.01$. The water density and sound speed are $\rho_0 = 1000$ kgm^{-3} and $c_0 = 1387$ ms^{-1}, respectively.

In a similar way to the previous example, an indirect VBEM is first used to calculate the blocked pressure field acting on the sphere, which is used as input in a direct FEM/VBEM approach to calculate the structural displacement of the sphere together with the sound pressure radiated outside the sphere. The structural mesh consists of 1536 thin shell quad-4 elements and the acoustical boundary element mesh is compatible. Figure 8.14 (respectively Figure 8.15) display the modulus of the total pressure and the blocked pressure field (respectively the magnitude of the displacement) at two opposite points on the surface of the sphere. It is seen that both direct and indirect VBEM for the coupled problem lead to an excellent agreement with the analytical solution for both the acoustic and the structural response.

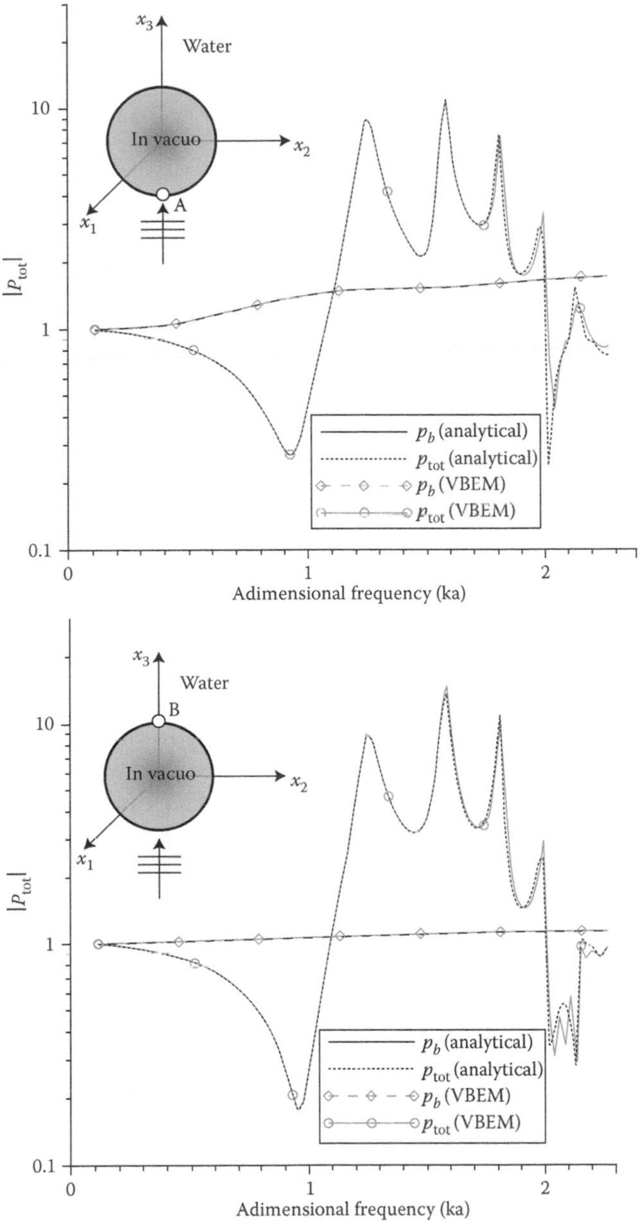

Figure 8.14 Vibroacoustic response of a free elastic sphere excited by a plane wave and radiating in an unbounded water medium as a function of adimensional frequency. Comparisons between two approaches (analytical and VBEM) for the modulus of the total and blocked pressures. Top—point A; Bottom—point B.

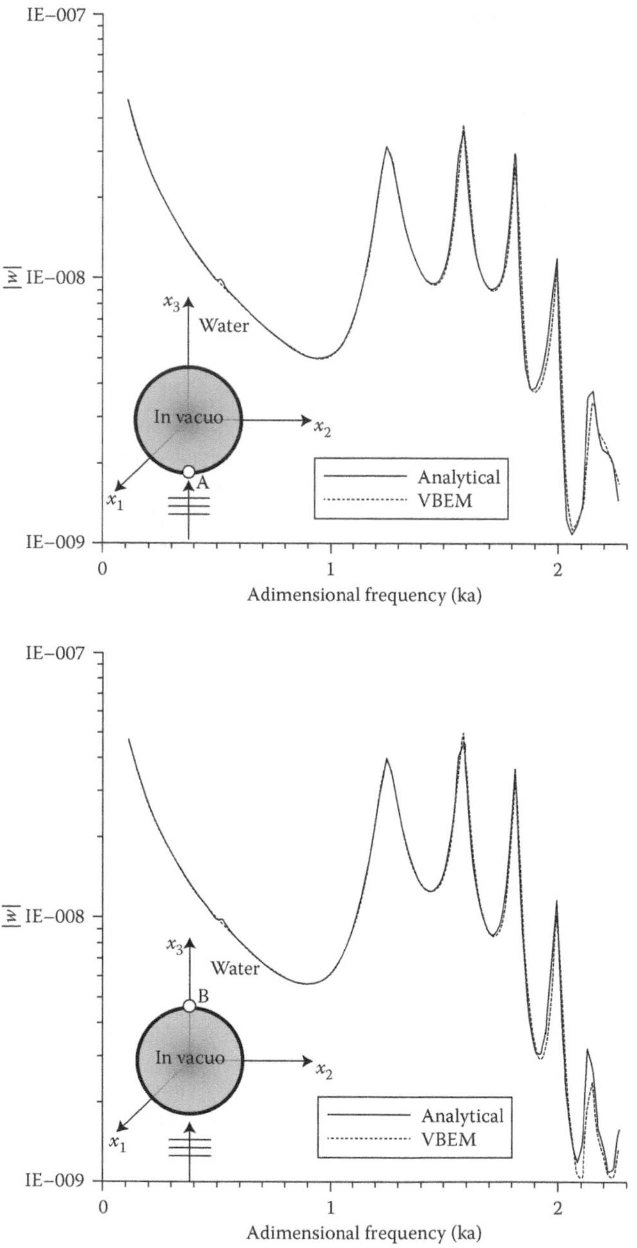

Figure 8.15 Vibroacoustic response of a free elastic sphere excited by a plane wave and radiating in an unbounded water medium as a function of adimensional frequency. Comparisons between two approaches (analytical and VBEM) for the magnitude of the point mobility. Top—point A; Bottom—point B.

8.8.7 Sound transmission through a curved structure

Let us consider now the case of a curved structure embedded in an infinite rigid baffle separating two semi-infinite fluid media. The interface between the exterior fluids $\Omega^1_{f,e}$ (excitation side) and $\Omega^2_{f,e}$ (reception side) and the wall are referred to as Σ^1_e and Σ^2_e, respectively. The configuration together with the physical and geometrical parameters are described in Figure 8.16.

The curved structure consists of an elastic half-cylinder. It is coupled to the exterior fluid $\Omega^1_{f,e}$ on one side through interface $\partial\Omega^1_{w,e}$ and to an interior fluid cavity $\Omega_{f,i}$ on the other side though interface $\partial\Omega_{w,i}$. The cylinder caps are acoustically rigid ($\hat{\underline{u}} \cdot \underline{n} = 0$ over the corresponding part of $\partial\Omega^1_{w,e}$ and $\partial\Omega_{w,i}$) and the curved part is elastic. The elastic structure is clamped on its edges. The cavity communicates with medium $\Omega^2_{f,e}$ through interface $\partial\Omega^2_{w,e}$. The system is either excited mechanically by a point force on $\partial\Omega^1_{F,e}$ or by an incident acoustic field in $\Omega^1_{f,e}$ acting on $\partial\Omega^1_{w,e}$.

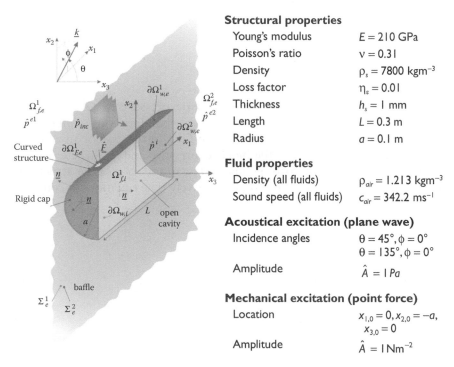

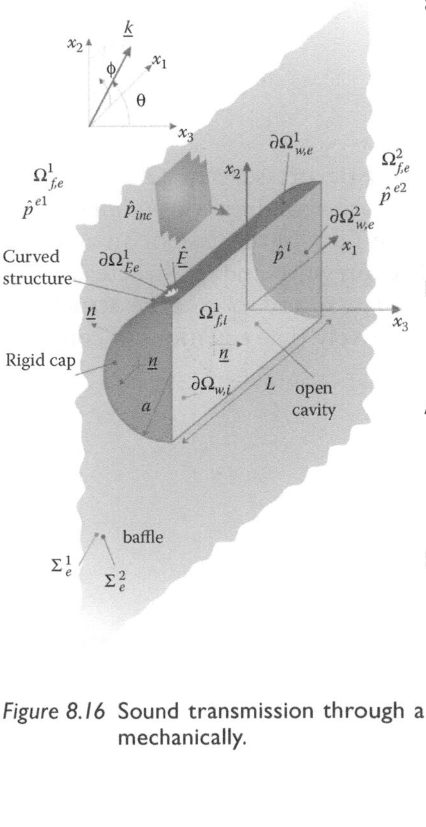

Structural properties	
Young's modulus	$E = 210$ GPa
Poisson's ratio	$\nu = 0.31$
Density	$\rho_s = 7800$ kgm^{-3}
Loss factor	$\eta_s = 0.01$
Thickness	$h_s = 1$ mm
Length	$L = 0.3$ m
Radius	$a = 0.1$ m
Fluid properties	
Density (all fluids)	$\rho_{air} = 1.213$ kgm^{-3}
Sound speed (all fluids)	$c_{air} = 342.2$ ms^{-1}
Acoustical excitation (plane wave)	
Incidence angles	$\theta = 45°, \phi = 0°$ $\theta = 135°, \phi = 0°$
Amplitude	$\hat{A} = 1$ *Pa*
Mechanical excitation (point force)	
Location	$x_{1,0} = 0, x_{2,0} = -a,$ $x_{3,0} = 0$
Amplitude	$\hat{A} = 1$ Nm^{-2}

Figure 8.16 Sound transmission through a baffled half cylinder excited acoustically or mechanically.

The problem of interest is, therefore, an exterior/interior/exterior coupled fluid-structure interaction. $\underline{n}$ is the normal to the system pointing into the exterior domains $\Omega^1_{f,e}$ and $\Omega^2_{f,e}$. The equations of motion for the coupled system are given by

$$\underline{\nabla} \cdot \hat{\underline{\underline{\sigma}}} + \rho_s \omega^2 \hat{\underline{u}} = 0 \quad \text{in } \Omega_s \tag{8.93a}$$

$$\hat{\underline{u}} = \underline{0} \quad \text{over } \partial\Omega_{s,u} \tag{8.93b}$$

$$\nabla^2 \hat{p}^i + k_i^2 \hat{p}^i = 0 \quad \text{in } \Omega_{f,i} \tag{8.94}$$

$$\nabla^2 \hat{p}^{e1} + k_{e1}^2 \hat{p}^{e1} = 0 \quad \text{in } \Omega^1_{f,e} \tag{8.95a}$$

$$\lim_{r\to\infty}\left(\frac{\partial \hat{p}_r^{e1}}{\partial r} + ik_{e1}\hat{p}_r^{e1}\right)r = 0 \quad \text{in } \Omega^1_{f,e} \tag{8.95b}$$

$$\nabla^2 \hat{p}^{e2} + k_{e2}^2 \hat{p}^{e2} = 0 \quad \text{in } \Omega^2_{f,e} \tag{8.96a}$$

$$\lim_{r\to\infty}\left(\frac{\partial \hat{p}^{e2}}{\partial r} + ik_{e2}\hat{p}^{e2}\right)r = 0 \quad \text{in } \Omega^2_{f,e} \tag{8.96b}$$

where $\hat{p}^{e1}, \hat{p}^{e2}$, and $\hat{p}^i$ are the total acoustic pressure in $\Omega^1_{f,e}, \Omega^2_{f,e}$, and $\Omega_{f,i}$, respectively. k_{e1}, k_{e2}, and k_i are the corresponding wavenumbers. $\hat{p}_r^{e1}$ is the pressure radiated by the structure in $\Omega^1_{f,e}$. In the case of an acoustic excitation, $\hat{p}^{e1}$ is the sum of the blocked pressure and $\hat{p}_r^{e1}$ whereas for a mechanical excitation it corresponds to $\hat{p}_r^{e1}$. $\hat{\underline{u}}$ and $\hat{\underline{\underline{\sigma}}}$ are the structural displacement field and stress tensor, respectively. The equations of coupling are given by

$$\hat{\underline{\underline{\sigma}}} \cdot \underline{n} = -\hat{p}^{e1}\underline{n} \quad \text{over } \partial\Omega^1_{w,e} \tag{8.97a}$$

$$\hat{\underline{\underline{\sigma}}} \cdot \underline{n} = -\hat{\underline{F}} \quad \text{over } \partial\Omega^1_{F,e} \tag{8.97b}$$

$$\hat{\underline{\underline{\sigma}}} \cdot \underline{n} = \hat{p}^i\underline{n} \quad \text{over } \partial\Omega_{w,i} \tag{8.97c}$$

$$\frac{\partial \hat{p}^{e1}}{\partial n} = \rho_{e1}\omega^2 \hat{\underline{u}} \cdot \underline{n} \quad \text{over } \partial\Omega^1_{w,e} \tag{8.97d}$$

$$\frac{\partial \hat{p}^{e1}}{\partial n} = 0 \quad \text{over } \Sigma_e^1 \tag{8.97e}$$

$$\frac{\partial \hat{p}^{i}}{\partial n} = \rho_{e2}\omega^2 \underline{\hat{u}} \cdot \underline{n} \quad \text{over } \partial\Omega_{w,i} \tag{8.97f}$$

$$\frac{\partial \hat{p}^{e2}}{\partial n} = \frac{\partial \hat{p}^{i}}{\partial n} \quad \text{over } \partial\Omega_{w,e}^2 \tag{8.97g}$$

$$\hat{p}^{e2} = \hat{p}^{i} \quad \text{over } \partial\Omega_{w,e}^2 \tag{8.97h}$$

$$\frac{\partial \hat{p}^{e2}}{\partial n} = 0 \quad \text{over } \Sigma_e^2 \tag{8.97i}$$

ρ_{e1}, ρ_{e2} are the fluid densities of $\Omega_{f,e}^1$ and $\Omega_{f,e}^2$, respectively.

In a similar way to Section 8.6, a variational approach is used to solve the problem. As before, for the acoustic excitation the coupled problem is solved in two steps. First, an indirect VBEM combined with the impedance coating regularization technique of Section 7.10.6 is used to calculate the blocked pressure field acting on the curved structure. Second, this calculation is used as input in a direct FEM/VBEM approach to calculate the structural displacement of the curved structure together with the sound pressure transmitted in medium $\Omega_{f,e}^2$. For the mechanical excitation, the problem is solved directly using a direct FEM/VBEM approach. In the FEM/VBEM approach, the mesh consists of 256 structural quad-4 shell elements, 256 fluid/structure coupling quad-4 elements, 1024 hexa-8 fluid elements for the internal cavity, 192 radiation impedance quad-4 elements (applied on the open part of the cavity), and 384 BEM quad-4 elements (applied on the half-cylinder part). The results are compared with those obtained with a commercial software (Virtual Lab, LMS© Siemens), which uses a full finite element description of the problem with an automatically matched layer (AML) approach to deal with the semi-infinite fluids. In this case, the mesh consists of 2312 structural 3-noded shell elements, 2312 tria-3 coupling elements, 96862 (respectively, 14702) tetra-4 elements for the convex domain coupled to the structure (respectively, the open part), 17728 tria-3 for the acoustic envelope coupled to the AML. In both configurations, the structure and inner cavity are modeled using the FEM without any modal reduction. Figure 8.17 shows the comparison for the sound power radiated by the open face of the half-cylinder in the case of the mechanical excitation. Figure 8.18 compares the sound transmission loss of the half-cylinder

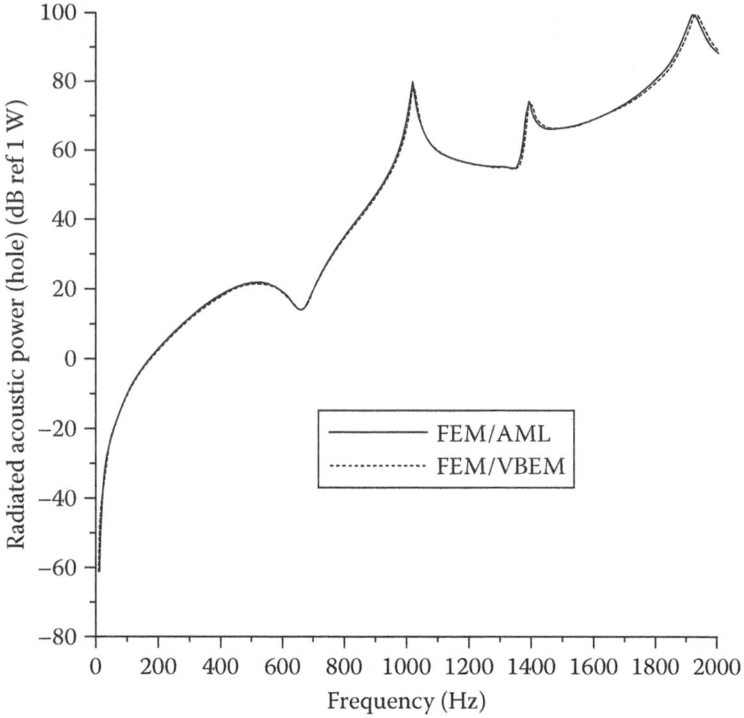

Figure 8.17 Sound power radiated by a baffled open half-cylinder excited mechanically and radiating in air—comparisons between a FEM/AML approach (Virtual Lab 12.0 © LMS) and a FEM/VBEM approach.

in two excitation configurations: (i) an oblique incidence plane wave in $\Omega^1_{f,e}$, and (ii) an oblique incidence plane in $\Omega^2_{f,e}$ which is symmetric from that in $\Omega^1_{f,e}$ with respect to the baffle. There is an excellent agreement between the two approaches for both mechanical and acoustical excitations. In the case of the transmission loss, it is interesting to note that the sound transmission loss is not symmetric.

8.8.8 Sound transmission through an aperture

In this section, we consider the sound transmission through an aperture. The geometry of the problem is depicted in Figure 8.19. An acoustic channel of axis x_3 and given cross section in the plane (x_1, x_2) is embedded into an acoustically rigid wall separating two semi-infinite fluids. The interface between the fluids $\Omega^1_{f,e}$ and $\Omega^2_{f,e}$ and the wall are referred to as Σ^1_e and Σ^2_e, respectively. The system is excited by an acoustic wave propagating in $\Omega^1_{f,e}$.

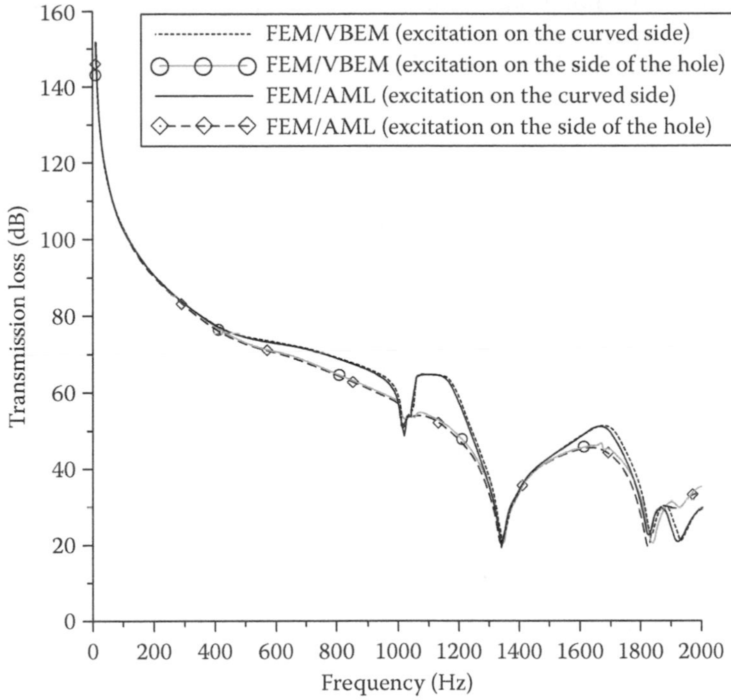

Figure 8.18 Sound transmission loss of a baffled elastic half-cylinder with an open rear cavity for an oblique incidence plane wave excitation on either the curved side ($\Omega^1_{f,e}$) or on the side of the aperture ($\Omega^2_{f,e}$)—comparisons between a FEM/AML approach (Virtual Lab 12.0 © LMS) and FEM/VBEM approach.

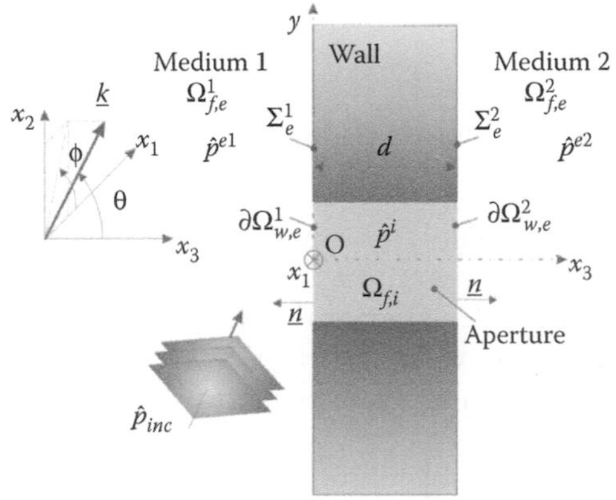

Figure 8.19 Sound transmission through an aperture in a rigid wall excited acoustically.

Using the same notations as in Section 8.8.7, the equations of motion are given by Equations 8.94, 8.95, and 8.96. The coupling conditions are given by Equation 8.97g, 8.97h and the following two additional boundary conditions

$$\frac{\partial \hat{p}^{e1}}{\partial n} = \frac{\partial \hat{p}^{i}}{\partial n} \quad \text{over } \partial\Omega^{1}_{w,e} \tag{8.98a}$$

$$\hat{p}^{e1} = \hat{p}^{i} \quad \text{over } \partial\Omega^{1}_{w,e} \tag{8.98b}$$

The problem can be solved using the method described in Section 8.6. The results are compared with two models available in the literature. The first is based on a modal approach and calculation of the impedance radiation matrix in the physical domain (Sgard et al. 2007) and the second on a modal approach and calculation of the impedance radiation matrix in the wavenumber domain (Park and Eom 1997). Figure 8.20 taken from Sgard et al. (2007) shows the oblique incidence $\theta = 45°, \phi = 0°$ transmission loss of (i) a rectangular aperture of dimensions $2a = 0.4\ m$, $2b = 0.2\ m$, $d = 0.2\ m$ and (ii) a circular aperture with geometrical characteristics $a = 0.1\ m$ and $d = 0.15\ m$ as a function of normalized frequency $k_{e,1}a_{eq}$. a_{eq} is the equivalent radius defined as the radius of the circular aperture or as $\sqrt{4ab/\pi}$ for a rectangular aperture. In the FEM/VBEM method, the aperture volume is discretized using acoustic hexa-8 elements and the coupling with the source and external fluids is taken into account with the admittance radiation matrix (see Section 8.6). For the rectangular aperture, results calculated using Park and Eom's model (Park and Eom 1997) are also shown. An excellent agreement between the FEM/VBEM approach and the other models is observed for both configurations.

8.8.9 Sound transmission through a structure inside a niche

Finally, we consider the sound transmission loss of a structure placed inside an opening that separates two rooms (also called a niche). One of the various reasons for the large deviations in sound transmission loss (STL) intralaboratory reproducibility tests is this so-called niche or tunneling effect. This opening behaves as an acoustic duct and alters the sound field on both sides of the sample.

The equations of the problem are very similar to the previous section, the difference being that there is a structure inside the aperture. The aperture is thus subdivided into two interior subcavities (see Figure 8.21). The

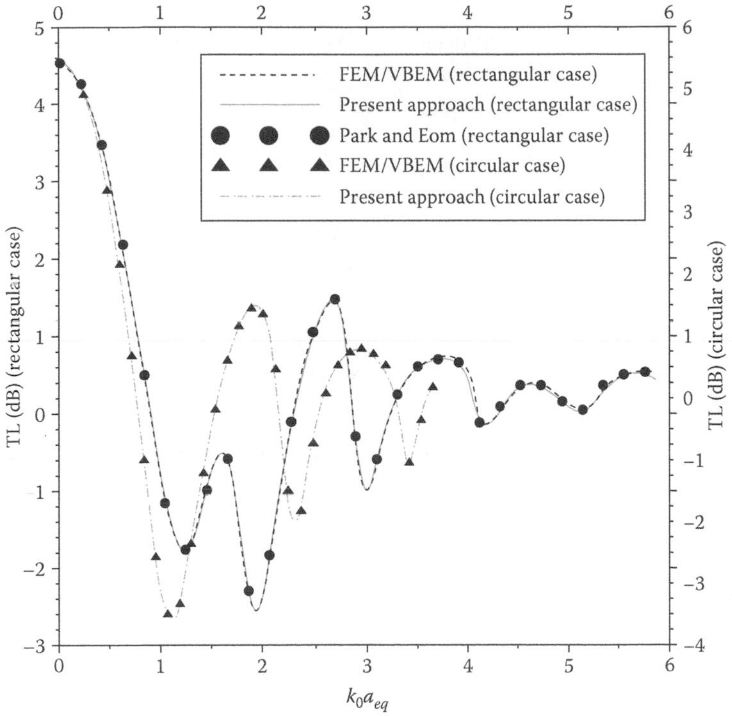

Figure 8.20 Oblique incidence transmission losses of a rectangular aperture *b*/*a* = 1/2, *d*/*a* = 1 and a circular aperture *d*/*a* = 2/3—comparison of the FEM/VBEM approach with other models. (Reproduced by permission of ASA taken from reference Sgard, F., H. Nélisse, and N. Atalla. 2007. On the modeling of diffuse field sound transmission loss of finite thickness apertures. *Journal of the Acoustical Society of America* 122 (1): 302–13.)

equations of motion that govern the problem consist of Equations 8.93, 8.95, and 8.96. Equation 8.94 becomes

$$\nabla^2 \hat{p}^{i1} + k_{i1}^2 \hat{p}^{i1} = 0 \quad \text{in } \Omega_{f,i}^1 \tag{8.99a}$$

$$\nabla^2 \hat{p}^{i2} + k_{i2}^2 \hat{p}^{i2} = 0 \quad \text{in } \Omega_{f,i}^2 \tag{8.99b}$$

where $\hat{p}^{i1}$ and $\hat{p}^{i2}$ are the acoustic pressure in $\Omega_{f,i}^1$ and $\Omega_{f,i}^2$ respectively. k_{i1} and k_{i2} are the corresponding wavenumbers. The coupling conditions are given by

$$\underline{\underline{\hat{\sigma}}} \cdot \underline{n} = \hat{p}^{i1} \underline{n} \quad \text{over } \partial\Omega_{w,i}^1 \tag{8.100a}$$

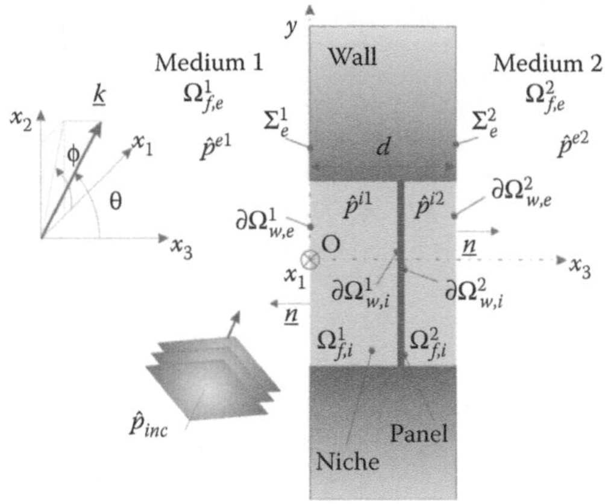

Figure 8.21 Sound transmission through a structure inside a niche excited acoustically.

$$\underline{\underline{\hat{\sigma}}} \cdot \underline{n} = \hat{p}^{i2} \underline{n} \quad \text{over } \partial\Omega^2_{w,i} \tag{8.100b}$$

$$\frac{\partial \hat{p}^{i1}}{\partial n} = \rho_{i1}\omega^2 \underline{\hat{u}} \cdot \underline{n} \quad \text{over } \partial\Omega^1_{w,i} \tag{8.100c}$$

$$\frac{\partial \hat{p}^{i2}}{\partial n} = \rho_{i2}\omega^2 \underline{\hat{u}} \cdot \underline{n} \quad \text{over } \partial\Omega^2_{w,i} \tag{8.100d}$$

$$\frac{\partial \hat{p}^{e1}}{\partial n} = \frac{\partial \hat{p}^{i1}}{\partial n} \quad \text{over } \partial\Omega^1_{w,e} \tag{8.100e}$$

$$\hat{p}^{e1} = \hat{p}^{i1} \quad \text{over } \partial\Omega^1_{w,e} \tag{8.100f}$$

$$\frac{\partial \hat{p}^{e1}}{\partial n} = 0 \quad \text{over } \Sigma^1_e \tag{8.100g}$$

$$\frac{\partial \hat{p}^{e2}}{\partial n} = \frac{\partial \hat{p}^{i2}}{\partial n} \quad \text{over } \partial\Omega^2_{w,e} \tag{8.100h}$$

$$\hat{p}^{e2} = \hat{p}^{i2} \quad \text{over } \partial\Omega^2_{w,e} \tag{8.100i}$$

$$\frac{\partial \hat{p}^{e2}}{\partial n} = 0 \quad \text{over } \Sigma^2_e \tag{8.100j}$$

The problem is solved using the method described in Section 8.6 for the exterior coupling. The internal structural and acoustic subcavity domains are modeled using FEM. The results are compared with a modal approach proposed by Sgard et al. (2015).

The test case (bare configuration) consists of an aluminum panel placed in the center of a niche (see Figure 8.22). A melamine foam slab can be attached to the panel (treated configuration). The system is excited by a diffuse acoustic field. All the fluids (external and internal) are identical (air). The physical parameters used in the calculation are given in Figure 8.22. Figure 8.23 shows the results obtained with the modal approach and with the FEM/VBEM. In the modal approach, modes up to a maximum frequency of $1.5 \times f_{max}$ for the structure and $2 \times f_{max}$ for both the cavities and the porous material were kept in the modal expansions where

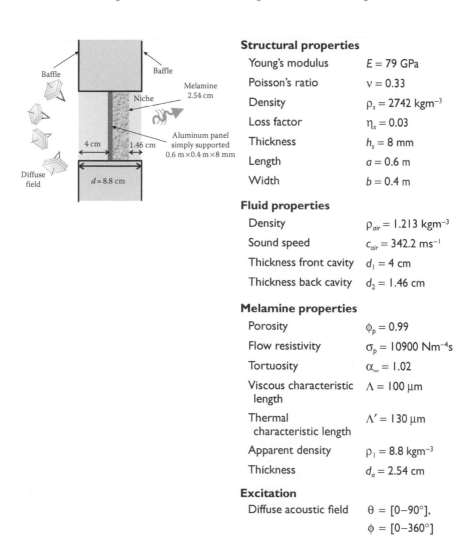

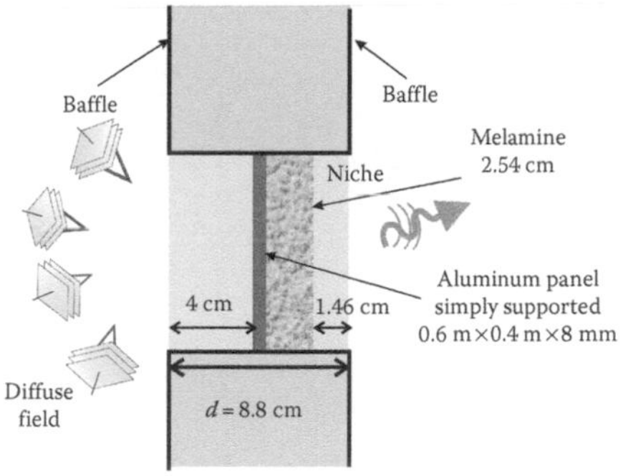

Structural properties	
Young's modulus	$E = 79$ GPa
Poisson's ratio	$\nu = 0.33$
Density	$\rho_s = 2742$ kgm^{-3}
Loss factor	$\eta_s = 0.03$
Thickness	$h_s = 8$ mm
Length	$a = 0.6$ m
Width	$b = 0.4$ m
Fluid properties	
Density	$\rho_{air} = 1.213$ kgm^{-3}
Sound speed	$c_{air} = 342.2$ ms^{-1}
Thickness front cavity	$d_1 = 4$ cm
Thickness back cavity	$d_2 = 1.46$ cm
Melamine properties	
Porosity	$\phi_p = 0.99$
Flow resistivity	$\sigma_p = 10900$ Nm^{-4}s
Tortuosity	$\alpha_\infty = 1.02$
Viscous characteristic length	$\Lambda = 100$ μm
Thermal characteristic length	$\Lambda' = 130$ μm
Apparent density	$\rho_1 = 8.8$ kgm^{-3}
Thickness	$d_a = 2.54$ cm
Excitation	
Diffuse acoustic field	$\theta = [0–90°]$, $\phi = [0–360°]$

Figure 8.22 Sound transmission through a structure inside a niche excited acoustically.

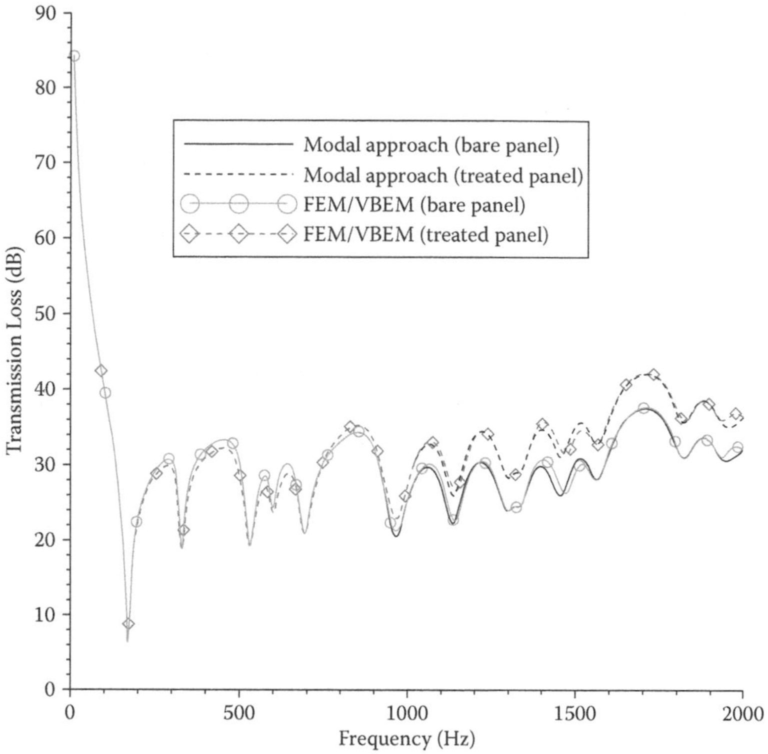

Figure 8.23 STL of a panel inside a niche with (treated) and without (bare) attached melamine foam—comparisons of a direct variational FEM/BEM approach and a modal-based approach.

$f_{max} = 2000\,Hz$. The diffuse field TL is calculated numerically using a 16×16 points Gauss quadrature. For the FEM/VBEM solution, the finite element mesh consists of 40 elements along x and 30 along y. Along the z-direction, the mesh consists of 4 elements for cavity 1, 2 elements for the panel, 10 elements for the melamine (modeled as limp), and 2 elements for cavity 2. Hexa-8 elements were used for the fluids, the porous material, and the panel (discretized using solid elements to account for pressure discontinuity along its thickness). Quad-4 BEM elements were used at the two ends of the tunnel to calculate the radiation impedance matrices and the transmitted acoustic field. Figure 8.23 displays the STL of the panel in the niche with and without the sound absorbing foam. In both cases, an excellent agreement is observed between the numerical FEM/VBEM approach and the analytical modal approach.

8.9 CONCLUSION

The solution of exterior fluid-structure interaction problems has been discussed in this chapter. Various formulations combining the FEM for the structure and the BEM for the fluid were presented and their advantages and drawbacks compared. Several examples were used to illustrate the accuracy of the presented methods by comparison with analytical or other FEM-based methods.

APPENDIX 8A: CALCULATION OF THE VIBROACOUSTIC RESPONSE OF AN ELASTIC SPHERE EXCITED BY A PLANE WAVE AND COUPLED TO INTERNAL AND EXTERNAL FLUIDS

The derivation of the equations can be found in Junger and Feit (1972) and are recalled here. To be consistent with their book, an exp($-i\omega t$) temporal dependency is assumed. Let us consider the axisymmetric vibrations (no torsion) of a thin isotropic elastic spherical shell of radius a and thickness h_s, excited by an incoming plane wave whose wave vector is directed along the x_3 axis in an external fluid of properties ρ_e, c_e and filled with an internal fluid of properties ρ_i, c_i (see Figure 8.24). Let ρ_s, E, and ν_s be the shell density, Young's modulus and Poisson's ratio, respectively. The incident acoustic pressure at point $\underline{x} = (r,\alpha)$ can be written as

$$\hat{p}_{inc} = \hat{A}\exp(ik_{e,3}x_3) = \hat{A}\exp(ik_{e,3}r\cos\alpha) \tag{8.101a}$$

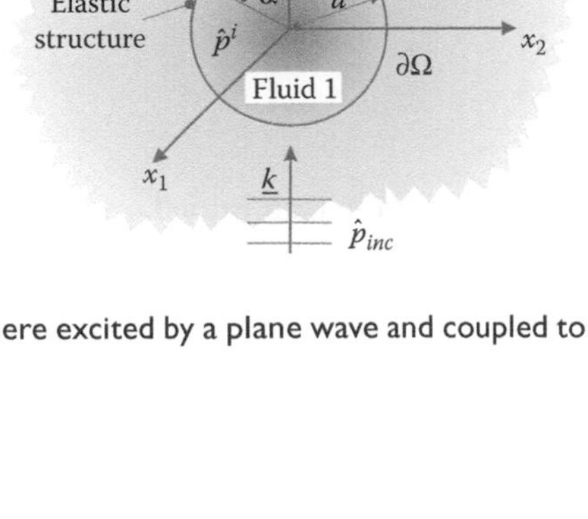

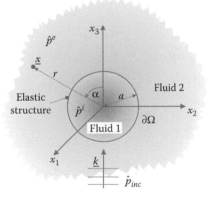

Figure 8.24 Elastic sphere excited by a plane wave and coupled to an internal and external fluids.

The shell displacement field ($r = a$) has two components $\hat{u}_t(\alpha)$ the tangential component and $\hat{u}_r(\alpha)$ the radial component which can be written as (Junger and Feit 1972)

$$\begin{cases} \hat{u}_t(\alpha) = \sum_{n=0}^{+\infty} \hat{u}_{tn}(1-\xi^2)^{\frac{1}{2}} \dfrac{dP_n}{d\xi}(\xi) \\ \hat{u}_r(\alpha) = \sum_{n=0}^{+\infty} \hat{u}_{rn} P_n(\xi) \end{cases} \tag{8.101b}$$

with $\xi = \cos\alpha$ and $P_n(\xi)$ is Legendre's polynomial of order n.

The sound pressure scattered by the sphere is given by

$$\hat{p}_{sc,e} = \hat{p}_{sc,\infty} + \hat{p}_r \tag{8.102}$$

with

$$\hat{p}_{sc,\infty}(r,\alpha) = -\hat{A}_i \sum_n (2n+1) i^n P_n(\cos\alpha) \frac{j_n'(k_e a)}{h_n'(k_e a)} h_n(k_e r) \tag{8.103}$$

where $j_n(x)$ and $h_n(x)$ denote the spherical Bessel and Hankel function of order n, respectively. The prime indicates the first derivative of the function with respect to its argument. It can be shown that (Junger and Feit 1972)

$$\hat{p}_r(r,\alpha) = \rho_e c_e \omega \sum_{n=0}^{+\infty} \hat{u}_{rn} P_n(\cos\alpha) \frac{h_n(k_e r)}{h_n'(k_e a)} \tag{8.104}$$

At $r = a$ we have

$$\hat{p}_r(a,\alpha) = \sum_{n=0}^{+\infty} \hat{p}_{rn} P_n(\cos\alpha) \tag{8.105}$$

with

$$\hat{p}_{rn} = -i\omega \hat{Z}_{en} \hat{u}_{rn} \tag{8.106}$$

where $\hat{Z}_{en}$ denotes the external radiation impedance of mode n. It is given by

$$\hat{Z}_{en} = i\rho_e c_e \frac{h_n(k_e a)}{h_n'(k_e a)} \tag{8.107}$$

Moreover, the sound pressure radiated inside the sphere can be written as

$$\hat{p}^i(r,\alpha) = \rho_i c_i \omega \sum_{n=0}^{+\infty} \hat{u}_{rn} P_n(\cos\alpha) \frac{j_n(k_i r)}{j_n'(k_i a)} \tag{8.108}$$

At $r = a$ we have

$$\hat{p}^i(a,\alpha) = \sum_{n=0}^{+\infty} \hat{p}_{in} P_n(\cos\alpha) \tag{8.109}$$

with

$$\hat{p}_{in} = i\omega \hat{Z}_{in} \hat{u}_{rn} \tag{8.110}$$

where $\hat{Z}_{in}$ denotes the internal radiation impedance of mode n. It is given by

$$\hat{Z}_{in} = -i\rho_i c_i \frac{j_n(k_i a)}{j_n'(k_i a)} \tag{8.111}$$

The total blocked pressure acting on the sphere constitutes the excitation term. It is given by

$$\begin{aligned} \hat{p}_b &= \hat{p}_{inc}(a,\alpha) + \hat{p}_{sc,\infty}(a,\alpha) \\ &= \hat{A}\sum_n (2n+1) i^n P_n(\cos\alpha) j_n(k_e a) - \\ &\quad \hat{A}\sum_n (2n+1) i^n P_n(\cos\alpha) \frac{j_n'(k_e a)}{h_n'(k_e a)} h_n(k_e a) \end{aligned} \tag{8.112}$$

Remembering that

$$j_n(k_e a) h_n'(k_e a) - j_n'(k_e a) h_n(k_e a) = \frac{i}{(k_e a)^2} \tag{8.113}$$

we get

$$\hat{p}_b = \frac{\hat{A}}{(k_e a)^2} \sum_n \frac{(2n+1) i^{n+1} P_n(\cos\alpha)}{h_n'(k_e a)} = \sum_n \hat{p}_{bn} P_n(\cos\alpha) \tag{8.114}$$

The displacement field of the shell of order n is thus solution of

$$\begin{pmatrix} \hat{a}_{11}^{(n)} & \hat{a}_{12}^{(n)} \\ \hat{a}_{21}^{(n)} & \hat{a}_{22}^{(n)} \end{pmatrix} \begin{Bmatrix} \hat{u}_{tn} \\ \hat{u}_{rn} \end{Bmatrix} = \begin{Bmatrix} 0 \\ \hat{F}_n \end{Bmatrix} \tag{8.115}$$

where

$$\begin{cases} \hat{a}_{11}^{(n)} = -\Omega_{red}^2 + (1+\beta_s^2)(\nu_s + \lambda_n - 1) \\ \hat{a}_{12}^{(n)} = \beta_s^2(\nu_s + \lambda_n - 1) + (1+\nu) \\ \hat{a}_{21}^{(n)} = \lambda_n a_{21}^{(n)} \\ \hat{a}_{22}^{(n)} = -\Omega_{red}^2 + 2(1+\nu_s) + \beta_s^2\lambda_n(\nu_s + \lambda_n - 1) - i(\hat{Z}_{in} + \hat{Z}_{en})\dfrac{(1-\nu_s^2)a^2}{Eh_s} \end{cases} \tag{8.116}$$

where $\beta_s^2 = h_s^2 / 12a^2$, $\lambda_n = n(n+1)$, $\Omega_{red} = \omega a / c_p$, $c_p = \sqrt{E / (1-\nu_s^2)\rho_s}$ and

$$\hat{F}_n = -\hat{p}_{bn}\frac{(1-\nu_s^2)a^2}{Eh_s} = -\frac{\hat{A}}{(k_e)^2}\frac{(1-\nu_s^2)}{Eh_s}\frac{(2n+1)i^{n+1}}{h_n'(k_e a)} \tag{8.117}$$

Let us define the mechanical impedance of order n as the ratio of the exciting pressure to the normal velocity of the sphere in vacuo:

$$\hat{Z}_{mecn} = \frac{-\hat{p}_{bn}}{-i\omega\hat{u}_{rn}} = -\frac{i\rho_s c_p}{\Omega_{red}}\frac{h_s}{a}\frac{\left[\Omega_{red}^2 - (\Omega_n^{(1)})^2\right]\left[\Omega_{red}^2 - (\Omega_n^{(2)})^2\right]}{\Omega_{red}^2 - (1+\beta_s^2)(\nu_s + \lambda_n - 1)} \quad n > 0$$

$$Z_{mec0} = \frac{-p_{b0}}{-i\omega\hat{u}_{r0}} = -\frac{i\rho_s c_p}{\Omega_{red}}\frac{h_s}{a}\left[\Omega_{red}^2 - (\Omega_0^{(2)})^2\right] \tag{8.118}$$

$\Omega_n^{(1)}$ and $\Omega_n^{(2)}$ are, respectively, the smallest and the largest real solution of the following equation:

$$\Omega_{red}^4 - \left[1 + 3\nu_s + \lambda_n - \beta_s^2(1 - \nu_s - \lambda_n^2 - \nu_s\lambda_n)\right]\Omega_{red}^2 + (1-\nu_s^2)(\lambda_n - 2) + \beta_s^2\left[\lambda_n^3 - 4\lambda_n^2 + \lambda_n(5-\nu_s^2) - 2(1-\nu_s^2)\right] = 0 \tag{8.119}$$

Equation 8.119 can be rewritten as

$$\left[\Omega_{red}^2 - (\Omega_n^{(1)})^2\right]\left[\Omega_{red}^2 - (\Omega_n^{(2)})^2\right] \quad n > 0 \tag{8.120}$$

We have

$$(\Omega_0^{(2)})^2 = 2(1 + \nu_s) \tag{8.121}$$

In the presence of the fluid loading on each side of the sphere, coefficient $\hat{a}_{22}^{(n)}$ of Equation 8.116 can be rewritten as

$$\hat{a}_{22}^{(n)} = -i\omega(\hat{Z}_{mecn} + \hat{\mathcal{Z}}_{in} + \hat{\mathcal{Z}}_{en})\frac{(1-\nu_s^2)a^2}{Eh_s} \tag{8.122}$$

The resolution of Equation 8.115 leads to $(\hat{u}_{tn}, \hat{u}_{rn})$

$$\begin{aligned}\hat{u}_{rn} &= \frac{-\hat{p}_{bn}}{-i\omega(\hat{Z}_{mecn} + \hat{\mathcal{Z}}_{in} + \hat{\mathcal{Z}}_{en})}\\ \hat{u}_{tn} &= -\frac{\hat{a}_{12}^{(n)}}{\hat{a}_{11}^{(n)}}\hat{u}_{rn}\end{aligned} \tag{8.123}$$

Consequently, substituting Equation 8.123 in Equations 8.108 and 8.104, we get

$$\hat{p}^i(r,\alpha) = -i\rho_i c_i \sum_{n=0}^{+\infty} \frac{\hat{p}_{bn}}{(\hat{Z}_{mecn} + \hat{\mathcal{Z}}_{in} + \hat{\mathcal{Z}}_{en})} P_n(\cos\alpha)\frac{j_n(k_i r)}{j_n'(k_i a)} \tag{8.124}$$

and

$$\hat{p}_r(r,\alpha) = -i\rho_e c_e \sum_{n=0}^{+\infty} \frac{\hat{p}_{bn}}{(\hat{Z}_{mecn} + \hat{\mathcal{Z}}_{in} + \hat{\mathcal{Z}}_{en})} P_n(\cos\alpha)\frac{h_n(k_e r)}{h_n'(k_e a)} \tag{8.125}$$

The acoustic power radiated by the sphere in the external fluid medium is given by

$$\Pi_{rad,e} = \frac{1}{2}\Re\left[\int_{\partial\Omega_w} \hat{p}_r(a,\alpha)\,\left(-i\omega\hat{u}_r(\alpha)\right)^* dS\right] \tag{8.126}$$

which simplifies to

$$\Pi_{rad,e} = \frac{1}{2}\Re\left[2i\rho_e c_e \omega^2 \pi a^2 \sum_{n=0}^{+\infty} |\hat{u}_{rn}|^2 \frac{h_n(k_e a)}{h_n'(k_e a)}\frac{2}{2n+1}\right] \tag{8.127}$$

The acoustic power radiated by the sphere in the internal fluid medium is given by

$$\Pi_{rad,i} = \frac{1}{2}\Re\left[\int_{\partial\Omega_w} \hat{p}^i(a,\alpha)\left(-i\omega\hat{u}_r(\alpha)\right)^* dS\right] \tag{8.128}$$

which simplifies to

$$\Pi_{rad,i} = \frac{1}{2}\Re\left[2\pi a^2 i\rho_i c_i \omega^2 \sum_{n=0}^{+\infty} \left|\hat{u}_{rn}\right|^2 \frac{j_n(k_i a)}{j_n'(k_i a)} \frac{2}{2n+1}\right] \tag{8.129}$$

If there is no damping in the internal fluid, we clearly see that the radiated power is zero (k_i is real).

The mean square velocity of the shell reads as

$$\begin{aligned}\left\langle V^2\right\rangle &= \frac{\omega^2}{2S}\int_{\partial\Omega_w} \hat{u}_{rn}(\alpha)\hat{u}_{rn}(\alpha)^* dS \\ &= \frac{\omega^2}{2}\sum_{n=0}^{+\infty}\left|\hat{u}_{rn}\right|^2 \frac{1}{2n+1}\end{aligned} \tag{8.130}$$

REFERENCES

Allard, J. F. and N. Atalla. 2009. *Propagation of Sound in Porous Media, Modelling Sound Absorbing Materials*, 2nd ed. Chichester, UK: Wiley-Blackwell.

Atalla, N., F. Sgard, and C. K. Amedin. 2006. On the modeling of sound radiation from poroelastic materials. *The Journal of the Acoustical Society of America* 120 (4): 1990–95. doi:10.1121/1.2261244.

Ding, H. J. and W. Q. Chen. 1996. Natural frequencies of an elastic spherically isotropic hollow sphere submerged in a compressible fluid medium. *Journal of Sound and Vibration* 192 (1): 173–98. doi:10.1006/jsvi.1996.0182.

Gaul, L., D. Brunner, and M. Junge. 2009. Simulation of elastic scattering with a coupled FMBE-FE approach. In *Recent Advances in Boundary Element Methods*. 131–45. Springer. http://link.springer.com/chapter/10.1007/978-1-4020-9710-2_10.

Guerich, M. and M. A. Hamdi. 1999. A numerical method for vibro-acoustic problems with incompatible finite element meshes using B-spline functions. *Journal of the Acoustical Society of America* 105: 1682.

Junger, M. C. and D. Feit. 1972. *Sound Structures, and Their Interaction*. Cambridge, MA: MIT Press.

Nelisse, H., O. Beslin, and J. Nicolas. 1996. Fluid-structure coupling for an unbaffled elastic panel immersed in a diffuse field. *Journal of Sound and Vibration* 198 (4): 485–506.

Olson, L. G. and K-J. Bathe. 1985. Analysis of fluid-structure interactions. a direct symmetric coupled formulation based on the fluid velocity potential. *Computers & Structures* 21 (1–2): 21–32. doi:10.1016/0045-7949(85)90226-3.

Park, H. H. and H. J. Eom. 1997. Acoustic scattering from a rectangular aperture in a thick hard screen. *Journal of the Acoustical Society of America* 101 (1): 595–98.

Selmane, A. and A. A. Lakis. 1997. Non-linear dynamic analysis of orthotropic open cylindrical shells subjected to a flowing fluid. *Journal of Sound and Vibration* 202 (1): 67–93. doi:10.1006/jsvi.1996.0794.

Sgard, F., N. Atalla, and H. Nélisse. 2015. Prediction of the niche effect for single flat panels with or without attached sound absorbing materials. *Journal of the Acoustical Society of America* 137 (1): 117–31.

Sgard, F., H. Nélisse, and N. Atalla. 2007. On the modeling of diffuse field sound transmission loss of finite thickness apertures. *Journal of the Acoustical Society of America* 122 (1): 302–13.

Sundqvist, J. 1983. An application of ADINA to the solution of fluid-structure interaction problems. *Computers & Structures* 17 (5): 793–807.

List of Symbols

$\underline{(.)}$	vector
$\underline{\underline{(.)}}$	second-order tensor
$\overline{(.)}$	prescribed value within parentheses
$\widehat{(.)}$	complex amplitude associated to a sinusoidal temporal dependency of the argument
$\dot{(.)}$	$\partial(.)/\partial t$ first temporal partial derivative
$\ddot{(.)}$	$\partial^2(.)/\partial t^2$ second temporal partial derivative
$(.)_y$	the operator is taken with respect to point $\underline{y}$ or the variable within parentheses is evaluated at point $\underline{y}$
$(.)_{,i}$	partial derivative of the quantity within parentheses with respect to x_i
$(.)^*$	complex conjugate of the quantity within parentheses
$(.)^e$	superscript indicating that the quantity within parentheses refers to an element
$\{.\}$	$\{a\} = \begin{Bmatrix} a_1 \\ a_2 \\ a_3 \end{Bmatrix}$ is a vector of components a_1,a_2,a_3 the use of the vector is also adopted to refer to a point a
$\langle . \rangle = \{.\}^T$	transpose of a vector
$\Vert . \Vert$	Euclidian norm of a vector
$\underline{\nabla}_y(.)$	gradient of the scalar within parentheses
$\nabla_y^2(.)$	Laplacian of the scalar within parentheses
$\underline{\nabla}_y \cdot (.)$	divergence of the vector within parentheses
$\underline{\nabla}_y \times (.)$	curl of the vector within parentheses
$\dfrac{\partial}{\partial n_y}(.)$	normal derivative $\underline{\nabla}_y(.) \cdot \underline{n}_y$
$\Re(.)$	real part of the quantity within parentheses
$\Im(.)$	imaginary part of the quantity within parentheses

a	lateral dimension or radius of a structure (m) real coefficient
A	amplitude of a plane wave
$\left(\underline{a}_1^e, \underline{a}_2^e, \underline{a}_3^e\right)$	vectors of the natural basis of the reference element Ω^r associated with Ω^e
α	angle (rad)
$\tilde{\alpha}$	dynamic tortuosity
α_1, α_2	Mellen decay coefficients along x_1- and x_2-axis, respectively, appearing in turbulent boundary layer excitation model
α_{bm}	arbitrary positive constant appearing in Burton Miller's method
α_h	exponent of Hölder condition ($\alpha \in]0;1]$)
α_{panich}	complex-valued coupling parameter appearing in Panich's method
α_∞	tortuosity
b	lateral dimension of a structure (m)
BEM	boundary element method
$B^2 = \dfrac{\eta C_p}{\kappa}$	Prandtl number
β	level of singularity real ≥ 0 constant
$\hat{\beta}$	specific normalized acoustic admittance defined as $\hat{\beta} = \rho_0 c_0 / \hat{Z}_n$
β_c	coefficient characterizing fluid–structure coupling defined as $\rho_0 c_0 / \rho_s h_s \omega_c$
β_s	$h_s / \sqrt{12} a$
$\underline{\underline{C}}$	elasticity tensor of components C_{ijkl}
$C^-(\underline{x})$	free coefficient for the interior problem
$C^+(\underline{x})$	free coefficient for the exterior problem
c_p	sound speed of flexural waves in a plate of infinite lateral extent defined as $\sqrt{E/(1-\nu_s^2)\rho_s}$ (ms^{-1})
C_p	heat capacity at constant pressure (JK^{-1})
c_0	sound speed in fluid domain (ms^{-1})
$\hat{c}_0$	complex sound speed in fluid domain accounting for dissipation defined as $c_0(1 + j\eta_a)^{1/2}$(ms^{-1})
C_v	heat capacity at constant volume (JK^{-1})
d	radius of sphere (m) depth of an aperture (m)
D	diameter of a pipe (m)
$\underline{D}$	dipole moment amplitude vector (kgm^4 s^{-2})
$\mathcal{D}(\phi, ka)$	pressure directivity function
d_B	distance of a point to the baffle (m)

$\delta(\underline{x}-\underline{y})$	delta function
δ_1, δ_2	spatial decay coefficients along x_1- and x_2-axis appearing in turbulent boundary layer excitation model (m^{-1})
δ_{DS}	distance by which the original boundary is shifted along the element inward normal in the Dual Surface method (m)
δ_{ij}	Kronecker symbol
δ_{tbl}	turbulent boundary layer thickness (m)
$\partial\Omega$	surface boundary
$\partial\Omega^*$	$\partial\Omega - e_\varepsilon$
$\partial\Omega_f$	fluid domain boundary
$\partial\Omega_\beta$	interior thin surface used in the prolongation by continuity technique
$\partial\Omega_c$	intersection of a domain Ω and an infinite rigid baffle
$\partial\Omega_e$	boundary on which essential boundary conditions are applied
$\partial\Omega_{f,N}$	fluid domain boundary with Neumann's condition
$\partial\Omega_{f,D}$	fluid domain boundary with Dirichlet's condition
$\partial\Omega_{f,R}$	fluid domain boundary with Robin's condition
$\partial\Omega_{im}$	image boundary of $\partial\Omega$ through the baffle
$\partial\Omega_n$	boundary on which natural boundary conditions are applied
$\partial\Omega_p$	poroelastic boundary or region of a fluid boundary $\partial\Omega$ where p is unknown
$\partial\Omega_{p,p}$	poroelastic boundary with imposed interstitial pressure
$\partial\Omega_{p,u}$	poroelastic boundary with imposed solid-phase displacement
$\partial\Omega_q$	region of a fluid boundary $\partial\Omega$ where $q = \partial p/\partial n$ is unknown namely $\partial\Omega/\partial\Omega_p$
$\partial\Omega_s$	structural boundary
$\partial\Omega_{s,F}$	structural boundary over which external force per unit area is applied
$\partial\Omega_{s,N}$	structural domain boundary with Neumann's condition
$\partial\Omega_{s,D}$	structural domain boundary with Dirichlet's condition
$\partial\Omega_w$	wetted structural boundary
$\partial\Omega_{w,e}$	wetted structural boundary in contact with external fluid medium
$\partial\Omega_{w,i}$	wetted structural boundary in contact with internal fluid medium
$\partial\Omega_\varepsilon$	boundary of Ω_ε defined as $(\partial\Omega - e_\varepsilon) \cup S_\varepsilon^+$
δp	kinematically admissible variation of the acoustic pressure (Pa)
$\left[\dfrac{\partial p}{\partial n}\right]$	normal pressure gradient jump defined as $\dfrac{\partial p^+}{\partial n} - \dfrac{\partial p^-}{\partial n}$
δS	small subregion of $\partial\Omega$ containing e_ε
$\delta\underline{u}$	kinematically admissible variation of the displacement vector
E	structural Young's modulus (Pa)

$\hat{E}$	structural Young's modulus accounting for structural damping defined as $E(1+i\eta_s)$ (Pa)
E_p	Young's modulus of a porous material solid phase (Pa)
e_ε	subregion of $\partial\Omega$ containing a singularity
$\underline{\underline{\varepsilon}}$	structural strain tensor of component ε_{ij} (l)
$\underline{\underline{\varepsilon}}^s$	solid-phase strain tensor
$\underline{F}$	contact force vector per unit area (Pa)
$\underline{F_b}$	body force vector per unit volume (component $F_{b,i}$) (Nm^{-3})
FEM	finite element method
f_{max}	maximum frequency of calculation spectrum (Hz)
f_v	volume loading term (Nm^{-3})
f_s	surface loading term (Nm^{-2})
$\hat{F}_s$	force per unit length (Nm^{-1})
$\varphi(\underline{x},t)$	acoustic velocity potential
ϕ	angle (rad)
ϕ_p	porous material porosity
$\phi_p\left(1+\dfrac{\tilde{Q}}{\tilde{R}}\right)$	Biot Willis coefficient
G	shear modulus (Pa)
$\tilde{G}(\omega)$	viscous correction factor
$\tilde{G}'(\omega)$	thermal correction factor
$\hat{G}(\underline{x},\underline{y})$	Green's function
$\hat{G}_b(\underline{x},\underline{y})$	baffled Green's function
$\hat{G}_{im}(\underline{x},\underline{y})$	image Green's function
$G_0(\underline{x},\underline{y})$	free space Green's function for Poisson's problem
$\hat{G}_\infty(\underline{x},\underline{y})$	free space Green's function
γ	ratio of specific heats defined as $\dfrac{C_p}{C_v}$
$\hat{\gamma}$	internal distributed heat source (Wm^{-3})
$\tilde{\gamma}$	coupling coefficient defined as $\phi_p\left(\dfrac{\tilde{\rho}_{12}}{\tilde{\rho}_{22}}-\dfrac{\tilde{Q}}{\tilde{R}}\right)$
Γ_R	fictitious outer sphere of radius R
Γ_∞	fictitious outer sphere of radius $R\to\infty$
h, h^e	size of a 1D finite element defined as $x_2^e - x_1^e$ (m)
h_b	beam height (m)
HBIE	hyper-singular boundary integral equation
h_c	depth of a rectangular cavity (m)
$h_n(x)$	spherical Hankel function of order n

h_s	thickness of structure (m)
I	area moment of inertia of the beam cross-section about the axis of interest (m^4) or mass moment of inertia of a rod around the pivot point (kgm^2)
$\underline{I}$	inertia force vector of component $I_i = -\rho_s \frac{\partial^2 u_i}{\partial t^2}$
$\hat{\underline{I}}$	sound intensity vector (Wm^{-2})
IBEM	indirect boundary element method
$[I_m]$	identity matrix of dimensions $m \times m$
IEN(:,:)	connectivity table
J	section torsional constant (m^4)
$j_n(x)$	spherical Bessel function of order n
$J_1(x)$	Bessel function of order 1
$k = \omega / c_0$	wavenumber in fluid domain (m^{-1})
$K = \frac{ES}{L}$	axial rigidity of a bar of cross section S, length L and Young's modulus E ($Pa \times m$)
$\hat{k}_c$	complex wavenumber of a wave propagating in a dissipative fluid (m^{-1})
k_{conv}	convective wavenumber defined as $\frac{\omega}{U_c}$ (m^{-1})
$\tilde{K}_e$	dynamic bulk modulus of the air in the pores of the equivalent fluid occupying the totality of a unit volume of porous material defined as $\tilde{R}/\phi_p$
k_p	$\frac{\pi D^4}{128\eta}$
k_{pqr}	modal wavenumber of a rectangular cavity modes defined as $\frac{\omega_{pqr}}{c_0}$ (m^{-1})
k_t	trace wavenumber (m^{-1})
k_{tc}	thermal conductivity ($Wm^{-1}K^{-1}$)
κ	thermal diffusivity (m^2s^{-1})
L	beam length or 1D domain length (m)
$\hat{\lambda}$	Lagrange's multiplier
Λ	viscous characteristic length (m)
Λ'	thermal characteristic length (m)
m_s	beam mass per unit length (kgm^{-1})
μ	double-layer potential density
N	solid-phase shear modulus (Pa)
$\underline{n}$	outward normal vector to a domain boundary
$\underline{\overline{n}}$	inward normal vector to a domain boundary
$\underline{n}_B$	normal vector to the baffle
η	fluid dynamic viscosity (Pa s)

η_a	fluid domain structural loss factor
η_p	porous material solid-phase damping loss factor
η_s	structural loss factor
(η_1,η_2,η_3)	local coordinates
ν_p	porous material solid-phase Poisson's ratio
ν_s	structural Poisson's ratio
ω	circular frequency (rads^{-1})
ω_c	characteristic frequency of the structure (rads^{-1})
ω_m	$\frac{m\pi c}{\ell}$, natural frequencies of a 1D open acoustic cavity (rads^{-1})
ω_{pqr}	natural frequencies of a 3D acoustic cavity (rads^{-1})
Ω	volume
$\tilde{\Omega}$	finite element discretization of volume $\Omega = \bigcup_{e=1}^{n_e} \Omega^e$
Ω_-	unbounded volume (exterior domain)
Ω_+	closed volume (interior domain)
Ω^e	finite element volume
Ω_f	fluid domain
Ω_p	poroelastic domain
Ω^r	reference element
Ω_{red}	reduced frequency defined as $\omega a/c_p$
Ω_s	structural domain
Ω_ε	$\Omega_- \cup \vartheta_\varepsilon$
$p(\underline{x},t)$	acoustic pressure (Pa)
$[p]$	pressure jump defined as $p^+ - p^-$
$\langle p^2 \rangle$	mean square pressure (Pa^2)
p_b	blocked pressure
p^f	fluid-phase pressure of a poroelastic material
FP	Hadamard finite part
p_{inc}	incident acoustic pressure
PML	perfectly matched layer
$P_n(\cos(\theta))$	Legendre polynomial of order n
p_r	radiated pressure
p_{ref}	reflected acoustic pressure
p_s	static pressure
p_{sc}	scattered acoustic pressure
$p_{sc,e}$	acoustic pressure scattered by an elastic structure
$p_{sc,\infty}$	acoustic pressure scattered by a rigid structure
p^+	acoustic pressure on positive side of a thin surface
p^-	acoustic pressure on negative side of a thin surface
P_0	ambient pressure
Π_d	power dissipated by structural damping (W)

$\Pi_{d,t}$	power dissipated by thermal effects in a porous material
$\Pi_{d,v}$	power dissipated by viscous effects in a porous material
Π_{exch}	power exchanged between a structure and a fluid cavity
Π_{in}	injected power
$\Pi_{inc}(\theta_i,\phi_i)$	acoustic power of an incident plane wave
$\Pi_{inc,d}$	acoustic power of a diffuse field
Π_{rad}	radiated power
Π_t	Transmitted sound power
$\psi(\underline{x},t)$	acoustic displacement potential
$\hat{\psi}_c(\underline{x},\underline{y},\omega)$	spatial correlation function of the excitation
$\Psi_{pqr}(\underline{x})$	modal shape of a rectangular fluid cavity
q	normal flux defined as $\partial p/\partial n$
$Q(\underline{x},t)$	volumic density of acoustic source
$\tilde{Q}$	coupling coefficient between the deformation of the solid phase and the fluid phase
q_i	generalized coordinates
$Q_s(\underline{x},t)$	volume velocity or source strength in $m^3\,s^{-1}$
r	Euclidian distance between points $\underline{x}$ and $\underline{y}$: $\lvert\underline{x}-\underline{y}\rvert = \sqrt{(x_i-y_i)(x_i-y_i)}$
r'	Euclidian distance between points $\underline{x}'$ the image of point $\underline{x}$ with respect to the baffle and $\underline{y}$: $\lvert\underline{x}'-\underline{y}\rvert = \sqrt{(x_i'-y_i)(x_i'-y_i)}$
$\tilde{R}$	dynamic bulk modulus of the fluid phase occupying a fraction ϕ_p of a unit volume of porous material
$\hat{\Re}(\theta)$	oblique incidence reflection coefficient
$\Re^3$	total space
$\rho(\underline{x},t)$	fluid density fluctuation (kgm^{-3})
$\tilde{\rho}$	Biot's complex-valued dynamic density defined as $\tilde{\rho}_{11}-\dfrac{\tilde{\rho}_{12}^2}{\tilde{\rho}_{22}}$
(ρ,ϕ)	polar coordinates
$\tilde{\rho}_e$	complex dynamic density of the equivalent fluid occupying the totality of a unit volume of porous material defined as $\tilde{\rho}_{22}/\phi_p$
$\tilde{\rho}_e'$	apparent dynamic complex density of the fluid phase of a limp material defined as $\left(\dfrac{1}{\tilde{\rho}_e}+\dfrac{\tilde{\gamma}^2}{\phi_p\tilde{\rho}}\right)^{-1}$
ρ_s	structural density (kgm^{-3})
ρ_{sk}	density of the material of the skeleton (kgm^{-3})
ρ_0	density of fluid domain
$\hat{\rho}_0$	density of fluid domain accounting for dissipation defined as $\rho_0(1-i\eta_a)$
$\tilde{\rho}_{11},\tilde{\rho}_{12},\tilde{\rho}_{22}$	Biot's complex-valued dynamic densities

S	cross section of a beam or area of a structure (m^2)
SBIE	standard boundary integral equation
$S_{E_c}(\omega)$	kinetic energy spectral density
$S_p(\omega)$	power autospectrum
$S_{pp}(\underline{x}, \underline{y}, \omega)$	power spectral density of the excitation
STL	sound transmission loss (dB) defined as $10\log\left(\frac{1}{\tau}\right)$
$S_{\langle v^2 \rangle}(\omega)$	mean square velocity spectral density
$S_1(x)$	first-order Struve function
$S_{\Pi_{in}}(\omega)$	injected power spectral density
$S_{\Pi_{rad}}(\omega)$	radiated power spectral density
S_ε^+	boundary of ϑ_ε. Commonly a half sphere of radius ε centered at point $\underline{x}$
σ	single layer potential density
$\underline{\underline{\sigma}}$	structural stress tensor of component σ_{ij} (Pa)
σ_p	flow resistivity (Nsm^{-4})
σ_{rad}	radiation efficiency defined as $\frac{\Pi_{rad}}{\rho_0 c_0 S \langle V_n^2 \rangle}$
$\underline{\underline{\tilde{\sigma}}}^s$	*in vacuo* solid-phase stress tensor
$\underline{\underline{\sigma}}^t$	total stress tensor (Pa)
$\underline{t}$	traction vector (Pa) defined as $\underline{\underline{\sigma}} \cdot \underline{n}$
T	acoustic temperature (K) or kinetic energy (J)
T_s	tension in a string
τ	transmission factor defined as $\frac{\Pi^t}{\Pi_{inc}}$
θ	angle (rad)
$\underline{u}(\underline{x}, t)$	structural displacement vector of component u_i (m)
$\underline{U}(\underline{x}, t)$	acoustic particle displacement
U_c	convective velocity (ms^{-1})
$\underline{U}^f$	fluid-phase displacement of a poroelastic material
u_r	radial displacement
$\underline{u}^s$	solid-phase displacement of a poroelastic material
u_t	tangential displacement
u_{tn}	nth coefficient of a modal expansion of the transverse displacement
U_∞	free stream velocity (ms^{-1})
$\underline{v}(\underline{x}, t)$	structural velocity vector of component v_i (ms^{-1})
$\underline{V}(\underline{x}, t)$	acoustic particle velocity (ms^{-1})
VBEM	variational boundary element method
V_D	Strain energy
V_{ext}	Potential energy variational boundary element method
$v_n(\underline{x}, t)$	normal structural velocity

$\left\langle V_n^2 \right\rangle$	normal mean square velocity (m^2s^{-2})
$\hat{\upsilon}$	volume flow rate (m^3s^{-1})
ϑ_ε	external small volume of arbitrary shape
W_i^g	Gauss weight
$\underline{x} = (x_1, x_2, x_3)$, $\underline{y} = (y_1, y_2, y_3)$	position of source or receiver point in physical coordinate system
$\underline{x}_0$	location of a point source
$x_{1,i}^g, x_{2,i}^g, x_{3,i}^g$	coordinates of the *i*th Gauss point in the three directions
XYZ(:,:)	table of node coordinates
(ξ_1, ξ_2, ξ_3)	local coordinates
$\hat{Z}_c$	characteristic impedance of a dissipative fluid (Pa m^{-1} s)
$\hat{Z}_n$	specific acoustic impedance (Pa m^{-1} s)

Index

A

B

C

D

E

F

G

H

I

J

L

M

Q

R

S